# Advances in Invertebrates and Fish Telemetry

Developments in Hydrobiology 130

*Series editor*
H. J. Dumont

# Advances in Invertebrates and Fish Telemetry

## Proceedings of the Second Conference on Fish Telemetry in Europe, held in La Rochelle, France, 5–9 April 1997

*Edited by*

Jean-Paul Lagardère, Marie-Laure Bégout Anras & Guy Claireaux

*Reprinted from Hydrobiologia, volumes 371/372 (1998)*

CENTRE NATIONAL
DE LA RECHERCHE
SCIENTIFIQUE

IFREMER

PROGRAMME
com'Science
CONSEIL REGIONAL
POITOU-CHARENTES

Kluwer Academic Publishers
Dordrecht / Boston / London

**Library of Congress Cataloging-in-Publication Data**

A C.I.P. Catalogue record for this book is available from the Library of Congress.

ISBN 0-7923-5211-4

---

Published by Kluwer Academic Publishers,
P.O. Box 17, 3300 AA Dordrecht, The Netherlands

Sold and distributed in North, Central and South America
by Kluwer Academic Publishers,
101 Philip Drive, Norwell, MA 02061, U.S.A.

In all other countries, sold and distributed
by Kluwer Academic Publishers Group,
P.O. Box 322, 3300 AH Dordrecht, The Netherlands

*Printed on acid-free paper*



Printed in Belgium

*Hydrobiologia* **371/372**, v–vii, 1998.
*J.-P. Lagardère, M.-L. Bégout Anras & G. Claireaux (eds), Advances in Invertebrates and Fish Telemetry.*

# Contents

## 4. Fish Migration and Stock Management

*Hydrobiologia* **371/372:** ix, 1998.
*J.-P. Lagardère, M.-L. Bégout Anras & G. Claireaux (eds), Advances in Invertebrates and Fish Telemetry.* 

# Introduction

The Second Conference on Fish Telemetry in Europe was held in La Rochelle, France, from 5 to 9 April 1997. The aim of this conference was to facilitate exchanges between groups of researchers working on different compartments of marine or freshwater ecosystems and sharing telemetry techniques as investigating tools. This kind of conference also allows us to be informed of the most recent technical advances and associated conceptual developments.

After the success of the First Conference on Fish Telemetry held in Liège, Belgium, April 4–6, 1995, the presence of 110 participants from 17 countries attending this second conference confirmed that telemetry is establishing itself as a major technique for the analysis and management of aquatic ecosystems. With 57 oral and posted contributions presented, we can observe:

- a fast instrumental progression based on electronic and signal treatment advances,
- an enlargement of continuous investigation on aquatic animals tracked over wider areas (lakes, estuaries and coastal areas) and longer periods of time,
- a developing use of telemetry for aquatic animal stock management and the conservation and protection of their environment.

Forty manuscripts were selected by the organizing committee after being reviewed by members of the Editorial Board. They provide the material for this book and are organized into four sections:

1. Methodology and New Developments,
2. Tagging Procedures,
3. Behavioural and Physiological Ecology,
4. Fish Migration and Stock Management.

The organization of the Second Conference on Fish Telemetry in Europe was made possible by the support of Maurice Héral, Director, and the staff of the Centre de Recherche en Ecologie Marine et Aquaculture de l'Houmeau (CREMA). Special thanks are due to Miss A. Bonpain, Mrs A. L. Monteragioni and Mrs E. Travers, who undertook the secretarial work. We also thank Henri J. Dumont, Editor-in-Chief of *Hydrobiologia* and René Mijs, Kluwer Academic Publishers, for their interest in the production of this special issue.

*CREMA (CNRS – Ifremer)*
*B.P. 5, 17137 L'Houmeau, France*

J.-P. LAGARDÈRE
M.-L. BÉGOUT ANRAS
G. CLAIREAUX

**List of sponsors**

*Tutorial organisms*

Centre National de la Recherche Scientifique, CNRS
Institut Français de Recherche pour l'Exploitation de la Mer, IFREMER
Région Poitou Charentes, Programme Com'Sciences

*Other sponsors*

- Institut National de la Recherche Agronomique, INRA
- Institut Français de Recherche Scientifique pour le Développement en Coopération, ORSTOM
- Conseil Supérieur de la Pêche, CSP
- Conseil Général de la Charente Maritime
- Ville de La Rochelle
- Muséum d'Histoires Naturelles de La Rochelle
- Aquarium de La Rochelle
- Fish Eagle
- Lotek
- Vemco Ltd.

*Hydrobiologia* **371/372:** xi, 1998.
*J.-P. Lagardère, M.-L. Bégout Anras & G. Claireaux (eds), Advances in Invertebrates and Fish Telemetry.*


# List of Participants

| | |
|---|---|
| Kim AARESTRUP | Denmark |
| Gary ANDERSON | Canada |
| Marie-Laure BÉGOUT ANRAS | France |
| John ARMSTRONG | U.K. |
| Bjorn Erik AXELSEN | Norway |
| Jean-Luc BAGLINIÈRE | France |
| Etienne BARAS | Belgium |
| Toni BEDDOW | Canada |
| Vincent BENECH | Mali |
| Stephen BLACK | Ireland |
| Drew BODALY | Canada |
| Gunnbjorn BREMSET | Norway |
| Andre BREUKELAAR | The Netherlands |
| Catherine BRUGEL | France |
| Eilif BRODTKORB | Norway |
| Patrice CAYRE | France |
| Guy CLAIREAUX | France |
| Stuart CLOUGH | U.K. |
| Johan COECK | Belgium |
| Sandra COLAZZO | Belgium |
| Martin COLLINS | U.K. |
| Gabriel EDEL | France |
| Bruno ERNANDE | France |
| Koos FOCKENS | The Netherlands |
| Frank FREDRICH | Germany |
| Mathieu GERLIER | France |
| Eduardo GONZALEZ-GURRIARAN | Spain |
| Juliette GUET | Norway |
| Eli Kyrkjebo HAUGLAND | Norway |
| Tor HEGGBERGET | Norway |
| Cyril HERISSE | France |
| Marianne HOLM | Norway |
| Martin HUBER | Switzerland |
| George JACKSON | Australia |
| Niels JEPSEN | Denmark |
| David JOHNSTON | Canada |
| Erwan JOSSE | French Polynesia |
| Adrian JOWITT | U.K. |
| Anders KOED | Denmark |
| Françoise LAGARDÈRE | France |
| Jean-Paul LAGARDÈRE | France |
| François LAUTERS | France |
| Cheryl LE DREW | Canada |
| Christel LEFRANÇOIS | France |
| Mario LEPAGE | France |
| Jeffrey LINES | U.K. |
| Vladimir LINNIK | Russia |
| Svein LOKKEBORG | Norway |
| Pierre LOONIS | France |
| Don MacCUBBING | U.K. |
| Ridha MALLEKH | France |
| Gerd MARMULLA | Italy |
| Francis MARSAC | Seychelles |
| Theresa MARTINELLI | U.S.A. |
| Andy MOORE | U.K. |
| George NIEZGODA | Canada |
| Mark John O'CONNELL | U.K. |
| Ron O'DOR | Canada |
| Odd Arild OLSEN | Norway |
| Dominique OMBREDANE | France |
| Michael OVIDIO | Belgium |
| Armin PETER | Switzerland |
| Pascal ROCHE | France |
| Ian RUSSELL | U.K. |
| Richard SABATIE | France |
| Stefan SCHMUTZ | Austria |
| Henrik SCHURMANN | Denmark |
| David SCRUTON | Canada |
| Mitchell SISAK | Canada |
| Per Olav SKJERVOLD | Norway |
| Gordon SMITH | U.K. |
| Philip SMITH | U.K. |
| Kim SUN | The Netherlands |
| Fredrik SUNDSTROM | Sweden |
| Dany SUREAU | France |
| John TAYLOR | U.K. |
| Eva THORSTAD | Norway |
| François TRAVADE | France |
| Gordon URQUHART | U.K. |
| Fred VOEGELI | Canada |
| Magnus WAHLBERG | Sweden |
| Dale WEBBER | Canada |
| Hakan WESTERBERG | Sweden |

Hydrobiologia 371/372: 1–8, 1998.
J.-P. Lagardère, M.-L. Bégout Anras & G. Claireaux (eds), Advances in Invertebrates and Fish Telemetry.



# Applications and performance of Radio-Acoustic Positioning and Telemetry (RAPT) systems

R. K. O'Dor[1], Y. Andrade[1], D. M. Webber[1], W. H. H. Sauer[2], M. J. Roberts[2], M. J. Smale[3] & F. M. Voegeli[4]
[1] *Biology Department, Dalhousie University, Halifax, Canada B3H 4J1 (odor@is.dal.ca)*
[2] *Sea Fisheries Research Inst., Private Bag X2, Roggebai 8012, Cape Town, South Africa*
[3] *Port Elizabeth Museum, P.O. Box 13147, Humewood, Port Elizabeth, South Africa*
[4] *VEMCO Ltd., 100 Osprey Dr., Shad Bay, Nova Scotia, Canada B3T 2C1*

**Abstract**

RAPT was developed to use systems of semi-autonomous buoys with hydrophones and radio transmitters to continuously monitor the positions and performance of multiple objects, animals and/or people tagged with miniature acoustic transmitters under water. Buoys communicate signal arrival times to shore, ship or aircraft based computers which triangulate positions in three dimensions and decode telemetered information such as heart rate, respiration rate, temperature, salinity and light encoded in pulse intervals. It is the only way of tracking with high-resolution (meters) at intermediate ranges (10's–1000's of meters) in seawater and the most accurate in freshwater. The technique is powerful and flexible with wide application, but is constrained by tradeoffs between electrical power and signal accuracy under extreme conditions. Technological solutions to some of these constraints are possible, but optimization of information gathering, in many cases, simply requires more experience and can be achieved by software, information sharing and a cadre of trained personnel.

## Introduction

Three dimensional sound source location for marine animals was described by Watkins & Schevill (1972) and a variety of wire-linked hydrophone systems have since been developed (Hawkins et al., 1974; Chapman et al., 1975; Maynard & Webber, 1987; Lagardère et al., 1990; Wartzok et al., 1992). Experimental radio-linked x-y plotting systems were described by McKibben & Nelson (1981) and acoustic-radio relay buoys by Solomon & Potter (1988). Commercial Radio-Acoustic Positioning and Telemetry (RAPT) systems using digital radio transfers were developed by VEMCO Ltd., Shad Bay, Nova Scotia, in collaboration with the Dalhousie group (senior contributors have included: principal investigators; R. K. O'Dor, R. G. Boutilier, S. Kerr, J. Hutchings; research associates, D. M. Webber, G. Claireaux, Y. Andrade) starting in 1985 (O'Dor et al., 1989), and commercial production began in 1990 (Bjordal et al., 1993).

This portable system was developed to allow biologists to track and monitor acoustically tagged animals in a wide range of applications without a team of engineers. Six systems are in use in Canada and eight internationally; these systems are listed in Table 1. The Dalhousie group has played an active role in demonstrating and/or advising many other users and proposers, linking them through its Internet site: http://is.dal.ca/~raptnet

This report recognizes wide use of VEMCO's RAP (radio-acoustic positioning) buoys, with VRAP (VEMCO RAP) software becoming a de facto standard for biologists using wire links and RAPT, and it encourages formalization of an international network with the goals of:

(a) improving the quality of data gathering by RAPT,

(b) accelerating its integration into marine monitoring/research,

(c) promoting common software and international standards,

*Table 1.* Radio-Acoustic Positioning and Telemetry Applications*

| | Location | Application | System |
|---|---|---|---|
| ANIMAL MONITORING | | | |
| 1) | Dalhousie Biology Dept., NS, Canada | | |
| | Laboratoire Arago, Banyuls-sur-Mer, France | octopus[1] | Wired |
| | Dept. Fisheries & Oceans, Moncton, Canada | snow crabs[2] | Wired |
| | DFO, Halifax Lab, Canada | lobsters[3] | 3B |
| | Sable Island, Canada | seal behavior | 4B, 3D |
| | Pt. Elizabeth Museum, S. Africa | shark feeding | 4B, 3D |
| | Rhodes University, S. Africa | squid mating[4] | 4B, 3D |
| 2) | U. Missouri, St. Louis, MO, USA | polar seals[5] | Wired/ice |
| 3) | Inst. Marine Research, Bergen, Norway | lobsters, cod[6] | 5B, 2C |
| 4) | DFO, Freshwater Inst., Winnipeg, Canada | seals, fish | 6B, 2C |
| 5) | University of New Hampshire, NH, USA | lobster behavior | 3B |
| 6) | DFO, NWAFC, St. John's, Canada | cod/trawls | 4B, 2C |
| 7) | Ano Neuvo I., CA, USA | shark behavior | 3B |
| HABITAT MONITORING & MARINE RESERVES | | | |
| 8) | University of Girona, Spain | Lake Banyoles | 4B |
| 9) | Sea Fisheries Research Inst., S. Africa | squid/physics[8] | 4B, 3D |
| 10) | Sumiyoshihama Park, Japan | horseshoe crabs[9] | 3B |
| 11) | Lincoln Marine Science Center, Australia | multispecies | 4B, 3D |
| 12) | Exuma Cay Land & Sea Park, Bahamas | coral reef fishes | 3B |
| 13) | Lizard I. Stn., James Cook U., Australia | reef fish, squid | 4B, 3D/RR |
| AQUACULTURE MONITORING | | | |
| 14) | CREMA, L'Homeau, France | multispecies[10] | Wired |
| DIVER MONITORING & VIDEO SURVEY | | | |
| 15) | DCIEM/DND, North York, ON, Canada | diver & search control | 3B |
| 16) | Dalhousie Biology Dept.,Halifax, Canada | | |
| | DFO, Halifax Lab, Canada | juvenile lobster habitat | 3B |

B = buoys, C = cells, D = dimensions, RR = Radio Relay

*For more proposals and contact addresses see: http://is.dal.ca/~raptnet. This table emphasizes the wide acceptance of systems built by VEMCO Ltd. or functioning on the same principles. Many are recent installations, but references describe some examples of biological observations: [1] Mather & O'Dor, 1991; [2] Maynard & Webber, 1987; [3] O'Dor & Webber, 1991; [4] Sauer et al., 1997; [5] Wartzok et al., 1992b; [6] Bjordal et al., 1993; [8] O'Dor et al., 1996; [9] O'Dor & Seino, 1997; [10] Lagardère et al., 1995. Several additional examples are included in the present volume.

(d) facilitating hardware sharing for cellular buoy arrays, and
(e) avoiding duplication of effort in overcoming problems.

Applications for the systems, to date, can be divided into the four categories outlined below.

*Animal Monitoring.* For purposes such as detailed resolution of the movements and activities of animals over spatial and temporal scales too large for visual observations by divers and submarines but too small for boat tracking, RAPT is the underwater equivalent of binoculars. Such data can, for example, provide information on breeding behaviour, energy costs of species and energy flow in ecosystems which may be critical to fisheries management. Continuous monitoring by divers is extremely expensive and limited to a few meters by visibility. Boat tracking typically resolves positions to 10's of meters and can only reliably follow one animal at a time. Triangulation from a mobile platform is technically possible, but maintaining a vessel of adequate size may be prohibitively expensive for long-term projects.

*Habitat Monitoring & Marine Reserves.* Rare or sentinel species can be continuously monitored in marine reserves or disturbed areas. Conservation of rare species requires knowledge of their ranges throughout life, which can best be established by continuous tracking. Public interest in conservation is heightened by RAPT displays on shore-based computers. This provides real-time information about animal activities even when weather or other factors limit direct observation; several public centres for preservation are interested in providing such indirect 'viewing' to their visitors, either to reduce disappointment when animals are not visible or to reduce direct visitor pressure. Continuous monitoring also reveals changes of behaviour which may give early warning of environmental degradation or recovery.

*Aquaculture Monitoring.* Both finfish and shellfish culture operations are occurring on increasing scales. Aquaculturalists have the disadvantage that they cannot see what their stock is doing, particularly in large enclosures. RAPT systems can reveal anomalous behaviours and environmental changes remotely, in real time. This can provide early warning of catastrophic events and improve basic husbandry practices.

*Diver Monitoring and Surveys.* SCUBA divers can be monitored by RAPT in the same way as animals, allowing remote supervision of diver safety and effectiveness from shore or a mobile platform. Objects and sites can be marked for rapid recovery. Divers or remote equipment can carry larger telemetry packages, allowing transponder contact via the buoys. In the biological context, bottom surveys to monitor habitat types and changes, such as kelp bed or coral reef damage and recovery, can be done by video recordings followed by onshore analysis, to minimize expensive dive and boat time. However, even then divers must spend time laying out an area for survey to relate images to locations. RAPT can move all analyses to shore, and perhaps even automate them. In marine reserves which combine conservation with public viewing by divers, RAPT can direct divers to unusual specimens that have been tagged, but can also keep track of tagged visitors who might move to areas where they are a conservation or safety risk.

This is by no means an exhaustive list of RAPT applications, but shows the scope of the associated information management challenge, as well as some of the common research problems that could benefit from shared solutions. To illustrate this, we report a detailed reconstruction of a diver exercise, which indicates the current limits of resolution for RAPT and some of the available approaches for optimizing information recovery and presentation. Although the example is not directly relevant to the conference theme of fish telemetry, we did ask the diver to behave like a pair of mating squid, and, unlike real squid (or fish), he was able to confirm that what the RAPT system recorded was, in fact, what happened. This example was chosen as an extreme test of the RAPT system with four buoys moored in heavy swell near the advertised radio range limit, using prototype VRAP-B software to calculate positions of two fast-moving pingers in 3-D from hyperbolic equations (VRAP-A requires a depth transducer to provide 3-D positions).

## Methods and materials

In November, 1994, we conducted diver tests of new VRAP-B RAPT base station software designed to calculate 3-D positions of mating pairs of squid tagged with the smallest possible transmitters in real time. Biological observations made with the system are reported elsewhere (O'Dor et al., 1995, Sauer et al., 1997). A four buoy array was located in St. Francis Bay, South Africa, 5 km from the base station, at a site studied extensively in relation to the fishery on the squid breeding grounds (Figure 1; Augustyn et al., 1994). The buoys were arranged in an approximate 40 m square in 25 m of water with hydrophones in a tetrahedral array. Two were located at the surface diagonally opposite each other, and two were on the mooring ropes at 20 m depth (See Figure 3). Analysis of the equation solutions with small errors in arrival times introduced randomly, indicated that this configuration provided minimum position error in the zone inside the array. Mooring ropes were attached to 5 m lengths of heavy chain, anchored at one end, to minimize horizontal movement of the buoys while allowing more than 3 m of vertical movement for swells. Each hydrophone includes a pinger that can be activated from the base station on shore for self-positioning.

The base station initiated acoustic sampling at the buoys and polled them via a 458.5375MHz radio link. It then forwarded arrival times to a Toshiba 4800CT computer which calculated and plotted positions using color to distinguish between animals (Figure 2). Vemco's VRAP software was originally designed for slow-moving benthic species and includes averaging and data rejection algorithms to improve position accuracy by rejecting noise. Because of power constraints, existing buoys listen to only one frequency, and optimal listening times vary with the number and speed of moving targets. The diver tests were conducted to optimize sampling for a project with a

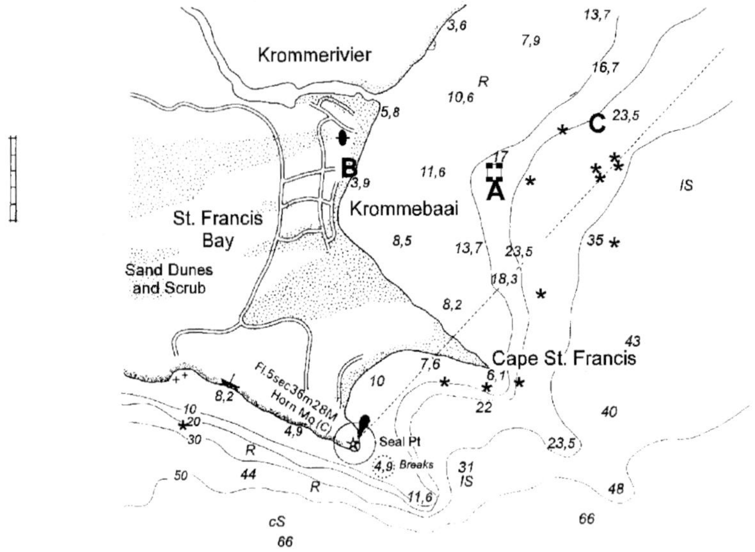


*Figure 1.* Positions of the square buoy array (A), base station (B) and permanent physical monitoring site (C) at 39E09.83' S, 24E53.00' E, relative to know spawning sites of squid (∗) in St. Francis Bay, South Africa. Scale bar is 1.85 km, soundings in meters.

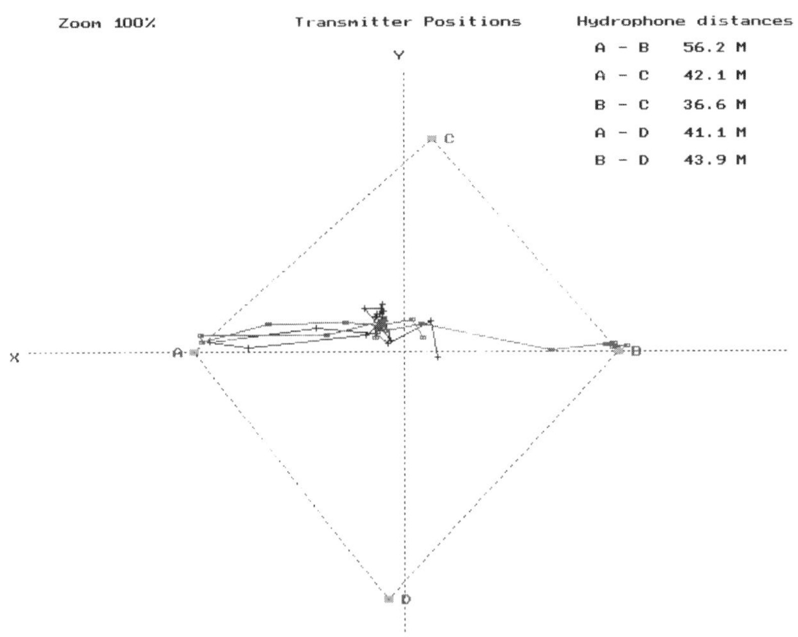


*Figure 2.* A standard VRAP (VEMCO radio-acoustic positioning software) screen display (x,y; A-B vs.C-D axes) of real-time tracking data, including self-positioning buoy distances. The data displayed show two pingers (+, 71 kHz; rectangle, 76.8 kHz) being carried by a diver across the array. The diver was instructed to: 1) enter the water at buoy B, 2) position himself approximately at the center of the array, 3) remove the pingers from his pocket, 4) hold a pinger in each hand, 5) swim to buoy A with arms outstretched, 6) bring his arms together, 7) return to a bottom mooring, which marked a squid egg bed, 8) wave his arms 'to simulate squid mating behavior', 9) tie the 71 kHz pinger to the mooring, 10) swim back to buoy B, 11) surface and return to the boat.

primary aim of tracking squid-squid interactions at speeds up to 1 m $s^{-1}$. To simulate this, a diver (MJS) carried a VEMCO V16-3H pinger (1.5 s pulse interval, 71 kHz), used for large male squid, in one hand and a V8-2L pinger (1.5 s pulse interval, 76.8 kHz), used for smaller females, in the other, while swimming inside the array at 'squid speeds'. The test with the best results alternately listened to frequencies for 8 s. This gave about five positions in each sample after filtering (See below.). Switching between frequencies requires two-way radio communication as well as gain adjustments, so that each frequency was actually updated every 25 s on average.

Real-time plotting is essential to allow onshore observers to optimize data gathering (i.e. Continuous monitoring of an animal that is present may be preferable to monitoring for absent animals.), but there are still tradeoffs in price, portability, power consumption and speed for field-ready computers. When targets are being tracked continuously, pulse intervals are too short to allow most current portable computers time for solving 3-D hyperbolic equations, choosing the best solution, completing a full statistical analysis and updating a 3-D plot display in real time. There is a need for software that will run on any computer, because few researchers will risk (or have backup) for high-end computers in the field. In any case, real-time analysis can never equal post-hoc analysis because computers will (presumably?) never actually be able to predict the future with absolute certainty.

Comparison of real-time to post-analysis plotting requires some understanding of data structure. When each buoy is polled at the end of a sampling interval, it returns a series of timed events which are compared to solve position equations. Noise near one buoy can add extra events and distance or sound absorbing objects (e.g. bubbles from a wave, a fish's swimbladder or a diver's suit) can differentially block events. Knowing the pulse interval and having quadruply redundant sampling allows noise to be filtered and missed pulses to be inserted with very little error, post-hoc. Because of real-time constraints, data rejection is an equally appropriate treatment except for fast moving targets where real information may be lost. After editing, VRAP software will calculate additional positions, which can be used to reconstruct improved trajectories.

We do post-analysis using Matlab (The Math Works, Inc., Natick, MA) software, including the Signal Processing Toolbox. The 'spline' function quickly provides a cubic interpolant curve to all positions to estimate the position of each pinger at every ping, which can be plotted in 3-D, statistically evaluated and even animated.

## Results

Figure 2 shows a typical real-time screen plot from VRAP of a 12.5 min sequence of diver movements in a pattern defined in the caption. Much of the pattern was recognizable from the standard x,y (A-B vs. C-D) plot in the figure, and, when he returned to the boat, shore base was able to immediately tell the diver by radio that the 76.8 kHz pinger had been in his right hand, to his surprise and our delight.

The details become clear in the x,y zoom in Figure 3a produced after Matlab post-analysis. This plot actually records a fourth dimension because the data for the 76.8 kHz pinger is displayed as splined points at 1.5 s intervals. The swim with arms outstretched averaged 0.23 m $s^{-1}$ and peaked at 0.37 m $s^{-1}$, but the return with arms together reached 0.55 m $s^{-1}$. The vertical aspect of the return to the boat shows up better in the x,z (A-B vs. Depth) plot in Figure 3b. The 3-D view in Figure 3c highlights the difference between the swim with arms outstretched and the return with the pingers together, as well as the squid-mating simulation. One problem with 3-D plots is that there are an infinite number of possible views, and whichever one you choose in real-time will likely be the worst one possible. The second problem is that no two brains seem to see them exactly the same, although color and perspective tricks available on powerful workstations can improve this.

Figures 3b and c also give an indication of a feature of the system that shows up most clearly when data are animated. Vertical movements on the order of 1–2 m of the 71 kHz pinger continue even after the diver swims away. The magnitude and timing of these oscillations is approximately that of the day's swell, a continuous feature of this part of the South African coast. Because the moorings allow the buoys to rise and fall on the swell, while the pinger remains fixed, the pinger appears to move vertically. Subsquent monitoring of the 71 kHz pinger 1 m off-bottom for 1 h using 30 s averaging gave standard deviations of 0.3 m in x,y and 1.0 m in z in 1–2 m swells. The resolution of transmitter separation on the diver's swim indicates overall accuracies of fitted tracks in the range of 1–2 m and x,y accuracy considerably better than this. This could be further improved by using the pinger

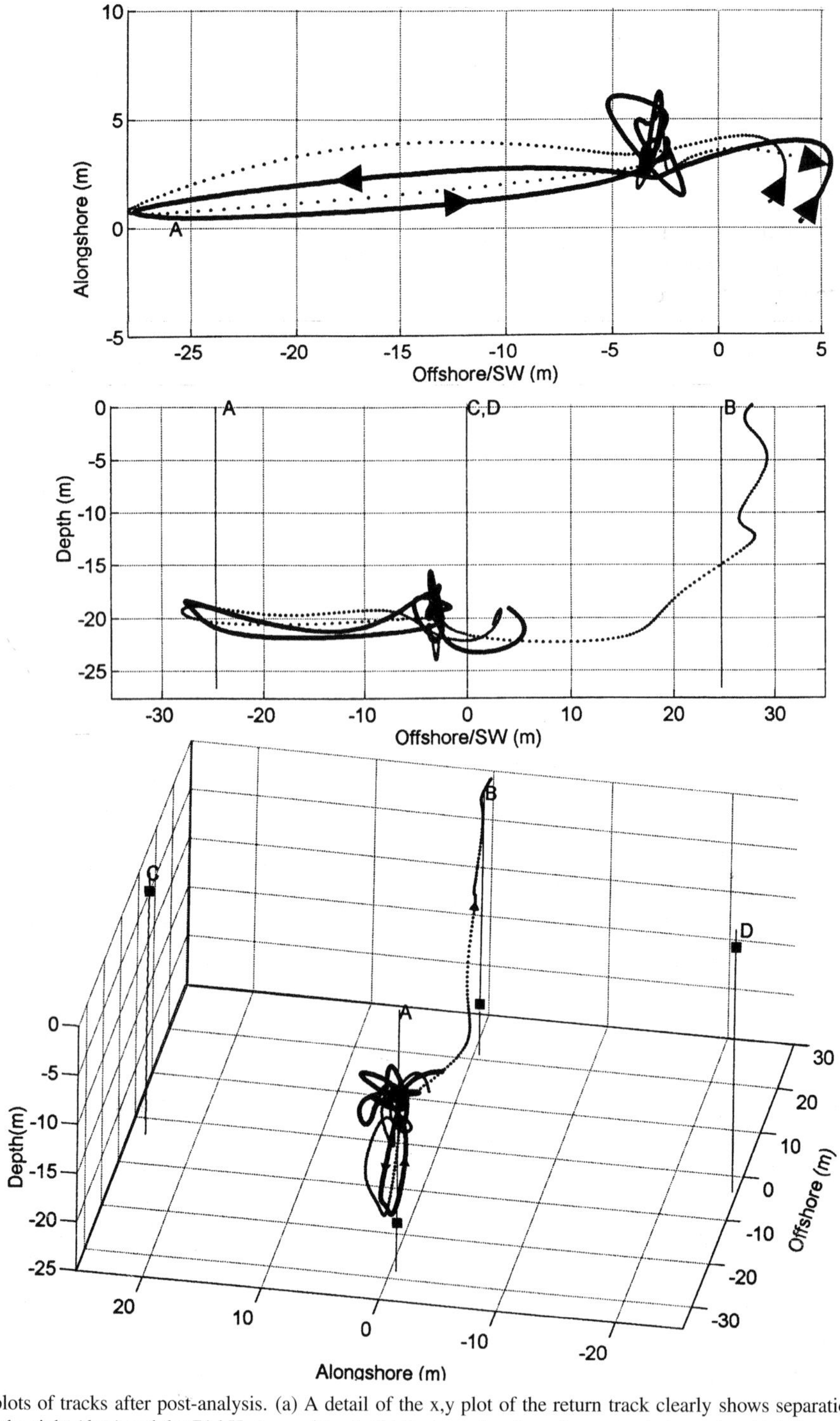


*Figure 3.* Matlab plots of tracks after post-analysis. (a) A detail of the x,y plot of the return track clearly shows separation between the 76.8 kHz transmitter on the right (dots) and the 71 kHz transmitter (solid line) moving from the center to buoy A, which then come together on the return. Arrowheads indicate the direction of swimming; the width of the line is approximately equal to the standard deviation in the x,y plane (0.3m) of a fixed position pinger. Data from Figure 2 were fitted with a cubic spline and positions reconstructed for each 1.5s pulse interval, so the dot graphs convey time for the 76.8 kHz transmitter. Distance between points indicates velocity. See details in Figure 2. (b) A vertical cross-section (x,z; A-B vs. Depth) illustrating the diver's return to a boat at buoy B with the 76.8 kHz transmitter after leaving the 71 kHz transmitter the mooring near. Note the apparent vertical motion of 71 kHz discussed in the text. (c) This 3-D view really displays 4-D, including the time based dots. The diver's swim, simulated squid movements and vertical movements due to swells all show up in this view.

in a fixed position inside the array as a reference. This would make it possible to correct for buoy movements due to swell or changes in wind and current on a moment-to-moment basis in post-analysis, but such corrections are not yet included in routine analysis.

Accuracy dropped off quickly outside the array, but probably averaged 3–5 m in a 100 m radius, except in the worst zones (Wartzok et al., 1992a). We are currently funded to conduct detailed tests of accuracies outside the array to the limits of reception in connection with studies of the use of the system to monitor diver safety. Hopefully, formation of a RAPT Network will accelerate the pace of analysis beyond that of the present 1994 data!

## Discussion

There are many different ways of using RAPT, and each application superimposes constraints on those that are fundamental to the approach. The squid study had four major constraints:

(a) Squid weighing as little as 70 g (sneaker males) are important in the mating process.
(b) Squid are pelagic and their behaviors had to be monitored in three dimensions.
(c) Even small squid can swim at speeds approaching 1 m $s^{-1}$ and average 0.17 m $s^{-1}$ over an entire day (Sauer et al., 1997).
(d) Rough weather and limited harbors along this coast make recharging buoys expensive, high risk and sometimes impossible, so power consumption was minimized.

Three-buoy systems and VRAP-A software can provide 3-D positions by using telemetered data from depth transducers (V16D-3H), but this would only have allowed us to look at large male squid (> 400 g) behavior. In this case, such data would certainly have been the least significant aspect of the mating process (Sauer et al., 1997), and in many situations small animal behavior can be critical. Even with larger animals, avoiding the requirement for a depth transducer makes it possible to monitor other variables, such as temperature, salinity and/or jet pressure in squid. However, the three-buoy system can work with buoys separated by as much as a half kilometer, depending on transmitter power and conditions. To use the four-buoy system for maximum accuracy within a maximum volume requires that all array dimensions be similar. Thus, the horizontal array dimensions should not exceed water depth by more than about 50%. Fortunately, because the squid were highly concentrated for mating, this was not a problem. The four-buoy system should theoretically perform optimally in deeper water, although there may be deployment problems.

With post-analysis, the present buoys worked better than we expected and will probably give adequate behavioral coverage with 8 s sampling of up to three targets traveling at speeds approaching 0.5 m $s^{-1}$. We are now applying the filter and splining techniques to extensive squid tracks with this sort of data outside the array and should soon be able to provide more details. We are also examining continuous, uninterrupted tracks of individuals which indicate little about interactions, but seem likely to give detailed records of individual behavior to much higher speeds. Acoustic systems capable of sampling multiple frequencies simultaneously are feasible, but there are power tradeoffs for increased onboard processing and time tradeoffs for radio relays. Increasing radio power to increase reliability not only increases power drain, but also increases the problems of licensing systems. One of the barriers to RAPT cooperation and collaboration is access to radio frequencies and this would likely be exacerbated by increased power. New radio designs allow a variety of frequencies to be synthesized, so that buoys can be moved between countries, but there is a price in power consumption for this convenience.

Only extensive experience in a wide range of applications by groups with a wide range of constraints and skills will allow us to discover the general rules for optimizing this approach. We hope that a network of users to facilitate communication and share equipment, experience and ideas will help this exciting new technology fulfill its promise for increasing our understanding of the underwater world. Since humans have demonstrated that we now have the power to destroy the marine environment, it is critical that we learn how to manage and conserve it.

## Acknowledgements

We acknowledge support from FRD (South Africa), SASMIA (South Africa), SFRI (South Africa), NSERC (Canada).

## References


Augustyn, C. J., M. R. Lipinski, W. H. H. Sauer, M. J. Roberts & B. A. Mitchell-Innes, 1994. Chokka squid on the Agulhas Bank: life history and ecology. S. Afr. J. Sci. 90: 143–154.

Bégout Anras, M. L., R. A. Bodaly, P. M. Cooley, R. J. P. Fudge & K. H. Mills, 1998. Habitat utilization and swimming behaviour of lake whitefish during spawning season in a boreal lake integrating acoustic telemetry and G.I.S. this volume, in press.

Chapman, C. J., A. D. F. Johnstone & A. L. Rice, 1975. The behaviour and ecology of the Norway lobster, *Nephrops noevegicus* L. Proc. 9th Eur. Mar. Biol. Symp.: 59–74.

Bjordal, A, A. Engas, A. V. Soldal & J. T. Ovredal, 1993. A new radio link telemetry system: experiences from tracking fish and crustaceans. ICES cm 1993/B:23, 6 pp.

Hawkins, A. D., D. N. MacLennan, G. G. Urquhart & C. Robb, 1974. Tracking cod, *Gadus morhua* L., in a Scottish sea loch. J. Fish. Biol. 6: 225–236.

Lagardère, J. P., M-L. Bégout, H. Breton & J. B. C. Claret, 1995. The effects of illumination, temperature and oxygen concentration on swimming activity of turbot *Psetta maxima* (Linne 1758). Fish. Res. 24: 165–173.

Lagardère, J. P., J. J. Ducamp, L. Favre, J. Mosneron Dupin & M. Sperandio, 1990. A method for quantitative evaluation of fish movements in salt ponds by acoustic telemetry. J. exp. mar. Biol. Ecol. 141: 221–236.

Mather, J. & R. K. O'Dor, 1991. Foraging strategies and predation risk shape the natural history of juvenile *Octopus vulgaris*. Bull. Mar. Sci. 49: 256–269.

Maynard, D. R. & D. M. Webber, 1987. Monitoring the movements of snow crab (*Chionoecetes opilio*) with ultrasonic telemetry. Oceans '87 Proc. 3: 962–966.

McKibben, J. N. & D. R. Nelson, 1981. A portable real-time, x-y plotting system for ultrasonic tracking of fish. In F.M. Long (ed.), Proc. 3rd Int. Conf. Wildlife Telemetry: 105–115.

O'Dor, R. K., Y. Andrade, M. Roberts, W. H. H. Sauer, M. Smale & D. M. Webber, 1996. Upwelling impact on spawning behaviour of tracked *Loligo vulgaris*. Benguela Dynamics Symp., Cape Town.

O'Dor, R. K. & S. Seino, 1997. Monitoring marine reserves with radio-acoustic positioning and telemetry (RAPT) systems. Univ. British Columbia. Fish. Cent. Res. Rpts. 5: 14–15.

O'Dor, R. K. & D. M. Webber, 1991. Ultrasonic telemetry. The Lobster Newsletter 4: 1–2.

O'Dor, R. K., D. M. Webber, W. H. H. Sauer, M. Roberts & M. Smale, 1995. High resolution, 3D tracking of squid on spawning grounds. In C. Cristalli, C. J. Amlaner, Jr. & M. R. Neuman (eds), Proc. 13th Int. Symp. Biotelemetry, Un. Arkansas Press, Fayetteville: 193–198.

O'Dor, R. K., D. M. Webber & F. Voegeli, 1989. A multiple buoy acoustic/radio telemetry system for automated positioning and telemetry of physical and physiological data. In C. J. Amlaner (ed.), Proc. 10th Int. Symp. Biotelemetry, Un. Arkansas Press, Fayetteville: 444–452.

Sauer, W. H. H., M. J. Roberts, M. R. Lipinski, M. J. Smale, R. T. Hanlon, D. M. Webber & R. K. O'Dor, 1997. Choreography of the squid's 'nuptial dance'. Biol. Bull. 192: 203–207.

Solomon, D. J. & E. C. E. Potter, 1988. First results with a new fish tracking system. J. Fish Biol. 33 (suppl. A): 127–132.

Wartzok, D., R. Elsner, H. Stone, B. P. Kelly & R. W. Davis, 1992b. Under-ice movements and the sensory basis of hole finding behavior in ringed and Weddell seals. Can. J. Zool. 92: 682–687.

Wartzok, D., S. Sayegh, H. Stone, J. Barchak & W. Barnes, 1992a. Acoustic tracking system for monitoring under-ice movements of polar seals. J. Acoust. Soc. Am. 92: 682–687.

Watkins, W. A. & W. E. Schevill, 1972. Sound source location with a three-dimensional hydrophone array. Deep-Sea Res. 19: 691–706.

*Hydrobiologia* **371/372:** 9–17, 1998.
J.-P. Lagardère, M.-L. Bégout Anras & G. Claireaux (eds), Advances in Invertebrates and Fish Telemetry.



# A comparison of theoretical estimates of the errors associated with ultrasonic tracking using a fixed hydrophone array and field measurements

G. W. Smith[1], G. G. Urquhart[2], D. N. MacLennan[2] & B. Sarno[2]
[1] *Freshwater Fisheries Laboratory Field Station, 16 River Street, Montrose DD10 8DL, Scotland*
*E-mail: g.w.smith@marlab.ac.uk*
[2] *Marine Laboratory, P.O. Box 101, Victoria Road, Aberdeen AB11 9DB, Scotland*

*Key words:* ultrasonic telemetry, fixed hydrophone array, position errors, multi-paths

## Abstract

Consideration of position fixing errors associated with fixed array acoustic tracking techniques has generally concentrated upon the consequences of inaccuracies in the timing of the arrival of the ultrasonic pulse at each hydrophone, the overall accuracy of the system being determined by combining timing errors with information on the position of the transmitter in the array. This paper presents a mathematical treatment which develops previous studies, allowing the distribution of position errors to be predicted from this same information and also making it possible to quantify multi-path effects and the resulting errors in position fixing.

These theoretical considerations are compared with the distributions of 1000 position fixes generated from each of five stationary transmitters using a fixed hydrophone array. In each case the transmitters were on the sea-bed within the boundary of the hydrophone array. Four of them were monitored in what would generally be considered unfavourable tracking conditions on an underwater reef characterised by deep crevices and a dense covering of *Laminaria* sp. Although the total area encompassing all estimated positions obtained during the field measurements varied considerably between tags, all exceeded 150 $m^2$. The position errors of the four tags associated with the reef were large compared to theoretical estimates based upon expected errors in time measurement.

It is concluded that multi-path effects and random noise can make a substantial contribution to the errors associated with fixing the positions of tagged individuals, particularly when tracking conditions are difficult and the signal level reaching any given hydrophone is low relative to the background noise. Thus timing errors and the general position within the hydrophone array are only contributory factors in the determination of the total error, which may be more strongly influenced by the micro topography in the vicinity of the tag.

## Introduction

Sound travels at a predictable speed through water. If a pulsed signal is detected at a series of hydrophones at known positions, the position of the sound source can be calculated from the differences in time taken for the sound to reach each of the hydrophones successively more distant from the source (Watkins & Schevill, 1972; MacLennan & Hawkins, 1977). A number of tracking systems based on this principle have been used to track sound producing animals (Freitag & Tyack, 1993) as well as acoustically tagged individuals (Hawkins et al., 1974; Lagardère et al., 1990; Glass et al., 1992; Keats & Steele, 1992; Juell & Westerberg, 1993; Engas et al., 1996). Several transmitters may also be tracked simultaneously using such techniques (Smith et al., 1993; Sarno et al., 1994).

Consideration of position fixing errors associated with fixed array acoustic tracking techniques has generally centred upon the consequences of inaccuracies in the timing of the arrival of the ultrasonic pulse at each hydrophone (MacLennan & Hawkins, 1977;

Juell & Westerberg, 1993). Such treatments conclude that errors in position estimates vary predictably with the position of the signal source in the hydrophone array, being generally smallest near the middle of the array and increasing towards the perimeter and beyond. Multi-path effects (reflections) and random noise also contribute to errors in estimating the positions of tagged individuals, however, particularly when the signal level reaching any given hydrophone is relatively low compared to the background noise, but such effects have previously not been taken into account.

This present study presents a mathematical analysis which allows the distribution of position errors to be predicted from inaccuracies in the timing of the arrival of the ultrasonic pulse at each hydrophone, together with information on the position of the transmitter in the array. The analysis also makes it possible to quantify multi-path effects and the resulting errors in position fixing. The predictions derived from such an approach are compared with the distribution of estimated positions of five stationary transmitters using a hydrophone array on the seabed.

## Theory

Following MacLennan et al. (1976) and Juell and Westerberg (1993), consider an array of three hydrophones at positions $H_1$, $H_2$ and $H_3$ (Figure 1). The pinger is at a point P in the plane of the hydrophones. The position of P may be calculated from the arrival times $t_i$ of pulses recorded at the respective hydrophones (I = 1, 2, 3). The time when the pinger transmits is unknown, but differences in the arrival times define hyperbolic coordinates which locate P. The lines joining any two pairs of hydrophones may be used as the baselines for the hyperbolic coordinates, for example:

$$a_1 = c\,(t_2 - t_3);\ a_2 = c\,(t_3 - t_1), \tag{1}$$

where c is the sound speed in water. The position of P is defined by the intersection of the hyperbolae defined by $a_1$ and $a_2$. For convenience we use the subscripting convention that $a_i$ is the coordinate for the hydrophone pair not including $H_i$.

As shown in Figure 1, $\varphi_1$, $\varphi_2$ and $\varphi_3$ are half the angles subtended by the hydrophone baselines at P, with the same subscripting convention ($2\varphi_1$ is the angle subtended by $H_2H_3$ etc.). These angles are key parameters in the analysis of errors in fixing positions.

In practice, there are various measurement errors to be considered, so let Q be the estimated position as calculated from the measured $a_1$ and $a_2$. Q is some distance $\Delta r$ from the true position P. Errors can arise in the time measurement, from multi-path effects or other causes, but the combined effect may be expressed as equivalent time errors $\Delta t_i$ in the measurements. Thus the true arrival times are $T_i = t_i + \Delta t_i$, and the correct hyperbolic coordinates are given by $(a_1 + \Delta a_1, a_2 + \Delta a_2)$ where

$$\Delta a_1 = c\,(\Delta t_2 - \Delta t_3);\ \Delta a_2 = c\,(\Delta t_3 - \Delta t_1). \tag{2}$$

The position error $\Delta r$ is defined by the area QAPB bounded by the four hyperbolae shown in Figure 2. Assuming that the errors $c\Delta t_i$ are small compared with the distances between the hydrophones, the error area is approximated by a parallelogram formed by the tangents to the hyperbolae, and $\Delta r$ is the length of the diagonal PQ.

An important property of hyperbolae in the present context is that each tangent at P (or Q) bisects the angle subtended by the relevant baseline. Thus the angle between the tangents to the hyperbolae defined by $a_1$ and $a_2$, opposite to $H_3$, is $(\varphi_1 + \varphi_2)$.

Furthermore, it can be shown that the perpendicular distance between the hyperbolae $(a_i)$ and $(a_i + \Delta a_i)$ is $\Delta a_i/(2 \sin \varphi_i)$. Noting that $\varphi_3 = \pi - (\varphi_1 + \varphi_2)$, we find that the sides of the error parallelogram are $\Delta a_1/(2 \sin \varphi_1 \sin \varphi_3)$ and $\Delta a_2/(2 \sin \varphi_2 \sin \varphi_3)$ respectively. The angle QAP $= \pi -$ AQB is simply $\varphi_3$, so the distance error $\Delta r$ is given by the equation

$$(2\,\Delta r \sin \varphi_3)^2 = (\Delta a_1/\sin \varphi_1)^2 + (\Delta a_2/\sin \varphi_2)^2 - 2\,\Delta a_1\,\Delta a_2 \cos \varphi_3/(\sin \varphi_1 \sin \varphi_2). \tag{3}$$

We require $\Delta r$ to be expressed in terms of the timing errors $\Delta t_i$. Substituting from (2) in (3), and noting that

$$2 \sin \varphi_1 \sin \varphi_2 \cos \varphi_3 = \sin^2 \varphi_1 + \sin^2 \varphi_2 - \sin^2 \varphi_3$$

some algebraic manipulation of the equations leads to the result

$$(2\,\Delta r \sin \varphi_1 \sin \varphi_2 \sin \varphi_3)^2 = (c\,\Delta t_1)^2 \sin^2 \varphi_1 + (c\,\Delta t_2)^2 \sin^2 \varphi_2 + (c\,\Delta t_3)^2 \sin^2 \varphi_3$$

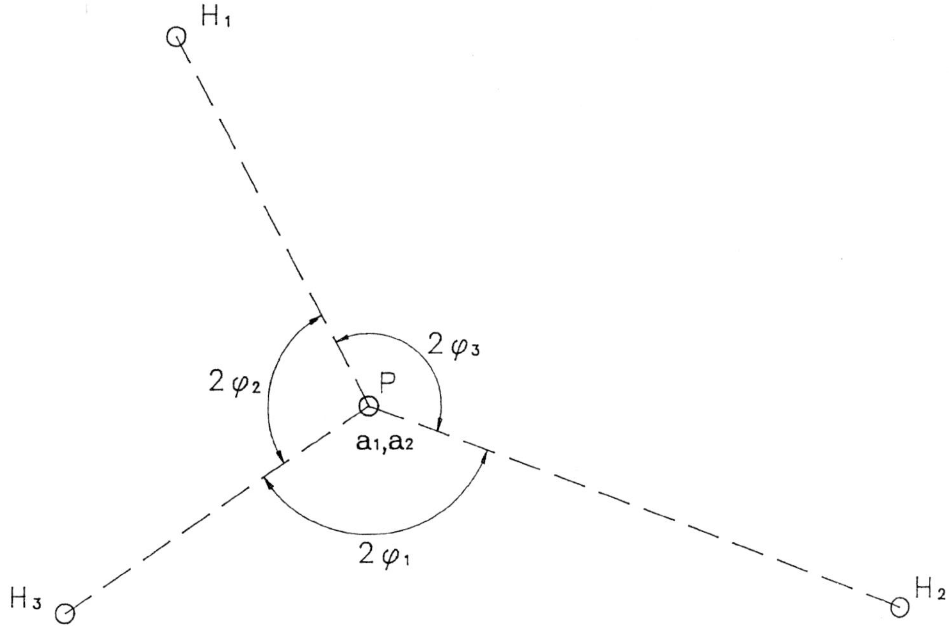


*Figure 1.* Theoretical geometry of a tracking array. The pinger, P, is in the plane of hydrophones $H_1$ to $H_3$ at a position determined by hyperbolae ($a_1$, $a_2$).

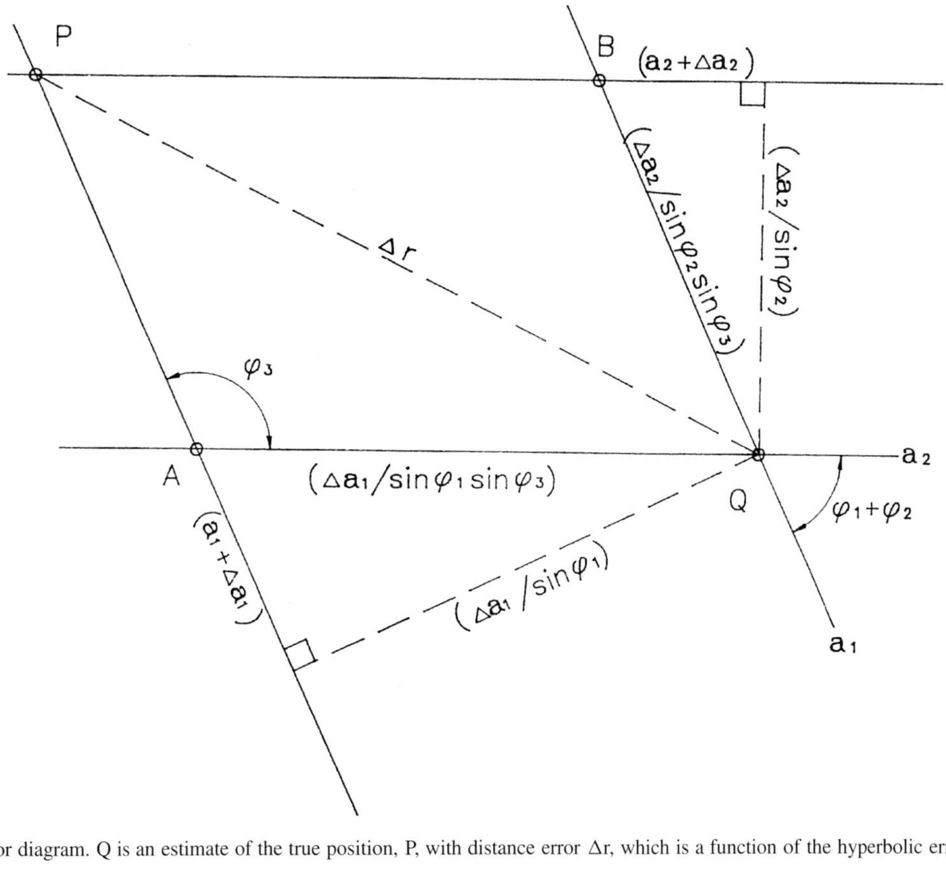


*Figure 2.* Error diagram. Q is an estimate of the true position, P, with distance error $\Delta r$, which is a function of the hyperbolic errors $\Delta a_1$ and $\Delta a_2$.

$$-2\sin\varphi_1 \sin\varphi_2 \cos\varphi_3 (c^2\ \Delta t_1\ \Delta t_2)$$

$$-2\sin\varphi_2 \sin\varphi_3 \cos\varphi_1 (c^2\ \Delta t_2\ \Delta t_3)$$

$$-2\sin\varphi_3 \sin\varphi_1 \cos\varphi_2 (c^2\ \Delta t_3\ \Delta t_1). \quad (4)$$

We suppose that each of the error terms ($c\ \Delta t_i$) has some statistical distribution, not necessarily Gaussian (normal) but with a mean of zero and variance $V_i$ which is the mean of $(c\ \Delta t_i)^2$. We further suppose that the $\Delta t_i$ are uncorrelated with each other. Thus for $I \neq j$, the mean of ($c^2\ \Delta t_i\ \Delta t_j$) is zero. The position error is usefully described by the root-mean-square (RMS) of $\Delta r$, say D. Taking means of each term in (4), the last three terms vanish and so

$$D = (V_1 \sin^2\varphi_1 + V_2 \sin^2\varphi_2 + V_3 \sin^2\varphi_3)^{1/2}/(2\sin\varphi_1 \sin\varphi_2 \sin\varphi_3). \quad (5)$$

The factors involving sines in (5) depend on the position of the pinger. Ignoring for the present any position-dependence of the $V_i$, D increases with distance from the array when P is outside the triangle of hydrophones. When P is inside the triangle, D varies slightly with position, except close to a hydrophone where one of the $\varphi_i$ is indeterminate. There is an optimum position for minimum error at the point defined by the condition

$$V_1 \sin^2\varphi_1 = V_2 \sin^2\varphi_2 = V_3 \sin^2\varphi_3 \quad (6)$$

which is generally within the array but some distance from any hydrophone. If the $V_i$ are all equal, then the error is minimum when the subtended angles $\varphi_i$ are all 60°. We note that the optimum position according to this criterion is not in general the same as the centroid of the triangle formed by the three hydrophones.

If the $V_i$ are not all equal, equations (5) and (6) still apply to the magnitude of the position error. However, the error will have directionality in that case. A scatter plot of repeated measurements (with the pinger stationary) would show more variation in some directions than others. In principle, the anisotropy of the scatter plot could be used to determine the relative magnitudes of the timing errors. For example, if all the error was in the arrival time at one hydrophone, then the estimated positions would be distributed along a hyperbola with foci at the other two hydrophones.

*Table 1.* The area encompassing a given proportion of fixes for each tag

| Proportion of fixes (%) | Area covered by the position fixes ($m^2$) | | | | |
|---|---|---|---|---|---|
| | Tag a | Tag b | Tag c | Tag d | Tag e |
| 50 | 23 | 107 | 15 | 23 | 20 |
| 75 | 55 | 278 | 37 | 47 | 56 |
| 90 | 107 | 428 | 74 | 82 | 117 |
| 100 | 205 | 528 | 160 | 155 | 216 |

## Materials and methods

Acoustic tags (designed at the MAFF Fisheries Laboratory, Lowestoft) transmitted at a nominal 75 kHz and produced short (2 ms) bursts of ultrasound repeated approximately once per second. The peak acoustic output was approximately 165 dB re 1 $\mu$Pa at 1 m. Estimation of the tag positions was carried out in an identical way to previously published studies (Hawkins et al., 1974), using a fixed hydrophone array linked to an onshore computer controlled system which timed the delays between each acoustic signal reaching hydrophones at known positions and logged the resulting cartesian coordinates (Hawkins et al., 1980; Hawkins & Urquhart 1983; Urquhart & Smith, 1992).

The study was conducted in Loch Ewe, a sea loch on the west coast of Scotland. An underwater reef, approximately 150 m long by 30 m at its widest point was situated in a depth of 12 to 15 m of water (MLWS) in the middle of the hydrophone array (Figure 3). It has a dense covering of *Laminaria* sp. and forms a large isolated feature surrounded by an expanse of flat sand. Four of the tags (a,b,c and e Figure 3) were located on the reef, the fifth (tag d) was located on a smaller outcrop of rock some 250 m away (Sarno et al., 1994). Further details of the site are given by Johnstone et al. (1991) and Glass et al. (1992).

## Results

Data from five tags which had been regurgitated by fish during the course of a tracking study were monitored and a total of 1000 position fixes were generated for each tag. Although the total area encompassing all estimated positions varied considerably between tags, all exceeded 150 $m^2$ (Figure 3, Table 1). 90% of fixes were contained within an area of between 74 and 117 $m^2$ in four out of five tags (Table 1), the area associated with the fifth tag being somewhat larger at 428 $m^2$.

The distribution of estimated positions varied between tags (Figure 4). Tag d generated a unimodal distribution of position estimates, variation being similar in all directions. Tag c, on the other hand, while still generating a unimodal distribution of position estimates, displayed greater variation along the vertical axis compared to the horizontal axis. Tags a and e showed distinct bimodal distributions in their estimated position fixes while there was no clear distribution of points in any direction for tag b. Consideration of the general direction along which most variation occurs can indicate which hydrophone contributes most to errors in estimating the position of the tag. For example, following the logic developed at the end of the Theory section, above, suppose hydrophones 1, 2 and 3 are used to fix the position of a transmitting tag and the timing at 3 is the least precise. Then the estimated positions will tend to lie on a hyperbola defined by the (more precise) timings at 1 and 2. Applying this argument to the observed distributions of fixes for each of the four tags on the reef suggests that hydrophone 3 (Figure 3) was responsible for most of the error associated with estimating the position of tags a,c and e. There was no clear evidence that any one specific hydrophone contributed most to the errors associated with the fixing the position of tag b.

Table 2 gives the predicted RMS position error for each tag (Formula 5), assuming equal variance in all directions and a $\Delta t$ of either 0.5 ms or 1.0 ms. The standard deviations (S.D.) of the observed position estimates of each tag from a calculated mean position are also shown. For tag d, which was positioned off the reef (Figure 3) and most closely satisfies the assumption of equal variance in all directions (Figure 4), the observed S.D. is very close to the predicted RMS calculated from a $\Delta t$ value of 1 ms. Predicted RMS values underestimate the observed S.D. for all other tags.

It is also possible to predict RMS values for tags whose position estimates may suffer from multi-path effects and thus which may show greater variance in some directions than others. Assume the position of tag c, for example, is computed using time delay data from hydrophones 1, 2 and 3 (Figure 3). The path from the tag to the furthest hydrophone (hydrophone 3) is a direct distance of approximately 136 m. A single reflection from the surface (depth 12 m) increases the path length by approx 2 m which effectively in-

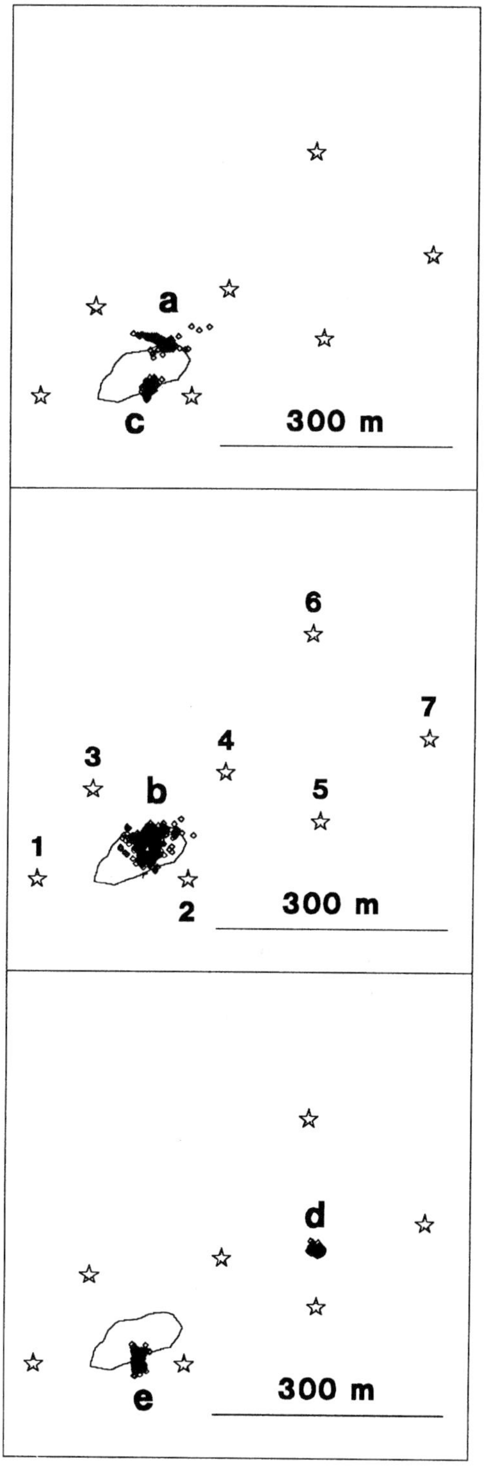


*Figure 3.* The distribution of position fixes estimated for each tag (labelled a to e) together with the hydrophones in the range (stars) and the outline of the underwater reef.

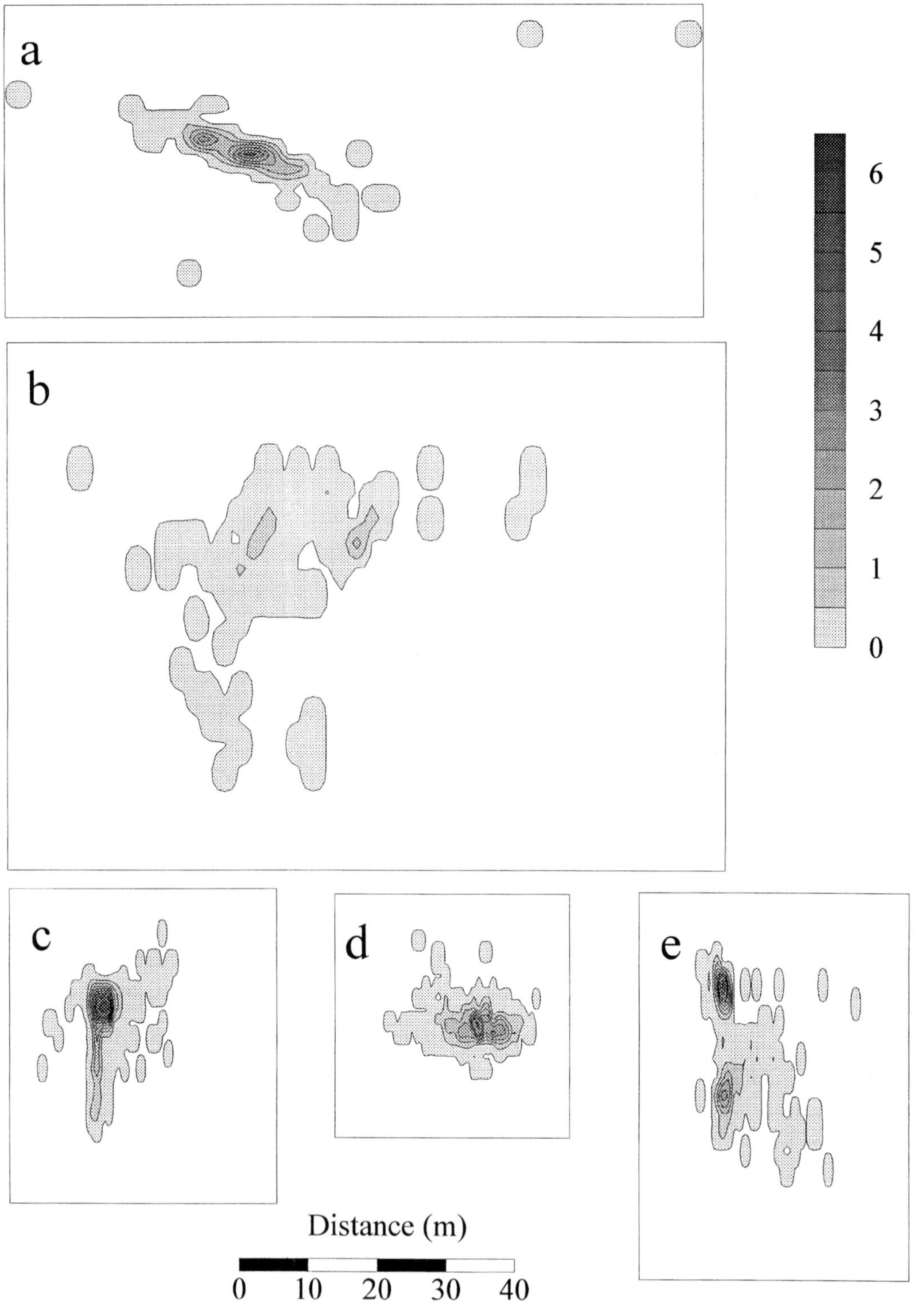


*Figure 4.* The distribution of estimated positions for each tag (labelled a to e). Contour lines enclose areas which account for equal proportions of the positions estimates generated. The horizontal bar indicates the scale on the ground while the vertical bar indicates the scale of the probability distributions as a percentage of the position estimates generated for each tag. Both scales are the same for all tags.

*Table 2.* Extent of the errors for each tag predicted by formula 5 (see text) together with the standard deviations of the observed distributions of position estimates from a calculated mean position. Also shown are the estimated notional mean speeds (± S.D.) between successive position fixes for each of the tags

| Tag[1] | RMS error, D (m) (formula 5) for $\Delta t = 0.5$ ms | RMS error, D (m) (formula 5) for $\Delta t = 1.0$ ms | Standard deviation obtained from sample variance[2] (m) | Observed mean speed (± S.D.) between successive position estimates ($m\ s^{-1}$) |
|---|---|---|---|---|
| a | | | | |
| 2,3,4 | 0.91 | 1.83 | 5.76 | 0.122 (±0.118) |
| b | | | | |
| 1,2,3 | 1.03 | 2.06 | 7.93 | 0.128 (±0.107) |
| c | | | | |
| 1,2,3 | 1.18 | 2.35 | 3.13 | 0.053 (±0.041) |
| d | | | | |
| 5,6,7 | 1.03 | 2.05 | 2.12 | 0.065 (±0.044) |
| e | | | | |
| 1,2,3 | 1.20 | 2.40 | 3.35 | 0.064 (±0.059) |

[1] Also shown are the 3 hydrophones used to estimate the tag positions (Figure 3).
[2] RMS distance from calculated mean position for each tag.

creases $\Delta t$ for hydrophone 3. Assume direct paths to hydrophones 1 and 2 with $\Delta t = 1$ ms (as above). The three variances $V_1$, $V_2$ and $V_3$ can then be calculated separately, and the predicted RMS increases from 2.35 m (Table 2) to 4.10 m.

As position estimates may be subject to significant error, even stationary transmitters appear to move and thus generate a notional speed of movement between position fixes. Mean speed can thus be estimated for each tag in the present study (Table 2) which can be thought of as the 'background noise' associated with the tracking system as it operates within that precise area of the tracking array.

## Discussion

Chapman et al. (1975) calculated the error associated with estimating the position of tagged Norway lobster, *Nephrops norvegicus* (L.) using a fixed array tracking system by considering the accuracy with which the arrival of pulses at any given hydrophone could be timed. An accuracy of ±0.5 ms obtained in the study was believed to result in an error of ±1 m within the array. Similarly Juell and Westerberg (1993), tracking Atlantic salmon (*Salmo salar* L.) in a sea cage, estimated an error of approximately 0.1 m in the middle of the array when timing the arrival of pulses by means of a computer with a clock speed of 12 Khz. In this case, however, Juell and Westerberg (1993) suggested that variation in the rise time of the envelope of the signal pulse may have been a greater source of uncertainty than limitations in the instrument used to time the arrival of the signal. For a tag frequency of 100 kHz, they estimated an error in timing the arrival pulses of between 0.1 and 1 ms, corresponding to an error of between 0.2 and 1.5 m in the middle of the array.

While these estimates may reflect the expected errors when signal to noise ratio is high and multi-path effects are at a minimum, they greatly underestimate the actual position fixing errors measured in four of the five tags in the present study, despite the fact that tags were in areas within the hydrophone array (Figure 3) where consideration of time of arrival inaccuracies would predict relatively small position fixing errors (MacLennan et al., 1976; Juell & Westerberg, 1993). Further, and also contrary to previous expectations, tags in similar general positions with respect to the overall geometry of the array also showed substantial differences in the overall distribution of their position errors (Figure 4).

Such findings arise because micro topography in the direct paths between the transmitters and the hydrophones may play a dominant role in determining the errors associated with position fixing, for example where dense vegetation and deep crevices contribute to a low signal to noise ratio and signal reflection (multi-paths). While the effects of multi-paths has received some attention from workers tracking radio tagged individuals by means of triangulation (Springer, 1979; Lenth, 1981; Garrott et al., 1986) they have received little systematic attention in relation to acoustic tracking systems based upon the measurement of time delays.

The developments in the mathematical theory of position fixing errors presented here has allowed errors to be calculated which reflect more closely those experienced in the field (Table 2). However, multi-path errors are to a large extent dependant on the exact conditions along the acoustic paths and it is therefore not possible to predict in advance either the overall error associated with tracking in a particular area or the precise distribution of position fixes. Previously, decisions on the combination of hydrophones which should be used to estimate the position of a given transmitter were generally made on the basis of the position of the tag in relation to the overall geometry of the array. The present analysis suggests that a more effective strategy may be to design the array in such a way that more than the minimum number of 3 hydrophones are in a position to collect time delay data from any given part of the range, and to analyse the data after they have been collected to identify and eliminate those hydrophones most associated with position fixing errors. In this respect, it is important to collect the raw time delay data to facilitate future analysis.

Animals moving over a study area may, of course, be expected to generate more complex distributions of position fixes than the stationary transmitters presented here. It may therefore be argued that the task of discriminating the actual movements of the animal from artifacts generated by the tracking system become unrealistically complex. Again, the solution may be found in collecting the raw time delay data. Analysis after the event may make it possible to disregard those distribution patterns which are dependant upon data from a particular hydrophone.

A number of species, because of a combination of their behaviour patterns and the habitat they occupy will inevitably be unsuitable candidates for ultrasonic tracking using fixed hydrophone array systems. Before undertaking such studies, limited trials using stationary transmitters to investigate the magnitude of the errors expected at a given site may be a quick and relatively inexpensive method of determining whether ultrasonic tracking techniques are capable of providing movement information at the levels of accuracy required.

## Acknowledgements

We would like to thank Tony Hawkins, Dick Shelton and Alan Youngson for comments on an earlier draft of this paper.

## References

Chapman, C. J., A. D. F. Johnstone & A. L. Rice, 1975. The behaviour and ecology of the Norway lobster, *Nephrops norvegicus* (L.). Proc. 9th Eur. mar. biol. Symp.: 59–74.

Engas, A., S. Lokkeborg & J. T. Ovredal, 1996. Natural and fishing-gear induced behaviour of tagged fish, studied by means of a stationary positioning system. In E. Barras & J. C. Philippart (eds), Underwater Biotelemetry. Proceedings of the First Conference and Workshop on Fish Telemetry in Europe. University of Liege, Belgium, 4–6 April 1995: 203–211.

Freitag, L. E. & P. L. Tyack, 1993. Passive acoustic localization of the Atlantic bottlenose dolphin using whistles and echolocation clicks. J. Acoust. Soc. Am. 93: 2197–2205.

Garrott, R. A., G. C. White, R. M. Bartmann & D. L. Weybright, 1986. Reflected signal bias in biotelemetry triangulation systems. J. Wildl. Mgmt 50: 747–752.

Glass, C. W., A. D. F. Johnstone, G. W. Smith & W. R. Mojsiewicz, 1992. The movements of saithe (*Pollachius virens* L.) in the vicinity of an underwater reef. In I. G. Priede & S. M. Swift (eds), Wildlife Telemetry: Remote Monitoring and Tracking of Animals. Ellis Horwood, Chichister: 328–341.

Hawkins, A. D., D. N. MacLennan, G. G. Urquhart & C. Robb, 1974. Tracking cod, *Gadus morhua* L., in a Scottish sea loch. J. Fish Biol. 6: 225–236.

Hawkins, A. D. & G. G. Urquhart, 1983. Tracking fish at sea. In A. G. MacDonald & I. G. Priede (eds), Experimental Biology at Sea. Academic Press, London: 103–166.

Hawkins, A. D., G. G. Urquhart & G. W. Smith, 1980. Ultrasonic tracking of juvenile cod by means of a large spaced hydrophone array. In C. J. Amlaner & D. W. MacDonald (eds), A Handbook of Biotelemetry and Radio Tracking. Pergamon Press, Oxford: 461–470.

Johnstone, A. D. F., C. W. Glass, W. R. Mojsiewicz & G. W. Smith, 1991. The movements of saithe (*Pollachius virens* L.) as revealed by acoustic tracking. Prog. Underwater Sci. 16: 61–73.

Juell, J.-E. & H. Westerberg, 1993. An ultrasonic telemetric system for automatic positioning of individual fish used to track Atlantic salmon (*Salmo salar* L.) in a sea cage. Aquacult. Eng. 12: 1–18.

Keats, D. W. & D. H. Steele, 1992. Diurnal feeding of juvenile cod (*Gadus morhua*) which migrate into shallow water at night in eastern Newfoundland. J. Northw. Atlan. Fish. Sci. 13: 7–14.

Lagardère, J.-P., J.-J. Ducamp, L. Favre, J. Mosneron Dupin & M. Spérandio, 1990. A method for the quantitative evaluation of fish movements in salt ponds by acoustic telemetry. J. exp. mar. Biol. Ecol. 141: 221–236.

Lenth, R. V., 1981. On finding the source of a signal. Technometrics 23: 149–154.

MacLennan, D. N. & A. D. Hawkins, 1977. Acoustic position fixing in fisheries research. Rapp. P.-v. Reun. Cons. Int. Explor. Mer. 170: 88–97.

MacLennan, D. N., G. G. Urquhart & A. D. Hawkins, 1976. Acoustic tracking of fish and fishing gears. International Council for the Exploration of the Sea, C.M. 1976/B:24.

Sarno, B., C. W. Glass, G. W. Smith, W. R. Mojsiewicz & A. D. F. Johnstone, 1994. A comparison of the movements of two gadoid species in the vicinity of an underwater reef. J. Fish Biol. 45: 811–817.

Smith, G. W., C. W. Glass, A. D. F. Johnstone & W. R. Mojsiewicz, 1993. Diurnal patterns in the spatial relationships between saithe, *Pollachius virens*, schooling in the wild. J. Fish Biol. 43 (Supplement A): 315–325.

Springer, J. T., 1979. Some sources of bias and sampling error in radio triangulation. J. Wildl. Mgmt 43: 926–935.

Urquhart, G. G. & G. W. Smith, 1992. Recent developments of a fixed hydrophone array system for monitoring movements of aquatic animals. In I. G. Priede & S. M. Swift (eds), Wildlife Telemetry: Remote Monitoring and Tracking of Animals. Ellis Horwood, Chichister: 342–353.

Watkins, W. A. & W. E. Schevill, 1972. Sound source location by arrival-times on a non-rigid three-dimensional hydrophone array. Deep Sea Res. 19: 691–706.

*Hydrobiologia* **371/372:** 19–28, 1998.
*J.-P. Lagardère, M.-L. Bégout Anras & G. Claireaux (eds), Advances in Invertebrates and Fish Telemetry.*



# Selection of optimal positioning intervals in fish tracking: an experimental study on *Barbus barbus*

E. Baras
*University of Liège, Laboratory of Fish Demography and Aquaculture*
*Mailing address: Chemin de la Justice, 10 B-4500 Tihange, Belgium; E-mail: E.Baras@ulg.ac.be*

*Key words:* telemetry, positioning frequency, home range, mobility, *Barbus barbus*

## Abstract

Selecting an optimum sampling frequency is a key problem in fish telemetry studies. High frequency positioning gives an accurate idea of the actual mobility pattern of the animal but may be impossible due to environmental constraints, prohibitive cost or when using long life transmitters operating on duty cycles. Low frequency positioning may prove effective, but only for a minimum loss of accuracy (i.e. decrease in the estimates of home range or mobility). This paper analyses the respective benefits and limitations of the 'high cost-high return' and 'low cost-unknown return' strategies. It is based on a day-by-day tracking study of the cyprinid fish *Barbus barbus* (L.) in Belgian waters (River Ourthe, seven fish, 112 tracking days each), and re-examines the estimates of home range and distance travelled, as if the fish had been located at longer time intervals (2–28 d). The estimates are compared to the original data set, in order to quantify the resulting loss of accuracy, and to provide predictive models for future studies.

Increasing the time interval between successive locations from 2 to 28 d introduced losses of accuracy, which were higher for mobility (21.1 to 70.0%) than for home range estimates (0.5 to 35.3%). Based on cost-effectiveness, the rationale for tracking *B. barbus* was determined as once a week. This positioning frequency produced consistent home range estimates (5% of inaccuracy) but rather poor estimates of mobility (*circa* 40% of inaccuracy). A simple predictive model of loss of accuracy depending on time interval, and a multiple regression model, integrating the heterogeneity between subsamples, were generated and tested with an independent data set. Both substantially reduced the inaccuracy of estimates (from 39.4 to 14.1% with 7 d intervals and from 19.7 to 5.7% when using four parallel subsamples at 7 d intervals).

These results clearly indicate that positioning fish at intervals longer than one day always generates a bias, but may prove effective, at least when the resulting inaccuracy can be predicted and (partly) corrected. The analysis and proposed predictive models open promising perspectives for long term studies relying on the use of transmitters operating on duty cycles (e.g. reproductive homing, role of experience on behaviour, ...).

## Introduction

Telemetry nowadays represents a performing set of methodologies enabling to improve our knowledge and understanding of the way wild or captive animals react and adapt to their environment (Winter, 1983; Priede & Swift, 1992; Baras & Lagardère, 1995). The problems most frequently addressed in animal telemetry refer to the innocuity of tagging procedures (Summerfelt & Smith, 1990; Baras et al., 1998), the accuracy of horizontal and vertical positioning techniques (Lagardère et al., 1988; 1996; Axelsen et al., 1997; Smith et al., 1997), the reception range and endurance of transmitters (Voegeli & Pincock, 1996), which are limited in weight and bulk by the size of the animal being tagged. A further key problem refers to the frequency of positioning the animal.

Day-night alternation represents a major zeitgeber which rules the use of time and space by most teleost fishes (Thorpe, 1978; Helfman, 1986; Boujard & Leatherland, 1992). Depending on species, age, size and environmental conditions, fish may use the same habitat all day long both for feeding and resting (e.g. *Salmo trutta*, Heggenes et al., 1993; Ovidio et al., 1998), or different habitats for different activities at different times of the 24 h cycle (e.g. *Cyprinidae*; Clough, 1997; Fredrich et al., 1997). Although this is not a rule of thumb, residence and feeding areas, when spatially distinct are frequently connected by functional links: fish exploit a feeding area from a precise resting place (or set of neighbouring resting places), move to another resting place as they select another feeding area and vice versa (Baras, 1992; 1996; Fredrich et al., 1997). For these reasons, the daily positioning of fish equipped with telemetry devices, coupled with continuous tracking over some 24-h cycles, may give a representative picture of the way they utilise their environment.

In many tracking studies, however, constraints by weather or environment, as well as increasing costs of manpower and travel, make it impossible to go and locate fish each and every day (e.g. tropical environment: Tito de Morais & Raffray, 1996; subarctic regions: Beddow et al., 1997). This limitation also applies to long term studies, which aim to test for the consistency of individual behavioural responses over a period of which the duration exceeds the battery life of continuously transmitting units. These studies rely on the use of 'intelligent' transmitters, which operate on programmed duty cycles (e.g. transmitting one day and resting six days each week) and operate over much longer periods than simple transmitters (Voegeli & McKinnon, 1996). What would be the reason for it, tracking studies may rely on a set of locations collected at *n* day intervals, with n supposed to correspond to the best compromise between accuracy, feasibility and cost. Basically, this tradeoff between 'high cost-high return' and 'low cost-unknown return' tracking strategies strongly depends on environment, animal behaviour and parameter examined (home range, mobility, habitat selection) and should be addressed by answering the following questions: 'How much accuracy do I lose when increasing the time interval between successive locations, and does this loss of accuracy outweigh the advantage granted by less frequent positioning?'. These questions somehow resemble the interrogations that caused scientists to re-analyse, at the light of tracking data, the apparent restricted movements produced by mark-recapture studies (Gowan et al., 1994; Philippart & Baras, 1996). In any case, no appropriate answer can be given without a proper reference set, consisting of daily locations.

This paper relies on a tracking study on the cyprinid *Barbus barbus* in Belgian waters (1989–1990; Baras, 1992), for which fish were positioned each and every day. In order to address the aforementioned questions, it re-examines the results as if they would have been collected at looser positioning frequencies (2 to 28 d intervals). It aims to:

(a) quantify the loss of accuracy depending on positioning frequency and behavioural parameter considered (home range and overall mobility)
(b) determine a rationale for a cost-effective positioning frequency
(c) test whether the loss of accuracy can be modelled from sampling frequency, and to evaluate the adequacy of this model by applying it to an independent data set (collected in 1991).

## Material and methods

In 1989, 1990 and 1991, 12 barbel (four, three and five, respectively) were equipped with surgically implanted radio transmitters and radio tracked in the River Ourthe (main tributary of the R. Meuse in Belgium), in the same study area, centred at Hamoir-sur-Ourthe. From late winter to early summer, the fish were positioned each and every day, at a time of the day when they usually rest in their residence area (10:00–14:00; Baras, 1995). This time of the year encompasses the spawning period (Hancock et al., 1976; Baras, 1994), when *B. barbus* occupies the largest seasonal home range and shows the highest mobility (Baras, 1992; 1993a; Lucas & Batley, 1996), thus the conditions when the risk of underestimating these parameters is maximum. A minimum of 112 daily positions (excluding the initial release point and final location) was collected for each fish during the two first tracking years.

In order to measure the loss of accuracy due to non daily positioning, the data sets of daily positions from 1989 and 1990 were sampled as if the fish had been located at two, four, seven, fourteen and twenty eight days intervals. For each sampling frequency, the number of data subsets generated by subsampling was equal to the duration (days) of the time interval between consecutive locations (e.g. 14 subsets of eight locations for a 2 week interval). Positions on the day

*Table 1.* Home range and mobility of the seven *B. barbus* radio tracked in the River Ourthe in 1989 and 1990, and loss of accuracy (LA, %) when increasing the time interval between consecutive locations. For each positioning frequency (PF, every n days), the values given are the mean loss and coefficient of variation (%) over the *n* subsets generated by the subsampling process.

| Fish | | | Home range over 112 d Loss of accuracy (%) if positioning every | | | | | | Distance travelled over 112 days (sum of net journeys). Loss of accuracy (%) if positioning every | | | | | |
|---|---|---|---|---|---|---|---|---|---|---|---|---|---|---|
| n° | FL (mm) | Sex | Size (m) | 2 d | 4 d | 7 d | 14 d | 28 d | Total (m) | 2 d | 4 d | 7 d | 14 d | 28 d |
| 1 | 316 | M | 1760 | 0.0 (0.0) | 0.0 (0.0) | 6.7 (18.9) | 33.0 (46.9) | 53.7 (65.6) | 9255 | 25.3 (19.7) | 41.8 (24.2) | 54.1 (20.5) | 74.3 (60.0) | 86.1 (86.4) |
| 2 | 403 | F | 2050 | 0.0 (0.0) | 0.0 (0.0) | 0.0 (0.0) | 0.0 (0.0) | 2.4 (5.5) | 5006 | 20.1 (9.6) | 34.1 (12.4) | 45.0 (15.6) | 49.4 (14.3) | 55.5 (9.8) |
| 3 | 271 | M | 1200 | 0.0 (0.0) | 0.4 (0.5) | 3.1 (3.5) | 4.9 (3.1) | 5.8 (2.4) | 3661 | 26.8 (27.4) | 44.3 (27.2) | 58.2 (5.7) | 63.8 (9.9) | 68.3 (5.7) |
| 4 | 530 | F | 9760 | 0.5 (0.7) | 1.1 (0.8) | 1.4 (0.7) | 47.0 (88.6) | 70.4 (135.6) | 26521 | 2.2 (3.3) | 13.2 (3.1) | 19.1 (4.5) | 58.8 (83.3) | 78.2 (136.6) |
| 5 | 378 | F | 14870 | 0.4 (0.6) | 4.6 (5.6) | 5.5 (4.4) | 13.6 (17.0) | 53.0 (88.3) | 37900 | 8.3 (8.9) | 21.8 (10.2) | 26.6 (4.1) | 33.9 (17.2) | 65.4 (93.5) |
| 6 | 342 | M | 730 | 2.7 (4.0) | 7.2 (7.4) | 10.0 (4.9) | 16.2 (8.3) | 32.0 (26.1) | 9119 | 45.3 (1.4) | 61.8 (18.0) | 72.1 (10.3) | 79.0 (20.8) | 87.7 (34.9) |
| 7 | 525 | F | 11210 | 0.2 (0.3) | 0.3 (0.2) | 2.8 (5.3) | 18.0 (19.0) | 29.5 (18.3) | 25854 | 14.2 (1.3) | 18.9 (1.7) | 24.4 (4.5) | 38.6 (22.6) | 49.1 (21.3) |
| Mean loss of accuracy (%) | | | | 0.5 (0.8) | 1.9 (2.1) | 4.2 (5.4) | 19.0 (25.4) | 35.3 (48.8) | | 21.1 (10.2) | 33.7 (13.8) | 42.8 (9.3) | 56.8 (32.6) | 70.0 (55.5) |
| Models | | | $LA = -3.210 + 1.398$ PF; $R^2 = 0.517$; $F = 35.3$, df $= 34$, $p < 0.0001$ | | | | | | LA (%) $= 7.835 + 42.684$ Log (PF); $R^2 = 0.554$; $F = 41.0$, df $= 34$, $p < 0.0001$ | | | | | |

of release (day 0) and final tracking day (day $n+1$) were added to each data set.

For each data set at each positioning frequency, two estimates of dispersion and mobility were calculated: the home range (i.e. longitudinal distance between the most upstream and downstream locations) and the overall distance travelled (sum of distances between consecutive locations). The home range and mobility calculated from subsamples were always inferior or equal to those produced by daily positioning, enabling to calculate a relative loss of accuracy (%), depending on the time interval between consecutive locations. The consistency between different subsamples at the same positioning frequency was deduced from the coefficient of variation of the mean loss of accuracy. For both mobility and home range, the relationships between the loss of accuracy, time interval between consecutive locations, and consistency of samples were examined by simple and multiple regression analyses. The resulting models were later applied to the independent data set (fish tracked over four weeks in spring 1991) to test for their reliability.

In order to determine a rationale for a cost-effective positioning frequency in barbel tracking studies, I compared the gain of accuracy (%) granted by an increase of positioning frequency to the corresponding increase of budget (%). Search time was not considered as a key factor in this study, as fish were dispersed over the study area, which was entirely surveyed on each tracking day. Budgets were calculated for standard tracking projects running five to thirty fish, equipped each with a 200 US $ radio transmitter. The consumables (use of receiver, antenna, batteries), labour and travel costs were estimated at 74 US $ per tracking day, based on the River Ourthe study (4 h of travel and tracking, one technician, 55 km of travel). Indicative project costs depending on positioning frequency and number of fish tracked are given in Table 2. Increasing the positioning frequency was empirically considered as cost effective as long as the gain of accuracy (%) granted by more frequent

*Table 2.* Comparison between the gain of accuracy in estimating the home range and mobility of *B. barbus*, resulting from an increase of positioning frequency, and the related increase of project cost, depending on the number of fish tracked. * indicates that the ratio between the gain of accuracy (%) and increase of project cost (%) is higher than 1 (rationale for cost-effective tracking).

| Interval between locations (days) | Labour and travel cost (US $) | Total cost of Project (US $) five fish | Total cost of Project (US $) 30 fish | Increase % of project cost when decreasing the interval between locations (from ...to) | five fish | 30 fish | Increase (%) of accuracy when decreasing the interval between locations: Home range | Distance travelled |
|---|---|---|---|---|---|---|---|---|
| 28 | 444 | 1444 | 6444 | | | | | |
| | | | | (28 to 14 d) | +17.73 | + 3.97 | +16.30* | +13.22* |
| 14 | 700 | 1700 | 6700 | | | | | |
| | | | | (14 to 7 d) | +37.18 | + 9.43 | +14.88* | +14.04* |
| 7 | 1332 | 2332 | 7332 | | | | | |
| | | | | (7 to 4 d) | +38.08 | +12.11 | + 2.10 | + 9.10 |
| 4 | 2220 | 3220 | 8220 | | | | | |
| | | | | (4 to 2 d) | +64.35 | +25.21 | + 1.40 | +12.57 |
| 2 | 4292 | 5292 | 10292 | | | | | |
| | | | | (2 to 1 d) | +78.31 | +40.26 | + 0.54 | +21.33 |
| 1 | 8436 | 9436 | 14436 | | | | | |

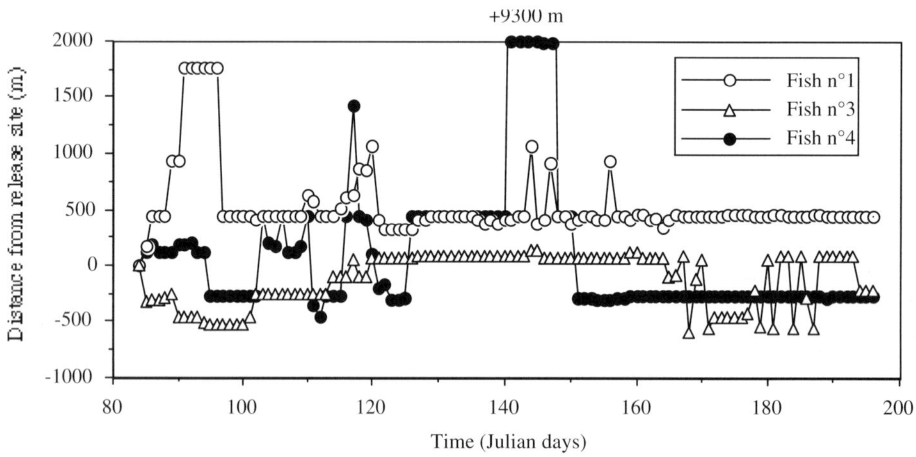


*Figure 1.* Illustration of mobility patterns of *Barbus barbus* in the River Ourthe in 1989 (fish n° 1, 3 and 4 positioned daily over 112 d).

positioning exceeded the related increase (%) of total project cost.

## Results

From late winter to early summer, the fish utilised home ranges with a longitudinal extension from 730 to 14 870 m. The running sum of their net daily journeys over 112 d ranged from 3 661 to 37 900 m. These values exceeded by far the corresponding home ranges (Table 1), and indicated that barbel did not exploit their home range in a sequential way. Homing movements, consecutively to short or long journeys, were frequent, especially during the spawning season (mid-May to mid-June, Figure 1).

### *Measures of inaccuracy resulting from non-daily positioning*

Positioning frequencies lower than once a day produced contrasted data sets, which either closely resembled the original set or differed from it by a substantial margin, depending on fish behaviour and parameter

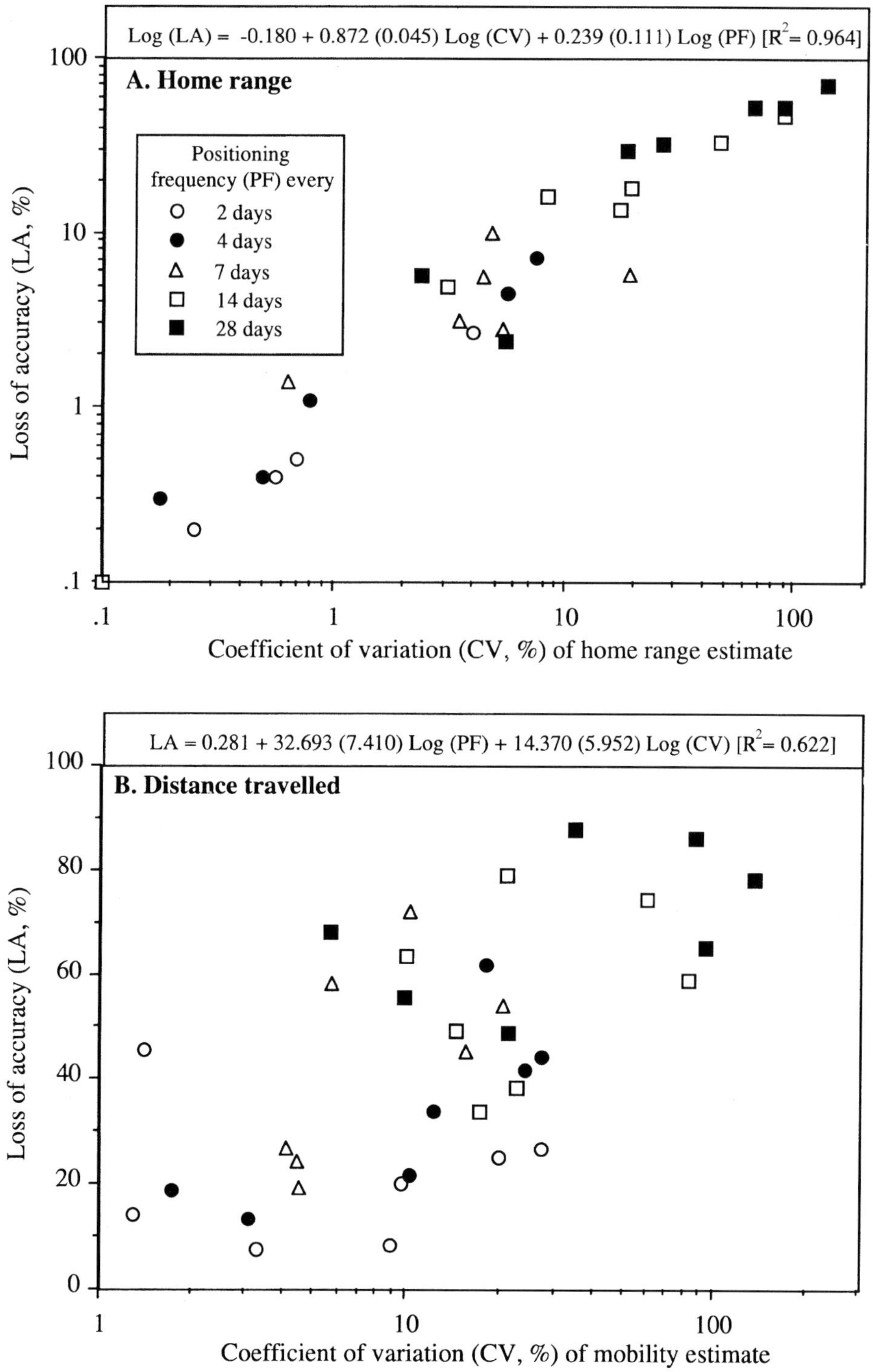


*Figure 2.* Comparison between the loss of accuracy in estimates of home range and mobility of *B. barbus* when increasing the time interval ($n$ days) between consecutive locations and the corresponding heterogeneity (coefficient of variation) of the $n$ data sets generated by the subsampling process. The models are generated by stepwise multiple regression analyses, with the variables (± standard error) presented in order of integration into the analysis.

considered (Table 1). For fish n° 2, which occupied a small home range, positioning at least every two weeks would have produced the same home range estimate as daily positioning, and a four week interval would only have generated a loss of accuracy of 2.4%. By contrast, the corresponding loss of accuracy for fish n° 4, which made a long upstream spawning excursion then homed, would have amounted to over 70%. Considering the home ranges of all seven fish, it becomes obvious that intervals shorter or equal to 7 d between successive locations generate acceptable losses of accuracy (always $\leq$ 10%, mean < 5.0%) whereas longer intervals cause unacceptable errors (means of 19.0 and 35.3 for 14 and 28 d intervals, respectively). For each fish and positioning frequency, increasing the interval between consecutive locations caused a much higher loss of accuracy for mobility than for dispersion. The distance travelled by the fish was underestimated by an average of 21.1% when the fish were positioned every two days and by more than 40% for time intervals longer or equal to one week. This loss of accuracy was essentially due to the mobility pattern of the fish, which frequently travelled in between the same residences (Figure 1).

*Cost effective tracking*

For both behavioural parameters, the rationale for cost-effective tracking was determined as once a week (Table 2). When the time interval between successive locations was reduced from 28 to 14 d, then from 14 d to 7 d, each percent of additional budget granted at least one additional percent of accuracy, at least for a project involving 30 fish. This assumption was no longer valid for more frequent positioning, especially for home range. The rationale was never reached for projects running five transmitters. Based on the cost of equipment, labour and travel, and on the observed gain of accuracy given in Table 2, the rationale for cost effective tracking would be reached with a minimum of 8 transmitters for bimonthly positioning and of 19 transmitters for weekly positioning.

*Predictive models of loss of accuracy*

Both for home range and mobility, the loss of accuracy of estimates given by non-daily positioning was significantly correlated with the time interval between successive locations, which accounted for 51.7 and 55.4% of the variation of accuracy (Table 1). As illustrated by Figure 2, the higher the loss of accuracy, the higher the heterogeneity between subsamples at the same positioning frequency. Both variables were significantly retained by a stepwise multiple-regression analysis and substantially improved the resolution of the models, especially for estimating home range (Figure 2).

These models were evaluated with the set of tracking data from 1991, using time intervals of 7 d (rationale for cost-effective positioning), 2 d and a corresponding sample of four sets of weekly intervals, enabling to test the multiple regression model. The results of this evaluation are given in Table 3. Weekly positioning resulted into losses of accuracy of 4.5% for home range and of 39.4% for mobility, which were similar to the losses presented in Table 1 (4.2 and 42.8%, respectively). Applying the simple predictive models, based on positioning frequency only, reduced the underestimation but introduced some overestimation, especially for home range. For this reason, the modelled estimates for home range were slightly less accurate than the estimates based on raw data (7.0% v. 4.2%, whereas mobility was more accurately estimated (14.1% v. 39.4%). As expected, positioning every two days gave more accurate estimates (0.0% and 19.7% of inaccuracy for home range and mobility, respectively). Modelling with the simple regression models produced similar trends as with weekly positioning: a marked reduction of inaccuracy for mobility (down to 10.7%) and a slight increase of inaccuracy for home range (0.4%). When the set of locations at 2 d interval was used as a group of four sets of locations at 7 d intervals, in order to apply the multiple regression models, relying on positioning frequency and heterogeneity between estimates, the mean accuracy was strongly improved: only 0.2% and 5.7% of inaccuracy, for home range and mobility, respectively. Furthermore, the multiple regression model strongly reduced the discrepancy between the accuracy of estimates for individual fish, especially for mobility. No error of more than 8.0% was introduced by this model, whereas observations or simple regression models generated much higher errors (respectively 37.2 and 20.8%; for positioning at 2 d intervals).

## Discussion

When reviewing the recent literature on fish tracking in rivers, I noticed that numerous studies relied on non-daily positioning. The spawning migrations of trout *Salmo trutta* (L.) have been documented using various positioning frequencies: daily (Evans, 1994),

*Table 3.* Relative loss of accuracy in estimating the home range (longitudinal extension, m; upper table) and mobility (distance travelled, m; lower table) of *B. barbus* depending on positioning frequency and use of correcting models. Fish A, B, C, D and E were located daily over 28 consecutive days in spring 1991. Fish A: male, 234 mm; B: male, 338 mm; C: male: 350 mm; D: female, 402 mm; E: male, 252 mm. Observed columns give estimates based on observations at 1 d, 2 d and 7 d intervals. Values given are the mean estimate (coefficient of variation, %) and loss of accuracy (%) against the values obtained by daily positioning. Modelled columns were generated by the correcting models obtained from the 1989–1990 data sets (simple models in Table 1; multiple linear models in Figure 2). Values given are the mean estimate and inaccuracy (under- or overestimation) against the values obtained by daily positioning.

| Home range (m) | Observed | | | Modelled | | |
|---|---|---|---|---|---|---|
| | 1-d interval | 2 d interval | 7 d interval | simple, 2 d interval | simple, 7 d interval | multiple, 7 d interval ($N=4$) |
| Fish A | 95 | 95 (0.0) | 95 (0.0) | 94.6 | 101.7 | 95.1 |
| | | 0.0% | 0.0% | −0.4% | +7.0% | +0.1% |
| Fish B | 560 | 560 (0.0) | 552.5 (0.9) | 557.8 | 591.4 | 557.8 |
| | | 0.0% | −1.3% | −0.4% | +5.6% | −0.4% |
| Fish C | 850 | 850 (0.0) | 668 (31.6) | 846.6 | 715.0 | 848.3 |
| | | 0.0% | −21.4% | −0.4% | −15.9 | −0.2% |
| Fish D | 1720 | 1720 (0.0) | 1720 (0.0) | 1713.2 | 1841.0 | 1718.3 |
| | | 0.0% | 0.0% | −0.4% | +7.0% | +0.1% |
| Fish E | 75 | 75 (0.0) | 75 (0.0) | 74.7 | 80.3 | 75.1 |
| | | 0.0% | 0.0% | −0.4% | +7.0% | +0.1% |
| Mean inaccuracy | | [0.0%] | [4.5%] | [0.4%] | [8.5%] | [0.2%] |
| Fish A | 385 | 310 (8.1) | 220 (21.7) | 390.8 | 392.0 | 415.0 |
| | | −19.5% | −42.9% | +1.5% | −1.8% | +7.8% |
| Fish B | 720 | 650 (5.5) | 555 (1.8) | 819.5 | 989.0 | 759.6 |
| | | −9.7% | −22.9% | +13.8% | +37.4% | +5.5% |
| Fish C | 2050 | 1580 (3.2) | 973 (41.9) | 1992.0 | 1733.9 | 1994.7 |
| | | −22.9% | −52.5% | −2.8% | −15.4% | −2.7% |
| Fish D | 2880 | 2620 (8.4) | 1720 (4.5) | 3303.2 | 3065.0 | 2741.8 |
| | | −9.0% | −40.3% | +14.7% | +6.4% | −4.8% |
| Fish E | 215 | 135 (0.0) | 132 (11.3) | 170.2 | 235.2 | 231.8 |
| | | −37.2% | −38.6% | −20.8% | +9.4% | +7.8% |
| Mean inaccuracy | | [19.7%] | [39.4%] | [10.7%] | [14.1%] | [5.7%] |

once every two days (Brown & Mackay, 1995), twice a week (Meyers et al., 1992), once to four times a month (Clapp et al., 1990; Young, 1994; Økland et al., 1996). Similarly, the seasonal movements of cyprinids have often been studied relying on different time intervals between successive fixes (from 1 to 14 d; Lucas & Batley, 1996; Clough, 1997; Fredrich et al., 1997; Huber & Kirchhoffer, 1997; Lucas & Frear, 1997). In some studies, the positioning frequency varied depending on time of the year (e.g. Meyers et al., 1992; Young, 1994; Lucas & Batley, 1996). The loss of accuracy may be low for fish making straightforward migrations, like salmonids, whereas it can be substantial for fish making home range movements, like cyprinids.

This study on *Barbus barbus* provided evidence that low positioning frequencies could substantially bias the results, and permitted to quantify the loss of accuracy resulting from increasing time intervals between successive locations. Based on cost-effectiveness, the optimum positioning frequency for *B. barbus* in the River Ourthe would be once a week, essentially because the search time did not increase when increasing the time interval between successive locations. This could be substantially different in larger environments or situations where the fish may travel into distant tributaries. Positioning the barbel once a week in the River Ourthe would have caused a minor and acceptable loss of accuracy for home range (ca. 5%) but a considerable loss of accuracy for mobility (ca. 40% for 7 d interval, and still ca. 20% for 2 d interval). These findings support the idea that non-daily positioning may be a major shortcoming when analysing a biological phenomenon, and may account for the discrepancies between the interpretations of the same phenomenon proposed by different authors relying on different positioning frequencies. A recent study where trout were positioned each and every day (Ovidio et al., 1998, this volume) provided evidence that the migrations of trout were ruled by a combination of variations of environmental factors (flow and temperature), which apparently acted at the daily level. Positioning the trout at a lower frequency would have caused them to put forward the influence of different variables, depending on the day where the fish would have been positioned, and to join the conclusions of other papers, which documented a preponderant influence of either flow or temperature.

This study also indicated that, when estimating home range or mobility, the loss of information was proportional to the positioning frequency. These relationships further suggest that any comparison between the movement patterns of fish tracked at different time intervals should be considered with extreme caution. As a corollary, I would strongly recommend to use the same positioning frequency, consistently throughout a study, in order to collect homogeneous results (i.e. results sharing a consistent loss of accuracy), of which the accuracy can be improved by using predictive correcting models.

In this study, day-by-day positioning permitted to establish such models, which substantially reduced the inaccuracy of estimates, especially when using multiple regression models taking into account the heterogeneity between subsamples. As the mobility pattern exhibited by the fish presumably corresponds to an adaptive response to environmental conditions, the predictive models presented in this paper may unfortunately not be transposed to barbel populations in other environments, at least without a preliminary evaluation. Studies on other barbel populations indeed revealed that the overall pattern of space utilisation could differ to some extent from the pattern described in the River Ourthe (Baras, 1992; 1993a), especially in environments where the mobility of fish is restricted by man-made obstacles (e.g. Lucas & Batley, 1996). Based on the results illustrated here, similar correcting models could be produced by using a similar subsampling methodology on other existing data sets, even if these do not consist of daily positions. This process would imply to consider the real $n$ day interval as a 100% basis, to measure the loss of information due to subsampling and to extrapolate the curve down to $n = 1$. In any case, it is most obvious that this process would require tracking durations proportional to the time interval and could hardly be attempted when relying on very loose positioning frequencies ($n > 7$ d). Throughout this paper, daily positioning was considered as the standard with highest accuracy for estimating the home range and mobility of fish. However, it should be reminded that daily positioning may only give a fuzzy picture of the animal's life, which should be completed by more detailed investigations of their foraging habits over 24 h cycles.

This study on *B. barbus* was voluntarily focused on the time of the year when the fish were most mobile (Baras, 1993a; Lucas & Batley, 1996), and thus when the risk of underestimating their mobility was maximum. In these conditions, weekly positioning and use of correcting models would grant home range and mobility estimates with an error margin of less than 10 and 15%, respectively. Relying on the annual cycle of mobility of *B. barbus* in the River Ourthe (Baras, 1993a), the loss of accuracy resulting from weekly positioning in winter or summer would be much lower. During these periods, the probability that the fish moves from a residence to another between consecutive days in the River Ourthe is less than 10%, as a result of low temperatures in winter (Baras, 1995) or as an attempt to exploit more efficiently its activity area (optimal foraging strategies, see Baras, 1993b). Similarly, movements in autumn are far less frequent than during spring time. Based on past knowledge and on the results presented here, future studies on the ecology of *B. barbus* in the River Ourthe could thus efficiently rely on intelligent transmitters programmed to transmit one out of seven days. This duty cycle would increase their endurance by 250–300%, still grant accurate estimates and permit to compare the utilisation of time and space by the same individuals over several years. This could prove crucial to our understanding of the way long term experience affects fish behaviour, and of population genetics, through the analysis of reproductive homing (*sensu* fidelity to the same spawning ground).

## Acknowledgements

I wish to thank Dr. J. C. Philippart (FNRS and University of Liège) for promoting the behavioural ecology study, as well as B. Cherry, J. M. Lambert and G.

Rimbaud for their assistance in field tracking surveys. The study was supported by the Belgian Fonds National de la Recherche Scientifique and by the Ministry of Nature Conservation and Fisheries of the Région Wallonne.

## References


Axelsen, B., M. Holm & M. Steinset, 1997. Developing a high-accuracy acoustic fish positioning system for monitoring migratory behaviour of small, non-schooling fish. In M. L. Bégout-Anras, G. Claireaux & J. P. Lagardère (eds), Abstracts of the Second Conference on Fish Telemetry in Europe, CREMA l'Houmeau, France, April 5–9, 1997: 4 (abstract).

Baras, E., 1992. Etude des stratégies d'occupation du temps et de l'espace chez le barbeau fluviatile, *Barbus barbus* (L.). Cah. Éthol. 12: 125–442.

Baras, E., 1993a. Etude par biotélémétrie de l'utilisation de l'espace, chez le barbeau fluviatile, *Barbus barbus* (L.). Caractérisation et implications des patrons saisonniers de mobilité. Cah. Éthol. 13: 135–138.

Baras, E., 1993b. A biotelemetry study of activity centres exploitation by *Barbus barbus* in the River Ourthe. Cah. Éthol. 13: 173–174.

Baras, E., 1994. Constraints imposed by high densities on behavioural spawning strategies in the barbel, *Barbus barbus*. Folia Zool. 43: 255–266.

Baras, E., 1995. Seasonal activities of *Barbus barbus* (L.) – Effect of temperature on time-budgeting. J. Fish Biol. 46: 816–828.

Baras, E., 1996. Selection of residence area selection and non-reproductive homing in a shoaling freshwater teleost, the barbel *Barbus barbus* (L.). In E. Baras & J. C. Philippart (eds), Underwater Biotelemetry, Proceedings of the First Conference and Workshop on Fish Telemetry in Europe. University of Liège, Belgium: 47–58.

Baras, E., C. Birtles, L. Westerloppe, X. Thoreau, M. Ovidio, D. Jeandrain & J. C. Philippart, 1998. A critical review of surgery techniques for implanting telemetry devices into the body cavity of fish. In Y. Le Maho (ed), Proceedings of the 5th European Conference on Wildlife Telemetry. Strasbourg, France, 26–30 August 1996: 10 pp, in press.

Baras, E. & J. P. Lagardère, 1995. Fish telemetry in aquaculture: review and perspectives. Aquacult. Int. 3: 1–26.

Beddow, T. A., C. Deary & R. S. McKinley, 1997. Migratory and reproductive activity of radio-tagged Arctic charr (*Salvelinus alpinus* L.) in Northern Labrador. In M. L. Bégout-Anras, G. Claireaux & J. P. Lagardère (eds), Abstracts of the Second Conference on Fish Telemetry in Europe, CREMA l'Houmeau, France, April 5–9, 1997: 8 (abstract).

Boujard, T. & J. F. Leatherland, 1992. Circadian rhythms and feeding times in fishes. Envir. Biol. Fishes 35: 109–131.

Brown, R. S. & C. Mackay, 1995. Spawning ecology of cutthroat trout (*Oncorhynchus clarki*) in the Ram River, Alberta. Can. J. Fish. aquat. Sci. 52: 983–992.

Clapp, D. F., R. D. Clark & J. S. Diana, 1990. Range, activity and habitat of large, free-ranging brown trout in a Michigan Stream. Trans. am. Fish. Soc. 119: 1022–1034.

Clough, S., 1997. Preliminary results of using external attached miniature radio-transmitters to track the movements of *Leuciscus leuciscus*, in the River Frome, Dorset. In M. L. Bégout-Anras, G. Claireaux & J. P. Lagardère (eds), Abstracts of the Second Conference on Fish Telemetry in Europe, CREMA l'Houmeau, France, April 5–9, 1997: 14 (abstract).

Evans, D. M., 1994. Observations on the spawning behaviour of male and female adult sea trout, *Salmo trutta* L., using radio-telemetry. Fish. Mgmt. Ecol. 1: 91–105.

Fredrich, F., S. Ohmann & B. Curio, 1997. Spawning migrations and daily movements of chub (*Leuciscus cephalus*) in the Spree River. In M. L. Bégout-Anras, G. Claireaux & J.P. Lagardère (eds), Abstracts of the Second Conference on Fish Telemetry in Europe, CREMA l'Houmeau, France, April 5–9, 1997: 19 (abstract).

Gowan, C., M. K. Young, K. D. Fausch & S. C. Riley, 1994. Restricted movements in resident stream salmonids: a paradigm lost. Can. J. Fish. aquat. Sci. 51: 2626–2637.

Hancock, R. S., J. W. Jones & R. Shaw, 1976. A preliminary report on the spawning behaviour and the nature of sexual selection in the barbel, *Barbus barbus* (L.). J. Fish Biol. 9: 21–28.

Heggenes, J., O. M. W. Krog, O. R. Lindas, J. G. Dokk & T. Bremnes, 1993. Homeostatic behavioural responses in a changing environment: brown trout (*Salmo trutta*) become nocturnal during winter. J. anim. Ecol. 62: 295–308.

Helfman, G. S., 1986. Fish behaviour by day, night and twilight. In T. J. Pitcher (ed), The Behaviour of Teleost Fishes. London: Croom Helm: 366–387.

Huber, M. & A. Kirchhoffer, 1997. Radio telemetry as a tool to detect habitat preference in medium-sized rivers. A case study on the nase (*Chondrostoma nasus*), an endangered rheophilic cyprinid. In M. L. Bégout-Anras, G. Claireaux & J. P. Lagardère (eds), Abstracts of the Second Conference on Fish Telemetry in Europe, CREMA l'Houmeau, France, April 5–9, 1997: 22 (abstract).

Lagardère, J. P., J. J. Ducamp, L. Frikha & M. Sperandio, 1988. Ultrasonic tracking of common sole juveniles (*Solea vulgaris* Quensel, 1806) in a saltmarsh: methods and fish response to some environmental factors. J. appl. Ichthyol. 4: 87–96.

Lagardère, J. P., M. L. Bégout-Anras & V. Buchet, 1996. The acoustic positioning system as a valuable tool for estimating the well-being of fishes in aquaculture. In E. Baras & J. C. Philippart (eds), Underwater Biotelemetry, Proceedings of the First Conference and Workshop on Fish Telemetry in Europe. University of Liège, Belgium: 177–186.

Lucas, M. C. & E. Batley, 1996. Seasonal movements and behaviour of adult barbel *Barbus barbus*, a riverine cyprinid fish: implications for river management. J. appl. Ecol. 33, 1345–1358.

Lucas, M. C. & P. A. Frear, 1997. Effects of a flow-gauging weir on the migratory behaviour of adult barbel, a riverine cyprinid. J. Fish Biol. 50: 382–396.

Meyers, L. S., T. F. Thuemler & G. W. Kornely, 1992. Seasonal movements of brown trout in Northeast Wisconsin. N. am. J. Fish. Mgmt. 12: 433–441.

Økland, F., J. A. Jensen & B. O. Johnsen, 1996. Winter habitat and seaward migration of a Norwegian brown trout population. In E. Baras & J. C. Philippart (eds), Underwater Biotelemetry, Proceedings of the First Conference and Workshop on Fish Telemetry in Europe. University of Liège, Belgium: 161–171.

Ovidio, M., E. Baras, D. Goffaux, C. Birtles & J. C. Philippart, 1998. Environmental unpredictability rules the autumn migration of brown trout (*Salmo trutta* L.) in the Belgian Ardennes. Hydrobiologia 371/372: 263–274.

Ovidio, M., C. Birtles, E. Baras, J. C. Philippart & F. Giroux, 1998. Daily and seasonal movements of radio-tagged brown trout (*Salmo trutta*) in the Belgian Ardennes: mobility versus residency. In Y. Le Maho (ed), Proceedings of the 5th European

Conference on Wildlife Telemetry. Strasbourg, France, 26–30 August 1996: 8 pp, in press.

Philippart, J. C. & E. Baras, 1996. Comparison of tagging and tracking studies to estimate mobility patterns and home range in *Barbus barbus*. In E. Baras & J. C. Philippart (eds), Underwater Biotelemetry, Proceedings of the First Conference and Workshop on Fish Telemetry in Europe. University of Liège, Belgium: 3–12.

Priede, I. G. & S. M. Swift (eds), 1992. Wildlife telemetry: remote sensing and monitoring of animals. Ellis Horwood Ltd., Chichester, UK, 708 pp.

Smith, G. W., G. G. Urquhart, D. N. MacLennan & B. Sarno, 1997. A comparison of theoretical estimates of the errors associated with ultrasonic tracking using a fixed hydrophone array and field measurements. In M. L. Bégout-Anras, G. Claireaux & J. P. Lagardère (eds), Abstracts of the Second Conference on Fish Telemetry in Europe, CREMA l'Houmeau, France, April 5–9, 1997: 45 (abstract).

Summerfelt, R. C. & L. S. Smith, 1990. Anaesthesia, surgery, and related techniques. In C. B. Schreck & P. B. Moyle (eds), Methods for fish biology. American Fisheries Society, Bethesda, Maryland: 213–272.

Thorpe, J. E. (ed), 1978. Rhythmic activity of fishes. Academic Press, London.

Tito de Morais, L. & J. Raffray, 1996. Behaviour of *Hoplias aimara* during the filling phase of the Petit-Saut Dam (Sinnamary river, French Guyana, South America). In E. Baras & J. C. Philippart (eds), Underwater Biotelemetry, Proceedings of the First Conference and Workshop on Fish Telemetry in Europe. University of Liège, Belgium: 153–160.

Voegeli, F. A. & D. G. Pincock, 1996. Overview underwater acoustics as it applies to telemetry. In E. Baras & J. C. Philippart (eds), Underwater Biotelemetry, Proceedings of the First Conference and Workshop on Fish Telemetry in Europe. University of Liège, Belgium: 23–30.

Voegeli, F. A. & G. P. McKinnon, 1996. Recent developments in ultrasonic tracking systems. In E. Baras & J. C. Philippart (eds), Underwater Biotelemetry, Proceedings of the First Conference and Workshop on Fish Telemetry in Europe. University of Liège, Belgium: 235–241.

Winter, J. D., 1983. Underwater biotelemetry. In L. A. Nielsen & J. D. Johnsen (eds), Fisheries Techniques. Am. Fish. Soc. Bethesda, Maryland: 371–395.

Young, M. K., 1994. Mobility of brown trout in south-central Wyoming streams. Can. J. Zool. 72: 2078–2083.

*Hydrobiologia* **371/372:** 29–33, 1998.
*J.-P. Lagardère, M.-L. Bégout Anras & G. Claireaux (eds), Advances in Invertebrates and Fish Telemetry.*



# Inland migration study of sea trout (*Salmo trutta*) into the rivers Rhine and Meuse (The Netherlands), based on inductive coupling radio telemetry

André W. Breukelaar[1], Abraham bij de Vaate[2] & Koos T. W. Fockens[3]
[1]*Institute for Inland Water Management & Waste Water Treatment, P.O. Box 9072, 6800 ED Arnhem, The Netherlands (e-mail: a.breukelaar@riza.rws.minvenw.nl)*
[2]*Institute for Inland Water Management & Waste Water Treatment, P.O. Box 17, 8200 AA Lelystad, The Netherlands*
[3]*Nedap N.V., P.O. Box 6, 7140 AA Groenlo, The Netherlands*

*Key words:* sea trout (*Salmo trutta*), Atlantic salmon (*Salmo salar*), migration, inductive coupling radio telemetry, transponders

## Abstract

A new telemetric method, the NEDAP TRAIL System® , has been developed to study sea trout migration in the delta area of the rivers Rhine and Meuse. This method is based on inductive coupling between an antenna loop on the bottom of a river or canal and ferrite rod antenna incorporated in a tag of the transponder type. Field tests showed that the detection system functions well with a maximum antenna length of at least 550 m, a water depth of 15 m and a passing speed of the tag with 5–6 m $s^{-1}$ at maximum. A first group of 18 sea trout (of the total number of 600 that will be tagged) was tagged in December 1996. By March 1997 two of them had already been detected at different detection stations.

## Introduction

An important aim in the ecological rehabilitation of the rivers Rhine and Meuse is the return of anadromous fish species such as the Atlantic salmon (*Salmo salar* (L.)) and twaite shad (*Alosa fallax* (Lacépède)), which were common in The Netherlands in the past (Bij de Vaate, 1989; De Groot, 1990). One of the preconditions for their successful return is the improvement of inland migration from the Dutch coastal zone. However, possibilities for anadromous fish species to migrate into the rivers Rhine and Meuse have been drastically reduced during the last decades (De Haas, 1991). Natural outlets in the Dutch coast, that could serve as a migration route, have been blocked by dams and locks. Nowadays physically unhindered inland migration is only possible through the Nieuwe Waterweg, a man-made canal connecting the harbours of Rotterdam with the North Sea (Figure 1). For the management of the sluices in the dams, locks and other obstacles in the migration route of salmonids, information is needed on the routes they follow in the delta area of the rivers Rhine and Meuse during their journey to the spawning areas.

## Study area

Originally, the Haringvliet (Figure 1) was the most important entrance for salmonids (Bij de Vaate, 1989). However, after the 1953 flooding, the Rhine/Meuse estuary in the south-western part of The Netherlands was dammed. Migration into the River Rhine was also possible in the northern part of The Netherlands through the former Zuiderzee (called lake IJsselmeer after its damming in 1932) into the River Rhine. Little is known about the importance of this migration route for Atlantic salmon and sea trout. Catches of both species in the Zuiderzee were relatively low which means that this migration is to be considered of minor importance.

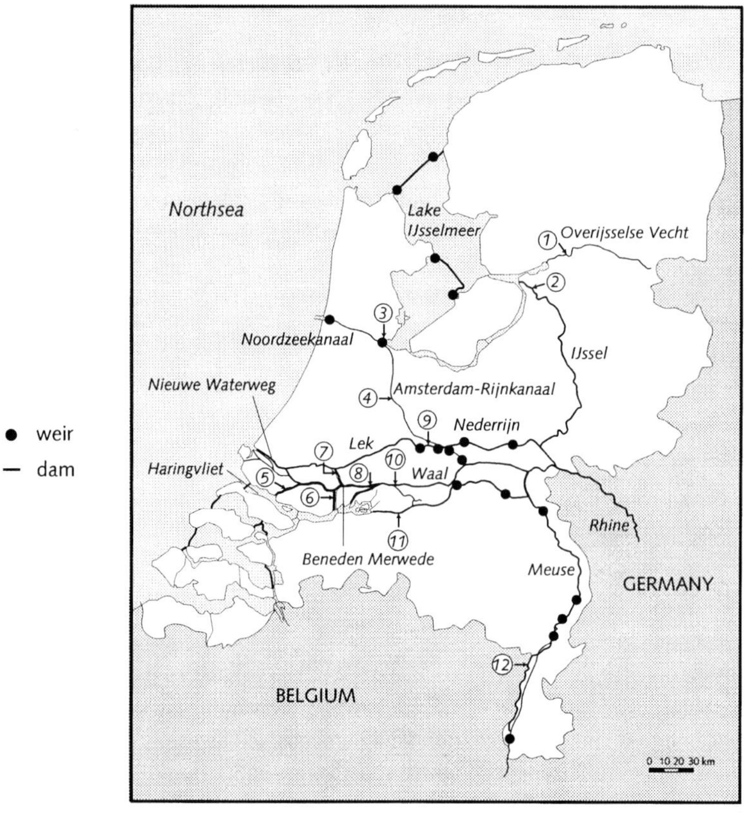


*Figure 1.* Locations of the detection stations and the physical barriers for sea trout migration in the Dutch part of the rivers Rhine and Meuse.

In general, migration, via sluices in the Afsluitdijk, the Haringvlietdam and in the outlet of the Noordzeekanaal (Figure 1) can only take place during two relatively short periods (15–20 min) at the beginning and the end of a tidel movement.

Further inland, there are barriers in the River Meuse, in one of the branches of the River Rhine, the Nederrijn/Lek (weirs provided with shiplocks), and in the Amsterdam-Rijnkanaal (shiplocks). Fish ladders of the basin type have been constructed to improve fish migration in the River Meuse. Improvements for fish migration to be taken in the Nederrijn/Lek are under study.

## Methods

The study consists of the following elements:

* Sea trout are caught in the coastal area, tagged with individually coded transmission tags, and released on the catching spot. Catching started in December 1996 and will continue till 1999.
* On fixed locations along possible migration routes automatic detection stations register the identification codes and the times of passage of the tagged sea trout. The locations were selected in such a way that the route followed can be unambiguously derived from the registrations.

Sea trout was selected in this study because the results are important for successfully reintroduction of the Atlantic salmon in the rivers Rhine and Meuse. The Atlantic salmon became extinct from both rivers in the first part of this century. Sea trout are quite common in the Dutch coastal waters and large rivers, although in limited numbers. Moreover, the species is closely related to the Atlantic salmon. The demands that both species make on migration routes, as well as on spawning and nursery areas, are largely identical.

A new telemetric method, the NEDAP TRAIL System®, has been developed. This method is based on the inductive coupling between an antenna loop and ferrite rod antenna within transponders. The identification system can be used in rivers, canals and estuaries. For the transmission mode a narrow Low Frequency

(LF) band (33.25 kHz), low bitrate, approach was chosen. This combines a high sensitive receiver with a low chance of interference from radio signals and other man made interference. The choice of the frequency is based on the consideration that coupling between the tag, in this study a transponder, and the antenna on the river bottom is a loosely coupled transformer. This means that no radio signals are emitted and thus no interference is caused to radio users. The system consists of a detection station and implantable transponders. A network of 12 detection stations (Figure 1), was built along possible migration routes.

Each detection station (Figure 2) incorporates an antenna loop, consisting of three parallel cables (distance 10 m) crossing the river or canal bottom, an antenna connection box, a receiver, and a transmitter. In the antenna connection box the antenna loop is tuned. The antenna itself is earthed by a 10 $\Omega$ electrode on the opposite bank.

Every 4 seconds the transmitter generates an interrogating signal. This signal triggers a transponder passing the antenna loop. The transponder responds by transmitting its unique signal which is demodulated by the receiver, decoded, and recorded by a microprocessor unit connected with a telephone modem. Transmission of a transponder signal takes two periods of 8 seconds, separated by 8 seconds of silence. In each 8 seconds transmission period the unique code number is transmitted 32 times. After the whole transmission period (24 seconds) the transponder is mute for two minutes to prevent the batteries from running down in the case where a tagged fish stops above an antenna.

The implantable transponder consist of a biocompatible glass tube with a diameter of 15 mm and a length of 85 mm. Inside is a ferrite rod with a length of 40 mm, a custom made integrated electronic circuit, and two batteries with diameter of 12 mm and a length of 10 mm each, ensuring a lifetime of at least two years. Weight of a transponder is 25 g (in water).

Field tests showed that the detection system still functions well with a maximum antenna length of 550 m, a water depth of 15 m and a passing speed of 5–6 m $s^{-1}$ at maximum.

Effects of ship engine noise are negligible. When the transponder is close to the hull of a vessel (less than 10 cm) the signal detection is interrupted. Maximum tolerable conductivity of the river water was not tested. However, calculations showed that a conductivity $<6000\ \mu S\ cm^{-1}$ does not affect transmission when the distance between the antenna and the transponder is $< 15$ m (Figure 3).

The transponders are implanted in the abdominal cavity of the animals (Lucas, 1989; Moore et al., 1990; McKinley et al., 1994). Rainbow trout (*Oncorhynchus mykiss* (Walbaum)), kept in captivity, did not show any significant difference in growth and behaviour during a half year period after implantation of 25 fishes with dummy transponders (Vriese, 1995). Histological analyses showed some encapsulation of the transponder in only two of them, most probably the effect of insufficient cleaning. None of the implanted fishes died as result of implantation.

During the field study approximately 150–200 sea trout will be tagged yearly for the period December 1996–December 1999.

## First results and discussion

A first group off 18 sea trout was tagged in December 1996. All these fishes were caught with fyke nets placed in the coastal area near the Haringvlietdam. The first registration was observed at the detection station in the Spui (station 5, Figure 1). Within 3 days after tagging the fish passed the sluices in the Haringvlietdam. In the period 19–12–96 till 03–03–97 this fish was detected several times at station 5.

A second sea trout was detected for the first time in the Lek (station 9, Figure 1) five weeks after tagging. This fish was not able to migrate further upstream because of the presence of a weir about four kilometres upstream of the detection station. After one week the fish was detected in the Noord (station 7, Figure 1) and after that in the Beneden Merwede (station 8, Figure 1). The last observation was made on 06–02–97 when the fish passed the detection station in the Waal (station 10, Figure 1) in upstream direction. It can be concluded that this fish swam from the catching site near the Haringvlietdam through the coastal zone in a northerly direction and entered the delta area via the Nieuwe Waterweg. It did not pass the weir at Hagestein (Figure 1), but swam about 40 km in a downstream direction, before swimming again in upstream direction to the Waal. Via this branch of the River Rhine spawning areas in some tributaries in Germany are reachable without passing any obstacle.

A survey of telemetric techniques used in migration studies was made by Priede & Swift (1992). Tags being used are generally based on two principles of signalisation: acoustically or via radio waves. Tags

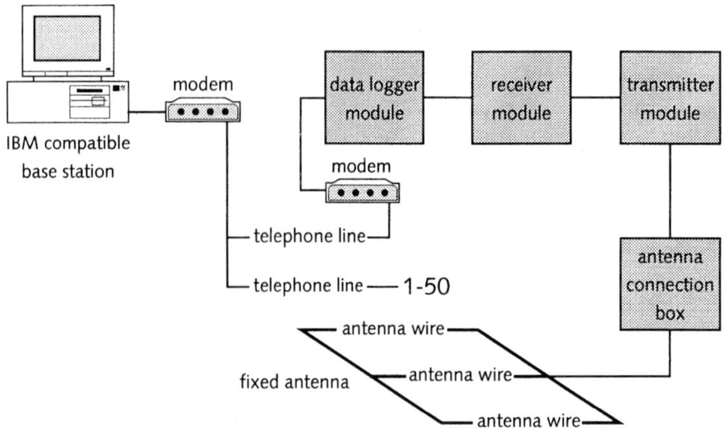


*Figure 2.* Scheme of a detection station.

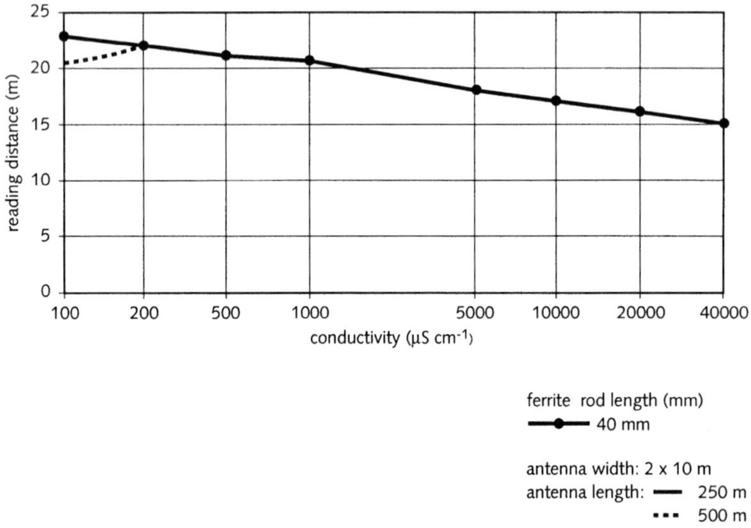


*Figure 3.* Relation between conductivity and maximum detection distance between transponder and antenna.

emitting radio waves are only applied in fresh water because signal strength is negatively determined by the conductivity of water. In the River Rhine conductivity varied between 320–1390 $\mu$S cm$^{-1}$ in the period 1987–1994, which was considered to be too high for the optimal functioning of radio tags. In acoustic signal transmission conductivity does not play any role. However, background noise can interfere with acoustic signals, which is specially the case in navigation areas like the River Rhine in which a ship is passing the German-Dutch border every 2 minutes on average. To avoid problems with high conductivity and shipping traffic, the NEDAP TRAIL System® has been developed for this study.

## References

Bij de Vaate, A., 1989. Rekolonisatiemogelijkheden van het stroomgebied van de Rijn voor riviertrekvissen en echte riviervissen: samenvatting van een literatuurstudie. Report no. 89.048, Institute for Inland Water Management & Waste Water Treatment, Lelystad, 17 pp.

De Groot, S. J., 1990. Herstel van riviertrekvissen in de Rijn een realiteit? 1. De Atlantische zalm (*Salmo salar*). De Levende Natuur 91: 82–92.

Haas, A. W. de, 1991. Inventarisatie van en verbeteringsplanning voor de fysieke belemmeringen voor de migratie van vis op de grote Nederlandse rivieren. Institute for Inland Water Management & Waste Water Treatment, Lelystad, report no. 31–1991 of the project 'Ecological rehabilitation of the River Rhine', 31 pp.

Lucas, M., 1989. Effects of implanted dummy transmitters on mortality, growth and tissue reaction in rainbow trout, (*Salmo gairdneri*, Richardson). J. Fish. Biol. 35: 577–587.

Moore, A., I. C. Russell & E. C. E. Potter, 1990. The effects of intraperitoneally implanted dummy acoustic transmitters on the behaviour and physiology of juvenile Atlantic salmon (*Salmo salar* L.). J. Fish. Biol. 37: 713–721.

McKinley, R. S., G. Power & H. E. Kowalyk, 1994. Transmitter attachment/implant: laboratory manual. Report Ontario Hydro Research, Toronto, and Univ. Waterloo, Biol. Dept., Waterloo (Canada), 18 pp.

Priede, I. G. & S. M. Swift, 1992. Wildlife telemetry. Remote monitoring and tracking of animals. Ellis Horwood Ltd., West Sussex (UK), ISBN 0 13 957994 X, 708 pp.

Vriese, F. T., 1995. Implantering van transponders in salmoniden. Report no. 26, Organisation for the Improvement of Inland Fisheries, Nieuwegein, 59 pp.

*Hydrobiologia* **371/372:** 35–46, 1998.
*J.-P. Lagardère, M.-L. Bégout Anras & G. Claireaux (eds), Advances in Invertebrates and Fish Telemetry.*



# Development of miniature pingers for tracking Atlantic salmon smolts at sea

F. A. Voegeli[1], G. L. Lacroix[2] & J. M. Anderson[3]
[1] *VEMCO Limited, 100 Osprey Drive, Shad Bay, NS B3T 2C1, Canada*
[2] *Department of Fisheries and Oceans, Biological Station, St. Andrews, NB E0G 2X0, Canada*
[3] *Atlantic Salmon Federation, P.O. Box 429, St. Andrews, NB E0G 2X0, Canada*
*E-mail address of corresponding author: lacroixg@mar.dfo-mpo.gc.ca*

*Key words:* Atlantic salmon, marine migration, smolts, ultrasonic telemetry

## Abstract

Miniature pingers and automated receivers were developed and tested over three years to track Atlantic salmon smolts of different size and origin as they migrated from fresh water and at sea. Pingers (8 mm diameter × 38 mm) with delayed activation were first tested and surgically implanted in large hatchery-reared smolts (31 cm). After improvements, these pingers were implanted in smaller hatchery smolts (23 cm) in a second study. They were detected by automated receivers at fixed sites and tracked at sea as far as 49 km. Range of detection was at least 400 m, and duration of improved pingers exceeded 75 d. Pingers were then reduced in size (8 mm × 26 mm) by using a custom integrated circuit, and they were implanted and tested in wild smolts (18 cm) in a third study. They were tracked over the same period and distances as the previous year. Power output was maintained and signal strength and range of detection were as good as for the larger pingers, and pinger duration was up to 86 d. In all tests, the frequency of transmitters was crystal controlled for decoding by the narrow-band automated receivers moored underwater at fixed sites. A laboratory study to examine the long-term effects of surgically implanting dummy pingers of different sizes (8 mm × 24, 28, and 32 mm) in juvenile salmon (15 cm) indicated that pinger shape needed modification to increase retention time past 5–6 mo and that pinger size should be reduced further to eliminate mortality. These studies have led to the development of miniature coded transmitters and small single-channel receivers which will make it possible to detect and track large numbers of small, wild salmon smolts over long distances and for several months at sea.

## Introduction

Increased natural mortality of Atlantic salmon (*Salmo salar*) in the marine environment has recently emphasized the need to determine the migratory success of smolts and post-smolts (Ritter, 1993; Reddin & Friedland, 1993). Until recently, only radio transmitters have been small enough (e.g., 7 mm × 7 mm × 18 mm and 1.0 g from Advanced Telemetry Systems, Isanti, Minnesota) for use in fish of the size of wild smolts (i.e., 12–18 cm), and this restricted tracking to fresh water. Tracking at sea has been limited by the need to use ultrasonic transmitters (i.e., pingers), and there have been few attempts to track salmon in the open sea (Potter & Moore, 1992). Most tracking studies of Atlantic salmon smolts have been carried out with large hatchery-reared fish because of the size of available pingers (Fried et al., 1978; LaBar et al., 1978; review by Potter & Moore, 1992). Smaller pingers recently developed and used in hatchery and wild smolts used a high frequency (300 kHz) and had limited power output, resulting in a low detection range (usually less than 50 m) and short duration (Moore et al., 1990, 1995). This restricted their use to rivers and narrow estuaries, and to the stage at which smolts make the transition from fresh to salt water. A new development in these estuarine studies was the use of a sonar buoy tracking system for automated detection of smolts (Potter, 1988; Solomon & Potter, 1988).

Under the initiative of the Atlantic Salmon Federation, a collaborative project involving the St. Andrews Biological Station (Department of Fisheries and Oceans) and VEMCO Ltd. was therefore started in 1994 to produce transmitters and receivers to detect and track Atlantic salmon smolts as they migrate from fresh water and at sea. At that time, pingers available with desirable features, such as a detection range of about 0.5 km and duration of 1–2 mo to allow tracking of post-smolts at sea, were too large to implant in or attach to fish of the size of smolts as they left fresh water. Larger post-smolts can rarely be captured at sea to implant the larger pingers. In this paper, we review the chronological sequence of development and testing that led to the production of low frequency ultrasonic pingers (65–77 kHz) suitable for use in wild smolts to track their migration and success in coastal areas in the initial stage of seaward migration and, thereafter, as post-smolts over longer distances at sea.

## Developments and tests in 1994

### *Delayed-start transmitters*

A new micro-controller ultrasonic pinger (V8 described in Table 1) was developed to extend the battery life by transmitting only after a preset delay (Voegeli & McKinnon, 1996). The delay was to allow pingers to be implanted in juvenile salmon before the final phase of smoltification when the fish change physiologically and become very sensitive to stress. At the time that smolts are ready to enter sea water their scales are also easily lost during handling and the risk of infection and death can be greater than when handled several weeks earlier. The flesh can be softer if the water has warmed, and surgery can then be more difficult. The transmission delay therefore saves valuable battery life for the later tracking effort while the fish are held for a recovery and observation period before release.

This first release of Atlantic salmon smolts used very large hatchery-reared fish (as shown in Table 1). This was a test of materials and methods with the view towards tracking a large number of smaller hatchery-reared smolts in coastal waters the following year using the same pingers. Performance of the new pingers (V8 with power output of 136 dB) was also compared to that of standard pingers (V2) without delayed start and with power outputs of 136 and 141 dB. Pingers were implanted surgically in the abdomen as described by Lacroix & McCurdy (1996). Details on surgical materials, procedures, and holding times before release are summarized in Table 2.

### *Automated ultrasonic receivers*

The ultrasonic receivers that we used were a combination of: (1) boat-mounted VR60 receivers, (2) shore-based VR60 receivers with automated pinger detection system and with long, weighted hydrophone cables, and (3) a newly designed, fully submersible VR20 receiver (described by Voegeli & McKinnon, 1996) with automated detection and with built-in hydrophone. All receiver models were from VEMCO Ltd. (Shad Bay, Nova Scotia). They were deployed as described by Lacroix & McCurdy (1996).

The detection algorithm developed by VEMCO Ltd. for automated monitoring used a pre-loaded matrix of pinger data (i.e., transmission frequency and pulse interval) and matching serial numbers (F. A. Voegeli, unpubl. data). This required that the pingers be crystal-controlled because of the need for very stable pinger frequency and pulse period. During a scan, the receiver listened alternately on each frequency for pinger pulses. If pulses were detected, then a time record of each pulse interval was saved for analysis. The analysis was performed *in situ* and the serial numbers of detected pingers were noted. The recording scheme was one which saved data to semiconductor memory, recording by exception. This method made an entry in memory only if something changed. A pinger which entered the reception zone and stayed for one hour would have its presence validated on each scan occurring at about 2 min intervals, but the data in memory would only hold two entries for this pinger, the time of arrival and the time of departure. This scheme allowed long-term monitoring with relatively small amounts of memory, the major benefit being the ease of downloading and handling the data for later analysis.

### *Lessons learned from 1994 study*

Surgical implantation of 38 mm long pingers (8 mm diameter) was relatively easy given the low transmitter to smolt weight and length ratios (as shown in Table 1). The use of a sedative to perfuse gills during surgery after MS-222 anaesthesia slowed the procedure slightly, but all fish recovered and the mortality of smolts with implanted pingers in the period before release was low.

*Table 1.* Characteristics of transmitters and of Atlantic salmon smolts used in tracking studies in Passamaquoddy Bay during 1994, 1995, and 1996. Data for 1994 from Lacroix & McCurdy (1996), and data for 1995 and 1996 from G. L. Lacroix (manuscripts in preparation)

| | 1994 | 1995 | 1996 |
|---|---|---|---|
| Transmitter | | | |
| Model[†] | V2B-2L and V8-2L/-3L Delayed-start | V8-2L Delayed-start | V8SC-1L |
| Size (mm) | 8 × 38 | 8 × 38 | 8 × 26 |
| Weight in water (g) | 3.5 | 3.5 | 2.7 |
| Power output (dB) | 136 and 141 | 136 | 136 |
| Smolt | | | |
| Source | Hatchery | Hatchery | Wild |
| Mean fork length (cm) | 31.8 | 23.4 | 18.7 |
| Mean weight (g) | 338.8 | 120.3 | 59.5 |
| N | 24 | 96 | 38 |
| Transmitter/smolt ratio | | | |
| Percent of fish length | 11.9 | 16.2 | 13.9 |
| Percent of fish weight | 1.0 | 2.9 | 4.5 |

[†] Transmitter models from VEMCO Ltd. (Shad Bay, Nova Scotia).

Fish held for the longer of the two post-surgery recovery periods (18 d vs. 3 d) had a higher mortality before and after release, but they were also significantly smaller ($t$-test, $P < 0.05$) than smolts held for the short period (as shown in Table 3). This difference in size, therefore, makes the comparison of mortality tentative.

Detection of pingers after release of smolts in sea water was good, but a review of the performance of pingers revealed some problems (as shown in Table 4). Six pingers failed to turn on successfully after a 1 wk delay, and six other pingers had abnormal pulse rates and shut down prematurely (i.e., before release). The problem was traced to the micro-controller transmitters in the reset circuitry of the new V8 pingers, and it was rectified in future pingers produced by replacing the minimum size resistor and capacitor delay circuit with an active reset circuit. Detection range of V8 pingers was about 250–500 m depending upon conditions and the type of hydrophone used (i.e., directional or omnidirectional), and it was similar to that of standard V2 pingers with similar power output (136 dB). The V2 pingers with high power output (141 dB) were not considered to be desirable for future test with smolts; although their signal strength was better than for pingers with regular power output, the range of detection was not increased and the duration was shorter (i.e., less than 1 wk).

The study pointed out the value and potential of the self-contained automated submersible receiver (VR20). Attempts at running long hydrophone cables from shore-based VR60 receivers to narrow channels around Ministers Island and from a mid-channel island in Letite Passage resulted in inadequate detection distance and failures due to broken cable conductors. The submersible receiver placed in the centre of a channel with strong tidal currents was effective in detecting most fish moving in and out through the channel. This resulted in a detailed analysis of smolt behaviour after entry in sea water in relation to tidal and diel cycles as described by Lacroix & McCurdy (1996).

## Developments and tests in 1995

### *Deployment of multiple automated submersible receivers*

In preparation for the release and tracking of smolts in 1995, we experimented with VR20 receivers in the passages from Passamaquoddy Bay to the Bay of Fundy (see Figure 1), where tidal currents are strong and depth ranges from 50–150 m (Forrester, 1958). Range tests were conducted with V8 pingers in areas with different current conditions to determine the spacing required between receivers to effectively detect all pingers passing through an area. A detection

*Table 2.* Surgical materials and methods used to implant transmitters in Atlantic salmon smolts in tracking studies in Passamaquoddy Bay during 1994, 1995, and 1996. Methodology used in 1994 is described in detail by Lacroix & McCurdy (1996), and methods for 1995 and 1996 are from G. L. Lacroix (manuscripts in preparation).

| | 1994 | 1995 | 1996 |
|---|---|---|---|
| Surgery location | Hatchery | Hatchery | Streamside |
| Anaesthetic brand† | MS-222 | Marinil | Marinil |
| Active agent | Tricaine methanesulfonate | Metomidate hydrochloride | Metomidate hydrochloride |
| Concentration (mg $l^{-1}$) | 90 | 10 | 10 |
| Induction time (min) | 5–7 | 3–4 | 3–4 |
| Surgery time (min) | 3–5 | 1–2 | 1–2 |
| Recovery time (min) | 3–5 | 2–3 | 2–3 |
| Sedative brand† | MS-222 | none used | none used |
| Concentration (mg $l^{-1}$) | 20 | | |
| Flow rate (ml $min^{-1}$) | 100 | | |
| Surgical supplies† | | | |
| Sterilizing solution | Germex | Germex | Germex |
| Therapeutant for fish | Furacin | Furacin | Furacin |
| Sutures | 4–0 Dermalon (CE-6 cutting needle) | 4–0 Dermalon (CE-6 cutting needle) | 4–0 Ethilon (FS-2 cutting needle) |
| Suture adhesive | Vetbond | Vetbond | Vetbond |
| Holding times | | | |
| Post-surgery (h) | 1 | 1 | 0.5 |
| Pre-release (h) | 72 ($N = 10$)<br>432 ($N = 14$) | 216 | 2 ($N = 8$)<br>24 ($N = 30$) |
| Holding location | Hatchery tanks | Hatchery tanks | River pens |
| Release method | Directly into 30°/‰ sea water | Into estuary (10–20°/‰ s.w.) | Estuary pen (1 h) (10–20°/‰ s.w.) |

† Brand suppliers are: MS-222 (Finquel®, Argent Chemical Laboratories), Marinil[Pr] (Syndel International Inc.), Germex (MTC Pharmaceuticals), Furacin (Austin), Dermalon® (Davis + Geck), Ethilon™ (Ethicon Inc.), Vetbond™ (3M Animal Care Products).

distance of 400 m was achievable even in the most turbulent areas, and this could be as high as 700 m in quieter conditions.

Attempts were made to suspend the receivers from moored floats, but the strong currents (up to 2.5 m $s^{-1}$) and the depths (often greater than 100 m) made it impossible to maintain receiver orientation and position, thereby affecting detection capability. The solution was to mount the receivers on heavy concrete moorings (100–150 kg) and lower them to the bottom. Cables with small surface floats were left attached to a harness on each mooring for later recovery on slack tides. Further tests with receivers placed on the bottom at depths of up to 150 m showed that the detection range was maintained even for detecting pingers near the water surface. An advantage of mounting receivers on the bottom was to place the hydrophone in a relatively quiet zone away from surface noise and turbulence.

Ten VR20 receivers were deployed at depths of up to 90 m to monitor all exits from Passamaquoddy Bay (see Figure 1). They were placed in a curved array rather than along a straight line across the two wide passages that required more than one receiver. This was done to increase the likelihood that a fish migrating through would pass within the detection range of more than one unit. The mean spacing between moored receivers was 664 m (as shown in Table 4), but

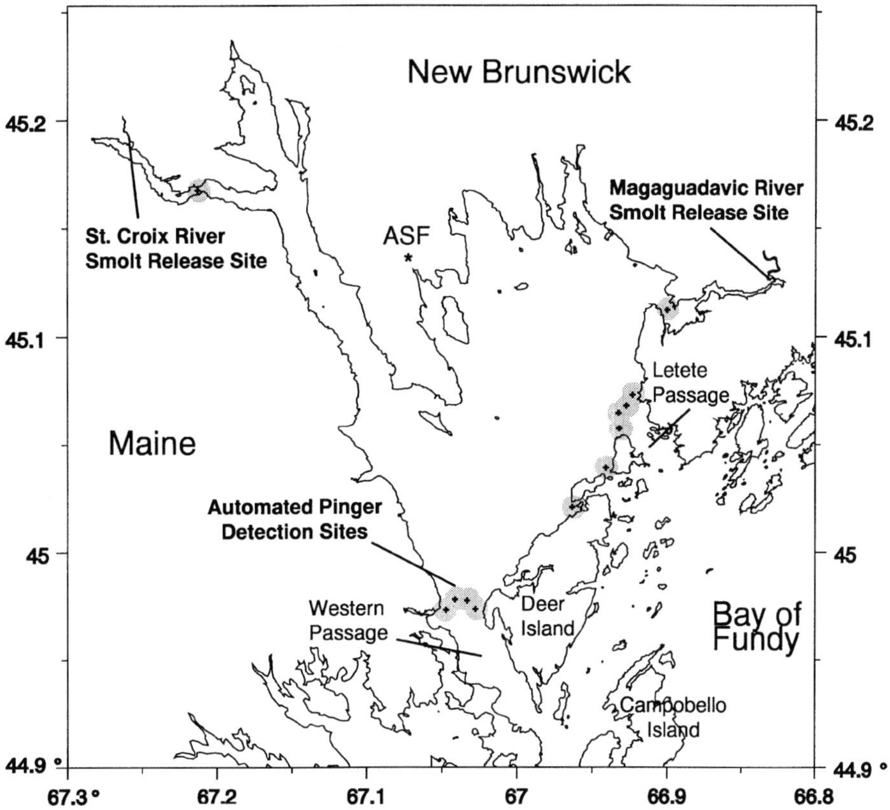


*Figure 1.* Map of Passamaquoddy Bay showing the locations of Atlantic salmon smolt release sites, of automated receivers deployed underwater in 1995 and 1996 (+ with shaded circle showing detection range), and of the Atlantic Salmon Federation hatchery (* ASF). Figure from G. L. Lacroix (manuscript in preparation).

*Table 3.* Performance of Atlantic salmon smolts with transmitters in relation to duration of holding period between surgery and release in tracking studies in Passamaquoddy Bay during 1994 and 1996. Mean times are given to nearest hour. n.a., not applicable. n.e., not estimated. Data for 1994 from Lacroix & McCurdy (1996), and data for 1996 from G. L. Lacroix (manuscript in preparation)

| | 1994 | | 1996 | |
|---|---|---|---|---|
| Post-surgery holding time (h) | 72 | 432 | 2 | 24 |
| Mean fork length (cm) | 38.6 | 26.9 | 19.7 | 18.4 |
| Mean weight (g) | 521.2 | 208.5 | 71.4 | 56.3 |
| Mortality (number of fish) | | | | |
| Pre-release | 1 | 4 | 1 | 17 |
| In river estuary | n.a. | n.a. | 1 | 0 |
| In Chamcook Harbour | 2 | 9 | n.a. | n.a. |
| In Passamaquoddy Bay | n.e. | n.e. | 0 | 6 |
| Mean time to estuary mouth (h) | n.a. | n.a. | 32 | 12 |
| Mean time to last detection (h) | 72 | 23 | 182 | 125 |

effective spacing perpendicular across a passage was smaller. Single VR20 receivers were also similarly deployed to seal off the mouth of two estuaries about 6–8 km from where smolts were released at the head of tide in each river (see Figure 1). Mean distance from release sites to the various exits from Passamaquoddy Bay was about 30 km.

The V8 pingers used in 1995 were effectively the same as the delayed-start pingers tested in 1994 (as shown in Table 1), but with modified reset circuitry. Four different frequencies (65–77 kHz) and pulse repeat intervals (i.e., periods) of 1500–3000 ms were used, which allowed for at least 96 pingers to be released at once without overlap in identification. These pulse periods could be tracked by surface receivers (VR60) as well as be identified by the VR20 monitoring receivers. Battery life could have been extended by increasing the transmitter pulse periods, but this could have reduced detection efficiency of the VR20 receivers. Longer pulse periods also become difficult

*Table 4.* Performance of transmitters used in tracking studies of Atlantic salmon smolts in Passamaquoddy Bay during 1994, 1995, and 1996. n.a., not applicable. Data for 1994 from Lacroix & McCurdy (1996), and data for 1995 and 1996 from G. L. Lacroix (manuscripts in preparation)

| | 1994 | 1995 | 1996 |
|---|---|---|---|
| Number of smolts | | | |
| Operated | 35 | 97 | 56 |
| Dead before release | 5 | 1 | 18 |
| Released | 24 | 96 | 38 |
| Number of transmitters | | | |
| Failing before release | 6 | 0 | 0 |
| Detected by single VR20 at estuary mouth | 13 | 88 | 37 |
| Detected by VR60 on boats in bay | 17 | 58 | 27 |
| Detected by VR20 array in bay passages | n.a. | 76 | 31 |
| Detected simultaneously by adjacent receivers | n.a. | 27 | 17 |
| Mean distance between adjacent VR20 in passages (m) | n.a. | 662 | 594 |
| Duration of longest lasting transmitter (d) | 10 | 75 | 86 |
| Maximum distance from release site pingers were detected (km) | 12 | 49 | 46 |

to decode by tracking with VR60 receivers on boats, especially in the presence of surface noise (e.g., waves, rain, wind, and other boats). The VR20 receivers were set to scan the four frequencies at about 2 min intervals and to identify the serial numbers of each passing fish. Computer routines were written to sort the data from the multiple receiving sites to create a file for each pinger with a chronological listing of all receivers that the pinger had passed.

*Large-scale release of hatchery smolts with transmitters*

Testing was part of a study involving the release of 96 hatchery-reared smolts with pingers; 48 in each of two rivers on opposite sides of Passamaquoddy Bay (see Figure 1). The surgical materials and methods were similar to those used in 1994, except that Marinil was used as an anaesthetic and no sedative was used during the surgery to speed up the process (as shown in Table 2). The Marinil rapidly produced a deep sedation facilitating surgery, and this was followed by a rapid recovery. The surgical procedure to implant pingers and close the incision with three sutures and Vetbond took about 1–2 min, and it included fish length and weight measurements. The complete procedure was therefore faster than in 1994.

Smolts were much smaller than those used in 1994, and the pingers therefore represented a larger percentage of the fish length and weight than the previous year (as shown in Table 1). Pinger insertion was more difficult than in 1994, especially in the smallest fish, but only one fish with an implanted pinger died before release (as shown in Table 4). The pingers were programmed for an activation delay of 8 d, and the fish were held in tanks to recover from surgery and for observation for 9 d before release. This was a 50% decrease in the long holding period used in 1994 (as shown in Table 2), because the 1994 study indicated that a short holding time could reduce pre-release mortality (as shown in Table 3). All pingers automatically started on day 8, and all were transmitting strong signals at the preset pulse period at the time of release, indicating that problems with pinger failure encountered in 1994 had been fully resolved.

The 96 smolts were released in the rivers at the head of tide and the subsequent detection of the pingers was excellent (as shown in Table 4). The single VR20 receivers at the mouth of each estuary detected 88 of the pingers. The VR20 receivers moored underwater in the exits from Passamaquoddy Bay detected 76 of the pingers, and 36% of these were detected simultaneously by two adjacent VR20 receivers. This figure shows a high degree of overlap in the coverage by the array of receivers, adding confidence in our estimate of the number of smolts that were detected leaving the bay. Daily searches in Passamaquoddy Bay aboard three boats with VR60 receivers detected 58 of the pingers at least once. Many of these were detected on a repeated basis until the fish left the bay, allowing for a more detailed analysis of behaviour in relation to currents and tidal cycles (G. L. Lacroix, manuscript in preparation). The furthest site where a pinger was detected and decoded from receivers on boats was 49 km at sea from the point of release. Two of the pingers were detected regularly after the smolts became trapped in weirs along the shoreline. One of these pingers was last detected 75 d after release, and provided a check on the battery life and duration at a power output of 136 dB. This was considered excellent and encouraging for planning long-term studies.

*Lessons learned from 1995 study*

This study demonstrated the effectiveness of the automated approach to sealing off the mouth of rivers or estuaries and large coastal embayments for monitoring fish emigration. The pinger detection range and battery life were also found to be adequate to work in the highly turbulent waters of the Bay of Fundy. The success rate for the boats searching daily for pingers within estuaries and Passamaquoddy Bay was significantly less than that of the automated receivers moored at fixed locations in areas of fish passage. However, the effort was considered worthwhile because of the information gathered on smolt behaviour and movements within the bay.

Underwater receivers powered by batteries were left to monitor unattended for up to 6 wk, and they were still operational when retrieved. However, we experienced some difficulty with O-ring seals on some of the VR20 underwater connectors leading to slow water seepage and shorting of the batteries in several units after several weeks underwater. Most fish had moved out of the bay by then. Nevertheless, the overlap in detection between adjacent units and the pattern of receiver deployment ensured that all fish passing through the array of receivers were probably detected. Future VR20 designs should minimize the number of seals used on the surface of the receiver casing. We also observed that it was possible for a VR20 to detect a pinger on an adjacent channel (i.e., frequency) if the fish passed very close to the hydrophone. This pinger then generated a false decode for another pinger that was not in the area. Such detections were readily identified and removed from the data because: 1) they only occurred at the same time as a long record of detections of a pinger with near identical pulse period (i.e., within 10 ms) on the correct channel, 2) the falsely decoded pinger was detected correctly at a location far away on the same day, and 3) most fish with pingers having similar pulse periods on adjacent frequencies had been released at different sites and the data could therefore be sorted on the basis that a fish released in one river could not be detected coming out of another river at the opposite side of the bay. In the future, these false detections could be avoided by selecting pulse periods that were more distinctly different (i.e., greater than 20 ms difference) for pingers on adjacent channels. However, this becomes difficult when a large number of fish are to be released at the same time and the range of pulse periods is to be kept within 1000–3000 ms to facilitate active tracking using surface receivers on boats.

The size of pingers used and surgically inserted in the abdomen of the hatchery-reared smolts was felt to be at the maximum limit for these fish based on the difficulty experienced with pinger insertion and closure of the incision on some of the smallest smolts. We considered that pingers smaller than 38 mm would be necessary to implant in even smaller wild smolts if we were to ensure their survival and the retention of the pinger. As a result, preliminary design of circuitry for a smaller transmitter to be used in wild Atlantic salmon smolts was undertaken for development of prototypes to be tested in 1996.

## Developments and tests in 1996

*Single-chip transmitters for implant in wild smolts*

In 1996, we performed tests using a new transmitter design with an Application Specific Integrated Circuit (ASIC). The ASIC as its name implies is a single-integrated circuit designed specifically to function as an acoustic pinger (Voegeli & McKinnon, 1996). This new device reduced the electronics to minimum size, leaving the choice of transmitter length up to a decision on balancing power output and duration of the pinger with size of the fish to be tracked.

Therefore, a laboratory experiment was first conducted to determine the effects of surgically implanted dummy pingers of different lengths on juvenile Atlantic salmon (G. L. Lacroix, manuscript in preparation). The length and weight of the salmon parr used were based on the expected size of wild smolt in nearby rivers. Dummies (8 mm diameter) of three different lengths (24, 28, and 32 mm) were implanted using the same procedure as in 1995, and the fish were kept in tanks for observation for 316 d along with control and sham-operated fish. The size of dummies was chosen to approximately represent pinger sizes achievable with the new circuitry and different power supplies.

The mortality of fish and the expulsion of some dummies through the wall of the abdomen over time indicated that the two largest sizes (28 and 32 mm) would be inappropriate for use in wild smolts, especially in a long-term study (as shown in Table 5).

*Table 5.* Long-term effects (316 d) of surgically implanting dummy transmitters (DT) of different length (8 mm diameter) in the abdomen of juvenile Atlantic salmon. Shams were handled and operated the same as treatment groups, but no DT was inserted. n.a., not applicable. Data from G. L. Lacroix (manuscript in preparation)

| | | | Group | | |
|---|---|---|---|---|---|
| | Control | Shams | 1 | 2 | 3 |
| Mean DT length (mm) | n.a. | n.a. | 24.2 | 28.3 | 31.8 |
| Mean DT weight in air (g) | n.a. | n.a. | 3.16 | 3.38 | 3.68 |
| Mean fish length (cm) | 14.6 | 14.4 | 16.6 | 19.5 | 14.8 |
| Mean fish weight (g) | 35.6 | 33.9 | 37.1 | 33.6 | 37.3 |
| Percent DT length/fish length | n.a. | n.a. | 14.6 | 14.5 | 21.5 |
| Percent DT weight/fish weight | n.a. | n.a. | 8.5 | 10.1 | 9.9 |
| Number of fish | | | | | |
| Total | 16 | 5 | 5 | 5 | 5 |
| Dead before expelling DT | 0 | 0 | 2 | 1 | 3 |
| Expelling DT | n.a. | n.a. | 1 | 4 | 2 |
| Dead after expelling DT | n.a. | n.a. | 0 | 0 | 0 |
| Retaining DT to 316 d | n.a. | n.a. | 2 | 0 | 0 |
| Alive after 316 d | 13† | 5 | 3 | 4 | 2 |
| Mean time | | | | | |
| To DT expulsion (d) | n.a. | n.a. | 166 | 164 | 181 |
| To death (d) | 0 | 0 | 139 | 33 | 66 |

† Three fish from control group jumped out of the tank and were lost.

Mortality occurred early in the smallest fish; at 19 d and 40 d in two fish with 32 mm dummies, and at 33 d in one fish with a 28 mm dummy. It should be noted that three other fish died during the week leading up to day 139, probably because of a failure of the tank cleaning system; two were from group 1 and one was from group 3. However, in the fish from group 3, the 32 mm dummy had started to protrude through the abdomen and it would have been expelled had the fish not died because of conditions in the tank. All other fish that expelled the dummies through the wall of the abdomen survived the event and lived to the end of the 316 d experiment; these included 4 fish in group 2 and 2 fish in group 3. Only one fish in group 1 expelled the 24 mm dummy. In all three dummy size groups, expulsion did not take place until approximately 5–6 mo after implant (as shown in Table 5). It usually took about 2–3 wk from the time a dummy was visible externally as starting to protrude through the abdomen to the time that it was expelled. No fish died after expelling the dummies. These were not expelled through the incision used for insertion, but rather at points on the abdomen where the sharp edged end of the dummy was pressing against. Examination at the end of the experiment showed that fish which had expelled dummies had external scars and internal irritation at points of contact with the ends of the cylinder. There was also a cavity surrounded by membrane where the dummy had been. Dummies that were retained in fish for the duration of the experiment had become completely encysted and surrounded by a membrane. It was felt that a smooth rounding of the end of the cylinder (i.e., a capsule shape) would reduce this irritation and possibly increase retention time.

When test pingers were produced with the ASIC, we found that the new circuit could be run on a single-cell silver oxide battery of 1.5 V which reduced the cylinder length to as short as 22 mm. However, this would have resulted in decreases in acoustic power output and detection range in comparison to pingers used in our previous studies. While this may be adequate for some studies in more restricted spaces, we felt that we could not compromise on detection range under conditions in Passamaquoddy Bay and with the design of the study. Therefore, pingers for the first release of wild smolts were produced to keep the acoustic power output at the same level (136 dB) as in the successful 1995 study. Two battery cells were used, and the first V8SC pingers produced had a cylinder length of about 25 mm. This was slightly longer than the smallest dummy tested (group 1, 24 mm), but the end edges of the cylinders were sanded to remove their sharpness to decrease the possibility of pinger expulsion.

*Release of wild smolts and lessons learned from 1996 study*

In 1996, wild Atlantic salmon smolts were captured at the mouth of the Magaguadavic River before entry into sea water. The fish were only half the weight and much smaller than the hatchery smolts released in 1995 (as shown in Table 1). The new V8SC pingers were implanted in wild smolts using the same surgical procedure as in 1995 (as shown in Table 2), except for the use of 4-0 Ethilon sutures which we found more supple and easier to handle. The wild smolts were very sensitive to capture and holding at this late stage of smoltification and migration in fresh water (late May–early June), and some fish died during the holding period. The holding period before release was therefore decreased considerably from previous years (as shown in Table 2), in an attempt to reduce mortality and also because of the absence of delayed activation in the new pingers. The effects of two holding periods (2 h and 24 h) were compared, and the shorter holding time resulted in less pre- and post-release mortality (as shown in Table 3). This was especially significant in the holding period before release. It seems that simply holding the smolts resulted in mortality. Wild smolts which were handled and held in the same way but on which no surgery was performed and no pingers were implanted had a similar level of mortality. The best procedure was therefore considered to probably be a rapid release after full recovery from surgery and anaesthesia because of the mortality apparently related to holding wild fish during advanced stages of migration. Fish were also held briefly (1 h) for acclimation to the brackish water in the estuary before release in case there was some initial mortality related to their transfer and handling. This allowed the pingers to be recovered and reused in another smolt in the event of unsuccessful transfer. Table 3 shows the mortality resulting from surgery and for different holding periods in 1996.

As a result of the handling mortality, 56 wild smolts were operated in order to release 38 fish with the small, single-chip pingers (as shown in Table 4). They were released from the Magaguadavic River at the same site as used the previous year with hatchery smolts. The methods used for tracking in the 1996 study replicated those used in 1995. In addition to a receiver at the mouth of the estuary, all exits from Passamaquoddy Bay were covered with the same pattern of receiver deployment used in 1995 (see Figure 1) to provide an accurate comparison between the behaviour and migration success of hatchery and wild smolts (G. L. Lacroix, manuscript in preparation). The mean spacing between moored receivers was 594 m and only slightly less than in 1995 (as shown in Table 4).

The performance of the new pingers and their detection was extremely successful; 37 pingers were detected by the single VR20 receiver moored at the mouth of the estuary, and daily searches in Passamaquoddy Bay by three boats equipped with VR60 receivers detected 27 pingers at least once. The moored array of ten VR20 receivers in the passages from the bay detected 31 pingers, of which 55% were detected simultaneously by two adjacent VR20 receivers. This again indicated a high degree of overlap in coverage by the array of receivers and added a high degree of confidence that most fish emigrating had been detected. A V8SC pinger that remained at the bottom of the estuary near the release site after a fish died was found to last 86 d, which is longer than was observed for the V8 pingers used in 1995.

A new problem surfaced that affected pinger detection during the 1996 study; a powerful acoustic seal scaring device (i.e., 4× scrammer, Ferranti-Thomson Sonar Systems UK Ltd.) was intermittently operated on a cage site for salmon aquaculture located near several of the underwater VR20 receivers. This device transmitted very high power levels of low ultrasonic noise (25.6 kHz) which contained harmonics on at least one of the standard V8 pinger frequencies used, and produced some false detections of pingers on that frequency. We were able to filter out the influence of this single seal scarer by post-processing the data, and by verifying fish positions with detections from boats. However, we have concerns for the continued ability to successfully detect miniature pingers if the use of these noise making devices increases and all available bandwidth is used. Regrettably, selection of pinger types and frequencies used in the future will probably be dictated by the type, intensity, and frequency of noise generated by such units when present. A prior reconnaissance is, therefore, highly recommended before designing a study in an area where sea cages for

## Code specifications

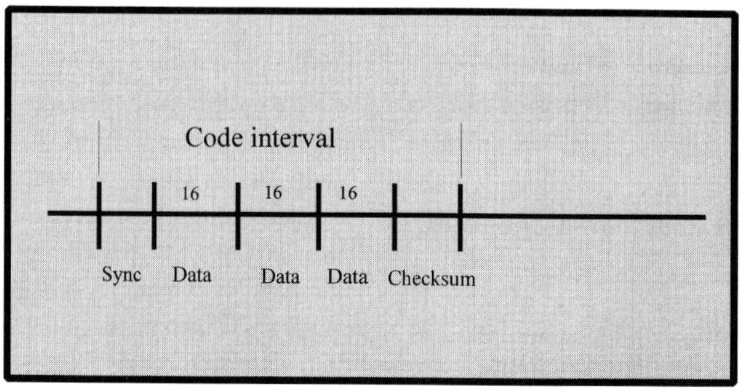


*Figure 2.* Example of code specifications with single error detecting checksum to produce 4096 unique code values.

## Random code repeat interval

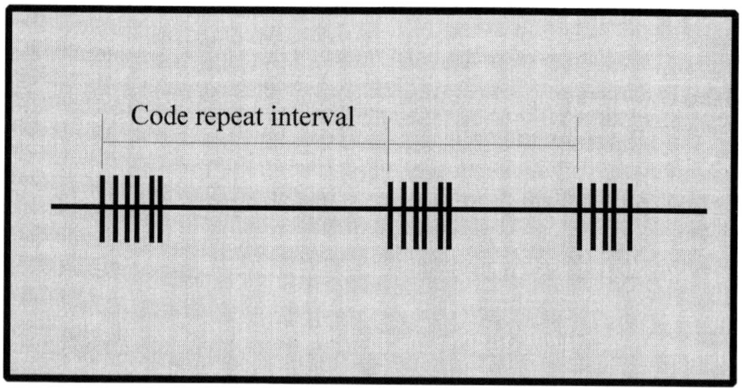


*Figure 3.* Example of random code repeat interval (RCODE) to avoid code collisions.

## ASIC BLOCK DIAGRAM

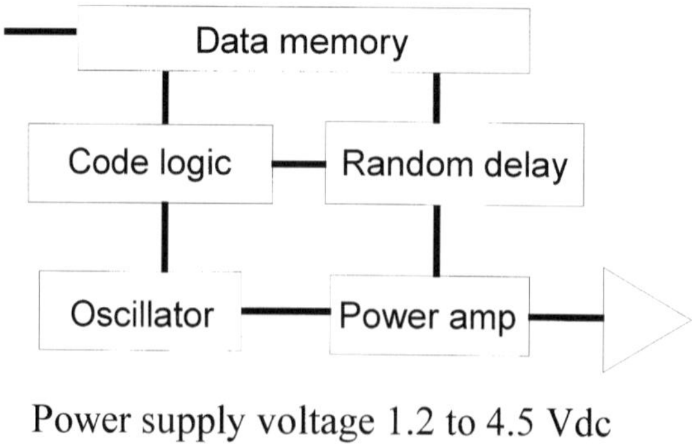


*Figure 4.* Details of new, small pinger designed with a single-integrated circuit (ASIC).

salmon aquaculture are present and the seal scarers are used. Recently, low frequency acoustic pingers (1–25 kHz) have also been attached to gillnet floatlines used in the groundfish fishery in the Bay of Fundy and tested as warning devices to reduce by-catch of harbour porpoise (*Phocoena phocoena*) (Trippel et al., 1996). If their use became widespread, these pingers could possibly also have an impact on our ability to detect or decode pingers in future studies of post-smolt migration at sea.

## Recent advancements to meet requirements for tracking at sea

In order to expand the number of transmitters which could be identified, tests were conducted with unique codes. V8-1L micro-controller transmitters were used to transmit complex codes of up to six pulses. A code was developed which allows identification of up to 4096 different fish on the same frequency and which has a single error detecting checksum for each unique code value (see Figure 2). For studies with 256 fish or less, a second checksum could be inserted in place of the unused data block. To place a large number of transmitters in the water at the same time, the code must be transmitted infrequently to reduce the probability of codes being transmitted at the same time (i.e., code collisions) from two or more fish. The repeat interval between the code of each transmitter should also be different, such that if a code collision does occur, then the next transmission will be at a different time and not collide. The problem that arises when making every code have a unique repeat interval is that some codes will be many minutes apart. This very infrequent code transmission rate would increase the probability of a fish swimming through an array of receivers without being detected. A pseudo-random code repeat interval (RCODE) was therefore developed to overcome this problem (see Figure 3). The RCODE system was implemented in a V8-1L micro-controller transmitter and tested in open water in 1996. A major advantage of the RCODE over fixed rate pingers is that the battery life of the transmitter can be extended because fewer transmitted pulses are required for automated decoding.

The RCODE system was first developed with the large V8-1L transmitter (8 mm × 33 mm), and a new ASIC therefore needed to be designed for smaller transmitters of this type to be produced and used in wild smolts. This second generation ASIC prototype was cast (see Figure 4 for ASIC block diagram), and it was tested successfully in the laboratory in 1997. Prototypes with this new circuit were also built and tested in the field. The extremely low current drain of this new ASIC between code transmissions should make it possible to implant a transmitter in a smolt which will last up to 1 yr. A new pinger capsule shape with smooth, rounded ends has also been designed for this pinger in the hope of increasing retention time from 5–6 mo to that of the new pinger life.

A new, single channel receiver (VR1) has also been developed and tested in the laboratory and in the field in 1997. The VR1 is much smaller than VR20 receivers, and it is designed for deployment of a large number of receivers in long lines or in other patterns at sea to build detection grids which could be picked up and moved easily to new locations as fish migrated further. Together, the development of the small coded transmitters and the single channel receivers make it possible to plan large-scale migration studies at sea. However, the deployment methodology and its effectiveness remain to be fully developed and performance of the gear needs to be tested with fish under open ocean conditions.

## Conclusions

With these studies, we demonstrated the effectiveness of using an automated monitoring approach to detect pingers in smolts migrating in the coastal marine environment. Pinger size, suitable for use in large hatchery-reared fish, was decreased to allow their use in the smaller wild smolts while in estuarine and coastal marine areas. The next step is to fully develop the capability to track these small fish over long distances or to locate them further out at sea over a long period of time. It should be noted that, even within the confines of Passamaquoddy Bay, the search method of locating transmitters was very labour intensive and weather dependent, and it was less successful than using a fixed array of receivers to monitor fish movements through specific areas. Tracking or searches for fish with pingers from boats are likely to be even more difficult in the open ocean.

A logical step towards successfully monitoring post-smolt migration at sea could be to deploy receivers with positioning capabilities over large areas at sea. This would require a large number of receivers given the present detection ranges for small pingers, and it would probably require the release of a large number of smolts (up to 500) to achieve a reasonable rate of detection. The existing VR20 receiver design, with its ability to scan multiple frequencies, was ideal for use in rivers, estuaries, and coastal areas because of the requirements for deployment and the high probability of multiple fish appearing at the same receiver site together. However, at sea fish would probably be distributed over a considerably wider area, and the use of a large number of easily deployed single channel receivers (i.e., non-scanning) such as the VR1 over a wide area would be an improvement. The recent development of small transmitters with a large number of unique codes now makes the release of large numbers of fish, all with transmitters on a single frequency, also possible. Coded transmitters on a well chosen frequency should also help reduce potential problems associated with the increasing presence of acoustic seal scarers and the use of other acoustic transmitters at large.

## Acknowledgements

We thank all personnel at the St. Andrews Biological Station, the Atlantic Salmon Federation, and VEMCO Ltd. that were involved in ensuring the success of these studies. Funding was provided by Fisheries and Oceans Canada and by the Atlantic Salmon Federation through special grants from the Molsen Family Foundation and the Robert R. McCormick Tribune Foundation.

## References

Forrester, W. D., 1958. Current measurements in Passamaquoddy Bay and the Bay of Fundy, 1957 and 1958. An Interim Report. Tidal Publication No. 40. The Canadian Hydrographic Service, Surveys and Mapping Branch, Department of Mines and Technical Surveys, Ottawa.

Fried, S. M., J. D. McCleave & G. W. LaBar, 1978. Seaward migration of hatchery-reared Atlantic salmon, *Salmo salar*, smolts in the Penobscot River estuary, Maine: riverine movements. J. Fish. Res. Bd Can. 35: 76–87.

LaBar, G. W., J. D. McCleave & S. M. Fried, 1978. Seaward migration of hatchery-reared Atlantic salmon (*Salmo salar*) smolts in the Penobscot River estuary, Maine: open-water movements. Journal du Conseil International pour l'Exploration de la Mer 38: 257–269.

Lacroix, G. L. & P. McCurdy, 1996. Migratory behaviour of post-smolt Atlantic salmon during initial stages of seaward migration. J. Fish Biol. 49: 1086–1101.

Moore, A., E. C. E. Potter, N. J. Milner & S. Bamber, 1995. The migratory behaviour of wild Atlantic salmon (*Salmo salar*) smolts in the estuary of the River Conwy, North Wales. Can. J. Fish. Aquat. Sci. 52: 1923–1935.

Moore, A., I. C. Russell & E. C. E. Potter, 1990. Preliminary results from the use of a new technique for tracking the estuarine movements of Atlantic salmon, *Salmo salar* L., smolts. Aquacult. Fish. Mgmt 21: 369–371.

Potter, E. C. E., 1988. Movements of Atlantic salmon, *Salmo salar* L., in an estuary in south-west England. J. Fish Biol. 33A: 153–159.

Potter, E. C. E. & A. Moore, 1992. Surveying and Tracking Salmon in the Sea. Atlantic Salmon Trust, Pitlochry, 51 pp.

Ritter, J. A., 1993. Changes in Atlantic salmon (*Salmo salar*) harvests and stock status in the north Atlantic. In D. Mills (ed.), Salmon in the Sea and New Enhancement Strategies. Fishing News Books, Oxford: 3–25.

Reddin, D. G. & K. D. Friedland, 1993. Marine environmental factors influencing the movement and survival of Atlantic salmon. In D. Mills (ed.), Salmon in the Sea and New Enhancement Strategies. Fishing News Books, Oxford: 79–103.

Solomon, D. J. & E. C. E. Potter, 1988. First results with a new estuarine fish tracking system. J. Fish Biol. 33A: 127–132.

Trippel, E. A., M. B. Strong, C. Hood, C. Richter & J. Lien, 1996. By-catch of harbour porpoise (*Phocoena phocoena*) in the lower Bay of Fundy gillnet fishery in 1995. DFO Atlantic Fisheries Research Document 96/110, 13 pp.

Voegeli, F. A. & G. P. McKinnon, 1996. Recent developments in ultrasonic tracking systems. In E. Baras & J. C. Philippart (eds), Underwater Biotelemetry. Proceedings of the First Conference and Workshop on Fish Telemetry in Europe, 4–6 April 1995, University of Liège, Belgium: 235–241.

*Hydrobiologia* **371/372:** 47–52, 1998.
*J.-P. Lagardère, M.-L. Bégout Anras & G. Claireaux (eds), Advances in Invertebrates and Fish Telemetry.*



# A dynamic combined acoustic and radio transmitting tag for diadromous fish

G. H. Niezgoda[1], R. S. McKinley[2], D. White[2], G. Anderson[2] & D. Cote[2]
[1] *Lotek Marine Technologies Inc, 114 Cabot Street, St. John's, NF A1C 1Z8, Canada (gniezgoda@lotek.com)*
[2] *Waterloo Biotelemetry Institute, Department of Biology, University of Waterloo, 200 University Av. W., Waterloo, ON N2L 3G1, Canada*

*Key words:* diadromous, conductivity, bio-telemetry, estuary, CART


## Abstract

A combined acoustic and radio transmitting (CART) tag employing a dynamic conductivity switch suitable for investigating the migratory behavior of diadromous fish is described. The unique feature of the transmitter is its ability to sense the electrical conductivity of the ambient water and therefore operate in the appropriate signal mode. Under fresh water conditions the transmitter operates in radio mode; in sea water it operates in acoustic mode. To optimize range performance the conductivity threshold setting at which the tag switches mode can be changed to suit the application. This paper describes a proposed criterion for threshold selection, the physical probe characteristics and the recommended tag attachment procedure.


## Introduction

Several species of fish, notably diadromous varieties including salmonids, sturgeon and the American eel, migrate at different periods of their life cycle between the ocean and inland bodies of fresh water. To monitor the behavior of these animals, biologists are faced with the task of making telemetry measurements under fresh, estuarine and sea water conditions. The large variations in water depth and electrical conductivity across these zones present significant challenges for radio and acoustic telemetry.

In most situations radio telemetry is restricted to fresh water applications since the conductivity levels associated with both sea water and brackish (estuary/coastal) waters, leads to severe radio signal attenuation through the water column. Also, significant attenuation may occur in fresh water depending on conductivity, transmitter frequency and depth. In the case of acoustic telemetry, severe signal loss under shallow water conditions commonly arises producing destructive interference at the receiving hydrophone. Consequently, acoustic telemetry is best applied to high conductivity, estuarine and sea water environments, with radio telemetry being most applicable in low conductivity, fresh water conditions. Therefore, for the most accurate assessment of the behavior of diadromous fish species during critical migratory phases as the animals move between fresh and saline waters, both acoustic and radio telemetry are required. Unfortunately, bio-compatibility issues such as weight and size restrictions prevent placing separate radio and acoustic tags on most animals. To address these concerns, combined acoustic and radio transmitting (CART) tags have been developed (Solomon & Potter, 1988; Armstrong et al., 1988).

Current CART configurations, (Solomon & Potter, 1988; Armstrong et al., 1988), rely upon non-dynamic switching between acoustic and radio modes based on a timer. Upon release of a tagged fish, the timer setting attempts to predict the time instant when the fish will transit between salt and fresh water. This approach has two drawbacks. First, since the timer setting is based on either historical data of fish movement or intuition, some level of measurement outage must be expected. Second, to ensure the accurate monitoring of fish movement, a large number of animals must be tagged to remove the effects of outage.

As a means of mitigating measurement outage and reducing the number of transmitters to an economical level without compromising the quality of measurements, a dynamic switching CART device has been developed. The present paper describes the development of such a device and suggests a suitable attachment procedure of the transmitter. Furthermore, the effects of the this attachment procedure on swimming performance of adult rainbow trout, *Onchorhynchus mykiss* is described.

## A criterion for conductivity threshold selection

In this section guidelines for selecting a conductivity threshold criterion for switching between radio and acoustic modes is proposed. The approach is based on range prediction formulae for radio (Velle et al., 1979; Lotek, 1994) and acoustic (Clay & Medwin, 1989; Coates, 1989) telemetry systems. It has been our experience that these range prediction formula provide range estimates comparable to data collected from controlled '*in-situ*' trials.

A prediction of radio range performance is derived by relating the receiver's minimum detectable signal sensitivity, to transmitter and propagation characteristics,

$$R_{\text{sen}} = \text{ERP} - L_{\text{water}}(d, C) - L_{\text{int}} - L_{\text{air}}(r) + G_{\text{ant}} - L_{\text{cable}}(\text{dBm}) \quad (1)$$

where ERP equals the effective radiated power of the transmitter in units of decibels referenced to a milliwatt (dBm). Receiver parameters include antenna gain in units of decibels referenced to a dipole antenna, $G_{\text{ant}}$ (dBd), and antenna cable loss, $L_{\text{cable}}$(dB). Transmission losses expressed in decibels (dB), include loss in water, loss at the water/air interface, and loss in air. Based on (Velle et al., 1979; Lotek, 1994) loss associated with radio propagation in water at 150 MHz is approximated by

$$L_{\text{water}}(d, C) \approx (0.0171 \cdot C + 1.5) \cdot d \ (\text{dB}) \ @ \ T = 10\,°\text{C}$$

where transmitter depth is $d$ (m) and $C$ is electrical conductivity measured in $\mu\text{S cm}^{-1}$ at a low frequency (i.e.: 100 kHz to 1 MHz). A detailed discussion of transmission loss at the water/air interface, $L_{\text{int}}$, can be found in Velle et al (1979) and Lotek (1994). The loss in air as a function of range, $r$ (m), is expressed as

$$L_{\text{air}}(r) = 17.7 + 20 \cdot \log_{10}(r/\lambda) \ (\text{dBd}),$$

where $\lambda = 2$ m equals the wavelength at 150 MHz.

A prediction of acoustic range performance follows by equating the sonar equation, (Clay & Medwin, 1989), to the receiver's minimum detection threshold level, DT (dB),

$$\text{DT} = \text{SL} - \text{TL}_{\text{spread}}(r) - \text{TL}_{\text{absorp}}(S) \cdot r + G_{\text{hydro}} + \text{NL}(\text{dB}) \quad (2)$$

where SL denotes transmitter source level in dB re $1\mu$ $Pa$ @ 1 m. Transmission loss due to spherical spreading is

$$\text{TL}_{\text{spread}}(r) = 20 \cdot \log_{10}(r) \ (\text{dB}), \quad (3)$$

and loss due to absorption is, where $S$ is salinity expressed in ‰. Hydrophone directional gain is given by $G_{\text{hydro}}$ (dB). The noise level, NL, will depend upon the receiver's effective bandwidth, BW(Hz); $\text{NL}_{\text{ambient}}$ (dB re $1\mu$ $Pa^2$ / Hz), the ambient noise level in a 1 Hz bandwidth; and $\text{NL}_{\text{wind}}$ (dB *re* 1 $\mu$ $Pa^2$ / Hz), the nominal noise level due to wind action in a 1 Hz bandwidth. It is assumed that the electronic noise level of the receiver is significantly smaller than the contributions from the ambient background and wind. To compute NL, the normalized noise levels $\text{NL}_{\text{ambient}}$ and $\text{NL}_{\text{wind}}$ are scaled by the receiver's bandwidth using the expression

$$\text{NL} = \text{NL}_{\text{ambient}} + \text{NL}_{\text{wind}} - 10 \cdot \log_{10}(\text{BW}) \ (\text{dB})$$

Using study site information, system characteristics of a radio telemetry receiver, and radio/acoustic specifications of the CART tag, a prediction of range is computed from Equations (1) and (2). Note that range values generated from Equation (2) should not be considered indicative of system performance under shallow water conditions (ie: a few meters). However, for the purpose of establishing guidelines, it is assumed that the water column depth is sufficient as to make range predictions from Equation (2) relevant. As an example, the following application scenario demonstrates an approach for establishing a conductivity threshold criterion for the CART device's dynamic conductivity switch.

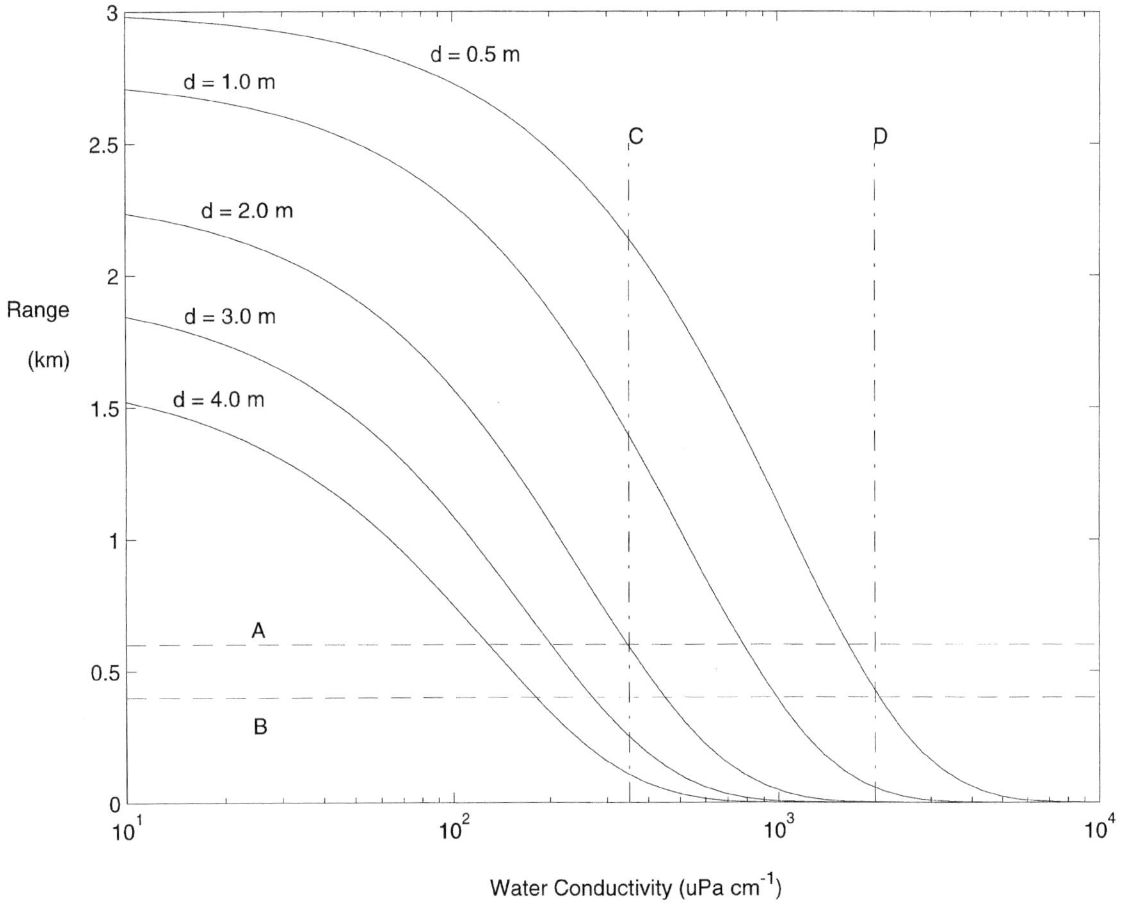


*Figure 1.* Acoustic/Radio Telemetry Ranges vs. Water Conductivity. Point A and B shows acoustic range predictions for water salinity values of 0.5‰ and 35‰, respectively. Points C and D separate three CART tag conductivity threshold regions. In between C and D transmission mode alternates between radio and acoustic; to the left of C the transmission mode is radio; and to the right of D the transmission mode is acoustic.

Consider an application scenario that monitors the movement of fish across three zones: a shallow fresh water river ($C < 200$ $\mu$S cm$^{-1}$), an estuary (200 $\mu$S cm$^{-1}$ $< C <$ 40000 $\mu$S cm$^{-1}$) and sea water ($C \approx$ 40000 $\mu$S cm$^{-1}$). The fish are assumed to transit the estuary within the top 2 meters of the water column. Across the estuary the nominal acoustic noise levels due to the ambient background and wind conditions are assumed to be known. A 150 MHz radio telemetry system to be used for the study is described by the following operating parameters:

$$\mathrm{ERP} = -20\ \mathrm{dBm},\ L_{\mathrm{int}} \approx 40\ \mathrm{dB},\ G_{\mathrm{ant}}$$

$$= 6\ \mathrm{dBd},\ L_{\mathrm{cable}} \approx 0\ dB,\ R_{\mathrm{sen}} = -136\ \mathrm{dBm}.$$

The study will also use a 76 kHz acoustic telemetry system. We model the estuary's noise conditions by $\mathrm{NL}_{\mathrm{ambient}}$ =40 dB re 1$\mu$ $Pa^2$ / Hz and $\mathrm{NL}_{\mathrm{wind}}$ =15 dB re 1$\mu Pa^2$ / Hz. In the open ocean, a typical ambient noise level is 20 dB re 1 $\mu Pa^2$ / Hz at 76 kHz (Clay & Medwin, 1989; Coates, 1989). The acoustic system's operating parameters are: SL = 158 dB re 1 $\mu$ *Pa @ 1* m, $\mathrm{G}_{\mathrm{hydro}}$ = 0 dB, BW = 7.5 kHz, and DT = 6 dB. Transmission loss due to absorption, (Clay & Medwin, 1989; Coates, 1989), in the estuary will range from

$$\mathrm{TL}_{\mathrm{absorp}}(S = .5‰) \approx 0.0024\ \mathrm{dB\ m^{-1}},$$

$$C \approx 200\ \mu\mathrm{S\ cm^{-1}} \quad @T = 12\ ^{\circ}\mathrm{C}$$

to

$$\mathrm{TL}_{\mathrm{absorp}}(S = 35‰) \approx 0.0027\ \mathrm{dB\ m^{-1}},$$

$$C \approx 40000\ \mu\mathrm{S\ cm^{-1}} \quad @T = 12\ ^{\circ}\mathrm{C}$$

Based on the radio and acoustic system parameters we compute a family of operating curves from (1), with each curve corresponding to a fixed depth. Figure 1 shows these curves for depths from 0.5 to 5 m. The dashed horizontal lines on the figure are the nominal range estimates of the acoustic telemetry system computed from Equation (2).

From Figure 1 we define a threshold window from 350 $\mu$S cm$^{-1}$ to 2000 $\mu$S cm$^{-1}$ corresponding to the intercepts between the $d=2$ m and $d=0.5$ m radio operating curves with the upper and lower limits of acoustic range. In this window the CART device is programmed to operate in a combined acoustic/radio mode. Below the window the tag transmits in radio mode, and above the window the mode is acoustic.

## Attributes of the CART device

The CART device consists of the following sub-systems: the radio section, acoustic projector section, battery pack, the microprocessor and conductivity probe assembly. A radio sub-system, including an external whip antenna, is derived from CFRT-3 type radio transmitters (Lotek Engineering Inc.). The acoustic projector assembly is taken from the CAFT family of acoustic transmitters (Lotek Marine Technologies Inc.). Battery packs for the CART device vary according to size and longevity requirements. A sketch of the CART tag is shown on Figure 2.

The unique feature of this transmitter is its ability to continually sense the electrical conductivity of ambient water and through a microcomputer algorithm, select the appropriate signal mode to maximize detection range. Conductivity measurements taken between two metal conductors exposed to the water is monitored by a programmable microprocessor that controls the transmission mode of the tag. The head of the conductivity probe as shown on Figure 2, is molded around a pair of metal conductors using a high durometer encapsulant. For bio-compatibility and flexibility the probe cable entering the head is coated with a flexible bio-compatible jacket. The strong bond between the head encapsulant and cabling jacket prevents any capillary action or water creep down the wire.

In addition to maximizing detection range, the continual sensing of conductivity is also motivated by the need to maximize tag life. Currently, the power consumption ratio between acoustic and radio modes of the CART tag is approximately four-to-one. Hence, a strategy of simply alternating between radio and acoustic modes would compromise the longevity of the device. Other attributes of this tag include; a programmable threshold setting; and an expandable threshold setting to cover a range of conductivity over which the device is set to operate in a combined transmission mode alternating between acoustic and radio. This latter feature allows for a 'gray' operating area where there is uncertainty as to whether acoustic or radio telemetry is optimal. The device can also be programmed to stop transmitting in a specified conductivity window and resume transmission outside this window thus providing a means of extending tag life.

*Table 1.* Mean Fork Length, weight and $U_{crit}$ of control and tagged groups. Results, expressed as a mean $\pm 1$ standard error of the mean value. Mean values were not found to be statistically different (ANOVA, systat)

| Group | Fork Length (cm) | Weight (g) | $U_{crit}$ (m/s) |
|---|---|---|---|
| Control | 41.73 ± 1.14 | 912.13 ± 92 | 0.804 ± 0.094 |
| Tagged | 46.2 ± 1.18 | 1103.5 ± 89 | 0.694 ± 0.087 |

## Attachment procedure

Rainbow trout were obtained from the Ontario Ministry of Aquaculture and Fisheries Research station, Alma, Ontario. Tests were conducted with two groups of rainbow trout, control and tagged fish. Both groups were acclimated to laboratory conditions for four weeks prior to experimentation in a 1400 l holding tank supplied with aerated well water ($T=11$ °C ± 0.5 °C) and lighting conditions of 16 h light: 8 h dark. Fish were fed *ad libitum* with commercial trout pellets (Martin Mills Inc., Elmira Ont.). However, food was withheld for a four day period prior to and during the experimental period to ensure a post-absorption state before the assessment of critical swimming velocities (Beamish et al., 1979).

All fish were anaesthetized to a 200 mg l$^{-1}$ solution of MS-222 neutralized with sodium hydrogen carbonate at a ratio of 1:1. Fork length and weight of each individual was recorded and the fish then placed into a 120 l Blaska type swim respirometer supplied with aerated well water ($T=11$ °C ± 0.5 °C). All fish responded to the conditions within the respirometer and swam under a shaded area at the anterior end of the chamber. Once placed in the respirometer, fish were given an acclimation period of 30 min at a water velocity of 1.5 body lengths prior to the assessment of

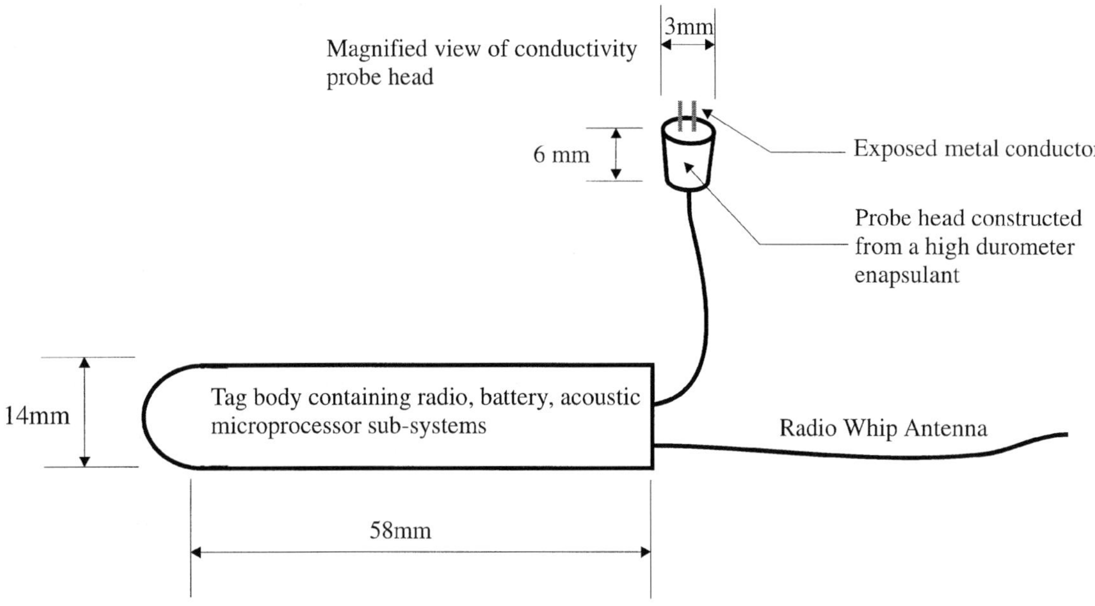


*Figure 2.* CART tag and Conductivity Probe.

the critical swimming speed. Following acclimation, fish were subjected to a forced swim by increasing the water velocity in a constant step-wise progression with a constant time increment of 5 m. Fish were deemed fatigued when they could no longer continue to swim against the current. Critical swimming speed ($U_{crit}$) was calculated for each individual using the formula described by Brett (1964),

$$U_{crit} = V + (t \cdot \Delta t^{-1})\Delta v \ \ (\mathrm{m\,s^{-1}}),$$

where $\Delta t$ is the time increment in minutes, $\Delta v$ ($\mathrm{m\,s^{-1}}$) is the velocity increment, $t$ (min) is the time elapsed at the final velocity, and $V$ ($\mathrm{m\,s^{-1}}$) is the highest velocity maintained for the prescribed time period.

Dummy CART transmitters (14 mm × 58 mm) were inserted into the abdominal cavity of the fish with anaethesia induced in the tagged individuals as described above (see Mellas & Haynes (1985, Figure 1)). These fish were weighted and measured then placed on a specially designed surgical table which allowed for the irrigation of a cold ($T$=6 °C ± 1.0 °C) solution of 100 mg $l^{-1}$ MS-222 neutralized with sodium hydrogen carbonate at a radio of 1:1. Attachment of the dummy transmitters involved insertion of the tag within the abdominal cavity of the fish, the antenna wire was led through the body wall at the posterior end of the cavity through a small puncture wound made by a 16 gauge needle. The conductivity probe head was brought through the body wall in a similar fashion at the anterior end of the cavity and protruded approximately 3.5–4.5 mm. Due to the dimension of the probe head (3 mm × 5 mm) a larger puncture wound was required to fit the probe through. This wound was closed with two small sutures to ensure that the probe head did not retract into the body of the fish during swimming. The main incision, through which the body of the tag was placed, was closed with 3–4 sutures. Total surgical time was a maximum of 4 minutes and fish were allowed to recover for a minimum of 16 hours prior to the assessment of critical swimming speeds. Tagged individuals were further anaesthetized in similar fashion as described above prior to being placed in the respirometer. All tagged fish were subject to the same protocol.

## Experimental results

Results are presented as a mean ± standard error. Statistical comparisons were made, using a $t$-test, between the control and tagged groups ($n$ = 6) (Table 1). No significant differences were found between each group demonstrating that the attachment procedure did not significantly disrupt the swimming performance of the tagged individuals 16 hours after insertion.

## Conclusion

The CART tag described in this paper is capable of continually sensing the electrical conductivity of ambient water. This feature permits the tag to operate in an appropriate signal mode that will maximize detection range performance. The strategy for switching modes can be modified through a microcomputer program to suit the application. An attachment procedure for the CART tag and conductivity probe is presented. Swimming performance results indicate that the attachment procedure does no significantly disrupt swimming performance.

## Acknowledgements

We gratefully acknowledge the assistance of Mark Ploughman, Ed Kennedy and Debbie Norman in the development, design and construction of the CART conductivity probe.

## References

Armstrong, J. D., M. Lucas, J. French, L. Vera & I. G. Priede, 1988. A combined radio and acoustic transmitter for fixing direction and range of freshwater fish (RAFIX), J. Fish Biol. 33: 879–884.

Beamish F. W. H., 1978. Swimming Capacity, Fish Physiology. Academic Press, Hoar and Randall (eds), New York, 7: 101–172.

Brett J. R., 1964, The respiratory metabolism and swimming performance of young sockeyed salmon. J. Fish. Res. Bd Can. 21: 1183–1226.

Clay C. & H. Medwin, 1977. Acoustical Oceanography: Principles and Applications, John Wiley.

Coates, R., 1989. Underwater Acoustic Systems. John Wiley, New York.

Lotek Engineering Inc., 1994, Telemetry Workshop Notes. Tondheim, Norway.

Mellas, E. J. & J. M. Haynes, 1985. Swimming Performance and Behavior of Rainbow Trout (Salmo gairdneri) and White Perch (Morone americana): Effects of Attaching Telemetry Transmitters. Can. J. Fish. aquat. Sci. 42: 488–493.

Solomon, D. J. & E. C. E. Potter, 1988. First results with a new estuarine fish tracking system. Fish Biol. 33 (supplement A): 127–132.

Velle, J., J. Lindsay, R. Weeks & F. Long, 1979. An Investigation of the Loss Mechanisms Encountered In Propagation From a Submerged Fish Telemetry Transmitter, 2nd International Conference of Wildlife, Biotelemetry: 228–237.

*Hydrobiologia* **371/372:** 53–59, 1998.
*J.-P. Lagardère, M.-L. Bégout Anras & G. Claireaux (eds), Advances in Invertebrates and Fish Telemetry.*



# Frequency choice for radio telemetry: the HF vs. VHF conundrum

Mitchell M. Sisak & James S. Lotimer
*Lotek Engineering Inc., Newmarket, Ontario, Canada*
*Correspondence should be sent to Mitchell M. Sisak (e-mail: mitch.sisak@lotek.com)*

*Key words:* biotelemetry, radio frequency, high frequency, very high frequency

## Abstract

The choice of operating frequency has always been the subject of considerable controversy amongst researchers. Many select the radio frequency at which they conduct their studies based solely upon the availability of equipment at hand or simply on tradition. Researchers are often not fully aware of the proximal impact of frequency choice on overall system performance and its ultimate impact upon the quality of the data that the study generates. In an attempt to provide researchers with the tools necessary to make informed choices regarding radio telemetry system configuration in general, and frequency choice in particular, this paper will present an overview of the 7 individual components of a radio telemetry system, examining the role played by each constituent in system optimization.

## Introduction

While direct observational techniques have long been available to researchers studying the movement patterns of terrestrial animals, the monitoring of aquatic organisms has always posed a challenge to the technology of the day. In the aquatic milieu, direct observation was an option for only a limited number of species and environments. Statistically significant results could only be realistically amassed through the use of remote monitoring techniques. One of the most widely employed techniques for the remote reception of data from aquatic organisms is through the detection of the active radiation from a transmitter which has been attached to the animal under study. This radiation can be in several forms, with radio and acoustic transmitters being amongst the most effective.

Both acoustic and radio telemetry are suitable for use in aquatic environments, but there are attendant costs and benefits associated with each. Such factors as water depth and its physico-chemical properties act in concert to maximize the utility of one or the other of these methods and this has been discussed at length in a number of survey papers by such authors as Winter (1983) and Priede (1992). For the purposes of this presentation we will limit ourselves to a discussion of radio telemetry systems.

## Radio telemetry

For the majority of freshwater studies, radio telemetry is the method of choice. The wide range of sophisticated transmitting, receiving and data logging systems available to researchers coupled with the relative ease with which radio telemetry equipment can be deployed and the studies conducted, has resulted in its widespread use in studies throughout North America and Europe.

While the suitability of radio over acoustic telemetry in many studies is rather clear-cut, the choice of operating frequency has always been the subject of considerable controversy amongst researchers. Through the course of dealing with researchers conducting radio telemetry studies the world over, it has come to our attention that many select the radio frequency at which they conduct their studies based solely upon the availability of equipment at hand or simply on tradition. They are often not fully aware of the proximal impact of frequency choice on overall system performance and its ultimate impact upon the quality of the data that the study generates. While this approach yields acceptable results in many cases, the far-ranging effects on the performance of the entire telemetry system posed by frequency choice deserves more careful eval-

*Table 1.* Radio frequency classification

| Band | Classification | Frequency Range |
|---|---|---|
| VLF | Very Low Frequency | 10 to 30 kHz |
| LF | Low Frequency | 30 to 300 kHz |
| MF | Medium Frequency | 300 to 3000 kHz |
| HF | High Frequency | 3 to 30 MHz |
| VHF | Very High Frequency | 30 to 300 MHz |
| UHF | Ultra High Frequency | 300 to 3000 MHz |

uation if systems are to be optimized for particular studies. On another front, choice of equipment for telemetry studies implies non-trivial capital costs associated with equipment acquisition as well as attendant training costs. The selection of telemetry frequencies results in a 'locking in' of future research to these frequencies, as equipment such as receivers and antennas are generally frequency specific.

In an attempt to provide researchers with the tools necessary to make informed choices regarding radio telemetry system configuration in general, and frequency choice in particular, this paper will present an overview of the 7 individual components of a radio telemetry system, examining the role played by each constituent in system optimization. Each component will be addressed in turn to afford newcomers to radio telemetry an appreciation from a system perspective. This exercise will also hopefully serve as a refresher course for those of you who are experienced radio telemetry users. Theoretical, practical and preliminary experimental data will be presented in a discussion whose aim will not be to recommend the optimal frequency of operation for all aquatic telemetry research, but rather to provide the tools necessary for researchers to select the optimal operating frequency for their particular study. We will conclude with a real-world example, permitting system performance to be evaluated.

### *Radio frequency bands*

Three divisions or bands within the radio frequency portion of the electromagnetic spectrum have been traditionally employed by fisheries researchers, these being designated as the HF, VHF and UHF bands. Table 1 demonstrates the relationship between these bands. The characteristics of these frequencies, which will be examined in more detail below, make them most suitable for aquatic applications.

## The radio telemetry system

In order to maximize the quality of the data collected in a radio telemetry-based fisheries monitoring program, the principal investigator must make wise system choices. A very significant variable in optimizing system performance is the choice of propagating frequency. Unfortunately, there is no one frequency that proves optimal in all cases. As is common in any science, an understanding of the system parameters and the trade-offs between them will ultimately lead the user to the best system for his application.

Seven system variables affect the choice of frequency. The system, as depicted in Table 2 is composed of the following variables:

### *Transmitter Radiated Power (ERP)*

The amount of radio energy that we can radiate from a transmitter is dependent upon two factors, these being:

*Energy Supply (battery)* – the amount of energy available from the battery. Regardless of the frequency chosen this variable will remain constant and can therefore be ignored in any frequency comparison.

*Antenna Radiating Efficiency* – the amount of radio energy the transmitter's antenna will transfer to the water relative to the input power. For fisheries transmitters, we have found that an 'efficient antenna' is one in which the antenna length is at least 25% of the frequency's wavelength ($\lambda$) as measured in free space; wavelength equals the speed of light (300 000 000 m $s^{-1}$) divided by the frequency (Hertz). For example; at a frequency of 50 MHz

$$\text{one wavelength} = 300\,000\,000/50\,000\,000$$

$$= 6\text{ m}$$

and an efficient antenna = $0.25 * 6 = 1.5\ m$ whereas at a frequency of 150 MHz

$$\lambda = 300\,000\,000/150\,000\,000$$

$$= 2\text{ m}$$

and an efficient antenna = $0.25 * 2 = 0.5\ m$.

It is important to note that antenna efficiencies in free space drop rapidly when their length is below one quarter wavelength.

*Table 2.* Major system variables

1. *Transmitter – Effective Radiated Power (ERP)* – the electromagnetic energy (radio waves) radiated from the transmitter's antenna.
2. *Propagation Loss in Water* ($L_{water}$) – the amount of radio energy lost as a result of traveling through water.
3. *Interface Loss* ($L_{int}$) – the amount of radio energy lost as a result of traveling through the air/water interface.
4. *Propagation Loss in Air* ($L_{air}$) – the amount of radio energy lost as a result of traveling through air.
5. *Receiving Antenna Gain* ($G_{ant}$) – the amount of radio energy gained in focusing the electromagnetic signal at the receiving antenna.
6. *Propagation Loss in the Cable* ($L_{cable}$) – the amount of radio energy lost as a result of traveling along the cable.
7. *Receiver Sensitivity* – the level at which the receiver will find and decode the radio signal. The receiver sensitivity can be limited by the *Ambient Noise Floor* – the level of radio energy in the surrounding environment.

*Propagation loss in water* ($L_{water}$)

Propagation loss in water for a given frequency is predictable, measured in decibels per meter (dB $m^{-1}$), and varies with the conductivity of the water. Figure 1 shows the relationship. Standard conductivity meters measure the conductivity of water at the frequency employed by the measuring instrument (approximately 1 MHz), so the conductivity value must be corrected before use at other frequencies as discussed by Velle et al. (1979).

As an example, from the graph in Figure 1, using water conductivity (measured at 1 MHz) of 200 $\mu$S $cm^{-1}$ and a temperature of 10 °C, the propagation loss at 50 MHz is approximately 3.6 dB $m^{-1}$ while at 150 MHz it is approximately 4.8 dB $m^{-1}$.

*Interface loss* ($L_{int}$)

Radio waves transiting the water/air interface exit the water in a cone directly above the transmitter, the sides of which form an angle of approximately 6 degrees from the vertical (Figure 2). Waves are also polarized vertically as they exit the water/air interface. Interface loss is frequency independent and thus does not affect the choice of frequency, but its effect is important when determining overall system performance.

*Propagation loss in air* ($L_{air}$)

Propagation loss in air is predictable, measured in dB and while the loss is frequency dependent the difference is constant over fisheries telemetry ranges. For example the difference in propagation loss between 150 MHz and 50 MHz is 10 dB regardless of the propagation range.

*Receiving antenna gain* ($G_{ant}$)

This situation has similar considerations to the transmitting antenna example. Roughly speaking, an efficient receiving antenna in fisheries telemetry studies will form a square with each side approximating 50% of a wavelength. Therefore, for 50 MHz you will need a receiving antenna 3 m by 3 m verses 1 m by 1 m at 150 MHz.

As to traditionally employed antennas for manual tracking, the diamond (or loop) antenna popular with researchers conducting radio telemetry in the HF band is not an ideal antenna as it represents a compromise between electrical characteristics and portability, and typically possesses a gain of –6 dBd. A typical 4-element Yagi antenna used in VHF band tracking has a boom length of 110 cm and a gain of +7 dBd.

Additionally, antenna height above the ground affects antenna efficiency. The lower the operating frequency the higher your antenna must be to realize any height advantage.

*Propagation loss in the cable* ($L_{cable}$)

Propagation loss in the cable running from the receiving antenna to the receiver is predictable and can introduce significant system losses when the cable length exceeds 50 meters. Below this length the losses in the cable will not affect the choice of frequency in any significant way. Indeed, even in a system requiring long cable runs to the receiver and data logger there are techniques that can be used to equalize and mitigate any differences in the frequency used, such as the inclusion of pre-amplifiers.

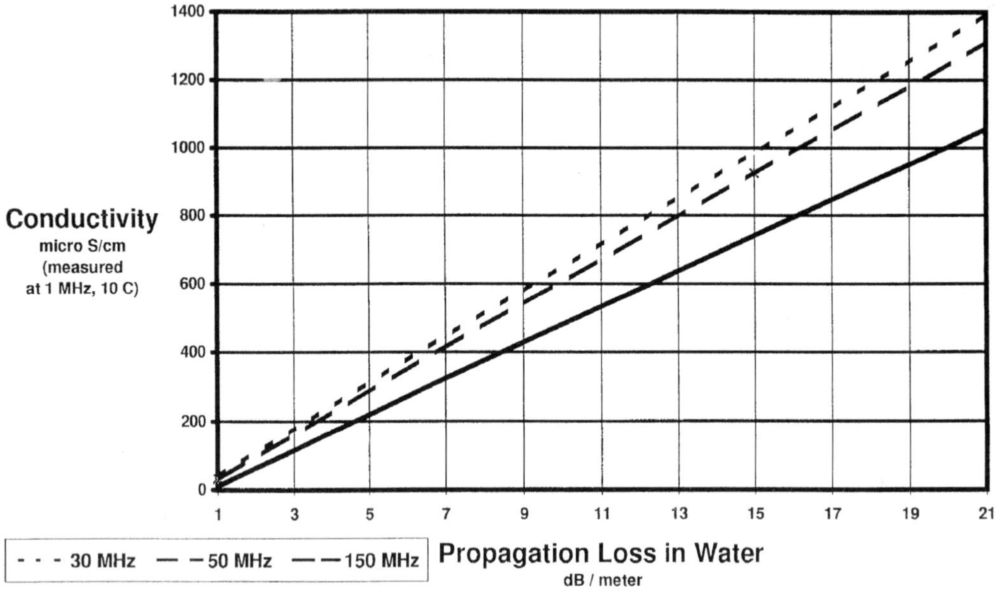


*Figure 1.* Effect of conductivity on propagation loss in water at 30 MHz, 50 MHz and 150 MHz (after Velle et al., 1979).

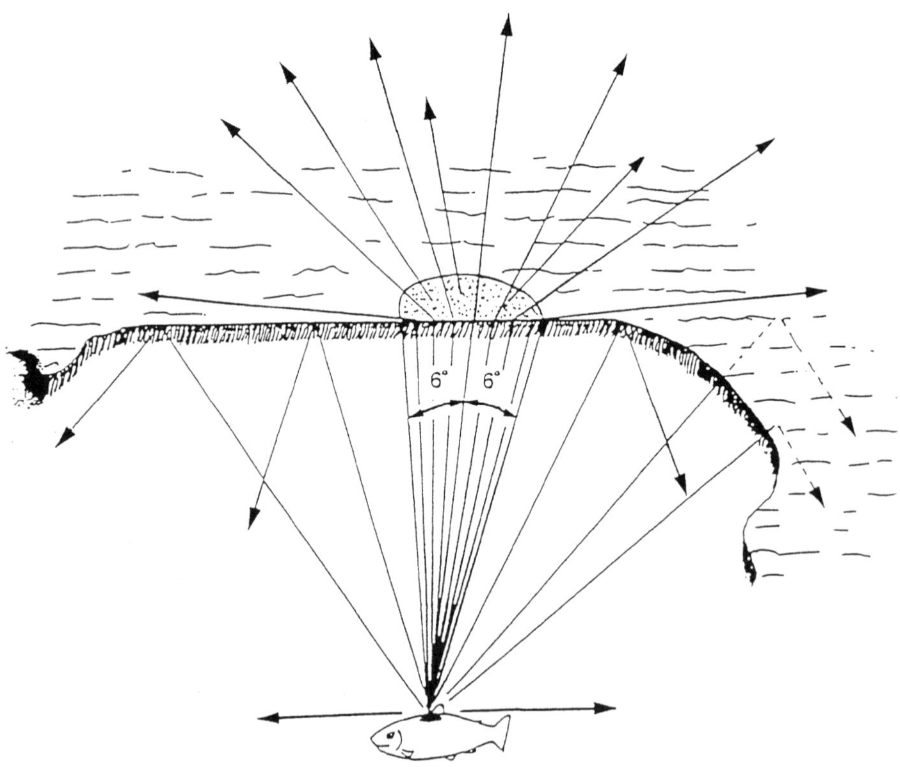


*Figure 2.* Idealized radiation pattern of an underwater radio transmitter (from Preide, 1992).

*Table 3.* Summary of system considerations affected by frequency selection

1. Transmitter Radiated Power: specifically Antenna Radiating Efficiency which is a function of transmitter antenna length. Lower frequencies require longer antennas to be efficient. How close to this ideal length can the studied species support?
2. Propagation loss in Water: losses increase with increasing frequency:how significant is this loss when other variables are considered?
3. Receiving Antenna Gain: Efficient antennas are longer at lower frequencies and physically manageable antennas at 50 MHz are inefficient. How realistic is it to use an efficient receiving antenna and will size compromises negatively impact on overall system performance?
4. Receiver Sensitivity: Receiver sensitivity can be limited by the ambient noise floor. This noise floor is generally higher at lower frequencies. How high is the ambient noise floor in your study area? Is it temporally or spatially heterogeneous?

*Receiver sensitivity*

The ability for your receiving system to 'hear' or decode a low level signal is known as the receiver sensitivity. This sensitivity is measured in dBm (dB re 1 mW), the same as the noise floor measurement. If the noise floor is lower than the receiver sensitivity, the receiver sensitivity becomes the limiting factor in the detection of the signal. On the other hand, if the noise floor is higher than the receiver sensitivity the noise floor becomes the limiting variable. It is important to note that while some automatic datalogging stations can receive and decode signals as low as –135 dBm, the processing power of the human brain permits a trained radio operator using narrow band equipment to detect signals at about –145 dBm.

*Ambient noise floor*

Electromagnetic noise comes from a variety of sources both natural and anthropogenic; the sun, electric transmission lines and generating plants, radio interference, automobile ignition, and motor boats to name a few. The interference or radiation from these sources combines to make up the total level of noise in the environment, and this level is measured in decibels relative to 1 mW. The effect of noise is especially pronounced in urban environments and around any man-made structure.

Electromagnetic noise floor is a significant system variable in a fisheries telemetry study. If your signal is buried in the noise (i.e. below the noise floor), when it reaches the receiving antenna you will not be able to 'hear' or decode the signal. The 'local' noise floor can be the single greatest influence on your system performance. The effect of this variable is demonstrated further in the example.

To summarize, we can list the system considerations affected by frequency selection (Table 3).

*Example*

To determine system capabilities one simply solves for the unknown variable by adding up the gains and losses in the following equation and comparing the signal available to the receiver sensitivity/noise floor limitation.

$$\mathrm{SL}_{\mathrm{received}} = \mathrm{ERP} - L_{\mathrm{water}} - L_{\mathrm{int}} - L_{\mathrm{air}} + G_{\mathrm{ant}} - L_{\mathrm{cable}}.$$

In many cases all the system variables are known to the user, and the equation can be employed to determine if the system will be capable of providing data. In addition, the user may wish to compare the performance of various frequencies to determine which frequency will provide the greatest margin of safety combined with ease of use.

Using the information discussed thus far we will end with an example. Table 4 depicts the appropriate values and typical results obtained using the concepts covered thus far for a typical freshwater telemetry study.

To begin, we will address the radiated power of the transmitter. The length of the transmitter antenna will be 35 cm long, a reasonable length for a majority of aquatic species. The radiated power of the transmitter will be typical for commercially available transmitters at 150 MHz and 50 MHz employing a 35 cm antenna. Environmental data suggest that the body of water under study has a maximum depth of 3 m and a conductivity of 200 $\mu$S cm$^{-1}$. Propagation loss in water is obtained from Figure 1. A typical interface loss value of 30 dB is employed. Propagation loss through

*Table 4.* Assumptions & measurements

| Frequency | 50 MHz | 150 MHz | Comments |
|---|---|---|---|
| Radiated power (ERP) | –30 dBm | +6 dBm | typical (similar) industry transmitters |
| Propagation loss in water ($L_{water}$) | 3.6 dB/m = 3 × 3.6 = 10.8 dB | 4.8 dB/m = 3 × 4.8 = 14.4 dB | from graph |
| Interface losses ($L_{int}$) | 30 dB | 30 dB | typical |
| Propagation loss in air ($L_{air}$) | 60 dB | 70 dB | calculated |
| Cable losses ($L_{cable}$) | 1 dB | 1 dB | short cable |
| Receiving antenna gain ($G_{ant}$) | –6 dB | +7 dB | note large difference |
| Received signal level ($SL_{received}$) | =–30–10.8–30–60–1–6=–137.8 dBm | =6–14.4–30–70–1+7 =–102.4 dBm | |
| Minimum signal detection | –135 dBm | –135 dBm | receiver (not noise floor) determined |
| Result | No Data Collected | Data Collected | |

the air is calculated for each frequency. We will assume a noise floor measurement of –150 dBm for both 50 MHz and 150 MHz. This value represents a very low level of ambient noise, suggesting a remote site. As this value is well below the minimum signal detection for the receiver, it will not directly impact on the calculations. Note that in an area where the noise floor is higher, it could seriously affect the ability of the receiver to detect the desired signal. Cable losses will be set at 1 dB implying a length of approximately 10 m of cable between the antenna and the receiver. Mechanical considerations with respect to the receiving antenna installation regarding such factors as weight and wind loading effects will limit the antenna size to 1 m × 1 m, again a reasonable figure. The installation will be an automatic data logging station, with a receiver sensitivity of –135 dBm.

As can be seen from Table 4, given the conditions presented above, the 150 MHz datalogging station would acquire and log signals as the received signal level is above the minimum signal detection level. As the received signal level for the 50 MHz station is below the minimum signal level, it would not be able to collect data. The suggested success of the 150 MHz system is due in large part to the greater radiated power of the transmitter and the greater efficiency of the receiving antenna, which is more easily attained with reasonably sized antennas at 150 MHz than at 50 MHz.

## Conclusion

This systems approach to radio telemetry is advanced to permit a rapid evaluation of expected system performance. It also serves to identify components which can be addressed in the event that improved performance is required to guarantee the success of a particular study. As the values of many components are estimated it is important to stress that this is a qualitative tool and should not be used quantitatively. This approach represents a good engineering approximation to the problem.

An appreciation for system optimization will permit you to make informed choices regarding the various components of your radio telemetry system. This system optimization will permit the collection of the maximal number and best quality data for the effort and expense that you put into your study.

In many cases, though propagation losses in water are higher at VHF frequencies, improvements in system variables and other system considerations favor the choice of VHF band frequencies.

Finally, with more and more studies tending toward an examination of meso- and -macro scale movement patterns of fish populations, researchers are finding it necessary to coordinate telemetry activities between groups and even countries. Frequency is one of the key issues that must be addressed in this coordination process. For example, the Scandinavian countries have recently adopted the 142 MHz band for their telemetry work, permitting as a side benefit the sharing of equipment amongst the various participating groups. In North America, the United States government have recently adopted the 162–174 MHz frequency band for all of their telemetry studies.

While choice of frequency is an important one, we must recognize that frequency is but one part of the telemetry system. Each component of the system is inter-related to the other constituents, the whole being the sum of the antagonistic and synergistic contributions of its parts.

## References

Velle, J. I., J. E. Lindsay, R. W. Weeks & F. M. Long, 1979. An Investigation of the Loss Mechanisms Encountered In Propagation From a Submerged Fish Telemetry Transmitter. In Proceedings of the Second International Conference on Wildlife Biotelemetry: 228–237.

Winter, J. D., 1983. Underwater Biotelemetry. In L. A. Nielsen & D. L. Johnson (eds), Fisheries Techniques. American Fisheries Society, Bethesda, MD: 371–395.

Priede, I. G., 1992. In I. G. Priede & S. M. Swift (eds), Wildlife Telemetry: Remote Monitoring and Tracking of Animals, Ellis Horwood, New York: 3–25.

*Hydrobiologia* **371/372:** 61–69, 1998.
*J. P. Lagardère, M.-L. Bégout Anras &G. Claireaux (eds), Advances in Invertebrates and Fish Telemetry.*



# Simultaneous observations of tuna movements and their prey by sonic tracking and acoustic surveys

Erwan Josse, Pascal Bach & Laurent Dagorn
*ORSTOM – B.P. 529 – Papeete – French Polynesia (Polynésie Française); e-mail address: josse@orstom.fr*

*Key words:* tuna, vertical and horizontal movements, sonic tracking, sound scattering layer, acoustic survey

## Abstract

This paper reviews results of some experiments conducted in French Polynesia on tuna behaviour. A method based on the simultaneous use of two techniques, acoustic tracking and acoustic surveys, was used. Experiments were conducted within the framework of the ECOTAP program, a joint program between two national research institutes (IFREMER and ORSTOM), and a territorial institute (EVAAM).

Acoustic tags equipped with pressure sensors were used in order to record horizontal and vertical movements of one yellowfin tuna (*Thunnus albacares*) and two bigeye tuna (*T. obesus*). Trackings lasted between 13 to 24 h. In the same time, echogram data were recorded between the surface and a depth of 500 m on board the tracking vessel. As the maximum range of the acoustic tags is small (a few hundred meters), vessel and tagged fish horizontal movements are therefore treated as equivalent. Echogram data from the sounder and data on the swimming depth of the fish given by the acoustic tag are then considered as having been obtained at the same time at the same place.

Comparison between the swimming depth of the tagged fish and the echogram data from the sounder clearly shows the important role of scattering layers, assimilated as food, on vertical and horizontal tuna movements, during daytime as well as during night-time.

The method used during these experiments allows to observe a new explanatory factor of tuna behaviour: the biotic environment. At small temporal and spatial scales, structure of the biotic environment and its dynamic appear to be a key factor to understanding the vertical and horizontal tuna movements. The simultaneous technique presented here must now be improved by using behavioural activities sensors. By this way, it would be possible to elucidate different tuna foraging phases in relationship with the dynamic of scattering layers.

## Introduction

Acoustic telemetry has been applied in a variety of studies on tuna behaviour in the world's oceans. Among the available sensors (pressure, temperature, tail beat frequency, electromyograms from epaxial and opercular muscles, electrocardiogram, body temperature), pressure sensors have been the most used if we consider the literature published on this subject. Horizontal movements have been interpreted in relation to the presence of anomalies in the environment such as Fish Aggregating Devices (Cayré & Chabanne, 1986; Holland et al., 1990; Cayré, 1991; Marsac et al., 1996) and seamounts (Yuen, 1970). Vertical movements detected by pressure sensor have been interpreted in relation to the vertical structure of abiotic variables such as temperature, oxygen, salinity (Laurs et al., 1977; Cayré & Marsac, 1993). Other experiments have examined changes in fish body temperature in relation to vertical movements and the vertical structure of the temperature (Holland et al., 1992).

Except experiments reported in Carey (1990), the biotic environment has been generally ignored as an explanatory factor in vertical and horizontal movements of tuna, as experiments did not produce relevant observations on this environment. This gap was already reported by Cayré & Chabanne (op. cit.): 'La relation éventuelle entre les nombreux déplacements verticaux et l'alimentation des thons est d'autant plus intéressante à explorer que celle entre ces mêmes mou-

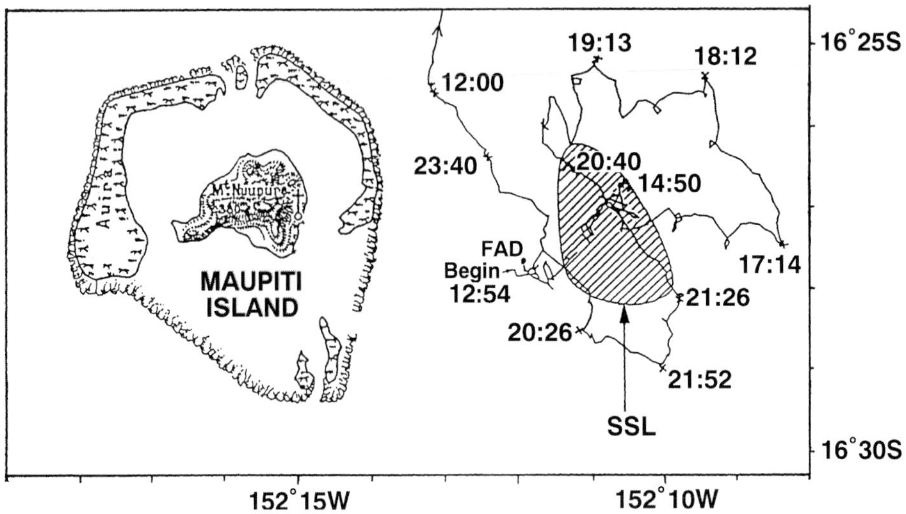


*Figure 1.* Experiment 1 (27/10/95). Horizontal movements of the tagged yellowfin tuna, with indication of the Sound Scattering Layer (SSL) position (hatched area) and the Fish Aggregating Device (FAD) position, near Maupiti Island (time is local time: UT – 10).

vements et les différents paramètres ou conditions hydrologiques n'est pas claire'. We present here a method based on the simultaneous use of two techniques: acoustic tracking and acoustic surveys. The simultaneous use of these techniques allows the interpretation of the vertical and horizontal movements of the fish in relation to the biotic factors of the environment. Three examples are used to illustrate the new observations on tuna behaviour we obtained with this method.

## Materials and methods

Our experiments were conducted in French Polynesia within the framework of the ECOTAP program. ECOTAP (Etude du Comportement des Thonidés par l'Acoustique et la Pêche / Studies of tuna behaviour using acoustic and fishing experiments) is a joint program between EVAAM[1], IFREMER[2] and ORSTOM[3]. The aim of this program is to study the distribution and the behaviour of subsurface tunas, *Thunnus albacares* (Bonnaterre, 1788), *T. obesus*, (Lowe, 1839), and *T. alalunga* (Bonnaterre, 1788), exploited by local longline and drop-stone (Moarii & Leproux, 1996).

Experiments were performed during cruises on board the ORSTOM research vessel 'ALIS'. The tracking equipment used during these experiments is a VEMCO system. Acoustic tags (model V16P, 50 kHz, 500 PSI) are equipped with a pressure sensor. Each tag transmits a 50 kHz pulse at a rate proportional to the pressure. For the first experiment, the signal was received through a VEMCO V10 directional hydrophone fitted on a V-fin towed depressor, then decoded and stored by a VEMCO VR60 receiver/decoder on board the R/V 'ALIS'. For the other experiments, a VEMCO V41 bearing hydrophone connected to VEMCO VR 28 receiver replaced the directional hydrophone. During acoustic survey, a SIMRAD EK 500 scientific sounder was used. This sounder was connected to a hull-mounted transducer SIMRAD ES 38 B Split Beam (frequency 38 kHz).

Fishing operations were conducted on board the R/V 'ALIS' using the drop-stone technique or trolling line, or on board longline fishing units. Two different methods were applied to attach the transmitter on the fish, according to its size. When the fish was small enough, it was pulled up on board without any injury, the fork length was measured and the sonic tag was attached on to its back, just behind the second dorsal fin, with two nylon tie-wraps. The tagging operation lasted between 45 to 90 sec. When the fish was too big, it was not moved out of the water but only drawn alongside the ship. The fork length was then estimated,

[1] EVAAM – Etablissement pour la Valorisation des Activités Aquacoles et Maritimes, B.P. 20, Papeete, French Polynesia.

[2] IFREMER – Institut français de recherche pour l'exploitation de la mer, Centre Océanologique du Pacifique, B.P. 7004, Taravao, French Polynesia

[3] ORSTOM – L'institut français de recherches scientifiques pour le développement en coopération, Centre ORSTOM de Tahiti, B.P. 529, Papeete, French Polynesia

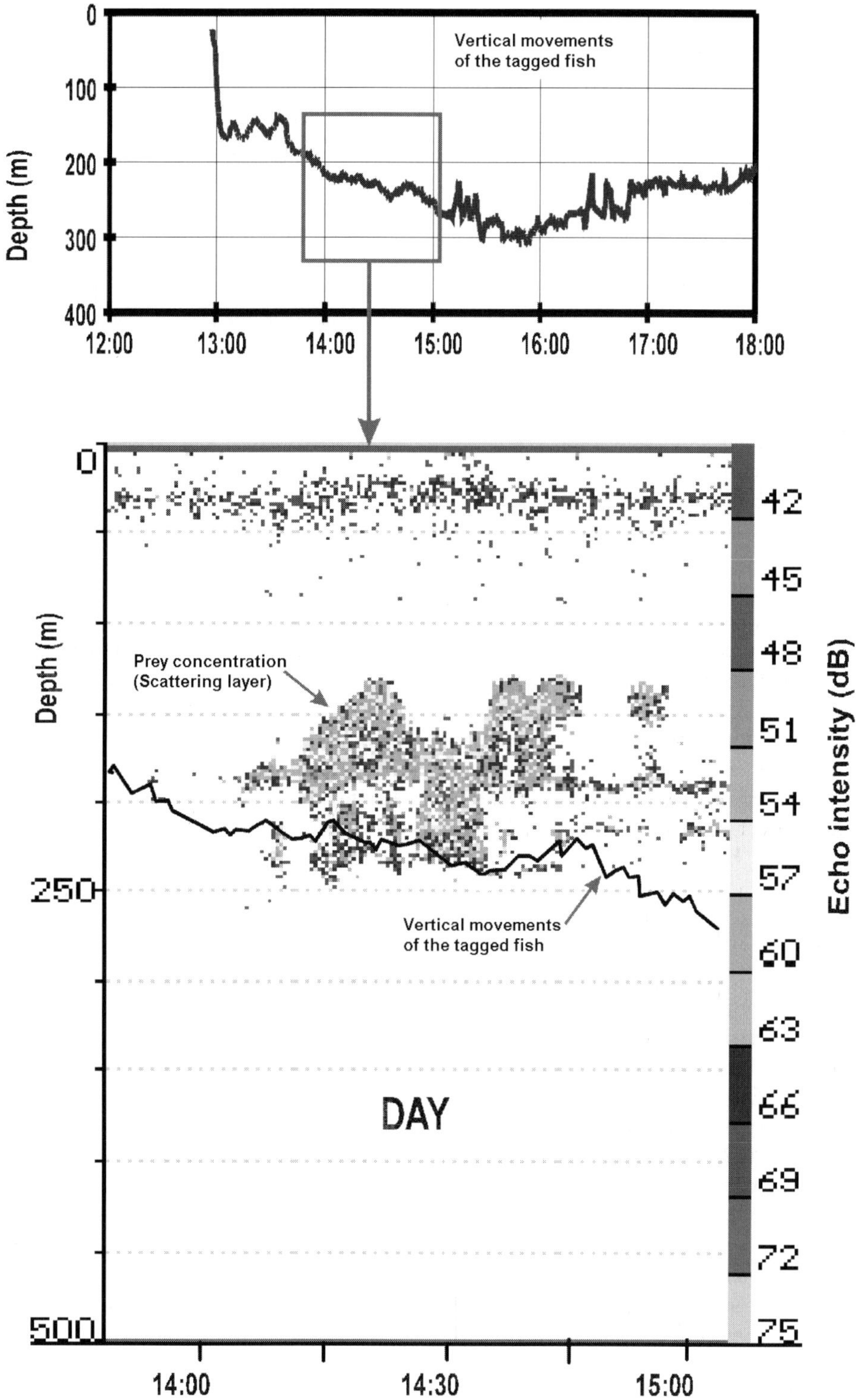


*Figure 2*. Experiment 1 (27/10/95). Vertical movements of the tagged yellowfin tuna and relation to the Sound Scattering Layer during daytime (time is local time: UT – 10).

and the sonic tag was fixed to the anterior dorsal musculature of the fish using a tagging pole. In this case, the tagging operation lasted about 20 sec. Immediately after the tagging operation, the fish was released at the same place and tracked. During tracking, simultaneous echosoundings were carried out between the surface and a depth of 500 m. The horizontal distance between the tracking boat and the tagged fish is generally short (between 500 and 800 m at the most). So, the horizontal movings of the vessel and those of the tagged fish are considered identical. The EP500 SIMRAD software was used to record echosounding data. At the same time, depth and geographical coordinates (GPS data) of the fish were recorded with a second recording unit using the VEMCO TRACK software. Interferences between the frequency of the tags (50 kHz) and the one used with the sounder (38 kHz) were very low and without effect on the quality of the data recorded.

During the tracking period vertical XBT profiles of temperature were performed between the surface and a depth of 460 m. At the end of each tracking, an hydrological station with a SeaBird SBE 19 probe was carried out. Pressure, temperature, salinity and oxygen were recorded between the surface and a depth of 600 m.

When possible, a fishing operation, using a pelagic trawl, was made before or after the tracking, in order to identify the species composition of scattering layers detected by the sounder.

## Results

### *Experiment 1*

A yellowfin tuna (*Thunnus albacares*), 60 cm fork length, was caught by the drop-stone technique at a depth of 120 m at midday, the 27th of October 1995, close to a Fish Aggregating Device (FAD) anchored near Maupiti Island (16°27′ S–152°17′ W), an island of the Leeward Islands of the Society archipelago. This tuna was tracked for 24 h. The results we present here concern only the first twelve hours of this tracking. Four periods can be distinguished for the horizontal movements: (i) a FAD association just after the release, (ii) a free-swimming phase directed offshore between 13:45 and sunset at 17:15, (iii) a progressive return to the FAD until 23:00, (iv) a free-swimming phase near the reef coast between 23:00 and midnight (Figure 1).

Vertical movements can be divided in two periods: (i) below the mixed layer during daytime, (ii) in the mixed layer during night-time.

Inside an oligotrophic area, an isolated Sound Scattering Layer (SSL) was observed in the vicinity of the Fish Aggregating Device (Figure 1).

The yellowfin tuna crossed the Sound Scattering Layer twice, first during daytime (Figure 2) and secondly during night-time (Figure 3). At each crossing, the tuna changed its movements. Horizontally it moved away from the Fish Aggregating Device during daytime although tuna are generally strongly associated with Fish Aggregating Devices during daylight (see experiment 2). Vertically, it swam in the mixed layer during night-time except during visits to the Sound Scattering Layer probably associated with foraging.

### *Experiment 2*

This experiment was realised in the open ocean around an anchored instrumented oceanographic buoy (5° S–140° W), the 20th of January 1996. Repeated echosoundings around this buoy were achieved every 4 h during a 24-h period just before the tagging.

These echosoundings confirmed the regular presence of a tuna school, essentially composed of juvenile bigeye tuna (*Thunnus obesus*), 35 to 84 cm fork length, during daytime (Figure 4). In the afternoon, an individual of this species (77 cm fork length) was caught by trolling line and equipped with an acoustic tag. This tracking lasted 24 h in all. We show here only some observations performed during this experiment.

During daytime, the tagged fish remained in the school. Depth and spread of the school did not seem to change while it appeared that the tagged fish exhibited vertical movements within the school (Figure 5). The maximum depth of the school and vertical movements of the fish seemed to be limited by the presence of a well-defined oxycline (1 ml $l^{-1}$ $0_2$) at a depth of 200 m. Because of this limit, the Sound Scattering Layer could be inaccessible to tuna during daytime.

At the end of the afternoon, the school structure disappeared and the tagged fish left the buoy. At dusk, the fish swam into the Sound Scattering Layer which migrated to the surface layer. The mean swimming depth of the tuna (60 m) corresponded to the maximum acoustic intensity in the Sound Scattering Layer (Figure 6). The first part of the vertical movements showed on Figure 6 is similar to the movements exhibited in the scattering layer by the yellowfin tuna tracked in experiment 1 (see Figure 3).

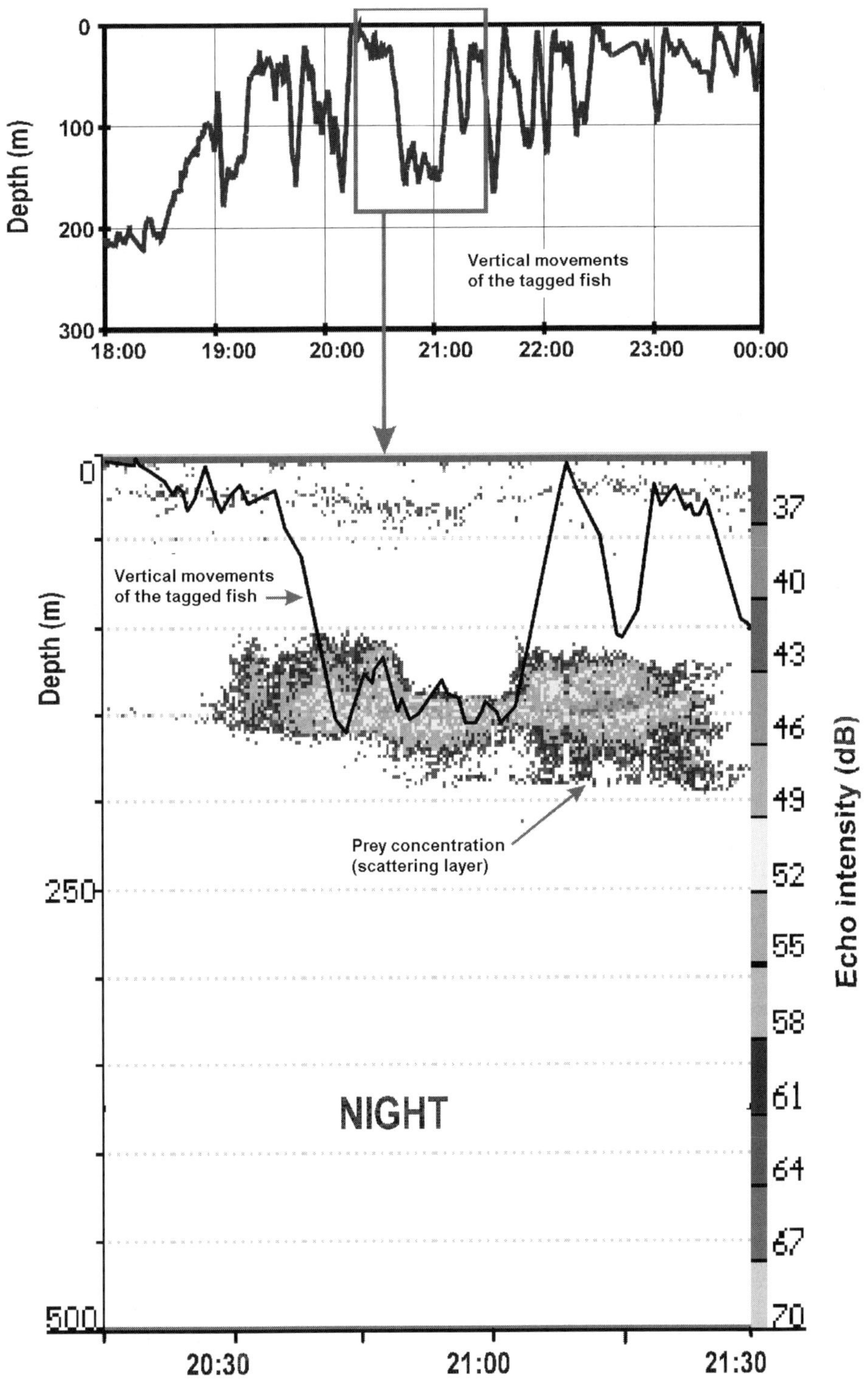


*Figure 3.* Experiment 1 (27/10/95). Vertical movements of the tagged yellowfin tuna and relation to the Sound Scattering Layer during night-time (time is local time: UT – 10).

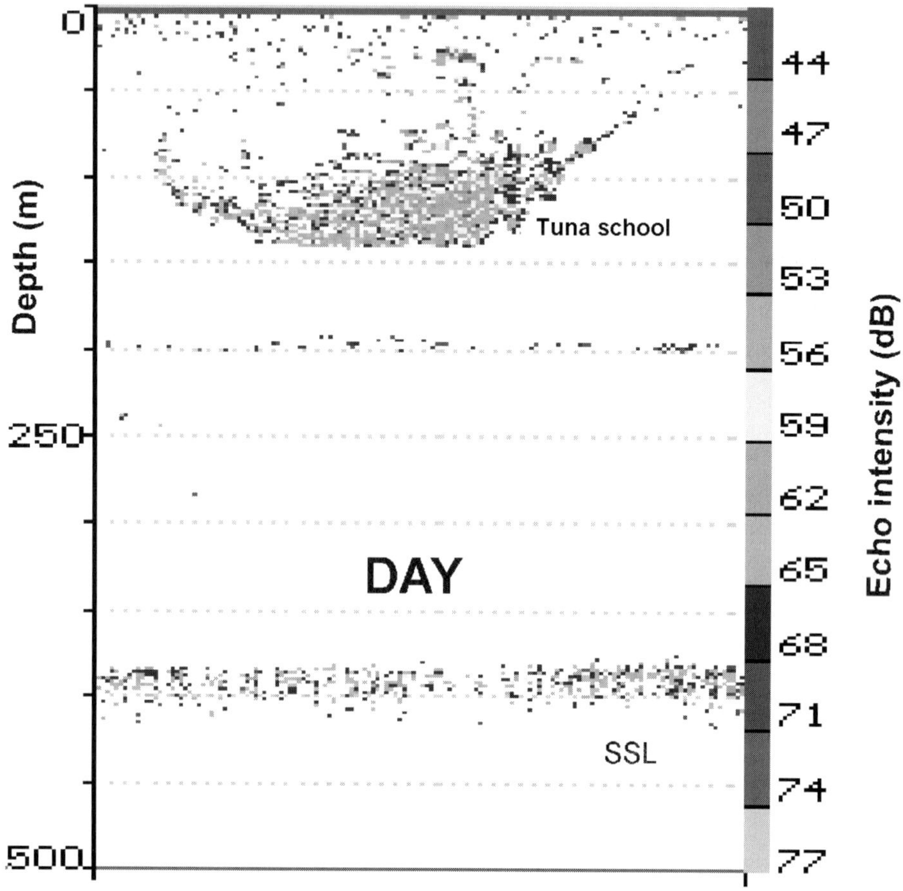


*Figure 4.* Experiment 2 (20/01/96). Tuna school (mainly bigeye tuna) and its biotic environment with indication of the Sound Scattering Layer (SSL) observed close to an anchored oceanographic buoy to the north (5° S–140° W) of the French Polynesian Exclusive Economic Zone.

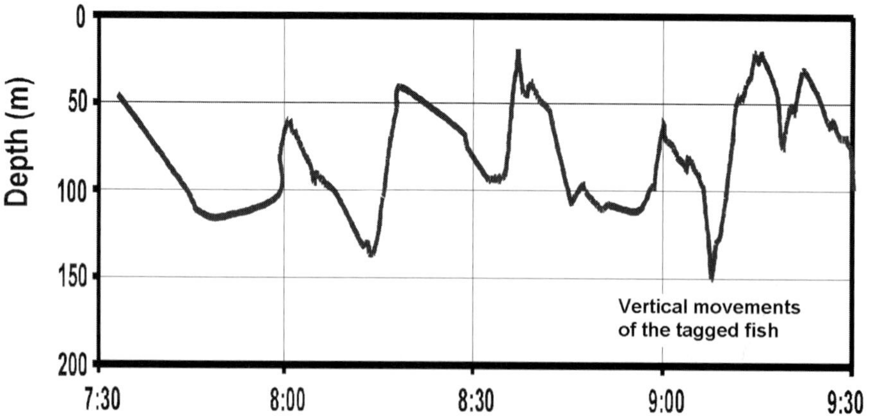


*Figure 5.* Experiment 2 (20/01/96). Detail of vertical movements of the tagged bigeye tuna inside the tuna school during daytime (time is local time: UT – 10).

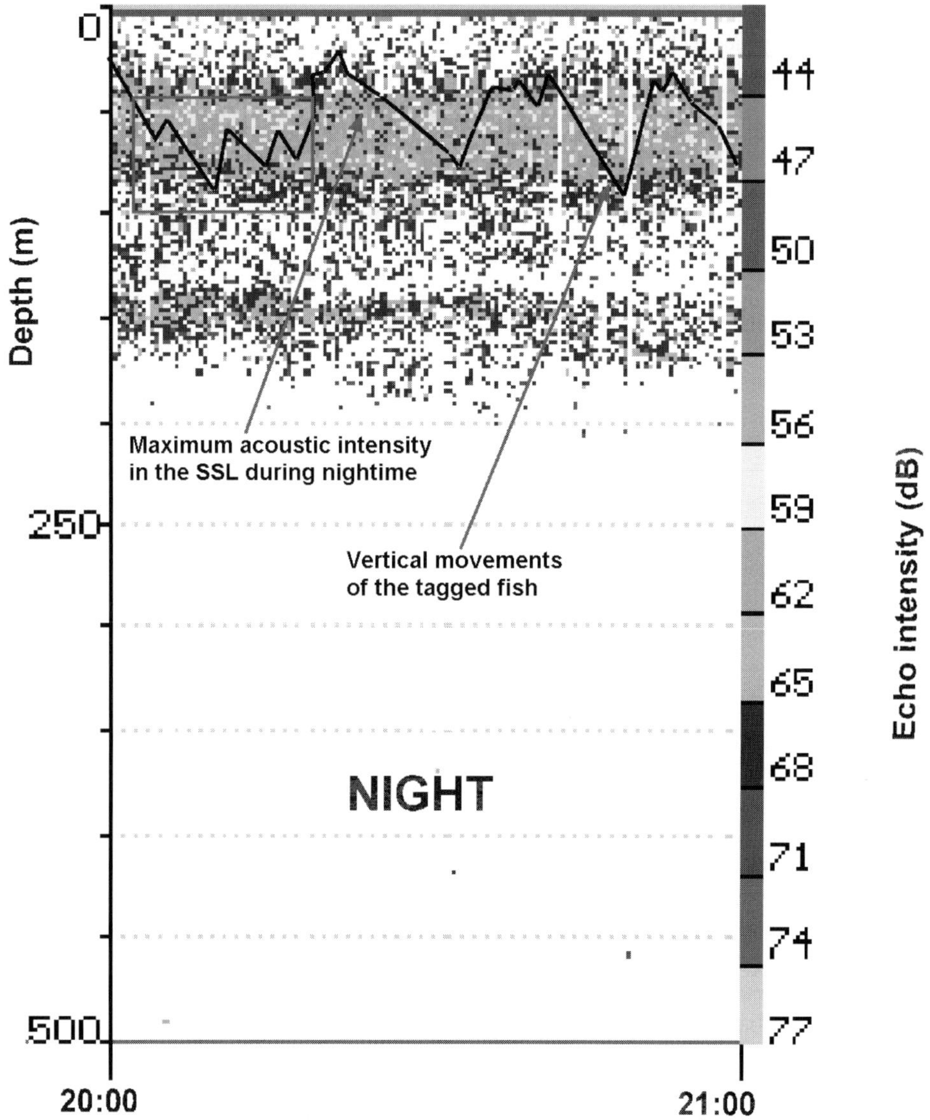


*Figure 6.* Experiment 2 (20/01/96). Vertical movements of the tagged bigeye tuna in relation to the Sound Scattering Layer (SSL) during night-time (time is local time: UT – 10).

*Experiment 3*

A bigeye tuna *(Thunnus obesus)*, 100 cm fork length, was caught in the open ocean by a longline fishing unit at 01:00, the 18th of December 1996 off Tetiaroa Island (17°00′ S–149°35′ W), an atoll of the Windward Islands of the Society archipelago. This experiment lasted at all 13 h. After tagging and release, the fish principally remained in the upper part of the Sound Scattering Layer between the surface and a depth of 40 m. Some rapid and small-scale changes in depth were observed down to 110 m deep corresponding to the lower limit of the Sound Scattering Layer (Figure 7).

At dawn (about 05:00), the tagged fish left the mixed layer and swam down slowly until a depth of 350 m. The simultaneous mapping of vertical movements and echosounding records observed during the tracking clearly shows that the tagged fish followed the Sound Scattering Layer migration at first light (Figure 7).

The fish stopped its descent at a depth of 350 m (water temperature: 15 °C) while the Sound Scattering Layer continued its migration to a depth of 550 m.

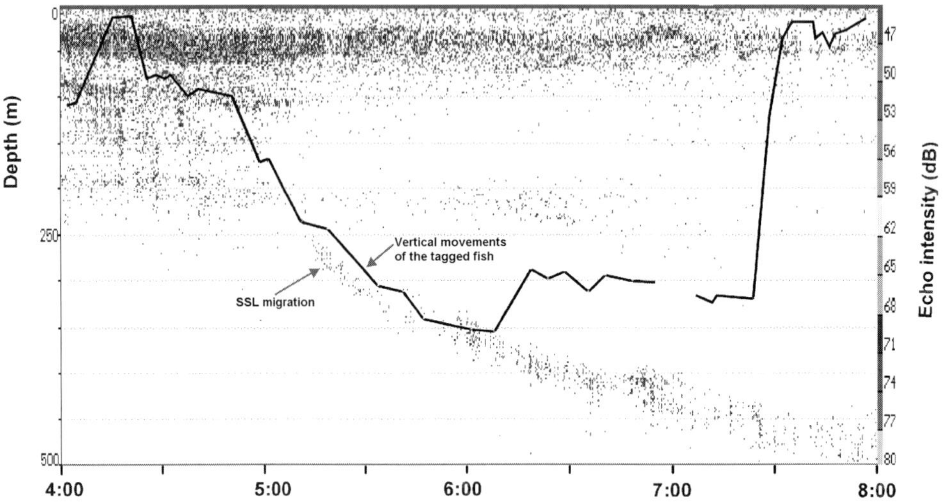


*Figure 7.* Experiment 3 (18/12/96). Vertical movements of the tagged bigeye tuna, with indication of the Sound Scattering Layer, off Tetiaroa Island (time is local time: UT – 10).

## Discussion

It is generally accepted that three-dimensional tuna movements are controlled by abiotic factors which determine habitat limits. Various studies have attempted to delineate the habitat requirements of tuna species, first by studying relationships between catch statistics and oceanographic conditions averaged over time and space (see Sund et al., 1981 for a review), more recently by developing researches on the physiological abilities and tolerances of tunas (see Brill, 1994, for a review).

Only one part of tuna movements observed during our experiments could be explained by limiting abiotic factors. For example during the experiment 2, the presence of a well-defined oxycline at 200 m deep (1 ml $l^{-1}$ $O_2$) seems to represent an impassable boundary, both for the tagged yellowfin tuna and for the tuna school associated with the anchored instrumented oceanographic buoy. In order to explain another part of tuna movements we observed, the presence of Fish Aggregating Devices must be considered (experiments 1 and 2). However, the structure of the abiotic environment and the presence of Fish Aggregating Devices do not allow to explain all of the observed vertical and horizontal movements of tunas. If we consider experiments 1 and 3, both conducted in the Society archipelago, abiotic factors do not allow to explain the limitation of tuna vertical movements. Oxygen concentrations are between 3 and 4 ml $l^{-1}$ from the surface to a depth of 500 m, whereas temperature (15 °C at 350 m depth) does not either represent a limitation according to the published literature.

Using simultaneous sonic tracking and acoustic surveys, it appears that tuna exhibited movements in relation to vertical and horizontal distributions of prey. Carey (1990) reported similar experiments conducted in the 80's on swordfish (*Xiphias gladius*) where a fish exhibited the same descent made by the bigeye tuna (experiment 3) at dawn with the migration of the scattering layer. Other studies used these two techniques together but not for the same topic (Cayré, 1991; Malinin et al., 1992).

Despite some evidences of the role of the biotic environment on tuna movements, some questions need further detailed observations. In experiment 1 as well as in experiment 3, why did tunas leave the scattering layers? Is it due to a behavioural feature (satiety, patch residence time, social behaviour)? Our experiments do not give us the answer. On the other hand, if tunas visit scattering layers during night-time, are they, for all that, able to eat on them, and with what efficiency? Finally, what are the relationships between scattering layers and tuna diet? This last question is, at the present time, investigated within the framework of the ECOTAP program, using comparative studies between stomach contents of longline tunas and the species composition of scattering layers sampled by means of pelagic trawling.

## Conclusion and perspectives

Simultaneous observations of tuna movements and the biotic environment show the important role of prey on tuna movements. At small temporal and spatial scales, and within habitat limits determined by abiotic factors, the three-dimensional structure of the biotic environment and its dynamic (prey density, patchiness, daily migrations, ...) appear to be the key factors in the understanding of tuna movements.

A foraging behaviour hypothesis can be suggested. But the simultaneous technique presented here must be improved by using behavioural activities sensors (existing or to be developed). By this way, it would also be possible to elucidate different foraging phases (search, attack, capture) at small temporal and spatial scales.

A behavioural model could be developed to use such small-scale results to predict large-scale movements of tuna. Then, records from archival tags equipped with behavioural activity sensors could be used in order to validate such a model.

Such results on tuna foraging behaviour would represent an interesting step to improve our knowledge on tuna catchability, and more particularly for longline fisheries studied as part of the ECOTAP program in French Polynesia.

## References


Brill, R. W., 1994. A review of temperature and oxygen tolerance studies of tunas pertinent to fisheries oceanography, movements models and stock assessments. Fish. Oceanogr. 3: 204–216.

Carey, F. G., 1990. Further acoustic telemetry observations of swordfish. In Stroud, R. H. (ed), Planning the future of billfishes – Research and management in the 90s and beyond, Proceedings of the Second International Billfish Symposium, Kailua-Kona, Hawaii, August 1–5, 1988. Mar. Recreat. Fish. 13: 103–122.

Cayré, P., 1991. Behaviour of yellowfin tuna (*Thunnus albacares*) and skipjack tuna (*Katsuwonus pelamis*) around fish aggregating devices (FADs) in the Comoros Islands as determinated by ultrasonic tagging. Aquat. Living Resour. 4: 1–12.

Cayré, P. & J. Chabanne, 1986. Marquage acoustique et comportement de thons tropicaux (albacore: *Thunnus albacares* et listao: *Katsuwonus pelamis*) au voisinage d'un dispositif concentrateur de poissons. Océanogr. trop. 21: 167–183.

Cayré, P. & F. Marsac, 1993. Modelling the yellowfin tuna (*Thunnus albacares*) vertical distribution using sonic tagging results and local environmental parameters. Aquat. Living Resour. 6: 1–14.

Holland, K. N., R. W. Brill & R. K. C. Chang, 1990. Horizontal and vertical movements of yellowfin and bigeye tuna associated with Fish Aggregating Devices. Fish. Bull. 88: 493–507.

Holland, K. N., R. W. Brill , R. K. C. Chang, J. R. Sibert & D. A. Fournier, 1992. Physiological and behavioral thermoregulation in bigeye tuna (*Thunnus obesus*). Nature 358: 410–412.

Laurs, R. M., H. S. H. Yuen & J. H. Johnson, 1977. Small-scale movements of albacore, *Thunnus alalunga*, in relation to ocean features as indicated by ultrasonic tracking and oceanographic sampling. Fish. Bull. 75: 347–355.

Malinin, L. K., V. I.. Kijasko & P. L. Vaaranen, 1992. Behavior and distribution of bream (*Abramis brama*) in oxygen deficit regions. In Priede, I. G. & S. M. S. Swift (eds), Wildlife telemetry: Remote Monitoring and Tracking of Animals. Ellis Horwood. Chichester: 297–306.

Marsac, F., P. Cayré & F. Conand, 1996. Analysis of small scale movements of yellowfin tuna around Fish Aggregating Devices (FADs) using sonic tagging. In Anganuzzi, A. A., K. A. Stobberup & N. J. Webbs (eds), Proceedings of the Expert Consultation on Indian Ocean Tunas, 6th session, Colombo Sri Lanka, 25–29 Sept. 1995. IPTP Coll. 9: 151–159.

Moarii, G. & F. Leproux, 1996. The drop-stone technique used by artisanal fishermen in French Polynesia. South Pacific Commission Fish Aggregating Device Bull. 1: 16–18.

Sund, P. N., M. Blackburn & F. Williams, 1981. Tunas and their environment in the Pacific Ocean: a review. Oceanogr. Mar. Biol. 19: 443–512.

Yuen, H. S. H., 1970. Behavior of skipjack tuna, *Katsuwonus pelamis*, determined by tracking with ultrasonic devices. J. Fish. Res. Bd Can. 27: 2071–2079.

*Hydrobiologia* **371/372:** 71–77, 1998.
*J.-P. Lagardère, M.-L. Bégout Anras & G. Claireaux (eds), Advances in Invertebrates and Fish Telemetry.*



# A data logger tag for the study of slaughter procedures in aquacultured salmon

O. A. Olsen[1]*, P. O. Skjervold[2] & S. O. Fjaera[2]
[1]*Prediktor AS, Postboks 296, N-1601 Fredrikstad, Norway*
[2]*Agricultural University of Norway, Dept of Agricultural Engineering, Postboks 5065, N-1432 Ås, Norway (e-mail: itfoao@itf.nlh.no)*
**Present address: Agricultural University of Norway, Dept of Agricultural Engineering, Postboks 5065, N-1432 Ås, Norway*

*Key words:* temperature logger, muscle temperature, Atlantic salmon, slaughter procedures, quality

## Abstract

This paper describes a micro controller based data logger designed to be externally attached to fish, in this case adult Atlantic salmon. The logger can measure ambient temperature and the temperature difference between ambient and dorsal white muscle, the latter with high resolution and accuracy. Communication with a computer for data transfer pass over a detachable transformer coupling. This transformer is also used to charge the internal battery. The logger can thus function without connectors, eliminating corrosion problems and simplifying encapsulation.

Tracking of the muscle temperature is important for the development of better slaughter procedures. But the principles described in the following can also be useful for conventional tags and other kinds of underwater equipment.

## Introduction

Numerous factors affect the term *quality* of fish products. One of the advantages of aquaculture is the possibility for control and manipulation of several major factors affecting this quality. There is a growing consciousness of how environmental conditions and handling of fish in both the pre- and post-slaughter period, affect product quality. The challenge is to obtain uniform product quality, despite of seasonal variations in climate and large biological variation in cultured fish.

### *Slaughtering procedures*

Both shelf life and sensorial qualities of farmed salmon are strongly affected by slaughtering and pre-slaughtering conditions (Andersen et al., 1994; Sigholt et al., 1997). Temperature is accepted to be the single most important factor (Wheaton & Lawson, 1985). To obtain a high product quality, the salmon should be packed before they enter rigor mortis, at a defined, low temperature, independent of fish weight. Time available for chilling before rigor mortis occurs is limited, and this often leads to unsatisfactory body temperature in the fish packet. In a traditional salmon slaughtering process, the fish are chilled after anaesthetisation and throat cut. A new alternative is to chill the fish while alive, before anaesthetisation (Skjervold et al., 1996). Control and documentation of actual body temperature at the time of packing are difficult due to variations in sea temperature and fish size. A systematic study of the internal temperature profile of fish passing the slaughtering process is therefore vital to the comparison and optimisation of slaughter procedures.

### *Pre-slaughter procedures*

Several studies have shown that pre mortal stress also affects the quality of fish muscle (Sikorski & Pan, 1994). Texture of fish muscle as well as development

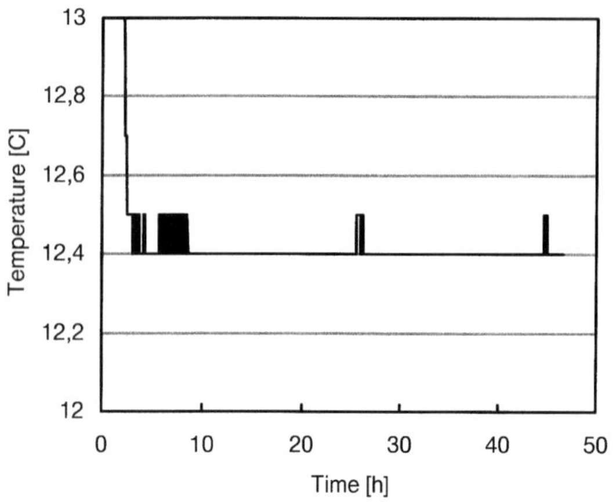


*Figure 1.* Increased body temperature of a fish during periods of increased activity at 26 and 45 h. In this case a commercial temperature logger was used. The resolution is too low to capture the details of the temperature increases. The fish entered the water 2 hours after start of the logger. During the experiment the water temperature varied less than 0.2 °C.

of rigor mortis (Azam et al., 1990; Abe & Okuma, 1991; Nakayama et al., 1992) and rate of fillet gaping (Love et al., 1969; Lavéty et al., 1988; Andersen et al., 1994) are affected by the level of muscle glycogen at death (Proctor et al., 1992). A high level of muscle glycogen at death is desirable. An elevated consumption of muscle glycogen occurs if stress induced swimming activity occurs before slaughtering. The handling procedures of the fish in the sea cage prior to slaughter and during transfer to the anaesthetisation basin are therefore important quality factors.

Salmonids lack a mechanism for body temperature control, and are therefore essentially thermal conformers. Rates of thermal equilibration in body temperature following a change in water temperature are affected by many factors including gill ventilation rate, blood flow and flow distribution, water movement and shape, size and activity of the fish (Stevens & Sutterlin, 1976; Kubb et al., 1980). Although many factors contribute to the body temperature, the swimming activity is one of them. The level of physical activity in fish may therefore be indicated by precise measurement of the difference between body core temperature and ambient water temperature (the excess temperature). A body temperature time series is shown in Figure 1. During this experiment the fish was kept at constant water temperature and the dorsal muscle temperature was measured by a commercial temperature logger. One can see that the body temperature is maintained at a constant value for long periods of time, but increases during intervals of higher activity due to stress. An important feature of this diagram is the low resolution of the logger, about 0.1 K, comparable to the increment in excess temperature. To do any further research on the relationship of activity and body temperature a logger with better temperature resolution is obviously needed.

### *Data logger, requirements*

Based on the above overview the study of the procedures pertaining to slaughter of salmon can be divided into two main phases: excess temperature before slaughter and body temperature during and after chilling. For the measurement of excess temperature the measurement range is small, but the accuracy and resolution must be high and independent of ambient temperature. During the second phase the important variable is the absolute body core temperature, but for the study of chilling efficiency the difference temperature is also of interest. During this phase the range is much larger and the demand on accuracy less.

To avoid problems due to variations between individuals, we want to follow individual fish through the total process. The logger must therefore handle both phases well. This means that the logger must simultaneously measure the differential temperature with high accuracy and the absolute temperature of the ambient water or air with moderate accuracy. We have not come across a commercial logger which satisfies these requirements, and therefore decided to develop our own for this application. However, the logger and most of the associated software are of general nature and can easily be adopted to the measurement and logging of other variables such as ambient light, salinity, heart rate, swimming direction, pressure and pH. A data logger can thus be of use also for the study of fish behaviour, a field mostly served by acoustic and electro-magnetic tags. A necessary condition being, of course, that the fish can later be caught and the logger retrieved.

Instead of a data logger we could also have used a conventional tag, but during the monitoring interval the fish stay both in sea water, air and on the slaughter line. A conventional acoustic or electro magnetic tag is therefore not suited for this task. When simultaneously monitoring 20–30 fish, tag interference is difficult to avoid too.

A data logger must have a way to communicate with a computer for the transfer of data. There are several principles that can be used for this, e.g. acoustic transducers and infra red light. We chose a magnetic

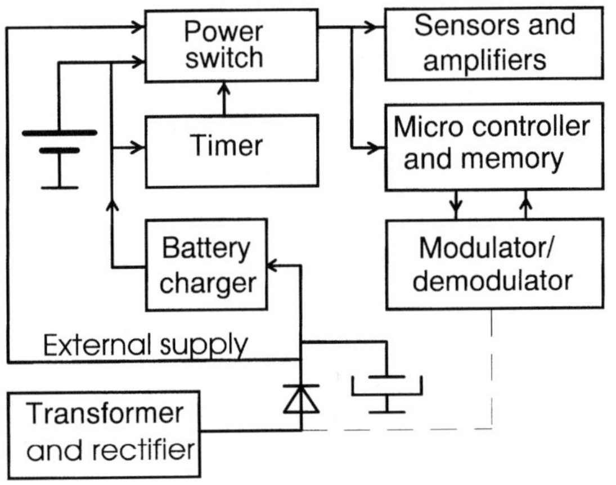


*Figure 2.* Block diagram of the data logger showing power distribution and communication signal paths. The signal from the transformer is split into modulation and power. The power switch selects the highest of the battery and external supply voltages. While running off the battery the power switch is enabled by the timer. External power also charges the battery.

coupling though, because this also can be used to transfer energy for charging of the logger's battery. Using a rechargeable battery eliminates the need for battery change. The logger can therefore be totally embedded in an encapsulation compound.

## Principles and results

### *Overview*

A simplified block diagram of the logger is shown in Figure 2. The lines connecting the boxes indicate the power distribution, control signals are left out. Most of the blocks in this figure take care of communication and power management. At the upper right is one block for all the analogue circuits, sensors and processor interface (analogue to digital converter) and one for the micro controller and data memory.

Communication and power supply is provided via a detachable transformer from an external power supply and communication link box (PSCL). The communication signals are sent and detected by a modulator and demodulator. See below for details on this. While connected to the PSCL the power supplied is fed to a power switch and the battery constant current charger. A power switch selects the highest voltage of the battery and the externally supplied power. When power is supplied from outside the micro controller and analogue circuits are always powered on. When powered from the battery these circuits are usually turned off, but turned on at regular intervals for measurements by the crystal controlled timer. After finishing the measurements the controller turns off the power again. The interval can be set from 1 s to 1 h. Up to one week of delay from configuration till the first measurement is taken can also be specified. Both these intervals can be set with 1 s resolution. Between measurements the battery is loaded only by the timer, drawing less than 1 $\mu$A of current.

The micro controller is presently a PIC16LC84 from Microchip Technology, but the next version will use an AT90S1200 from Atmel. Measured data are stored in a non-volatile memory of 16 kByte capacity.

The temperature difference measurement sensor is a pair of thermocouples, one sensing the ambient temperature and one the body temperature. The diameter of the wires are 0.07 mm, their teflon insulation outside diameter 0.2 mm. The wires are very flexible, one of the reasons why we use thermocouples. The ambient temperature is measured by a NTC resistance sensor. In addition to this, the battery voltage can also be measured. This can be used by the processor to stop logging should the voltage drop too low, but can also be used to check the battery state during charging. The analogue signals are converted to their digital representation by a 10 bit converter.

All circuits of the logger operate at a battery voltage of down to 2.8 V. The availability of analogue components specified to work at this low voltage is limited.

In the following description of the logger we will concentrate on the principles of the thermocouple sensor and on communication and power supply.

### *Differential temperature measurement*

The objective is to measure the temperature difference between a point within the fish body and the ambient water or air. The difference temperature can be calculated by subtracting the signals from two independent temperature sensors, e.g. NTC-resistors. With the resolution we are trying to obtain, this puts high demands on the calibration of each sensor since a significant temperature difference may only be some hundreds of a K. We therefore use two thermocouples as sensors.

Description of thermocouple principles can for example be found in Kinzie (1973), we will only mention the most relevant points here. Each couple consists of a pair of wires made from different alloys, and the principle utilises the thermoelectric Seebeck effect. The wires at the end of the thermocouple acting as sen-

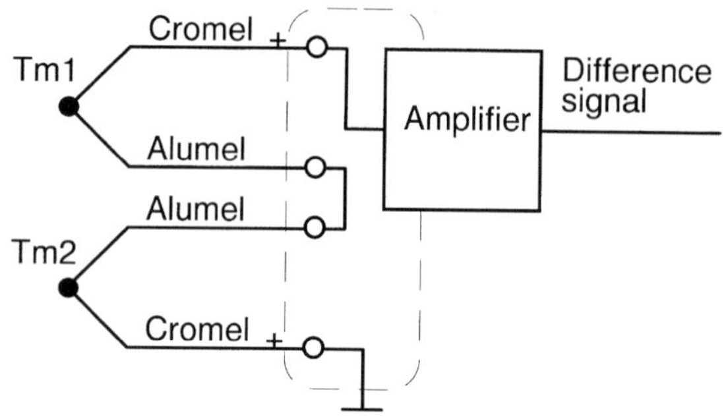


*Figure 3.* Thermocouple configuration for measurement of temperature difference. The dashed line indicates the part of the circuit where thermal gradients must be avoided.

sor are welded or soldered together and the voltage developed at the other open end is then a function of temperature

$$U = f(T_m - T_r) \approx \alpha(T_m - T_r).$$

Here $T_m$ is the measured temperature and $T_r$ the temperature at the open end. For small differences between them the constant $\alpha$ can be substituted for the non-linear function $f$. We can measure the temperature difference of two points by connecting two such thermocouples together back to back. This is illustrated in Figure 3. The voltage seen by the amplifier is then

$$\Delta U = f(T_{m1} - T_r) - f(T_{m2} - T_r) \approx \alpha(T_{m1} - T_{m2})$$

A thermoelectric voltage is developed at any junction of dissimilar materials along the signal path. If these junctions are kept at the same temperature, however, their sum is zero. The dashed box in the figure indicates the parts of the circuit that must be kept at uniform temperature. As long as this is maintained, the measured voltage is a function of the temperature differences only.

We used cromel/alumel alloy thermocouple wires, also known as *type K*. The sensitivity of this type is 40 $\mu$V/K or 25 mK/$\mu$V from −10 to +30 °C, see (IEC 584-1). The measured voltage is small and requires an amplifier with very small input offset drift and low noise.

*Error sources*

The most important error sources affecting the temperature difference measurement accuracy are the following

*Temperature gradients* As stated above, any temperature difference between pairs of joints of dissimilar materials will cause an error voltage, which size depends on both temperature and the joined materials. As a rule of thumb the error caused by any thermal differences along the signal path will be about the same size as the differences. To minimise thermal differences the sensitive circuitry should be contained within a small, thermally well-coupled volume, and the power dissipated should be small.

*Thermocouple wire variations* The non-linear function and $\alpha$ depend on the alloys used for the wires. The value can vary along the wire and between wires from different batches. To minimise this error source we therefore cut both thermocouples at adjoining sites along the same wire pair.

*Amplifier offset drift* The amplifier used is of a 'zero-drift' type, see (Linear Technology, 1995), which means that the amplifier's input offset is automatically and constantly zeroed. The resulting offset error is specified to be less than 10 $\mu$V and its drift typically less than 10 nV/K, equivalent to 0.25 K and 0.25 mK/K respectively. The static offset error can be compensated for by calibration.

*Noise* In this application the measurement noise is dominated by the noise produced within the amplifier, typically 130 nV/$\sqrt{\text{Hz}}$. This translates to 1.3 $\mu$V for a 100 Hz bandwidth and 0.4 $\mu$V for a 10 Hz bandwidth. The noise can thus be reduced by lowering the bandwidth, but then the settling time will also increase. Due to this increase the power must be turned on for a longer time before the measurement can be taken. The bandwidth is therefore a compromise between battery capacity (weight and volume) and measurement noise. For 10 mK accuracy, the noise must be as low as 0.4 $\mu$V.

*Representativity* In common with almost all principles of temperature measurement the temperature within the sensor itself is the one measured. To obtain representative measurements one must provide a good thermal coupling between the sensor and the object which temperature one will measure. In this case this means that the sensor should have only a thin layer of coating to ease heat transport from flesh to sensor and thin wires to attenuate transport of heat along the wires to the external medium. The thermocouple wire we use is 0.07 mm diameter. By inserting the wire tip 15 mm into the fish we get a length/diameter ratio of more than 200. Taking into account the low tempera-

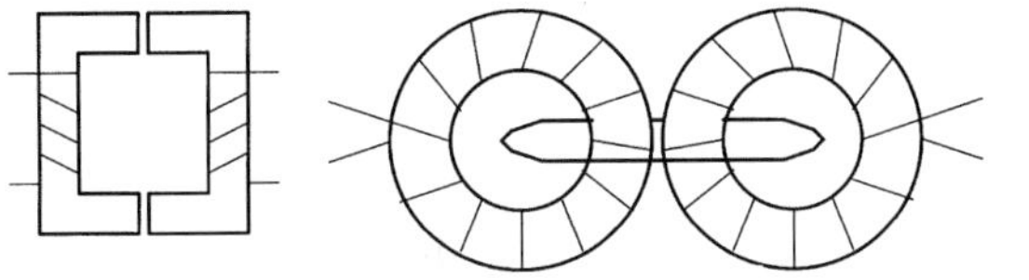

*Figure 4.* Inductive coupling for communication and power supply. Left: Half-transformers. This is a conventional transformer configuration transforming power from the primary to the secondary side. Right: Two toroids do not interact, but when coupled by a common loop they act as other transformers.

ture difference (1 degree or so), practically no heat can be transported along the wire compared to what flows radially into the sensor tip.

Another possible source of error which can degrade the representativity is a local increment in temperature due to inflammation of the wound.

*Power and communication*

The logger must be able to communicate with a computer for set-up before activation and for transfer of data after the experiment. To avoid connectors we use inductive coupling for this, see Figure 4. The simplest solution to this is to use two half-transformers, one embedded into the logger and the other external. When they are put together they become a transformer like we are used to. Alternating current can then be transformed from the windings on one side to the other. In our case we found it easier to obtain toroidal cores of the small size we wanted. They are 6.3 mm outer diameter made of 3F3 material by Philips. One is encapsulated into the logger and one is external. They are coupled by a single turn going through both of them, in practice we use a modified crocodile clip for this. The toroid can be totally embedded in the encapsulation compound as long as a hole for the detachable wire is provided.

Power is transferred from the PSCL into the logger by an alternating square current. To avoid saturation of these small cores the frequency must be fairly high, in our case we use about 1.2 MHz. The diagram in Figure 5 shows the secondary voltage of a toroid pair as a function of the secondary load current. The current consumed by the logger when externally powered is about 5 mA.

The communication over the transformer is half duplex: the computer always initiates communication by sending a message. The logger then responds. The PSCL-box connects to the RS232 serial port of the computer. Any zero-bit in the bit stream from the computer cuts off the power to the transformer for about 5 $\mu$s, see Figure 6. At the secondary the output of the rectifier thus drops for a short time for every zero-bit. This is recognised by the micro controller and the bit stream is reconstructed.

The data logger transmits messages by short-circuiting the rectifier output for about 5 $\mu$s for any zero-bit of its reply. This is detected as a short increase in the alternating current delivered to the logger. The pulse is stretched and the bit stream is thereby converted into a standard RS232-format for the computer.

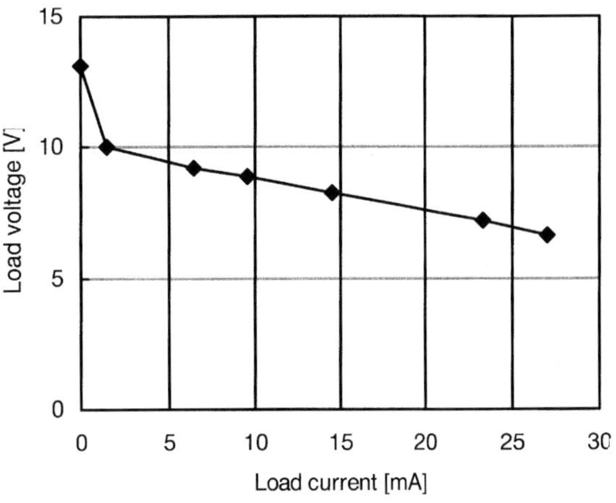


*Figure 5.* An example of secondary voltage over the load as a function of load current for coupled toroids. Even the small toroids used in this case can transfer a substantial current. The voltage obtained can be varied by changing the turn ratio of the two toroids and the drive voltage of the primary.

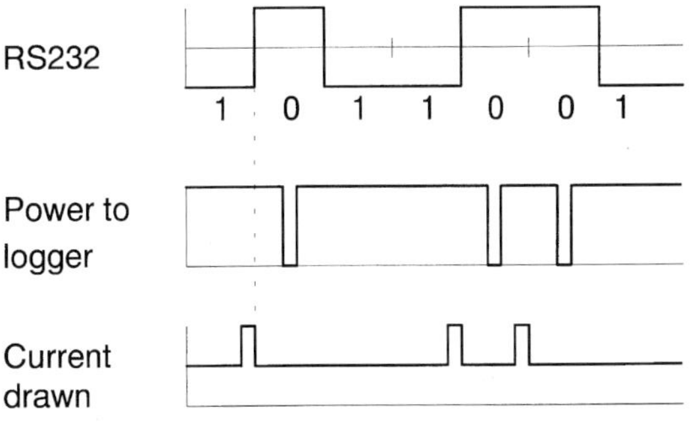


*Figure 6.* Principle of modulation used for communication. Upper trace is a conventional RS232 serial bit stream to or from a PC serial port, altering between +5 and –5 V. The unit of the abscissas of these diagrams are time. The middle trace indicates the output voltage from the rectifier when the PC transmits the above bit sequence. This voltage alters between 0 and about 10 V. The lower trace illustrates the increase in current drawn at the PC side of the transformer when the logger transmits the same bit sequence. The current peaks are then converted into a RS232 signal equal to the top trace and fed to the PC.

Using this principle a 9600 bit/s communication link is established and a transfer of 8 kBytes of logged data takes about 90 s. A check sum is embedded in every message to enable transfer error detection.

The external supply to the logger electronics is isolated from the intermittent voltage drops during communication by a series diode and a storage capacitor, see Figure 2.

The external power supplied is also used to charge the battery. We use a constant current charger and a stack of nickel-metal-hydride button cells, theoretically delivering 11 mAh at a nominal voltage of 3.6 V. The diameter of the stack is 12 mm and its height 10 mm. Such button cells can be overcharged without any harm and their self-discharge is about 1% per month at room temperature.

When awake the logger consumes about 3 mA, when turned off between measurements less than 1 $\mu$A. The total charge consumed is therefore well approximated by the product of the active current, the time taken for each measurement and the number of measurements. Assuming 3 mA, 200 ms duty and 4000 samples this translates into 2400 mAs or 0.7 mAh. The capacity of the battery should therefore be more than enough even at low temperatures and after many charge/discharge cycles.

## Conclusion

We have developed two prototypes of the logger, the second a near replication of the first, but more compact. This logger consists of two 16 × 32 mm and two 16 × 12 mm printed circuit boards. Including battery and toroid, the total encapsulated size is about 65 × 13 × 17 mm. In air the encapsulated weight is 20 g and in water 6 g. The tags are attached beside the dorsal fin by two sutures. Apparently, the fish of the size we use are not affected by the tag. See Claireaux & Lefrançois (1998) for a discussion of the attachment of even larger tags to fish of about the same size. Our tag is constructed using conventional surface mount technology. One might desire an even smaller size, especially to enable surgical or gastric implantation. This is of course possible if using naked chips directly bonded to a thick film circuit and/or by using custom chips. Although being technically possible, the financial support for this is an open question. The size and weight of the batteries and transformer set a lower, unrealistic limit of miniaturisation, but it would help much if the volume was reduced to about 25% of present.

The measurement error is a sum of several contributions as described above. We are most concerned about the zero point error of the difference temperature signal. The main contribution to this is inherent differences of the thermo-couple wires, this is a mostly static, but ambient temperature dependent error which can be calibrated for. Drift of the zero point is dominated by variations in ambient temperature, which can partly be compensated for by calibration. These errors are smaller than the short time error. The dominant short term error source is low frequency noise, having a standard deviation of about 10–20 mK. This is more than we would like to see, and forthcoming work will be concentrated on decreasing this. A varying battery voltage will also move the amplifier zero points, but this error source can be eliminated by stabilising the amplifier power supply.

The user interacts with the logger through a program running under Windows 95. The tag can be read and reconfigured while attached to the fish. This enables many or long term experiments with the same fish, the only disturbance being occasional anaesthetisation to keep the fish still while transferring data, and initiating a new logging session. In principle the batteries could be charged too, but that would take an hour or more of anaesthetisation. Long term attachment may of course not be possible due to infections of the wound and we do not have any experience with long term attachment yet.

The logger makes possible the accurate measurement we need of excess temperature caused by swimming activity. This will give new insight into the relationships between temperature/stress and quality. The tag was developed for studying slaughter procedures. But accurate measurement of excess temperature is fast, nearly continuous and simple to apply compared to methods involving blood sampling and chemical analysis. One may therefore envision other applications, e.g. in the study of time responses in fish energetics and metabolism.

A single circuit board of the smaller size (16 × 12 mm) carries all sensor amplifiers and up to three analogue signals of 0–2.5 V from this board can be read by the micro controller. Redesigning the logger to accept other types of sensors is therefore simple and do not affect the rest of the construction. Converting the logger to record other states, like pressure and ambient light, is therefore a small undertaking.

The use of inductive coupling for external power supply and communication allows the use of a rechargeable battery and communication without any connectors. Therefore the logger can be permanently encapsulated with no need for battery exchange and without any risk of leaks or corrosion. One disadvantage of using chargeable batteries instead of primary batteries is the lower capacity/weight ratio of the former. This is a point related to miniaturisation, and the chargeable batteries used here do have ample capacity given the limited number of samples the data memory can hold.

Data loggers are restricted to studies where the fish can be recaptured, as opposed to telemetry tags. The advantages of data loggers are foremost the ability to start the experiment and leave the fish alone until recapture, it is simple to use many tags simultaneously, and the data is not lost if secondary equipment (receiver and data storage on land) malfunction or the transmitted signal is lost or corrupted. The fish can also roam far without the need to move the secondary equipment. Examples of the use of data loggers to measure states of free-ranging animals are found in Bevan et al. (1994) and Woakes et al. (1995), where body temperature and heart rate of birds and seals were logged over a period of over five months. Recapture occurred up to a year later.

The disantvantages of a logger compared to a transmitting tag are also obvious. Especially when the measured data must be manually correlated with other data, such as observed behaviour, or when the measurements are used as a base for action, e.g. to sample blood at instances of characteristic changes in the measured state. Using a tag one will also immediately become aware of malfunctions in the system, e.g. a sensor drop out, while when using a logger such problems will not be evident before after the experiment.

## References


Abe, H. & E. Okuma, 1991. Rigor mortis progress of carp acclimated to different water temperatures. Nippon Suisan Gakkaishi 57: 2095–2100.

Andersen, U. B., A. N. Strømsnes, K. Steinsholt & M. S. Thomassen, 1994. Fillet gaping in farmed Atlantic salmon. Norw. J. Agricult. Sci. 8: 165–179.

Azam, K., N. J. C. Strachan, I. M. Mackie, J. Smith & P. Nesvadba, 1990. Effect of slaughter method on the progress of rigor of rainbow trout (Salmo gairdneri) as measured by an image processing system. Int. J. Food Sci. Technol. 25: 477–482.

Bevan, R. M., A. J. Woakes & P. J. Butler, 1994. The use of heart rate to estimate oxygen consumption of free-ranging black-browed albatrosses Diomedea melanophrys. J. exp. Biol. 193: 119–137.

Claireaux, G. & C. Lefrançois, 1998. A method for the external attachment of acoustic tags on roundfish. Hydrobiologia 371/372: 113–116.

IEC 584-1. Thermocouples, Part 1: Reference tables. Geneva: International Electrotechnical Commission.

Kinzie, P. A., 1973. Thermocouple temperature measurement. New York: John Wiley & Sons.

Kubb, R. N., J. R. Spotila & D. R. Pendergast, 1980. Mechanisms of heat transfer and time-dependent modeling of body temperatures in the largemouth bass (Micropterus salmonides). Physiol. Zool. 53: 222–239.

Lavéty, J., O. A. Afolabi & R. M. Love, 1988. The connective tissues in fish. IX. Gaping in farmed species. Int. J. Food Sci. Technol. 23: 23–30.

Linear Technology, 1995. LTC1152 Data sheet. Milpitas, CA: Linear Technology Corporation.

Love, R. M., J. Lavéty & P. J. Steel, 1969. The connective tissues of fish. II. Gaping in commercial species of frozen fish in relation to rigor mortis. J. Fd. Technol. 4: 39–44.

Nakayama, T., D. J. Liu & A. Ooi, 1992. Tension change of stressed and unstressed carp muscles in isometric rigor contraction and resolution. Nippon Suisan Gakkaishi 58: 1517–1522.

Proctor, M. R. M., M. Dorgan & J. V. McLoughlin, 1992. The concentration of adenosine triphosphate, creatine phosphate, glucose-6 phosphate, lactate and glycogen in skeletal muscle of marine and freshwater fish species anaesthetised with MS-222. Proceedings of the Royal Irish Academy. Section B 3: 45–51.

Sigholt, T., U. Erikson, T. Rustad, S. Johansen, T. Nordtvedt & A. Seland, 1997. Handling stress and storage temperature affect meat quality of farmed-raised Atlantic salmon (Salmo salar). J. Food Sci. 62: 898–905.

Sikorski, Z. E. & B. S. Pan, 1994. Preservation of seafood quality. In F. Shahidi & J. R. Botta (eds), Seafoods, Chemistry, Processing, Technology and Quality. London: Blackie Academic & Professional: 168–189.

Skjervold, P. O., S. O. Fjæra & K. Christoffersen, 1996. Pre-mortal chilling of farmed salmon. Bordeaux Aquaculture 1996, France, March 20–22. International colloquium 'refrigeration and aquaculture'. Refrigeration science and technology. Proceedings of the conference and IRR Commission C2.

Stevens, E. D. & A. M. Sutterlin, 1976. Heat transfer between fish and ambient water. J. exp. Biol. 65: 131–145.

Wheaton, F. W. & T. B. Lawson, 1985. Processing aquatic food products. New York: John Wiley & Sons: 181–222.

Woakes, A. J., P. J. Butler & R. M. Bevan, 1995. Implantable data logging system for heart rate and body temperature: its application to the estimation of field metabolic rates in Antarctic predators. Med. & Biol. Eng. & Comput. 33: 145–151.

*Hydrobiologia* **371/372:** 79–87, 1998.
*J.-P. Lagardère, M.-L. Bégout Anras & G. Claireaux (eds), Advances in Invertebrates and Fish Telemetry.*



# Growth and physiological responses to surgical and gastric radio transmitter implantation techniques in subyearling chinook salmon (*Oncorhynchus tshawytscha*)

T. L. Martinelli[1], H. C. Hansel & R. S. Shively
*U.S. Geological Survey, Biological Resources Division, Columbia River Research Laboratory, 5501A Cook-Underwood Rd., Cook, WA 98605, USA*
[1] *Author to whom correspondence should be addressed: Theresa_Martinelli@usgs.gov*

*Key words:* chinook salmon, radio-telemetry, surgery, growth

## Abstract

We examined the effects of surgical and gastric transmitter implantation techniques on the growth, general physiology and behavior of 230 subyearling chinook salmon (*Oncorhynchus tshawytscha*, Walbaum) (100 mm–154 mm fork length). The transmitter weighed 1.3 g in air (0.9 g in water) and comprised, on average, 6% of the body weight of the fish (in air). Individuals were randomly assigned to an experimental group (control, surgical or gastric) and a sampling period (day 5 or day 21). Relative growth rate was expressed as% body weight gained/day. General condition was assessed by necropsy. Physiological response variables included hematocrit, leucocrit and plasma protein concentration. The mean relative growth rates of control, surgical and gastric fish were not significantly different at day 5. By day 21, the gastric group had a significantly lower relative growth rate (1.3%) as compared to the surgical group (1.8%) and the control group (1.9%) ($P = 0.0001$). Mean hematocrit values were significantly lower in the surgical (41.8%) and gastric (42.2%) groups as compared to controls (47.3%) at day 5 ($P = 0.01$), but all were within normal range for salmonids. No significant differences in hematocrit values were detected at day 21. Leucocrit values for all groups were $\leq 1\%$ in 99% of the fish. Both tagged groups had significantly lower mean plasma protein levels as compared to controls at day 5 ($P = 0.001$) and day 21 ($P = 0.0001$). At day 21 the gastric group (64.4 g 100 $ml^{-1}$) had significantly lower mean plasma protein levels than the surgical group (68.8 g 100 $ml^{-1}$) ($P = 0.0001$). Necropsies showed decreasing condition of gastrically tagged fish over time, and increasing condition of surgical fish. Paired releases of surgically and gastrically implanted yearling chinook salmon in the lower Columbia River in spring, 1996 revealed few significant differences in migration behavior through two reservoirs. We conclude that gastrically implanted fish show decreased growth and condition over a 21 d period. We recommend a surgical implantation method for long-term studies of juvenile salmonids, however, gastric implantation may be suitable for short-term studies.

## Introduction

Telemetry studies of fishes are valuable in that they can provide insights into the movements and distribution, home range and habitat preferences of the study species without direct observation or recapture. Many fish telemetry studies place the transmitter in the stomach or surgically implant it into the peritoneal cavity. The placement of a transmitter into the stomach of a fish (gastric implantation) can be accomplished quickly, however, the transmitter may be regurgitated, rupture the esophagus or the stomach or may alter the feeding of the fish (Winter, 1983, 1996; Nielsen, 1992). Gastric implants have been commonly used with success in both juvenile and adult salmonids (*Oncorhynchus*, Walbaum) (Stier & Kynard, 1986; Wilson et al., 1991; Eiler et al., 1992; Ward et al., 1994; Burger et al., 1995). Gastric implantation is especially common for juvenile salmonids because it requires only minimal anesthesia and handling of the fish.

Surgical implantation of transmitters can be time consuming, requires some practice, and exposes the fish to the stress of prolonged handling and the potential for infection (Summerfelt & Smith, 1990). This technique has been used successfully in a large variety of fish species (Clapp et al., 1990; Moore et al., 1990; Guy et al., 1994; Carl, 1995; Isaak & Bjornn, 1996; Knights & Lasee, 1996). Surgical implantation of radio transmitters has been used, most commonly, in adult fish, however, the miniaturization of transmitters has made it a viable technique in juveniles as well (see Moore et al., 1990).

Ultimately, the choice of transmitter implantation method for a given species or life stage should be the method that affects the fish the least. We compared surgical and gastric methods for implanting a radio transmitter with an externally trailing antenna into subyearling chinook salmon (*O. tshawytscha*, Walbaum). We measured survival, growth, general condition and blood physiology parameters over the course of a 21 d laboratory experiment. In addition, we compared these methods by monitoring the migration behavior of paired releases of surgically and gastrically implanted fish in the Columbia River. We hypothesized that, although both groups of tagged fish would be affected by the presence of the transmitter, gastrically implanted fish would be chronically affected by the implantation method, while surgically implanted fish may be more acutely affected, but would make a rapid recovery and quickly return to normal behaviors.

## Methods

Two experiments were conducted from October-December, 1995, using subyearling chinook salmon from the Little White Salmon National Fish Hatchery (see Figure 1) reared to a mean size of 120 mm fork length (FL) (range 100–149 mm). During fish rearing and throughout both experiments fish were held under natural photoperiod conditions at water temperatures of 14 ± 1 °C and without chemical modification to the tank water. Fish were fed three times daily to satiation with 2.5 mm pellets. Food was withheld from fish 12–15 h prior to tagging and 20–24 h after tagging. Six tanks (76.2 cm diameter, 228 l) were used for each experiment, and 20 fish were held in each tank, giving a mean density of 1.7 g $l^{-1}$. For both experiments, each tank held two experimental groups: one tagged group (surgical or gastric) and a control group. Treatment and control fish were held together in the tanks to control for any possible tank effects. Experiments ran for 21 d with sampling at day 0, day 5 and day 21. At day 0, 20 control fish were sampled directly from the rearing tank to provide baseline condition data. In each experimental tank, two fish from each group were sampled at day 5, and the remaining fish were sampled at day 21. The response variables measured included relative growth rate (calculated at 5 d and 21 d), general condition, hematocrit, leucocrit and plasma protein concentration. The relative growth rate, expressed as percent body weight gained per day, was calculated according to Busacker et al. (1990); the weight difference was divided by the initial weight multiplied by the experimental period and was then transformed into a percent.

General fish condition was assessed by necropsy. The condition of external structures (gills, skin, fins, eyes and opercles) and internal structures (stomach, spleen, kidney and liver) were recorded. In surgical fish we examined the incision and the exit site of the antenna (both internally and externally) for redness, swelling or dehiscion. The mouths and stomachs of gastric fish were examined for irritation caused by the transmitter. In the mouth, we noted any irritation at the site where the transmitter antenna is bent to exit the mouth. The stomachs were examined for any instances of redness or rupture. For tagged fish, we graded abnormalities as mild or severe.

The transmitter used in the study was a replicate (no working components) of a 1.35 g (in air), 0.9 g (in water), 17.7 mm × 5 mm tag that is commonly used in juvenile chinook salmon. The transmitter trailed a 30 cm antenna, 0.5 mm in diameter, coated in a flexible plastic material. The tag represented, on average, 6% of the body weight of the fish (in air). We purposely choose the smallest fish that could accommodate the transmitter, anticipating that our response variables would show the greatest changes over time in smaller fish. We tagged fish smaller than would typically be used in normal field operations, where the tag would likely only represent about 2% of the body weight of the fish (Winter, 1983).

### *Tagging and sampling procedures*

Fish were removed from the rearing tank, anesthetized in 70 mg $l^{-1}$ tricane methanesulonate (MS-222) for no more than 3 min, weighed (to the nearest 0.1 g),

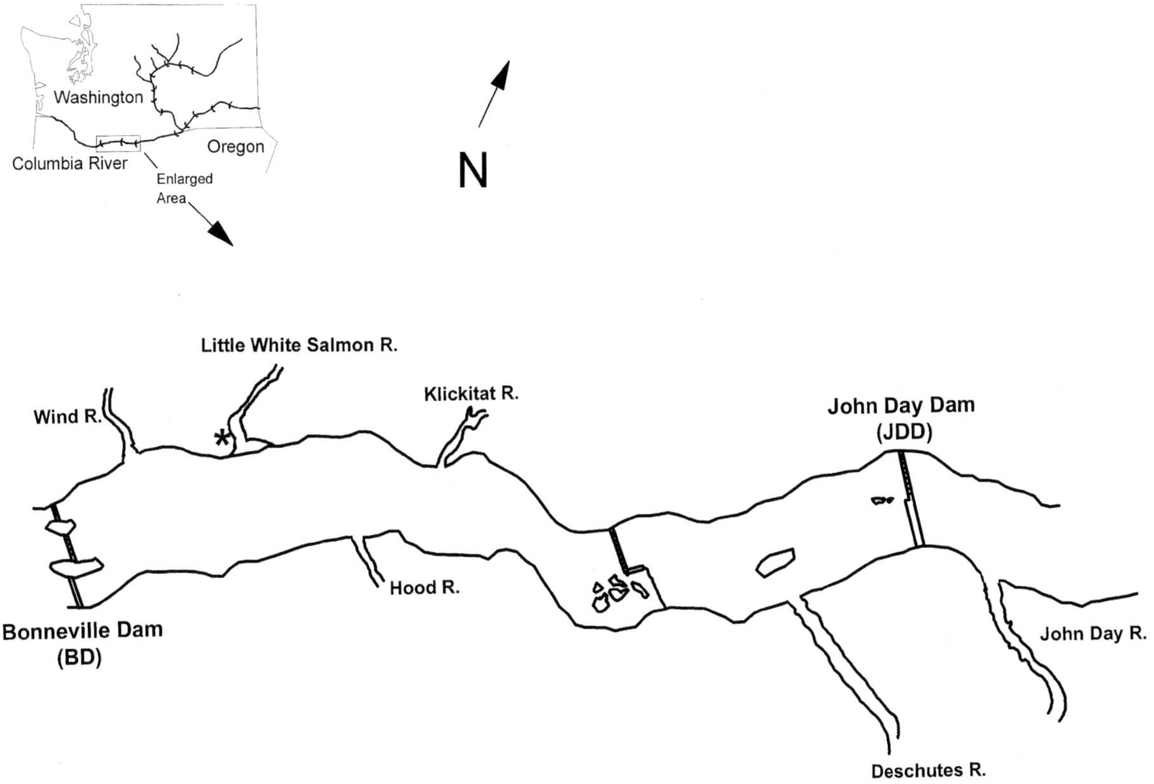


*Figure 1.* Map of the study area where yearling chinook salmon were surgically and gastrically implanted with radio transmitters and released upriver of John Day Dam (JDD), The Dalles Dam (TDD) and Bonneville Dam (BD) to monitor migration behavior. The asterisk indicates the site of the Little White Salmon National Fish Hatchery, the source of fish for the laboratory experiments comparing surgically and gastrically implanted fish. Map distances not to scale.

measured (FL in mm), freeze branded with a unique brand, tagged if appropriate, and placed into a 19 l recovery bucket. Oxygen was bubbled into the recovery bucket to facilitate the return to equilibrium. An electrolyte-mucus restorer was added to all containers and surfaces that fish contacted. When the experimental period was complete (5 or 21 d), all fish in a tank were killed in 90 mg $l^{-1}$ MS-222, weighed, measured, necropsied and sampled. A blood sample was taken by severing the caudal peduncle and collecting blood into a capillary tube. The blood components were separated by centrifugation, and the hematocrit and leucocrit were recorded as a percent of the packed column length. Plasma protein concentrations were determined using a protein refractometer.

For gastric implantation (gastric fish), fish were held in a shallow tray of water with the dorsal side upward. The transmitter was inserted into the mouth using a 10–15 cm plexiglass tube (diameters: 3 mm internal, 5 mm external) and gently pushed into the anterior portion of the stomach. Based on observation (no direct measurements were made), the tag consumed approximately 80–90% of the empty stomach volume. The transmitter antenna was bent where it exited the mouth to allow it to trail back along the body. This tagging procedure was very rapid (< 45 s) and the fish was never completely removed from the water. Henceforth fish gastrically implanted with a radio transmitter will be referred to as gastric fish.

For surgical implantation (surgery fish), the fish was placed ventral side up in a V-shaped plexiglass trough. Throughout the procedure the gills were perfused with a light anesthetic (20 mg $l^{-1}$ MS-222) through a tube in the mouth. Approximately 1 min prior to the completion of the procedure the gills were perfused with fresh water. The incision and insertion of the transmitter were similar to the protocol of Moore et al. (1990). After the transmitter was inserted, oxytetracycline was injected intraperitoneally at a dosage of 50 mg $kg^{-1}$ fish weight (Summerfelt & Smith, 1990). We used a 6 cm catheter and needle with a modification of Ross & Kleiner's (1982)

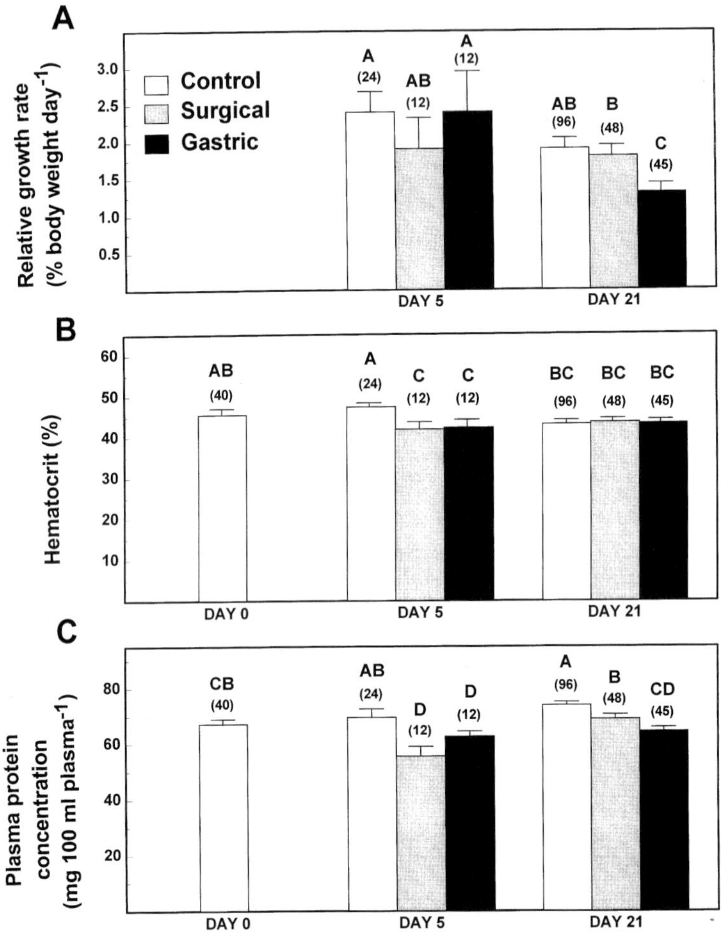


*Figure 2.* Relative growth rate (% body weight day$^{-1}$) (A.), hematocrit (%) (B.) and plasma protein concentration (mg 100 ml$^{-1}$ plasma) (C.) for control, surgically (Surgical) and gastrically (Gastric) implanted subyearling chinook salmon used in a laboratory experiment comparing tagging methodologies. Controls were sampled at day 0, day 5 and day 21. Groups of tagged fish were sampled at day 5 and day 21. Bars represent the mean, error bars are the standard error of the mean, and sample size is noted in parentheses above the bar. Bars without letters in common are significantly different by analysis of variance ($P < 0.05$).

shielded needle technique to provide a lateral body wall exit site for the transmitter antenna. The antenna was secured to the body wall with a single suture through the musculature, approximately 10–15 mm posterior to the antenna exit site. The incision (15–20 mm long) was closed with three individual coated Vicryl absorbable sutures (5–0) and covered with a cyanoadhesive compound. The surgical implantation procedure required that fish be removed from the water throughout the procedure, and took an average of 3.5±0.4 min to complete (range: 2.7–5.2 min). Henceforth fish surgically implanted with a radio transmitter will be referred to as surgery fish.

### *Migration behavior*

During April-June, 1996, we released pairs of fish surgically and gastrically implanted with radio transmitters and monitored their behavior as they migrated downriver past three dams on the lower Columbia River (Figure 1). Tagged fish were released 8–12 km upriver of John Day (JDD), The Dalles (TDD) and Bonneville (BD) dams. We collected hatchery reared yearling chinook salmon at JDD (for releases at JDD and TDD) and BD. Fish tagged in the field were larger, on average, than those used in the laboratory study: surgical fish 161.5 ± 19.9 mm FL, 43.3 ± 17.6 g and gastric fish 159.5 ± 22.6 mm FL, 43.5 ± 19.0 g. For both tagged groups the mean transmitter to body weight ratio was 3.1% in air and 2.1% in water. No significant differences were detected in the mean size of surgical and gastric fish. Following implantation, fish were held in 120 l containers (with holes) in the river for 24 h prior to release. The rationale behind holding fish prior to release was to allow for some recovery from tagging and handling. The holding period was set at 24 h because it was long enough to allow some recovery, and short enough to minimize the stress associated with holding actively migrating fish.

Tagged fish were monitored within the immediate dam areas by fixed receiving/data logging stations mounted on the upriver face of the dam. Each station comprised a series of four-element yagi antennas connected to a single data logging receiver. Receivers were programmed to scan and combine signals from up to eight individual antennas. Between five and six stations were established at each dam in order to provide as complete coverage of the area as possible. Individual yagi antennas were oriented such that they monitored an area approximately 100 m out from the dam, with some overlap with the neighboring antennas. Antennas were placed between 18 m and 35 m apart along the periphery of the dam, allowing fairly complete antenna coverage of the immediate dam area. This design allowed continuous monitoring, and gave us reliable information on fish passage time. Manual tracking with hand held yagi antennas was conducted from the face of the dams to verify fixed receiver data. Fish were monitored at the dam where they were released, and at all downriver dams in the study area.

Migration behavior was assessed with three variables, all of which were based on radio tagged fish being contacted by fixed receiving equipment on the dams. Travel time to the dam was defined as the travel time from the release site to the immediate dam area. Residence time was defined as the time between the first and last contact with a fish in the immediate dam area. Travel time between dams, the third migration behavior variable, was defined as the time elapsed between passage at one dam to arrival in the immediate dam area of the next downriver dam. All calculations were done using fixed receiver data and were confirmed with manual tracking data. We present medians, ranges and sample sizes, with nonparametric methods for analysis. Since releases were conducted over a variety of river flow and dam operating conditions which may have influenced the migration variables, each release was analyzed separately. For brevity, we present a tabular summary of overall medians (all releases combined) for each migration variable.

### *Analysis*

For all analyses, statistical findings were reported as significant if $P < 0.05$. Tanks within an experiment as well as experiments were pooled due to nonsignificant differences in all response variables. For the gastric implantation group, any fish that regurgitated its transmitter was omitted from analysis. Analysis of Variance (ANOVA) was used to make comparisons between treatments.

## Results

In both experiments we recorded 100% survival of fish. There was no evidence of expulsion of transmitters in the surgical group, either through the incision or an alternate route. Following 21 d, the incisions were well healed, most of the sutures over the antenna exit site were absorbed, and the transmitter was encapsulated in fibrous tissue. Three fish in the gastric group regurgitated their transmitters between day 5 and day 21. Throughout the experimental period gastric fish exhibited a 'coughing' type behavior (flared opercula) that appeared to be an attempt to forcefully expel the transmitter. Based on observational data, both groups of tagged fish had returned to normal feeding approximately 24 h after tagging.

### *Relative growth rate*

The initial size of fish included in the three experimental groups was not significantly different (see Table 1). All control and surgical fish showed positive weight gain throughout the study; several gastric fish had lost weight by day 21. The mean relative growth rate of the control, surgical and gastric groups could not be distinguished statistically at day 5 (see Figure 2A). By day 21, however, the relative growth rate of gastric fish was significantly lower than the control or surgical groups (see Figure 2A). Within each group, day 5 growth rate was compared to day 21 growth rate. For the control and surgical groups, the temporal influence on growth rate was not significant. Gastric fish, however, had significantly reduced growth at day 21 as compared to day 5 (see Figure 2A).

*Table 1.* Summary of initial fork length (FL), initial weight, sample size ($N$) and transmitter to body weight ratio (TBWR) in air and in water of subyearling chinook salmon used to compare surgical and gastric tagging methodologies. Measurements are presented as mean ± SD. No significant differences in initial FL or weight were detected among groups ($P > 0.05$)

| Group | $N$ | FL (mm) | Weight (g) | TBWR in Air (%) | TBWR in Water (%) |
|---|---|---|---|---|---|
| Control | 120 | 119 ± 11 | 21.7 ± 6.7 | | |
| Surgery | 60 | 119 ± 8 | 20.5 ± 5.9 | 7.0 ± 1.5 | 4.7 ± 1.0 |
| Gastric | 57 | 123 ± 14 | 21.6 ± 4.4 | 6.5 ± 1.6 | 4.4 ± 1.0 |

*General condition*

Due to the high density of fish in the rearing tank, the skin and fins of all experimental fish were in poor condition at the time of tagging. No additional damage to the skin or fins was noted consecutive to tagging in any of the experimental or control groups. Other than the skin and fins, no abnormalities were noted in control fish.

The condition of surgical fish, most notably the incision and antenna exit site wound, improved from day 5 to day 21. Five days after surgery 50% ($N = 6$) of the fish had slightly red or swollen incisions and antenna exit wounds. Of these six fish with abnormalities, four were rated as mild and two were rated as severe. By day 21, 98% ($N = 47$) of surgical fish had no irritation at these sites.

Gastric fish showed a decrease in general condition over time. The antenna exit site in the mouth was mildly irritated in one of twelve fish at day 5. By day 21 there was a distinct groove, rated as a severe irritation, in the mouth of 80% ($N = 36$) of the fish.

*Physiological response variables*

Hematocrit values for control fish were significantly higher than for either tagged group at day 5 (see Figure 2B), but the means for all groups remained within the normal range for salmonids (30–65%) (Randall & Wright, 1995). At day 21, the mean hematocrit values for the three groups were not significantly different (see Figure 2B). Hematocrit values remained within normal limits for 100% of surgical fish at both day 5 and day 21, indicating that the surgical procedure did not significantly affect the blood volume of the fish.

Leucocrit values were $\leq 1\%$ in 99% of the fish sampled, and $\leq 2\%$ in all fish. The normal range for leucocrit is $< 4\%$ (Adams et al., 1993), and elevated leucocrit levels may indicate a bacterial infection (Wedemeyer & McLeay, 1981). Since all leucocrit values were within the normal range, no signs of infection were noted for any group.

The mean plasma protein concentration for the controls at day 5 was significantly higher than either of the tagged groups (see Figure 2C), and the surgical and gastric groups were not significantly different (see Figure 2C). At day 21, the mean plasma protein concentration of the control group was significantly higher than the surgical group, and the mean concentration of the surgical group was significantly higher than the gastric group (see Figure 2C). The only temporal trend detected was within the surgical group where there was a significant increase in plasma protein concentration between day 5 and day 21 (see Figure 2C). Plasma protein concentrations between 30 and 70 g 100 $ml^{-1}$ are in the normal range (Adams et al., 1993); elevated levels may be due to hyperphagia and reduced levels could be the result of emaciation. None of the groups had any fish with plasma protein concentrations below the normal range. Elevated levels were detected for control fish at day 5 and day 21 and for surgical fish at day 21 (see Figure 2C). Gastric fish were apparently not eating as vigorously as control fish as their mean plasma protein concentration was always below 70 g 100 $ml^{-1}$. Surgical fish were apparently eating vigorously between day 5 and day 21 since the mean plasma protein concentration was elevated by day 21.

*Migration behavior*

Travel time to the dam was compared using seven paired releases; only one release showed significant differences. For the release of 23 May at JDD, surgical fish had significantly faster median travel times to the dam (4.5 h, range: 3.2–6.5 h) than gastric fish (6.0 h, range: 3.4–10.3 h). The overall median (where

*Table 2.* Migration behavior variables for paired releases of surgically and gastrically implanted yearling chinook salmon released into the Columbia River during April-June of 1996. The sample size (*N*), median and range of each variable and each dam is presented. Tagged fish were monitored at Bonneville (BD), The Dalles (TDD) and John Day (JDD) dams. Travel time to dam is defined as time in hours from release into the river until contact in the immediate dam area of the dam specified. Residence time at dam is defined as the time between the first and last contact with a tagged fish in the immediate dam area of the dam specified. Travel time between dams is defined as the time between the last contact in the immediate dam area of one dam and the first contact in the immediate dam area of the next downriver dam

| Migration Behavior Variable | Surgically Implanted Fish | | | Gastrically Implanted Fish | | |
|---|---|---|---|---|---|---|
| | *N* | Median | Range | N | Median | Range |
| **Travel Time to Dam (h)** | | | | | | |
| BD | 82 | 3.2 | 1.8–41.1 | 18 | 3.4 | 2.7–7.2 |
| TDD | 106 | 9.0 | 1.0–22.4 | 21 | 6.3 | 1.0–16.3 |
| JDD | 81 | 5.4 | 2.4–50.9 | 31 | 5.3 | 2.4–12.8 |
| **Residence Time at Dam (h)** | | | | | | |
| BD | 199 | 0.08 | 0–10.63 | 62 | 0.10 | 0–9.17 |
| TDD | 178 | 0.13 | 0–14.55 | 51 | 0.13 | 0–3.87 |
| JDD | 81 | 0.70 | 0–34.45 | 31 | 1.27 | 0–27.08 |
| **Travel Time Between Dams (h)** | | | | | | |
| JDD to TDD | 64 | 11.2 | 7.9–133.8 | 24 | 10.8 | 7.9–51.7 |
| TDD to BD | 107 | 24.2 | 9.4–153.9 | 37 | 27.1 | 19.7–94.1 |
| JDD to BD | 44 | 35.3 | 25.7–166.0 | 23 | 35.3 | 26.4–82.4 |

all releases are combined) for travel time to the dam was only different between surgical and gastric fish at TDD (see Table 2).

Residence time at the dam was compared using eight paired releases. The 16 May release at JDD had surgical fish with significantly lower median residence times (0.1 h, range: 0–0.7 h) than gastric fish (0.2 h, range: 0.02–6.2 h). The overall medians, summarized by dam, were similar between the surgical and gastric groups (see Table 2).

Travel time between dams was calculated for three routes, each using a different number of releases: four releases for JDD to TDD, four releases for TDD to BD, and six releases for JDD to BD. None of the releases, analyzed separately, revealed any significant differences between surgical and gastric fish. When releases were pooled for each route, the overall medians were similar for surgical and gastric fish (see Table 2).

## Discussion

The laboratory and field components of this study appear to support our hypothesis that gastric fish would be chronically affected by the implantation method and surgical fish would be acutely affected. At day 5 the surgical fish were, in some cases, more adversely affected than the gastric or control fish, but they steadily improved until they were not significantly different from the controls at day 21. Gastric fish showed the opposite trend, with some signs of being affected at day 5, and significant decreases in response variables by day 21. Few significant differences were detected in migratory behavior between surgical and gastric fish, possibly because most fish moved through the study area within five days, the time interval over which the lab study detected few differences between tagging methods. Another possible explanation would be similarities in physiological preparedness for migration between the gastrically and surgically implanted fish. For the two releases where differences were detected, surgical fish had significantly reduced travel times to the dam as well as residence time. Delays in travel time or residence time for migratory species are considered detrimental and call into question either the river flow, the operating conditions at the dam, (both of which were identical for surgical and gastric fish) or the condition of the fish.

Surgical fish were feeding normally within 24 h and steadily improved their condition over time. Expulsion of transmitters through the incision and by other routes has been reported in high percentages (9–71%) for several species, some of which were observed for extended periods (Marty & Summerfelt, 1986; Lucas, 1989; Moore et al., 1990; Helm & Tyus, 1992; Knights & Lasee, 1996). In other species transmitter retention over extended periods is high (see Thoreau & Baras, 1997). For example, northern squawfish (*Ptychocheilus oregonensis*) implanted with radio transmitters and released into the Columbia River were captured up to three years after implantation without any tag loss (T. Martinelli, unpublished data). In the present study we observed no transmitter loss and no indication that the process of expulsion had begun. Although the study period was relatively short, juvenile salmonids surgically implanted with radio transmitters and held in the laboratory (14 °C) for 75 d showed no signs of tag expulsion or serious infection (T. Martinelli, unpublished data). Tissue inflammation responses can serve as stimuli for tag expulsion in some species, therefore, the risk of transmitter expulsion can be reduced through efforts to minimize bacterial infection (Marty & Summerfelt, 1986). The aseptic procedures employed in our surgical technique, including the use of intraperitoneal antibiotics, may have minimized the risk of infection. Additionally, the use of a cyanoadhesive compound over the incision and sutures likely gave these vulnerable areas short-term protection from exposure to pathogens (Nemetz & MacMillan, 1988), and therefore, may have contributed to the observed lack of tag loss.

Few of the juvenile salmonids gastrically implanted with radio transmitters regurgitated their transmitters; three tags (5%) were regurgitated between day 5 and day 21. This regurgitation rate is relatively low as compared with other studies (Mellas & Haynes, 1985; Moser et al., 1990). We noted an obvious coughing behavior in gastrically implanted fish that has been recorded in other species (Mellas & Haynes, 1985), and does not indicate good general condition. The groove that developed in the mouth of most gastric fish, due to the irritation of the transmitter antenna, increased in severity over time. Although gastric implantation has been used successfully in studies of juvenile salmonids (Moser et al., 1990; Wilson et al., 1991; Ward et al., 1994), our findings support Winter's (1983) concept that this method is not appropriate for extended telemetry studies.

Based on the findings of this study, we recommend the use of a surgical implantation method for long term telemetry studies (between 5 and 75 d) of juvenile salmonids. Gastric implantation, however, appears to be suitable for short term ($\leq$ 5 d) monitoring. Although we have only observed surgically implanted fish in the laboratory for up to 75 d, these fish were in good condition, and we feel, had recovered from any harmful effects of surgery, including infection. Further evaluation of the extended (> 75 d) effects of surgical implantation of radio transmitters is required.

We would like to emphasize some considerations for the use of the surgical implantation technique on juvenile salmonids in the field. Care must be taken to minimize the potential for infection. As much as possible, an aseptic environment should be maintained, and the use of an intraperitoneal antibiotic agent is recommended (Summerfelt & Smith, 1990). The surgical technique should be practiced, especially the suturing, to minimize the amount of handling and air exposure time, and to ensure that no internal organs are damaged. The fish should be monitored closely during anesthesia induction so that they reach mid-stage three or early stage four (Jolly et al., 1972; Summerfelt & Smith, 1990) but are not allowed to go to deeper stages. Using the lightest anesthesia level that allows completion of the procedure will greatly speed recovery time. As for any tagging method, the buoyancy of tagged fish should be considered. Fish allowed access to the water surface will more readily compensate for the weight of the transmitter (Fried et al., 1976). Finally, we suggest that, although fish may be released immediately, data collection be delayed for at least 24 h to allow tagged fish time to recover from the immediate effects of surgery. If the study goals allow it, we recommend allowing at least five days of recovery time before data collection begins. Consideration should be given, however, to the potential stress associated with the delay of migration for actively emigrating fish.

## Acknowledgements

This work was funded, in part, by the U.S. Army Corps of Engineers, Portland, Oregon District, contracting officer Blaine D. Ebberts. We thank Tom Poe for review of earlier drafts of this manuscript. Field data was collected with the help of G. Holmberg, B. Liedtke, L. Blythe, J. Hardiman and D. Marks.

## References


Adams, S. M., A. M. Brown & R. W. Goede, 1993. A quantitative health assessment index for rapid evaluation of fish condition in the field. Trans. am. Fish. Soc. 122: 63–73.

Burger, C. V., J. E. Finn & L. Holland-Bartels, 1995. Pattern of shoreline spawning by sockeye salmon in a glacially turbid lake: evidence for subpopulation differences. Trans. am. Fish. Soc. 124: 1–15.

Busacker, G. P., I. R. Adelman & E. M. Goolish, 1990. Growth. In C. B. Schreck & P. B. Moyle (eds), Methods for Fish Biology. American Fisheries Society, Bethesda, Maryland, USA: 363–387.

Carl, L. M., 1995. Sonic tracking of burbot in Lake Opeongo, Ontario. Trans. am. Fish. Soc. 124: 77–83.

Clapp, D. F., R. D. Clark & J. S. Diana, 1990. Range, activity, and habitat of large, free-ranging brown trout in a Michigan stream. Trans. am. Fish. Soc. 119: 1022–1034.

Eiler, J. H., B. D. Nelson & R. F. Bradshaw, 1992. Riverine spawning by sockeye salmon in the Taku River, Alaska and British Columbia. Trans. am. Fish. Soc. 121: 701–708.

Fried, S. M., J. D. McCleave & K. A. Stred, 1976. Buoyancy compensation by Atlantic salmon (*Salmo salar*) smolts tagged internally with dummy telemetry transmitters. J. Fish. Res. Bd Can. 33: 1377–1380.

Guy, C. S., D. W. Willis & J. J. Jackson, 1994. Biotelemetry of white crappies in a South Dakota glacial lake. Trans. am. Fish. Soc. 123: 63–70.

Helm, W. T. & H. M. Tyus, 1992. Influence of coating type on retention of dummy transmitters implanted in rainbow trout. North amer. J. Fish. Mgmt 12: 257–259.

Isaak, D. J. & T. C. Bjornn, 1996. Movement of northern squawfish in the tailrace of a lower Snake River dam relative to the migration of juvenile anadromous salmonids. Trans. am. Fish. Soc. 125: 780–793.

Jolly, D. W., L. E. Mawdesley-Thomas & D. Bucke, 1972. Anesthesia of fish. Vet. Rec. 91: 424–426.

Knights, B. C. & B. A. Lasee, 1996. Effects of implanted transmitters on adult bluegills at two temperatures. Trans. am. Fish. Soc. 125: 440–449.

Lucas, M. C., 1989. Effects of implanted dummy transmitters on mortality, growth, and tissue reaction in rainbow trout, *Salmo gairdneri* Richardson. J. Fish Biol. 35: 577–587.

Marty, G. D. & R. C. Summerfelt, 1986. Pathways and mechanisms for expulsion of surgically implanted dummy transmitters from channel catfish. Trans. am. Fish. Soc. 115: 577–589.

Mellas, E. J. & J. M. Haynes, 1985. Swimming performance and behavior of rainbow trout (*Salmo gairdneri*) and white perch (*Morone americana*): effects of attaching telemetry transmitters. Can. J. Fish. aquat. Sci. 42: 488–493.

Moore, A., I. C. Russell & E. C. E. Potter, 1990. The effects of intraperitoneally implanted dummy acoustic transmitters on the behavior and physiology of juvenile Atlantic salmon, *Salmo salar* L. J. Fish Biol. 37: 713–721.

Moser, L. M., A. F. Olson & T. P. Quinn, 1990. Effects of dummy ultrasonic transmitters on juvenile coho salmon. Am. Fish. Soc. Symp. 7: 353–356.

Nemetz, T. G. & J. R. MacMillan, 1988. Wound healing of incisions closed with a cyanoacrylate adhesive. Trans. am. Fish. Soc. 117: 190–195.

Nielsen, L. A., 1992. Methods of marking fish and shellfish. American Fisheries Society Special Publication 23, Bethesda, Maryland, USA.

Randall, D. J. & P. A. Wright, 1995. Circulation and gas transfer. In C. Groat, L. Margolis & W. C. Clarke (eds), Physiological Ecology of Pacific Salmon. University of British Columbia Press, Vancouver, British Columbia, Canada: 441–458.

Ross, M. J. & C. F. Kleiner, 1982. Shielded needle technique for surgically implanting radio-frequency transmitters in fish. Progressive Fish Culturist 44: 41–43.

Stier, D. J. & B. Kynard, 1986. Use of radio telemetry to determine the mortality of Atlantic salmon smolts passes through a 17-MW Kaplan turbine at a low-head hydroelectric dam. Trans. am. Fish. Soc. 115: 771–775.

Summerfelt, R. C. & L. S. Smith, 1990. Anesthesia, surgery and related techniques. In C. B. Schreck & P. B. Moyle (eds), Methods for Fish Biology. American Fisheries Society, Bethesda, Maryland, USA: 213–272.

Thoreau, X. & E. Baras, 1997. Evaluation of surgery procedures for implanting telemetry transmitters into the body cavity of tilapia *Oreochromis aureus*. Aquat. Living Res. 10: 207–211.

Ward, D. L., A. A. Nigro, R. A. Farr & C. J. Knutsen, 1994. Influence of waterway development on migrational characteristics of juvenile salmonids in the lower Willamette River, Oregon. North am. J. Fish. Mgmt 14: 362–371.

Wedemeyer, G. A. & D. J. McLeay, 1981. Methods for determining the tolerance of fishes to environmental stressors. In A. D. Pickering (ed.), Stress and Fish. Academic Press, New York, New York, USA: 247–275.

Wilson, J. W., A. E. Giorgi & L. C. Stuehrenberg, 1991. A method for estimating spill effectiveness for passing juvenile salmon and its application at Lower Granite Dam on the Snake River. Can. J. Fish. aquat. Sci. 48: 1872–1876.

Winter, J. D., 1983. Underwater Biotelemetry. In L. A. Nielsen & D. L. Johnson (eds), Fisheries Techniques. American Fisheries Society, Bethesda, Maryland, USA: 371–395.

Winter, J. D., 1996. Advances in underwater biotelemetry. In B. R. Murphy & D. W. Willis (eds), Fisheries Techniques, second edition. Techniques. American Fisheries Society, Bethesda, Maryland, USA: 555–590.

*Hydrobiologia* **371/372:** 89–97, 1998.
J.-P. Lagardère, M.-L. Bégout Anras & G. Claireaux (eds), *Advances in Invertebrates and Fish Telemetry.*



# Use of miniature radio-transmitters to track the movements of dace, *Leuciscus leuciscus* (L.) in the River Frome, Dorset

Stuart Clough* & William R. C. Beaumont
*Institute of Freshwater Ecology, River Laboratory, East Stoke, Wareham, Dorset, BH20 6BB, England*
*(* Author for correspondence; e-mail s.clough@ife.ac.uk)*

*Key words:* radio-tracking, telemetry, migration, home range, dace, *Leuciscus*

## Abstract

Advances in radio-tag design have resulted in the production of tags suitable for use with relatively small fish. A method of release intended to minimise stress and increase validity of early tracks was designed. A technique of accurately locating radio-tags was developed and tested. Preliminary tracking results show that dace are highly mobile, and are capable of extensive, and often rapid migrations, both up and downstream. Most of the dace studied exhibited similar behaviour patterns, remaining at selected sites for extended periods of time, before rapidly relocating within the river.

## Introduction

Reviews of fish migration contain few references to the migration of cyprinids (Smith, 1991), and apart from a few notable exceptions (Lindsey & Northcote, 1963; Langford, 1974; L'Abée-Lund & Vollestad, 1987; Tyus, 1990), the study of daily and seasonal migrations of non-salmonid river fishes have been largely neglected. Historically, studies using telemetry techniques have been limited, in general, to the adult stages of medium sized and large salmonids owing to both the commercial importance of salmonids and the size of the tags available. Recent advances in the design, and in particular the encapsulation of radio-transmitters, have resulted in the production of smaller tags, that are sufficiently compact to be used with both small fish species, and the juveniles of larger species. An increase in the relative importance of non-salmonid fishes, in particular for recreational purposes, has resulted in a growing number of telemetric studies of their migrations (Baras, 1995; Lucas & Batley, 1996; Lucas & Frear, 1997).

The dace, *Leuciscus leuciscus* (L.), is a common European teleost, and in terms of population biomass, is the most important cyprinid in the larger chalk-streams of southern England (Mann & Mills, 1985). A shoaling, rheophilic, lithophilous species, the dace is considered to be particularly susceptible to habitat degradation (Mills, 1981, 1982). The length of adult dace is usually 150–250 mm (Maitland, 1972; Beaumont, 1982), with individuals in the upper half of this range being large enough to carry a miniature radio-tag. The preliminary study described was carried out between April 1992 and April 1994, during which period 25 adult dace were tagged with radio-transmitters.

## Study area

All of the dace in this study were caught, tagged and released in either the River Frome at East Stoke or the East Stoke Millstream.

The River Frome, situated in Southern England, is 'a typical medium-sized chalk stream' (Crisp et al., 1982), having most of the characteristic properties of such a river, as described by Ladle and Westlake (1995). East Stoke is situated in the lower portion of the Frome basin, approximately 6 km upstream of the tidal limit at Wareham. At East Stoke, widths range from 10 to 20 m, and the river meanders extensively. Maximum depth within the study area is approximately 3 m under normal summer river levels. Instream macrophytes, particularly *Ranunculus penicillatus* and *Potamogeton* spp. grow extensively except in the deepest pools. Sand and gravel are the predominant components of the substratum. Winter floods are a common occurrence, and surrounding fields are frequently inundated following heavy rain. Within the study area several tributaries enter, and the river splits into a main channel and carriers, known as millstreams, at regular intervals. The millstreams vary in size, carrying up to one quarter of the total discharge of the river, and are typically shallow and fast flowing, with a gravel substratum.

The East Stoke Millstream is a side branch of the River Frome 1.2 km long, and is divided into two sections by a flow gauging Crump weir. The upstream section (the Millhead) is generally deeper and slower flowing than the section below the weir, which is unmanaged and follows a well developed pool and riffle sequence. Historical management, including instream weed-cutting, has resulted in the Millhead being generally more even in depth with a trapezoidal cross-section. The width ranges from 4 to 7 m, and the maximum depth is 1.5 m. There are also several interconnecting side channels and branches, offering a range of habitat types. The entire East Stoke Millstream passes through a Fluvarium, which consists of two glass sided channels, (20 m × 1.5 m × 1.5 m) allowing visual observations of tagged fish in a near-natural environment. During the summer, adult dace in the East Stoke Millstream have been observed to occupy the same small area of channel for extended periods, and appear to select 40–60+ cm deep sites with no instream cover, a sand substratum and a flow velocity of between 10 and 40 cm $s^{-1}$ (Garner & Clough, 1996).

Environmental variables (air and water temperature, incident light level, turbidity and discharge) were monitored throughout the study period at the East Stoke flume, in the vicinity of the capture and release sites.

## Materials and methods

### *Fish tagging*

Biotrack SS-2 radio-tags (16 mm × 8 mm × 8 mm) were used throughout the study. The external tags were attached to the dorsal musculature of adult dace over 200 mm (fork length), using the technique described by Beaumont et al. (1996). Biotrack SS-2 tags weigh 0.9 g in water, and represent 0.84% and 0.75% of mean fish body weight, for 200 mm male and female dace respectively, and are well within the 1–1.25% of fish body weight recommendation suggested by Winter et al. (1978).

### *Release method*

In the earliest releases, radio-tagged dace were placed directly into the river within an hour after regaining balance following anaesthesia, and tag attachment. Relatively large and rapid post-release migrations were frequently observed, individuals were often seen darting away from the release site, and when several tagged fish were simultaneously released in this manner they rarely stayed together.

In later studies, a release cage was used. This allowed the tagged fish to acclimatise for a predetermined period in the presence of other fish, and then be released remotely, with a minimum of disturbance. It was thus possible to test whether or not the observed rapid post-release migrations were a result of the release method. The rectangular release cage (1000 mm × 500 mm × 500 mm) was constructed of black plastic mesh (1 mm × 5 mm slots), with one end being left open. A wooden frame was built into the open end, and a weighted, hinged, mesh door attached to the bottom of the frame. A ten metre long string was then attached to the door, in order that after the predetermined settling period, the door could be opened remotely with a minimum of disturbance to the fish. Clearly, a settling period which was as short as possible, but allowed the fish to have completely recovered from the stresses of tagging and handling and regained equilibrium, was required. Beaumont et al. (1996) suggest that recently radio-tagged dace exhibit a tail down posture initially, but show normal stability three hours after tagging, and behave normally by the following day. Similarly, neutral buoyancy was regained in bluegills (*Lepomis macrochirus*) less than 24 h after the insertion of a weight to mimic a tag (Gallepp & Magnuson, 1972). In consequence, 24 hours was selected as a suitable settling period.

### *Tag location*

Mobile radio-tracking was carried out using a modified Yaesu FT290R receiver and a hand held three element Yagi antenna. The receiver did not have a manual gain control, and fish positions were determined using a combination of signal triangulation, and de-tuning of the receiver. When quite close to the transmitter a substantial deflection was seen on the signal strength meter within a wide arc of antenna positions, and ascertaining the bearing of the strongest signal became difficult. By de-tuning the receiver away from the exact frequency of the transmitter a deflection of the meter needle was only seen when the antenna was pointing directly at the tag. A study of the accuracy of tag location from the river bank was carried out using hidden transmitters, with the exact position of the hidden transmitter being recorded on a detailed sketch map. The tracker searched for the hidden transmitter using the de-tuning technique, and recorded on a detailed map his interpretation of its position. The two maps were then compared, and the accuracy of location determined.

*Fish movements*

Tagged fish were tracked continuously after release in order to monitor immediate post-release behaviour. When several fish were released together the protocol was to follow the fish moving furthest away from the release point, should the tagged individuals separate. Continuous monitoring was terminated when the fish appeared to have become stationary, and attempts were made to locate each tagged fish once per hour for the next four hours. Thereafter, tagged fish were accurately located at least twice per day where possible. In addition, visual observations were carried out in order to establish whether or not the radio-tagged fish was part of a shoal.

*Effects of pike predation on a radio-tag*

Dace constitute the main prey items of pike (*Esox lucius* L.) in the River Frome (Mann, 1982; Mann & Mills, 1986). It was considered that during the study a radio-tagged dace could be eaten by a pike, and it was important therefore to assess the effects of such an event on both the tag and radio-signal. The damage incurred by the tag during the pike attack, both crush damage and puncture of the waterproof coating, was considered to be down to chance, and was likely to differ with each event, ranging from no effect to complete destruction. In consequence, it was decided to bypass this stage of the predation event and examine the effect on the signal characteristics by introducing the tag directly into the stomach of a pike.

In order to expose the tag to the same gastric conditions as would naturally occur following such an event, the transmitter was attached to a dead dace (245 mm, fork length), which was then inserted by means of a pipe into the stomach of an anaesthetised pike (1020 mm, fork length). Prior to insertion a small balsa float was attached to the tag using 1100 mm of nylon monofilament, with the float left trailing from the mouth of the pike. By examining the distance between the float and the mouth of the pike, the progress of the tag through the digestive system could be monitored. It was anticipated that should the tag cease to function, the presence of the float would facilitate the retrieval of the transmitter even in the event of it being regurgitated. The pike was retained in an experimental channel (20 m long $\times$ 2 m wide $\times$ 0.75 m deep) with a throughflow of river water for 7 days.

## Results

*Release method*

Prior to the use of the cage release method, tagged dace usually made immediate, rapid movements away from the release point. Early results from the use of a release cage did suggest a reduction in the incidence of these rapid, erratic post-release migrations. The data are however insufficient to provide statistical proof, as the release cage studies were carried out independently, and at a different time of year from the immediate releases. Cage releases in subsequent studies at comparable times of year have however shown that the fish often remain in the vicinity of the cage following release for up to 48 hours (S. Clough, pers. com.).

*Tag location*

The precision of location of hidden transmitters using triangulation and de-tuning showed that operatives were consistently able to pinpoint tags to within a 2 m transverse strip of river, and the position could be further narrowed down to near bank, mid-channel or far bank. The resulting precision corresponds to an area of approximately 4 $m^2$ in the East Stoke Millstream, and an area of 8 $m^2$ in the River Frome. The same de-tuning process enabled the recovery of three 'lost' transmitters from the river bed, two of which were buried in the substratum. It is not known how these tags became detached from the fish. One tag was recovered with both sutures and saddle intact, however visual observations over a number of days prior to recovering the tag suggested that the tagged dace had in fact been eaten by a pike, and it seems likely that this tag was either passed through or regurgitated. Each of the other two transmitters had one broken suture, suggesting that the increased pressure exerted on the remaining, intact suture had eventually resulted in the tag being shed, presumably by migration through the dorsal musculature.

*Fish movements*

During the study 25 dace, tagged with radio-transmitters, were located on 944 occasions between 1 April 1992 and 1 April 1994. Within this period, the movements of the radio-tagged dace were tracked for a total of 516 fish days, during which time they covered an accumulated minimum distance of 57 km. Radio-tagged fish released into the river were frequently observed visually, and on all occasions were associated with un-tagged dace, in shoals of between 20 and 1000 individuals. The longest known duration of tag attachment was 72 days, with the mean total track duration being 21 days. The net movement of 52% of radio-tagged dace was in a downstream direction from the point of release, however 88% of fish moved upstream immediately after release, with 16 out of 25 making large excursions on the day of release. In addition three fish were lost shortly after release, and it seems likely that these fish moved out of the study area. The maximum recorded distance from the release point was 3.3 km upstream and 9.1 km downstream, however the mean maximum distance was 1.3 km upstream and 1.6 km downstream. Full details of all 25 dace tracks are summarised in Table 1.

Many of the dace showed a similar pattern of behaviour, spending extended periods in one area, before rapidly moving to a new site, sometimes crossing weirs in the process (Figure 1). These relocation's were generally rapid with recorded speeds reaching 1.5 km $h^{-1}$. The mean maximum speed was 0.45 km $h^{-1}$, however the actual speed of the relocation was likely to be faster than the calculated value, because each fish was generally discovered to be stationary in its new position, and could have been at that location for some time. Two fish released together in the River Frome on 19 August moved rapidly upstream immediately after release, and both had crossed a weir, the East Stoke Flume, by 25 August. Each fish used a number of positions upstream of the weir during the next four weeks, sometimes simultaneously, however the tagged fish were located at different sites to one another more frequently than they were located together. Each fish displayed both upstream and down-

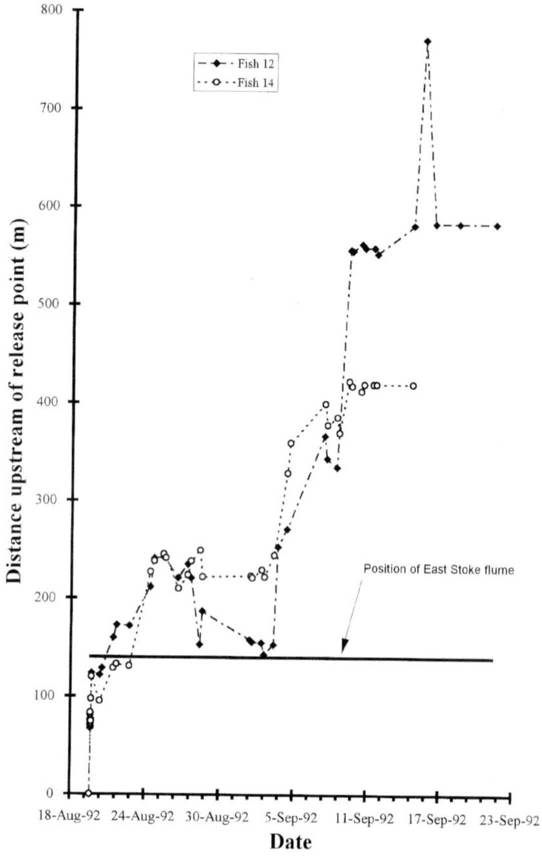


*Figure 1.* Example of the behaviour pattern displayed by many of the radio-tagged dace, showing residence periods interspersed by rapid relocation's. Both fish crossed the East Stoke flume within five days of release.

stream movements during this period, however the net movement of both dace was upstream.

When two or more tagged fish were released at the same time they rarely stayed together, suggesting that dace shoals are loose, constantly changing aggregations of individuals. Six fish released together in the River Frome on 6 August separated almost immediately into a group of three and three individuals (Figure 2). By the following day one of the individual fish had re-joined the group of three in a pool 0.4 km upstream of the release point, however by 8 August all six tagged dace were at different locations. Although the paths of some of the fish subsequently crossed on several occasions, tagged dace never stayed in the same shoal for long. Each of the fish used a number of different positions after release, and rapid relocations were observed in at least four fish.

Two radio-tagged dace (fish 2 & 3) released together on 20 July 1992, both moved upstream initially,

*Table 1.* Summary data of all 25 radio-tagged dace released between April 1992 and February 1994

| Fish number | Release date | First movement | Track duration (h) | Number of Records | Total distance (minimum) (km) | Net movement (km) | Maximum range (km) | | Maximum rate (km $h^{-1}$) |
|---|---|---|---|---|---|---|---|---|---|
| | | | | | | | Upstream | Downstream | |
| 1 | 27-Apr-92 | Upstream | 668 | 71 | 3.27 | –0.11 | 0.30 | –0.12 | 0.12 |
| 2 | 20-Jul-92 | Upstream | 214 | 29 | 1.39 | –0.27 | 0.24 | –0.43 | 0.06 |
| 3 | 20-Jul-92 | Upstream | 239 | 23 | 2.97 | –0.26 | 1.05 | –0.40 | 0.20 |
| 4 | 30-Jul-92 | Downstream | 144 | 17 | 1.46 | –0.61 | 0.35 | –0.62 | 0.21 |
| 5 | 30-Jul-92 | Downstream | 27 | 7 | 0.15 | –0.14 | 0.00 | –0.14 | 0.14 |
| 6 | 6-Aug-92 | Upstream | 915 | 49 | 2.30 | 1.10 | 1.10 | 0.00 | 0.04 |
| 7 | 6-Aug-92 | Upstream | 99 | 17 | 1.15 | 0.94 | 0.97 | 0.00 | 0.045 |
| 8 | 6-Aug-92 | Downstream | 279 | 16 | 0.94 | –0.81 | 0.00 | –0.84 | 1.50 |
| 9 | 6-Aug-92 | Upstream | 169 | 19 | 1.65 | –0.01 | 0.41 | –0.01 | 0.39 |
| 10 | 6-Aug-92 | Upstream | 76 | 15 | 0.98 | 0.96 | 0.96 | 0.00 | 0.66 |
| 11 | 6-Aug-92 | Upstream | 836 | 50 | 2.19 | –1.03 | 0.05 | –1.27 | 0.38 |
| 12 | 19-Aug-92 | Upstream | 738 | 45 | 1.38 | 0.58 | 0.77 | 0.00 | 0.40 |
| 13 | 19-Aug-92 | Upstream | 50 | 13 | 0.20 | 0.06 | 0.11 | 0.00 | 0.44 |
| 14 | 19-Aug-92 | Upstream | 576 | 42 | 0.73 | 0.42 | 0.42 | 0.00 | 0.42 |
| 15 | 10-May-93 | Upstream | 1724 | 107 | 2.55 | –0.47 | 0.90 | –0.47 | 0.94 |
| 16 | 10-May-93 | Upstream | 912 | 69 | 4.38 | 3.63 | 3.63 | 0.00 | 0.94 |
| 17 | 2-Jun-93 | Upstream | 408 | 37 | 1.82 | 1.82 | 1.82 | 0.00 | 0.18 |
| 18 | 2-Jun-93 | Upstream | 533 | 38 | 4.60 | 2.74 | 2.77 | 0.00 | 0.12 |
| 19 | 6-Jul-93 | Upstream | 984 | 37 | 9.57 | –8.80 | 0.04 | –9.15 | 0.24 |
| 20 | 16-Jul-93 | Upstream | 768 | 22 | 8.18 | –8.17 | 0.00 | –8.17 | 0.07 |
| 21 | 21-Feb-94 | Upstream | 667 | 64 | 0.31 | –0.01 | 0.01 | –0.05 | 0.60 |
| 22 | 21-Feb-94 | Upstream | 2 | 5 | 0.16 | 0.16 | 0.16 | 0.00 | 0.12 |
| 23 | 24-Feb-94 | Upstream | 574 | 54 | 1.03 | –0.06 | 0.28 | –0.06 | 1.14 |
| 24 | 25-Feb-94 | Upstream | 4 | 7 | 0.29 | 0.29 | 0.29 | 0.00 | 0.27 |
| 25 | 25-Feb-94 | Upstream | 789 | 91 | 4.10 | 3.26 | 3.30 | 0.00 | 1.14 |
| Total | | | 12395 | 944 | 57.72 | | | | |

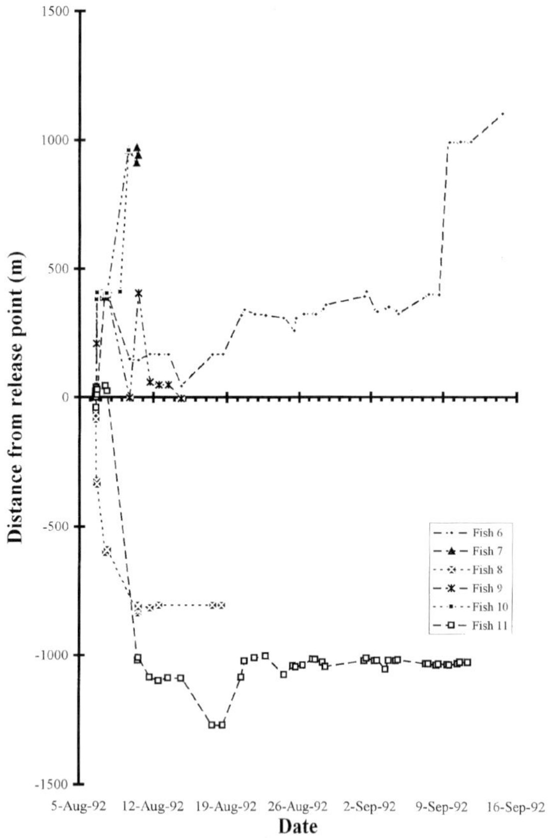


*Figure 2.* Post-release movements of six radio-tagged dace released on 6 August 1992. Negative distances are downstream of the release point. Although fish paths cross on a number of occasions, radio-tagged dace rarely remained together for long.

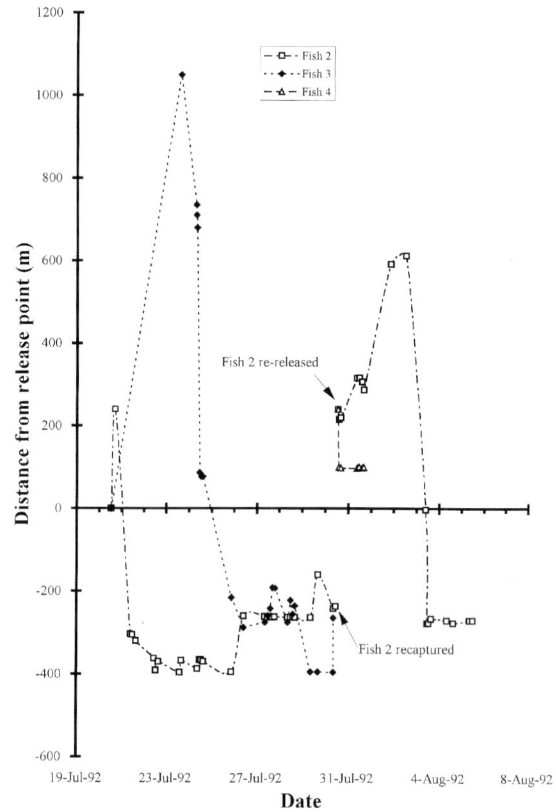


*Figure 3.* Movements of three radio-tagged dace released where the East Stoke Millstream splits from the River Frome. Negative distances are downstream of the release point, and correspond to fish positions in the East Stoke Millstream; Positive distances show fish positions upstream of the release point, and relate to locations in the River Frome. Fish 2, recaptured from the East Stoke Millstream and re-released in the River Frome returned to the exact position occupied immediately prior to recapture.

but separately (Figure 3). Fish 2 was found the next day 0.3 km downstream of the release site, and had entered the East Stoke Millstream. When fish 3 was located on 23 July it was over 1 km upstream of the release point, and 1.5 km upstream of the position of fish 2. By 25 July both fish were in the East Stoke Millstream, where over the next 5 days they occupied several positions, sometimes together. On 30 July both fish were captured by electrofishing, with fish 2 being re-released, along with another radio-tagged fish (fish 4) and a shoal of un-tagged individuals, 0.24 km upstream of the original release point. The two radio-tagged fish quickly separated, and by 3 August, fish 2 had returned to the location from where it was recaptured. Visual observation showed the tagged fish had taken up a position, among a shoal of about 50 other dace, within 1 $m^2$ of a position used on several occasions by both fish 2 & 3 during July, the same position it occupied immediately prior to recapture.

No significant trends in the behaviour of dace in connection with discharge were evident. Significant changes in discharge were arbitrarily considered to be either increases or decreases of 10% or over. Of the 57 single movements over 0.3 km exhibited by the 20 fish released between the months of April and July, only 13 were associated with significant changes in discharge.

Fish 25, a ripe female dace, was released on 25 February 1994 and moved upstream immediately after release covering 0.67 km in the first two hours (Figure 4). The following day the fish was located 1.1 km upstream of the release point, and was recorded at this location several times each day for the next five days. On 5 March, fish 25 was unable to be located in the River Frome, however a thorough search the following day revealed that the fish had entered the Bindon Millstream, and was occupying a position 3.2 km upstream

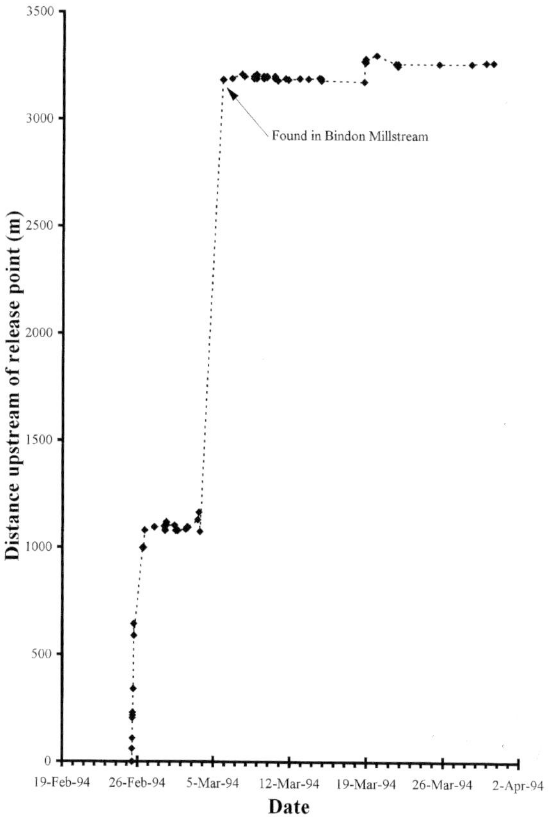


*Figure 4.* A radio-tagged female dace released prior to spawning moved upstream over 3 km and entered the Bindon Millstream.

of the release point (Figure 4). Visual observations showed that the radio-tagged fish was in a shoal of at least 1000 dace, in a pool immediately upstream of a shallow gravel bedded riffle. Fish 25 remained in the Bindon Millstream until 30 March, but was unable to be located thereafter, and visual observations suggested that the majority of the other dace had also left the area.

### *Effects of pike predation on a radio-tag*

No change in any of the signal parameters were observed following the insertion of a radio-tagged dace into the stomach of a pike (*E. lucius*). The balsa float moved approximately 100 mm on each of the first two days, however no further movement was observed, indicating that the tag had come to rest somewhere within the digestive tract. By applying gentle pressure to the nylon monofilament the radio-tag was easily removed from the stomach of the pike, and out through the mouth. The external antenna was un-kinked, both sutures intact, and examination under a microscope suggested that the coating of the transmitter was undamaged.

## Discussion

The timing and method of release of radio-tagged fish is likely to affect their subsequent behaviour, at least in the short term, and is worthy of further investigation. In this study, the release cage served two functions. It allowed the dace an appropriate period, in this case 24 hours, to regain normal posture following the tagging procedure, and it facilitated the remote release of the undisturbed fish, in a shoal, which were then free to swim out of the cage at will. Other species of fish including roach (*Rutilus rutilus* (L.)), are known to compensate for increased density (e.g. the addition of a tag) by tilting upwards and swimming with the caudal fin, with the upward vector of the thrust balancing their weight (Gallepp & Magnuson, 1972). This is probably the reason for the tail down posture observed in recently radio-tagged dace by Beaumont et al. (1996), and may explain why dace released immediately after tagging behave erratically. However, rapid relocations appear to be a normal feature of dace behaviour, and need not necessarily suggest an unnatural reaction to external stimuli.

On all occasions that radio-tagged fish were visually observed they were associated with other dace, suggesting that the radio-tagged fish were behaving in a similar way to un-tagged dace. When two or more tagged fish were released at the same time they rarely stayed together, and although their paths may have subsequently crossed on several occasions, time spent at the same location was generally limited. The reasons for this behaviour are unclear, but one possibility is that shoals of dace are loose aggregations, from which smaller groups break off, and others join.

Dace spend extended periods, often several days, in selected areas of the river. The fish periodically relocate within the river system, often moving several kilometres, and may pass potential obstructions, such as weirs, in the process. The relocation's are generally rapid, and could be in response to dwindling resources, could follow a disturbance (e.g. a perceived predation event), or may be a response to changes in abiotic factors such as turbidity.

Linfield (1985), in a discussion of home range theory, proposes that cyprinids form totally mobile populations. This is in contrast to previous studies (e.g. Stott, 1967) where cyprinid populations are considered to be based on home range territories, with intermixing of individuals arising only from the activity of mobile components of otherwise static populations. Overall movement patterns observed in this study support Linfield's alternative concept of home range theory, however individual fish do take up residence in defined areas for periods of days or weeks. At certain times of the year dace are known to carry out diel migrations between daytime and night-time habitats, with the timing of the movements closely linked to dawn and dusk (Clough & Ladle, 1997). These diel migrations may be superimposed upon these residence periods, with individuals consistently 'homing' to previously occupied habitats The area encompassed by such 'homing' migrations could be considered to be the fish's home range for that period of time. Following a period of residence, the fish relocate to a different area of the river, where they again take up residence for a variable period, establishing another 'home range' at the new site. In this study, a radio-tagged fish (fish 2) recaptured from the East Stoke Millstream, and re-released in the River Frome returned to the position it had occupied immediately before recapture. As no night time observations were carried out during this study, it is possible that fish 2 could have joined a shoal which were carrying out diel migrations between the East Stoke Millstream and River Frome (Clough & Ladle, 1997), and returned to the Millstream with the shoal around dawn. Indeed three of the four releases of fish during July 1992 resulted in dace entering and becoming resident in the East Stoke Millstream.

In the River Frome, dace are known to spawn from mid to late March (Mann, 1974), and a known spawning site, a gravel bedded riffle in the East Stoke Millstream, has been studied in detail (Mills, 1981). The discovery during March 1994 of a radio-tagged fish, along with large numbers of other dace at a similar site in Bindon millstream, over 3 km upstream of the release point, suggests that such sites are particularly suitable for the spawning of dace in the River Frome. Dace spawn on gravel (Mills, 1981; Mann & Mills, 1985), however extensive areas of clean gravel are available within the main River Frome between the release point and the entrance to the Bindon Millstream. In other rivers, dace are known to carry out 'spawning migrations into smaller streams' (Lobon-Cervia et al., 1996), and channel size may be an important factor in the selection of a suitable spawning site. It seems possible that millstreams, with similar characteristics to naturally braided sections of rivers, may provide specific combinations of habitat that are favoured by spawning dace, and further investigation is warranted.

No change in the signal emitted by a radio-tag was seen over 7 days inside the stomach of a pike (*E. lucius*). Predation of a radio-tagged dace by a pike during the study, could therefore result in the movements of a pike being tracked, with no remote indication to the tracker. The inserted transmitter was neither regurgitated or defecated during the study, and it seems possible that a naturally ingested tag could be similarly retained. During the study one radio-tagged dace was suspected to have been eaten by a pike, and the transmitter was found on the river bed two weeks later. It seems likely that this tag was either passed or regurgitated by the pike. A more detailed study is required before suggesting how long an ingested tag is likely to remain inside a pike, however both events indicate that transmitters can continue to work inside a pike, and this eventuality should be considered when interpreting the results.

## Acknowledgements

The authors wish to thank M. Ladle and J. S. Welton for their comments and assistance with all aspects of the radio-tagging study, and P. Garner for comments on an earlier draft of the manuscript. Thanks also to A. Wood and S. Randall for their help with tracking. S. Clough was funded by a doctoral studentship.

## References

Baras, E., 1995. Seasonal activities of *Barbus barbus*: effect of temperature on time-budgeting. J. Fish Biol. 46: 806–818.

Beaumont, W. R. C., 1982. An aberration in the scale formation of the dace *Leuciscus leuciscus* (L.) J. Fish Biol. 21: 321–324.

Beaumont, W. R. C., S. Clough, M. Ladle & J. S. Welton, 1996. A method for the attachment of miniature radio tags to small fish. Fish. Man. Ecol. 3: 201–207.

Clough, S. & M. Ladle, 1997. Diel migration and site fidelity in a stream-dwelling cyprinid, *Leuciscus leuciscus*. J. Fish Biol. 50: 1117–1119.

Crisp, D. T., A. M. Matthews & D. F. Westlake, 1982. The temperatures of nine flowing waters in southern England. Hydrobiologia 89: 193–204.

Gallepp, G. W. & J. J. Magnuson, 1972. Effects of negative buoyancy on the behaviour of the bluegill, *Lepomis macrochirus* Rafinesque. Trans. am. Fish. Soc. 3: 507–512.

Garner, P. & S. Clough, 1996. Habitat use by dace, *Leuciscus leuciscus* (L.), in a side channel of the River Frome, England. Fish. Man. Ecol. 3: 349–352.

L'Abée-Lund, J. H. & L. A. Vollestad, 1987. Feeding migration of roach, *Rutilus rutilus* (L.), in lake Arungen, Norway. J. Fish Biol. 30: 349–355.

Langford, T. E., 1974. Trials with ultrasonic tags for the study of coarse fish behaviour and movements around power station outfalls. J. Inst. Fish. Mgmt 5, 3: 61–62.

Ladle, M. & D. F. Westlake, 1995. River and stream ecosystems of Great Britain. In C. E. Cushing, K. W. Cummins & G. W. Minshall (eds), River and Stream Ecosystems. Elsevier: 343–388.

Lindsey, C. C. & T. G. Northcote, 1963. Life history of redside shiners, *Richardsonius balteatus*, with particular reference to movements in and out of Sixteenmile Lake streams. J. Fish. Res. Bd Can. 20: 1001–1030.

Linfield, R. S. J., 1985. An alternative concept to home range theory with respect to populations of cyprinids in major river systems. J. Fish Biol. 27 (Sup. A): 187–196.

Lobon-Cervia, J., Y. Dgebuadze, C. G. Utrilla, P. A. Rincon & C. Granado-Lorencio, 1996. The reproductive tactics of dace in central Siberia: evidence for temperature regulation of the spatio-temporal variability of its life history. J. Fish Biol. 48: 1074–1087.

Lucas, M. C. & E. Batley, 1996. Seasonal movements and behaviour of adult barbel *Barbus barbus*, a riverine cyprinid fish: implications for river management. J. appl. Ecol. 33: 1345–1358.

Lucas, M. C. & P. A. Frear, 1997. Effects of a flow gauging weir on the migratory behaviour of adult barbel, a riverine cyprinid. J. Fish Biol. 50: 382–396.

Maitland, P. S., 1972. Key to British freshwater fishes. Freshwater Biological Association scientific publication No. 27, 139 pp.

Mann, R. H. K., 1974. Observations on the age, growth, reproduction and food of the dace *Leuciscus leuciscus* (L.) in two rivers in southern England. J. Fish Biol. 6: 237–253.

Mann, R. H. K., 1982. The annual food consumption and prey preferences of pike (*Esox lucius*) in the River Frome, Dorset. J. anim. Ecol. 51: 81–95.

Mann, R. H. K. & C. A. Mills, 1985. Variations in the sizes of gonads, eggs and larvae of the dace, *Leuciscus leuciscus*. Envir. Biol. Fishes 13: 277–287.

Mann, R. H. K. & C. A. Mills, 1986. Biological and climatic influences on the dace *Leuciscus leuciscus* in a southern chalk-stream. Rep. Fw Biol. Assoc. 54: 123–136.

Mills, C. A., 1981. The attachment of dace eggs to the spawning substratum and the influences of changes in water current on their survival. J. Fish Biol. 19: 129–134.

Mills, C. A., 1982. Factors affecting the survival of dace in the early post-hatching period. J. Fish Biol. 20: 645–655.

Smith, R. J. F., 1991. Social behaviour, homing and migration. In I. J. Winfield & J. S. Nelson (eds), Cyprinid Fishes: Systematics Biology and Exploitation. Chapman & Hall: 509–529.

Stott, B., 1967. The movements and population densities of roach (*Rutilus rutilus*, (L.) and gudgeon (*Gobio gobio*, (L.) in the River Mole. J. anim. Ecol. 36: 407–423.

Tyus, H. M., 1990. Potamodromy and reproduction of Colorado squawfish in the Green River basin, Colorado and Utah. Trans. am. Fish. Soc. 119: 1035–1047.

Winter, J. D., V. B. Keuchle, D. B. Siniff & J. R. Tester, 1978. Equipment and methods for radio tracking freshwater fish. University of Minnesota, Inst. Agric. Misc. Rep. 152, 45 pp.

*Hydrobiologia* **371/372:** 99–106, 1998.
*J. P. Lagardère, M.-L. Bégout Anras &G. Claireaux (eds), Advances in Invertebrates and Fish Telemetry.*



# The effects of Passive Integrated Transponder tags on survival and growth of juvenile brown trout (*Salmo trutta* L.) and their use for studying movement in a small river

Dominique Ombredane[1,2], Jean Luc Baglinière[2] & Frédéric Marchand[2]
[1] *Ecole Nationale Supérieure Agronomique de Rennes, Département Environment et Exploitation des Ressources naturelles – 65 Rue de St Brieuc, 35042 Rennes cedex, France; e-mail: ombredan@roazhon.inra.fr*
[2] *Institut National de la Recherche Agronomique, Laboratoire d'Ecologie Aquatique, 65 Rue de St Brieuc, 35042 Rennes cedex, France*

*Key words:* PIT tag, survival, growth, movement, brown trout, freshwater

## Abstract

From 1993 to 1995, more than 2000 of 0+ brown trout (*Salmo trutta* L.) were captured in a small brook in October and released after clipping the left pelvic fin. More than 1700 of them were also tagged with a small Passive Integrated Transponder (PIT) tag (11 mm long) implanted in the body cavity. Tagged juveniles were recaptured in May in the brook but also one year later in the river Oir, and continuously in a trap during spring downstream migration in the river Oir.

The combined marking experiment (PIT tag and fin-clipping) carried out during the two first years showed an average transponder loss rate of 3.38% after seven months for 0+ juveniles tagged in October. Pelvic fin regeneration occurred at a average low rate of 2.37% during the same period. Futhermore, PIT tag showed no significant effect on growth, survival in 0+ juveniles (55 to 127 mm fork length), tagged in October 1995 and recaptured in May 1996.

PIT tagging has been used to analyse brown trout movement inside the river. Recapture rate in the brook from tagged 0+ juveniles after seven months decreased while size increased. The medium sized juveniles migrated in the main system Oir and the largest ones out of the main river.

Results showed (1) PIT tagging is a reliable technique to analyse the movement and population dynamics of brown trout by taking into account the individual life history, (2) recruitment of juvenile trout in the river depends on fish size.

## Introduction

The use of high performance techniques for individual tagging is turning out to be indispensable for obtaining precise data on growth, survival, migration and others aspects of fish behaviour in the natural environment (McFarlane et al., 1990). For brown trout (*Salmo trutta* L.), the use of individual tags early in the juvenile stage should enable us to develop models of population dynamics and to estimate production on a catchment basin scale. Individual tagging on this species will be also very useful for studies on the determinism of the migration and smolting in juveniles of brown trout in relation to environmental parameters and biological processes, such as growth, sexual maturation, genetic inheritance.

For a given study, the tagging technique chosen may depend on the number and the size of the fish to be marked, on the impact of the tag on survival and growth, on how long the tag lasts and often on how much it costs. For the individual tagging of salmonids at the juvenile stage (trout with a mean fork length of 85 mm) several techniques which are cheaper than PIT tagging were first tested in an aquarium in our laboratory. After 33 days, the retention rate of visible implants ($1.0 \times 2.5 \times 0.1$ mm) was below 50%,

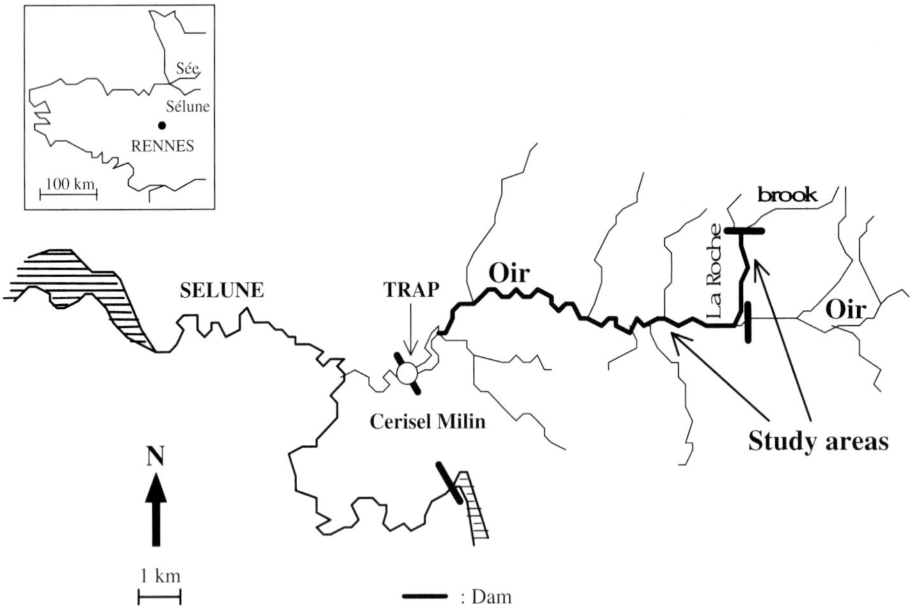


*Figure 1.* Hydrographic network and study areas where juvenile brown trout were captured, recovered and trapped from 1993 to 1996.

while those of small vinyl tapes through the base of the dorsal fin and of small Floy T-bar anchor tags were below 85%. These observations concerning the visible implants corresponded with the work of Bryan & Ney (1994) who obtained a retention rate of about 50% for 130–160 mm trout. Concerning the anchor tag, Brewin et al. (1995) founded loss rates of 15.0% and 44.3% for females and males brown trout (> 300 mm) respectively. These loss rates were too high to justify the use of these techniques on a large scale in the natural environment. Moreover, while the loss rate of coded wire tags on salmonids juveniles was low, 1.1% to 5.3% (Blankenship, 1990), the fish had to be killed for individual recognition, which is not compatible with studies of population dynamics.

Therefore, PIT tags, which seem to be highly efficient (Prentice et al., 1990), were used for our studies on trout despite their cost. However, before using the ecological data obtained, the reliability of the tagging technique in our conditions had to be ensured, in the natural environment. For 0+ juveniles of trout, several aspects were tested, including the loss rate of the tag, its impact on growth and its impact on mortality according to the size of the fish at the time of tagging. A experiment was then carried out using this tagging technique to study the movements of juvenile trout on a small catchment basin scale.

## Material and methods

### *Study site*

The study was carried out on the Oir which flows into the Sélune, at about 8 km from the sea and on one of the Oir tributaries, the La Roche Brook (Figure 1). In the downstream part of the Oir river, there is a double trapping system which allows control of the downstream and upstream migrating fishes. In the La Roche Brook, which is 2 m wide on average, the 2 km long study area is located downstream of a small dam which is impassable for brown trout juveniles. This area was subdivided into 20 sectors of 110 m long on average ($\sigma = 34$ m). The 12 km long Oir study area is bounded downstream by a trap and upstream by the confluence with the Roche Brook.

### *Capture and tagging of fish*

Over period of three years (1993–1995), about 2000 trouts aged 0+ were captured in October by electrofishing in the La Roche Brook. Fishes were anaesthetised with 2-phenoxy-ethanol, measured at the fork length ($\pm$ 1 mm) and then tagged.

All the fishes were then released in their captured sector within 30 to 60 min.

*Table 1.* Loss rate of PIT tags and whole regeneration rate of the pelvic fin after seven months. Mean characteristics of juvenile brown trout tagged in the La Roche brook in October 1993 and 1994. Standard deviations are given in parentheses.

| | 1993 Cohort | 1994 Cohort |
|---|---|---|
| Number of fishes tagged in October | 577 | 432 |
| Number of fishes recaptured in May (with PIT tag and/or fin clipped) | 186 | 172 |
| Mean (*SD*) fork length (mm) in October | 86.4 (*14.9*) | 82.6 (*12.3*) |
| Range of fork length (mm) in October | 55–129 | 54–124 |
| Mean (*SD*) fork length (mm) in May | 134.0 (*21.1*) | 138.9 (*21.1*) |
| Range of fork length (mm) in May | 78–202 | 83–185 |
| PIT tag loss (%) | 3.80 | 2.96 |
| Left pelvic fin regeneration (%) | 0.00 | 4.73 |

In order to evaluate the PIT tag retention rate and its effects on growth, survival, and migration behaviour, combined tagging was implemented on every cohort. In October 1993 and 1994, all juveniles, respectively 577 and 432, had their left pelvic fin clipped and had a PIT tag injected into their ventral part of the body cavity with a modified hypodermic syringe and a 12 gauge needle. In October 1995, this combined tagging was applied on only 74.8% of the 977 juveniles captured in the brook study area. The others 25.2% had only their left pelvic fin clipped.

The PIT tags used were 11 mm long with a 2.2 mm diameter (TROVAN system).

*Fish-tracking*

For a period of more than three years, the whole study area of the La Roche Brook was monitored twice a year – in May and in October – by electrofishing. Two successive fishings were made in 3 control sectors to estimate the average trout catchability according to the Seber-Lecren method (Seber and Lecren, 1967; Laurent and Lamarque, 1974). Over the whole study period, the efficiency of fishing was always above 80% regardless of the time of the year, the sector or the age of the fish.

Moreover, in October 1996, about 70% of the study area in the Oir River was monitored by electrofishing. Finally, the downstream movement of the juveniles trout was recorded by continuous downstream trapping in the Oir river throughout the year from 1994 to 1996.

For each capture by electrofishing or trapping, the fork length of every trout was measured and a routine monitoring of the tags was carried out. Scales were removed from some individuals in order to estimate their age. All the fish of the population could therefore be assigned to an age group.

## Results

### *Retention and effects of the PIT tags*

*Tags loss rates after 7 months*

These rates were evaluated for the 1993 and 1994 cohorts, when all the juvenile trout captured in The Roche brook were tagged in October with a PIT tag and by left pelvic fin clipping. When tagged, the trout were between 54 and 129 mm fork length (Table 1).

Seven months later, in May, three categories of 1+ trout were recaptured in the brook:

- individuals with a PIT tag and the left pelvic fin cut, (a)
- individuals with only a PIT tag, (b)
- individuals with only the left pelvic fin cut. (c)

Considering that the recapture rate of these three categories is identical, the loss rate of the PIT tags (including failed tags) and the whole regeneration rate of the left pelvic fin can be calculated. The formulae used to estimate these rates are the following:

Loss rate (%) of the PIT tags = $[c/(a+b+c)] \cdot 100$
Whole regeneration rate (%) of the left pelvic fin = $[b/(a+b+c)] \cdot 100$

This analysis shows (Table 1) that the loss rate of the PIT tags after 7 months varies only slightly and is very low (average 3.38%). The whole regeneration rate of the left pelvic fin is also very low, an average of 2.37%.

### *Effects of PIT tags on growth*

In order to investigate the influence of PIT tag on juvenile growth, the 977 trout 0+ of the 1995 cohort, with fork length ranging from 55 to 127 mm in October (Table 2), were tagged in two different ways:

- the $A_0$ group: PIT tag and left pelvic fin clipped,
- the $B_0$ group: left pelvic fin clipped only.

The average size (Table 2) of the $A_0$ and $B_0$ groups are not significantly different (*t* test, $P > 0.36$). In the

*Table 2*. Number and fork length of juvenile brown trout tagged in the La Roche brook in October 1995 and recaptured in May 1996. Standard deviations are given in parentheses.

| Tagging groups in October 1995 | $A_0$: PIT tagged and fin clipped | $B_0$: fin clipped | |
|---|---|---|---|
| Number of fishes | 731 | 246 | |
| Mean (SD) fork length (mm) | 88.5 (*14.9*) | 87.5 (*14.6*) | |
| Range of fork length (mm) | 55–127 | 57–126 | |
| **Recaptured groups in May 1996** | **$A_m$: PIT tagged** | **$B_m$: fin clipped** | **$C_m$: without mark** |
| Number of fishes | 149 | 55 | 67 |
| Mean (SD) fork length (mm) | 131.1 (*20.9*) | 129.3 (*19.1*) | 125.6 (*22.1*) |
| Range of fork length (mm) | 92–181 | 92–174 | 95–195 |

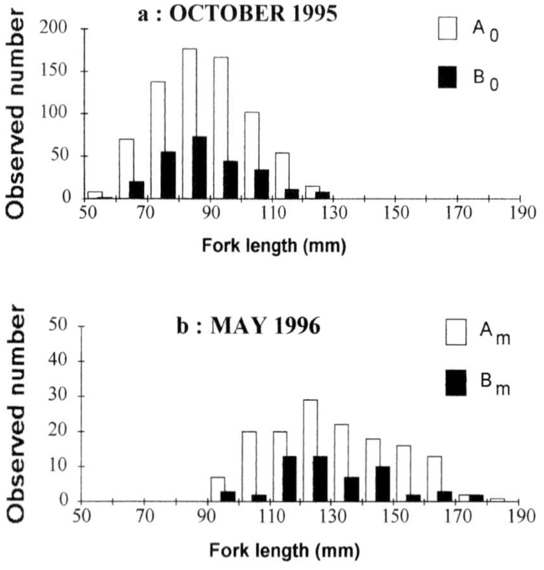


*Figure 2*. Size distributions of brown trout tagged with ($A_0$) or without ($B_0$) PIT tags in October 1995 (a) and recovered with ($A_m$) or without ($B_m$) PIT tags in May 1996 (b).

same way, the fish size distributions of the two groups (see Figure 2a) were identical ($\chi^2$ test, $P = 0.27$).

The fish recaptured in May were divided into three groups:

- $A_m$ group: with PIT tag,
- $B_m$ group: left pelvic fin clipped only,
- $C_m$ group: no PIT tag, no fin clipped

As the tag loss rates were low, the size distributions of the $A_m$ and $B_m$ groups recaptured in May were considered as representative of these of the $A_0$ and $B_0$ groups tagged in October.

In May, the average length (Table 2) of group $A_m$ and $B_m$ (with or without tags) did not differ statistically ($t$ test, $P = 0.56$) nor did the size distribution of both groups ($t$ test, $P = 0.49$) (see Figure 2b). In the same way, the average sizes of groups $A_m$ and $B_m$ were not statistically different ($t$ test, $P > 0.10$) to that of group $C_m$ (mainly fish not catched in October).

*Effects of PIT tags on fish mortality:*

*Survival at the total population scale*

The effect of PIT tag marking on the apparent survival for the total population in the brook was analysed on the basis of the protocol used for the 1995 cohort. The following approach (see Figure 3) was used:

- Firstly, the fish numbers were corrected in order to take into account the tag loss. The tag loss rates used, 3.38% for the PIT tags and 2.37% for the left pelvic fin-clipping, represent the averages calculated for 1993 and 1994 cohorts. The loss rate of both tags together, 0.08% was obtained by the product of the loss rates of each marking method.

- Secondly, the group $A_0$ apparent survival rate (21.1%) was calculated upon the basis of PIT tagged fish recovered in May. This procedure makes it possible to estimate both the number of fish from the $B_0$ group captured in May and the apparent survival rate (20.8%) of this group.

The apparent survival of each group is similar and the proportions of PIT tagged trout were not statistically different ($t$ test, $P = 0.94$) in October (75.5%) and in May (75.3%).

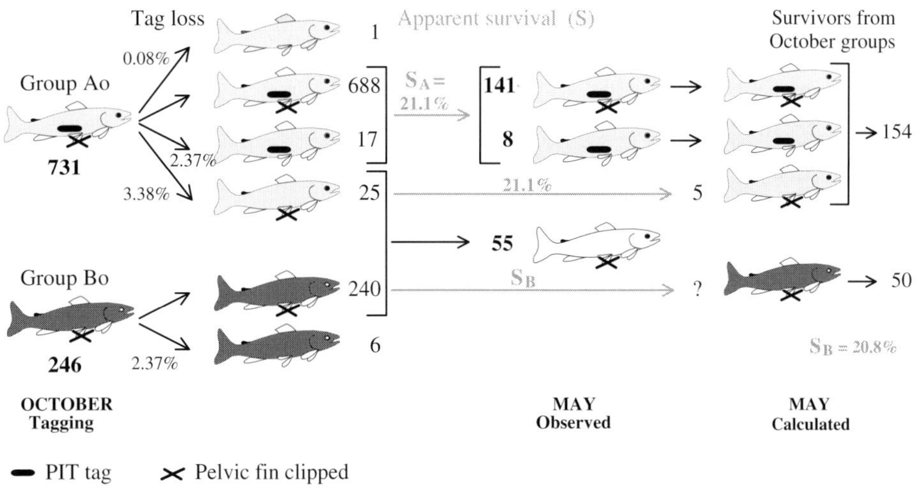


*Figure 3.* Procedure used to calculate survivor numbers after seven months of fish with ($A_0$) and without ($B_0$) PIT tags.

*Table 3.* Number (*N*) and percentage (%) by size class of juvenile brown trout with PIT tag captured in October 1995 and recovered in May 1996 in the La Roche brook. Percentage were compared with a *t* test (NS = Non Significant).

| Size class in October (mm) | October 1995 | | May 1996 | | Comparison with *t* test |
|---|---|---|---|---|---|
| | *N* | % with PIT tag | *N* | % with PIT tag | |
| 50–69 | 99 | 78.8 | 28 | 85.7 | NS ($p = 0.37$) |
| 70–89 | 443 | 71.1 | 116 | 69.8 | NS ($p = 0.76$) |
| 90–109 | 347 | 77.5 | 56 | 73.2 | NS ($p = 0.44$) |
| 110–129 | 73 | 78.6 | 3 | – | – |

*Survival according to the size of the fish*

Using the fact that PIT tagging had no effect on fish growth, fish survival according to size at tagging was analysed with the following procedure:

- A first step consisted in estimating the 'theoretical' fork length of $B_m$ group in October by using the linear regression ($r^2 = 0.82$) that exists between fish length in May 1996 and October 1995 for the $A_m$ group. The theoretical size distributions in October did not statistically differ (test $\chi^2$, $P = 0.19$) in both groups.
- As a second step, the proportions of fish with PIT tag by theoretical class size of 20 mm at marking in October were compared with these in May. There are no significant differences for all class sizes by *t* test (Table 3).

*Movements in the Oir system*

The tagged fish recapture rates in the La Roche brook after 7 months were calculated for the three years of the study. This was done both for the total population and by size classes at tagging. This recapture rate expressed cumulated effects of natural death and immigration on the capture location at age 0+. Global population recapture rates were 26.0%, 36.2% and 20.4% respectively for May 1994, 1995 and 1996. Although these rates are significantly different (test *t*, $P < 0.05$), the distribution patterns of recapture rates by size class can be compared. As regards the size classes comprised between 60 and 109 mm, distributions did not differed for any of the three years of the study (test $\chi^2$, $P = 0.10$). However, in order to remove inter-annual variations and to take account of the size groups where marked and recaptured numbers of fish were low, recapture rates for cumulative numbers during the study period were calculated (Figure 4a). The same procedure was used for migration rates out of the Oir system (Figure 4a) from trapping recaptures, which were very low (26 trouts over the three years period).

The recapture rates by size classes after one year for the 1995 cohort could also be calculated. Distributions of recapture rates by size classes in the La Roche Brook, in the Oir river and in the trap (Figure 4b) are significantly different ($\chi^2$ test, $P < 0.001$).

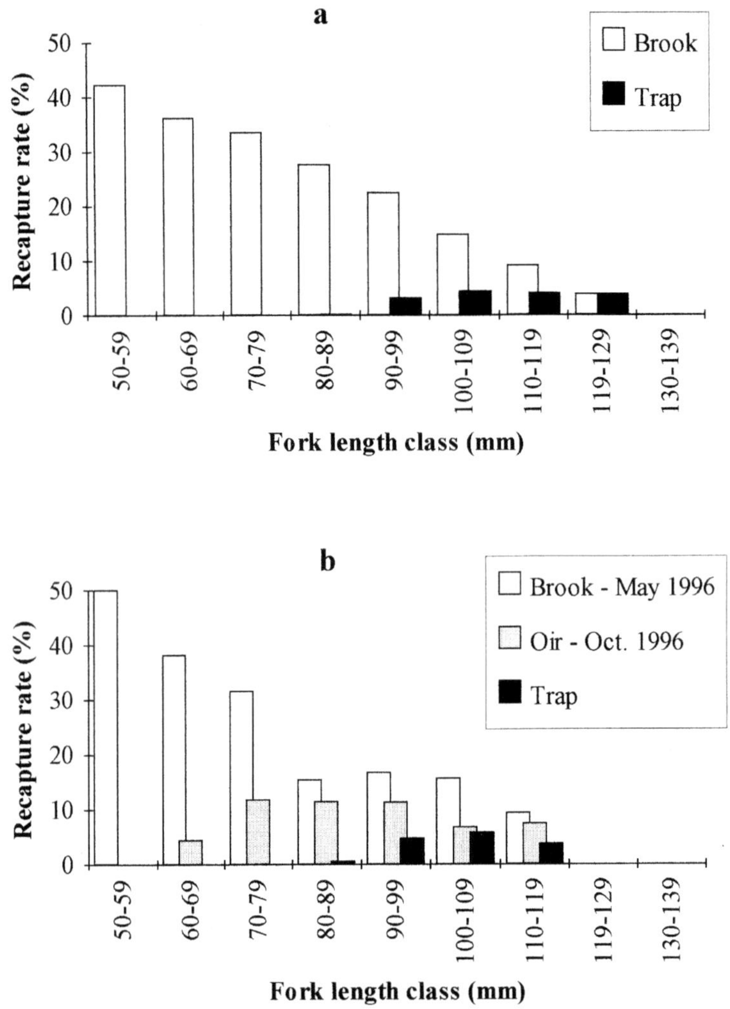


*Figure 4.* a – Mean percentage (1994–1996) of PIT tagged juvenile brown trout recaptured after seven months by electrofishing in the La Roche brook and by trapping. b – Percentage of PIT tagged (in October 1995) juvenile brown trout recaptured in 1996 after seven months by electrofishing in the La Roche brook or by trapping and after one year in the Oir by electrofishing.

The results show that the juvenile attachment for the brook decreases as the fish grow, and that only medium and large-sized trout migrate out of the La Roche Brook. Among these fish leaving the brook, only the largest juveniles migrate from the Oir as smolt or parr. Nevertheless, the overall degree of attachment for the brook vary from one year to another for all size classes studied.

## Discussion

### *Tag loss rate*

The tag loss rate recorded in brown trout after 7 months was low (an average of 3.38%). For several salmonids species tagged at the same size, Prentice et al. (1990) estimated loss rates which were even lower (0 to 2% for 3 to 4 month periods). The whole regeneration rate of the left pelvic fin was also very low (an average of 2.37%). However, Coombs et al. (1990) showed that after 3 months, 46.4% of the pelvic fins in atlantic Salmon were regenerated almost entirely, but the marking was recognisable on careful examination. Our observations confirm the rapid regeneration of this fin but this marking was still recognisable after 7 months.

*Influence of PIT tag on survival and growth*

In order to evaluate the effects of PIT tags in the natural environment, it was necessary to combine them with another external method of marking. This type of procedure had already been used to test the effects of coded wire tag in juveniles Chinook and Coho Salmon, but clipping the adipose fin, the best fin as its regeneration rate is lower than 0.1% (Blankenship, 1990). However, it was not possible for us to use the adipose fin clipping because this method has already been used for juvenile trouts implanted in the river. Therefore, the pelvic fin-clipping whose regeneration rate was higher (Coombs et al., 1990) was chosen.

As far as fish with a pelvic fin clipped were concerned, the comparison between fish with and without a PIT tag showed that this individual type of tag had no great effect on growth after 7 months for juveniles whose fork length was more than 55 mm. Moreover, in comparison with untagged fishes caught, neither the pelvic fin-clipping nor the PIT tag seemed affect this process. According to Prentice et al. (1990), PIT tagging induced a lower growth only during the first 20 days in juvenile chinook Salmon. Afterwards, the size was approximately equal for the control and tagged fishes. They also showed that within the same species, there was no effect of the PIT tag marking on the fish growth, whatever the size of the fish (between 55 and 107 mm) at the tagging time.

In comparison with fish marking only by a pelvic fin-clipping, PIT tagging seemed to have no great impact on mortality, whatever the fish size (55–127 mm) at the time of the tagging. These results agreed with these of Prentice et al. (1990) who showed that the survival of juvenile Chinook Salmon after 4 months in a controlled environment with a natural stream water supply ranged from 95 to 98%. Futhermore, the fork length of the fishes at the tagging time (ranging from 56 to 120 mm), seemed to have no influence on this survival rate.

In other studies in the natural environment, PIT tags and coded wire tags in juvenile Coho Salmon showed no difference in growth and survival after 7 months (Peterson et al., 1994). In fact, our results tended to show that mortality is influenced more by the different handling at the tagging time (capture, anaesthesic, . . .) than by the tag itself. Moreover, survival rates were similar (about 21%) for fish with both types of marking and fish with only the left pelvic fin clipped. Consequently, there was no cumulative effect of the two marking methods, either on mortality or on growth.

*Migration at the level of a catchment area*

The analysis of the results seemed to show that the recruitment into the river of juvenile brown trout from the brook depended on the fish size with only the largest fish migrated out of the system.

Elliot (1994) reported that the territorial behaviour of juvenile trout changes during the first winter, the fish being likely to look for deeper areas which better suit their size. The high sedentarisation rates in the La Roche brook observed for small fish show that they still found habitats adapted to their size in the brook after winter. However, in the Kernec brook which is of the same size, it was demonstrated that there was no relation between the migration rate out of the system and the fish size (Baglinière et al., 1989). This difference might be due to the geographical origin of the migrating fish and also to the habitat conditions. Indeed, in the Kernec brook, juveniles leaving the system mainly came from areas located upstream where the habitat conditions were not favourable to the maintenance of 1+ trouts.

The recapture rates in the Oir river obtained for the 1995 cohort were under-estimated, as only one part of the river was monitored by electrofishing. Nevertheless, the two sets of values could be compared and showed that medium-sized juveniles from the brook (70 to 100 mm at age 0+ in October) had a particularly important role in the recruitment of the river. These first results need to be confirmed by at least another experiment. The migration of the largest fish out of the Oir system confirmed the results obtained in other systems in Brittany and in Lower Normandy on migration-smolting-growth relationships (Ombredane et al., 1996). Theses two observations tended to show that the whole migrations of juvenile brown trout are correlated directly with size: the larger the fish are, the further they migrate downstream.

## Conclusion

PIT tags seem to provide a reliable method for ecological studies of salmonids in the natural environment by taking the life history of each fish into account. These tags, which last more than 10 yr (Prentice et al., 1990) should enable to study the migrations and the development of fish from the juvenile until the reproduction stage, as well as for the sea migratory form of the trout and the interactions between brown and sea trout cohabiting in the same stream. Futhermore, the use of PIT tags combined with antenna continuously recording fish numbers (Armstrong et al., 1996) will allow the collection of reliable data on movement, survival and migration rates at the different life stages of brown trout which are important parameters in population dynamics models. The method could also give information on the reliability of the capture trap in order to evaluate precise migrating juvenile flows.

## Acknowledgements

We would like to thank all the persons and all the organisations who have participated on the site and/or financially in these costly experiments: Conseil Supérieur de la Pêche, Electricité de France, the staff and the students of our ENSAR and INRA laboratories.

## References

Armstrong, J. D., V. A. Braithwaite & P. Rycroft, 1996. A flat-bed passive integrated transponder antenna array for monitoring behavior of atlantic salmon parr and other fish. J. Fish Biol. 48: 539–541.

Bagliniere, J. L., G. Maisse, P. Y. Lebail & A. Nihouarn, 1989. Population dynamics of brown trout, *Salmo trutta* L., in a tributary in Brittany (France): spawning and juveniles. J. Fish Biol. 34: 97–110.

Blankenship, H. L., 1990. Effects of time and fish size on coded wire tag loss from chinook and coho salmon. Am. Fish. Soc. Symp. 7: 237–243.

Brewin, M. K., L. L. Stebbins & J. S. Nelson, 1995. Differential losses of floy anchor tags between male and female brown trout. North Amer. J. Fish. Mgmt. 15: 881–884.

Bryan, R. D. & J. J. Ney, 1994. Visible implant tag retention by and effects on condition of a stream population of brook trout. North Amer. J. Fish. Mgmt 14: 216–219.

Coombs, K. A., J. K. Bailey, C. M. Herbinger & G. W. Friars, 1990. Evaluation of various external markings techniques for atlantic salmon. Am. Fish. Soc. Symp. 7: 142–146.

Elliot, J. M., 1994. Quantitative Ecology and the Brown Trout, Oxford University Press, Oxford (UK), 286 pp.

Laurent, M., P. Lamarque, 1974. Utilisation de la méthode des captures successives (De Lury) pour l'évaluation des peuplements piscicoles. Ann. Hydrobiol 5: 121–132.

McFarlane, G. A., R. S. Wydoski & E. D. Prince, 1990. Historical review of the development of external tags and marks. Am. Fish. Soc. Symp. 7: 9–29.

Ombredane, D., L. Siegler, J.,L. Bagliniere & P. Prunet, 1996. Migration et smoltification des juvéniles de truite (*Salmo trutta*) dans deux cours d'eau de Basse-Normandie. Cybium 20 (3) suppl: 27–42.

Peterson, N. P., E. F. Prentice & T. P. Quinn, 1994. Comparison of sequential coded wire and passive integrated transponder tags for assessing overwinter growth and survival of juvenile coho salmon. North am. J. Fish. Mgmt. 14: 870–873.

Prentice, E. F., T. A. Flagg & C. S. McCutcheon, 1990. Feasibility of using implantable Passive Integrated Transponder (PIT) tags in salmonids. Am. Fish. Soc. Symp. 7: 317–322.

*Hydrobiologia* **371/372:** 107–111, 1998.
J.-P. Lagardère, M.-L. Bégout Anras & G. Claireaux (eds), Advances in Invertebrates and Fish Telemetry.



# Evaluation of surgery procedures for tagging eel *Anguilla anguilla* (L.) with biotelemetry transmitters

E. Baras & D. Jeandrain
*University of Liège, Laboratory of Fish Demography and Aquaculture*
*Mailing address: Chemin de la Justice, 10 B-4500 Tihange, Belgium (e-mail: E.Baras@ulg.ac.be)*

*Key words:* tagging, surgery, cyanoacrilate adhesive, fish, eel, *Anguilla anguilla*

## Abstract

Externally attached telemetry transmitters are unsuitable to tag yellow eels *Anguilla anguilla* (L.), in streams where they exhibit cryptic life habits and hide in narrow cavities between rocks. We evaluated the adequacy of surgical implantation and closing procedures for tagging eels with biotelemetry transmitters. Epoxy dummy transmitters (18 × 8 mm, 1.6–1.7 g) were implanted in eels anaesthetised with 2-phenoxy-ethanol (0.9 ml $l^{-1}$), through a 20 mm mid ventral incision made in the posterior quarter of their body cavity. The incision was either left open, or closed in different ways: stitches (absorbable or non absorbable suture material) or commercial-grade cyanoacrilate adhesive (Loctite$^{TM}$). Fish were stocked in a 4 $m^2$ flow through tank (15–17 °C), controlled daily for mortality and weekly for evaluating the healing process.

No transmitter was expelled over a 12-week period, even in eels with unclosed incisions, of which 50% healed within 28 days ($t_{50}$). Regardless of the nature of the filament, suturing induced skin and muscle necrosis, caused significantly higher mortality rates (60% after 10 weeks) and paradoxically slowed down the healing rate (40 and 45 d, respectively). Cyanoacrilate suppressed the inflammatory response and granted higher survival rate (90%), but did not permit to speed up the closing process ($t_{50} = 52$ d), as eels actively bit and removed the adhesive within hours. This behaviour was suppressed when we applied a freshly cut fragment of the eel dorsal fin as a biological bandage over the drying cyanoacrilate. The adhesive remained in place for one to two days and permitted to substantially increase the healing rate ($t_{50} = 15$ d). These results substantiate the efficiency of surgery techniques for tagging eels with radio transmitters, at least for units of small weight and bulk.

## Introduction

Among the range of techniques available for marking or tagging fish, telemetry tags undoubtedly represent the most powerful tools as they permit to track the movements of the fish, and provide information on its environment or physiology, when appropriate sensors are coupled to the transmitter (Winter, 1983; Priede & Swift, 1992; Baras & Lagardère, 1995; Baras & Philippart, 1996). However, telemetry tags have a substantial size and weight (> 0.8 g and > 1.3 g in the air, for radio and acoustic tags, respectively), due to the electronic circuitry and battery required to power the acoustic or radio signal. These features limit the weight range of fish to be tagged with telemetry tags, and imply a fine tailoring of tagging procedures to avoid any major bias resulting from transmitter attachment or presence on, or inside the fish (syntheses in Summerfelt & Smith, 1990; Baras, 1991; Baras et al., 1998).

These considerations may account for the relatively small number of telemetry studies on eels (*Anguilla* spp., see review in Nielsen, 1988). Externally attached transmitters have been frequently used to tag large eels during their movements in lakes, estuaries, tidal creeks and the first stages of their marine migration on the continental shelf (Tesch, 1974, 1989; Helfman et al., 1983; McGovern & McCarthy, 1992). They are known to modify the hydrodynamism of the fish (Tesch, 1974) and may be unsuitable to study resident life stages in streams, where eels exhibit cryptic life habits and hide in narrow cavities in between rocks. Intragastric transmitters are prone to regurgitation by eels (Stasko & Rommel, 1974) and, when retained, might interfere with the appetite of eels as they modify the degree of stomach fulness. Surgically implanted tags may prove the best technique for long term tagging of yellow eels, although this technique has rarely been used and evaluated. LaBar et al. (1987) tracked European eels *Anguilla anguilla* (L.) with internal tags in a Spanish lake but gave no information on the tagging procedure and effects in the long run. Ernande (1995) noted that acoustic transmitters, with weight in the air *circa* 3% of the body weight of *A. anguilla*, might caused a deformation of the body wall and were systematically expelled within two weeks. This expulsion process may be the consequence of inappropriate incision closing procedures in a species with a narrow body cavity and anguilliform swimming mode, which may both cause outwards pressure and cause the expulsion of the transmitter before the incision has healed.

As a preliminary step to a tracking study on the behavioural ecology of yellow eels *A. anguilla* in Belgian streams, we evaluated the feasibility of tagging eels with radio transmitters implanted into the intraperitoneal cavity, and the relative efficiency of different closing procedures.

## Material and methods

The study was conducted on wild eels, captured in fish passes on the River Meuse. We used epoxy (Epon 812) constructed dummy transmitters that were similar in weight and bulk to commercially available motion-sensitive transmitters (selected model: ATS, Inc. 377, $16 \times 8$ mm in diameter, 1.6–1.7 g in air). This model weighed less than 0.6% of the body weight of the eels in this experiment (285–958 g), and its short length (comparatively to the body length of eels: 562–780 mm) was supposed to impose no major restriction on their swimming behaviour.

Eels were anaesthetised with 2-phenoxy-ethanol (0.9 ml $l^{-1}$), reached anaesthesia stage III.2 (MacFarland & Klontz, 1969) after 6 to 8 minutes then were placed ventral side up into a half cylindrical support, using wet paper to adjust the position of their body into the support. A 18–20 mm long incision was made with a scalpel on the mid-ventral line, in the posterior quarter of the body cavity. The dummy transmitter was inserted cranially and positioned 3 to 4 cm forth of the incision with a plastic plunger, in order to minimise the risk of inside out pressure during the healing process. Two closing procedures were evaluated against a control procedure, where the incision left open, as it is frequently for small implants such as passive integrated transponder (PIT) tags (Prentice et al., 1990).

In two groups of eels (2a and 2b), the incision was closed up with two stitches 6–7 mm apart, using 2 Dec filaments (absorbable plain catgut or non absorbable polyamide monofilament) that were fixed to 16 mm cutting needles. Cutting needles were preferred to atraumatic needles accordingly with the conclusions of Thoreau and Baras (1996) for species with strong body walls, like eels. In group 3a, the incision was closed with a commercial grade cyanoacrilate adhesive (Loctite$^{TM}$). Accordingly with the recommendations of Nemetz and MacMillan (1988), the two edges of the incision were blotted dry and maintained together by exerting a lateral pressure over the body wall of the eel during the application and drying of the adhesive (*ca.* 1 min).

All tagged eels were stocked together with a group of 15 control fish in the same flow-through tank ($4.0 \times 1.0 \times 0.5$ m$^3$), in the Fish Research Station of the University of Liège, in Tihange. In order to mimic the spring thermal regime of Belgian waters, the water temperature in the tank was maintained at $16 \pm 1$ °C throughout the experiment, by mixing waters from the River Meuse and from the effluent of the Tihange nuclear power plants. Dry food pellets were distributed by an automatic feeder. The tank was checked daily for fish survival. Eels were checked weekly for transmitter retention and for evaluating the healing process. It was initially programmed to analyse the growth of eels depending on the tagging procedure, but it rapidly turned out that all eels, including those of the control group, ate very little in captivity at $16 \pm 1$ °C and were gaining no weight. Transmitter retention was checked by a gentle pressure over the abdomen of the fish and empirically by search for lost transmitters on the bottom of the emptied tank. The healing stage was evaluated by testing, with a metal probe, which layers of the body wall had begun tissue reconstruction. The incision was considered as healed when the dermis and epidermis had started closing up over the whole length of the incision.

Transmitter retention rates and fish survival rates were compared by contingency table analyses. Healing rates were compared by Mann-Whitney *U*-tests. Null hypotheses were rejected at the 0.05 level of significance.

## Results

No transmitter was expelled over a 12 week period, even in the group where the incision was not closed. Regardless of the nature of the filament, suturing caused deep cuts into the body wall of most eels. From the second week onwards, most cuts degenerated into necrotic tissue, and infections were observed at these sites. The absorption of catgut filaments during the third week, or the removal of the permanent polyamide filaments after two weeks, limited the extents of the cuts and necrotic tissue, but about 50% of the sutured eels died within the five first weeks. Surviving eels healed their incision within 6–7 weeks (Table 1).

Eels with incision left open healed much faster and their survival rate was similar as in the control group. Inflammatory responses were observed on the edges of the incision but vanished within the first three weeks. The use of cyanoacrilate apparently suppressed the inflammatory response at the incision site and slightly further improved the survival rate (90% after ten weeks). However, it did not speed up the closing process, as eels required an average time of 52 days to heal their incision. Additional observations in aquarium on the behaviour of tagged eels closed up with cyanoacrilate revealed that the adhesive was shed within hours, not as a consequence of tensions caused by swimming, but because eels were actively removing it by biting. In order to possibly modify this behaviour, we refined the cyanoacrilate closing procedure by applying a freshly cut fragment of the eel's dorsal fin over the drying cyanoacrilate (group 3b, Table 1). Presumably due to the presence of this biological bandage over the incision, the eels did not actively remove the adhesive, which remained in place one or two days, then was shed. This procedure permitted to reduce the healing time to 15 days. Only one eel from group 3b died, after its incision had healed. The dissection of tagged eels three to four months after tagging indicated that most (90%) implants had moved posterior to the incision but had not caused internal damage.

## Discussion

The results indicated that surgical implantation was an adequate procedure for tagging eels with biotelemetry tags, at least when adequate closing procedures were used. Eels with unclosed incisions had a high survival, maximum retention rate and healed their incision within a month. These observations may somehow question the relevance or necessity of closing procedures, especially when dealing with short incisions comparatively to the body length of the animal. Additional experiments with PIT tags surgically implanted into the body cavity of eels also provided evidence that the non closing procedure was adequate, as 5–6 mm incisions completely healed within 17 days at $16\pm1$ °C. Although this procedure would undoubtedly be initially challenged by those responsible for enforcing animal welfare legislation, this study indicated that it granted far better survival rates than with suturing, which is the most popular and recommended technique for closing abdominal incisions in fish (e.g. Hart & Summerfelt, 1975; Baras et al., 1998). This finding is consistent with studies on liver biopsy in channel catfish *Ictalurus punctatus* Rafinesque (Carmichael, 1991), on abdominal implants of radio tags in the European barbel *Barbus barbus (L.)*, (Baras, 1992) and of PIT tags in juvenile fishes (salmonids: Prentice et al., 1990; tilapia *Oreochromis niloticus* (L.): Baras et al., 1996), which proposed that closing incisions with stitches did not substantially speed up the healing process. In this study, it appeared that suturing induced skin and muscle necrosis, caused significantly higher mortality rates and slowed down the healing process comparatively to unclosed fish, presumably as a result of general low condition. This effect of stitches on the recovery of eels is a further example to the potential detrimental effect of transcutaneous foreign bodies that have been reported by several authors (Roberts et al., 1973; Marty & Summerfelt, 1990; Baras, 1992; Knights & Lasee, 1996), and which presumably reaches its climax in fish species of which the skin is not covered by scales.

In comparison to the above procedures, the use of cyanoacrilate in eels suppressed the inflammatory response at the incision site, as observed in other fish species (Nemetz & MacMillan, 1988; Petering & Johnson, 1991). However, the healing rate of eels closed with cyanoacrilate was about twice as long as in unclosed eels, presumably because the removal of the adhesive by biting, within the hours following the surgery, caused additional lesions in the incision

*Table 1.* Survival and healing of surgically implanted eels, depending on the closing procedure. $t_{50}$ is the time of complete healing of 50% of the surviving fish. Values sharing at least one common upperscript label do not differ at the 0.05 level of significance (contingency table analyses for survival rates; Mann-Whitney *U* tests for healing)

| Group | *N* | Closing procedure | Survival (%) after *n* days | | | | Healing (days) |
|---|---|---|---|---|---|---|---|
| | | | 7 | 28 | 49 | 70 | $t_{50}$ (range) |
| Control | 15 | – | $100^{a}$ | $100^{a}$ | $93^{a}$ | $87^{a}$ | – |
| 1 | 10 | none | $100^{a}$ | $80^{a,b}$ | $80^{a,b}$ | $80^{a,b}$ | 28 (18–35)$^{a}$ |
| 2a | 10 | suture, catgut | $100^{a}$ | $60^{b}$ | $50^{b}$ | $40^{b}$ | 40 (24–46)$^{a,b}$ |
| 2b | 15 | suture, polyamide | $73^{b}$ | $53^{b}$ | $47^{b}$ | $40^{b}$ | 45 (36–48)$^{b}$ |
| 3a | 10 | cyanoacrilate | $100^{a}$ | $90^{a}$ | $90^{a}$ | $90^{a}$ | 52 (16–70)$^{a,b}$ |
| 3b | 10 | cyanoacrilate + bandage | $100^{a}$ | $100^{a}$ | $90^{a}$ | $90^{a}$ | 15 (14–28)$^{c}$ |

zone. The addition of a biological bandage caused the adhesive to remain in place for a few days, which was a sufficient delay to permit the adjustment of the edges of the incision and to speed up the first steps of the closing process. This 'bandage' technique could be worth testing in other fish species, especially in species without scales, where suturing could prove detrimental.

No transmitter expulsion was observed in this study, even in eels of which the incision had not been closed at all. In a recent field study (Baras et al., 1998), all radio-tagged eels released in a small stream with rock substrate survived and retained their 1.6 g transmitter until the end of the battery life. This study further documented that tagged eels could clear small weirs within their home range, suggesting that the presence of the tag did not cause any major impairment of their mobility or swimming capacities. These observations suggest that incision exits observed in past studies (Ernande, 1995) originated from the use of tags with excessive weight or bulk (transmitter to body weight ratios in the air from 2.4 to 3.0% in Ernande's study). A ratio of less than 0.6% obviously limits the risks of expulsion and permits to study the behaviour of eels in all types of ecosystems and environments. However, it reduces the weight range of eels which could be tagged with intraperitoneal transmitters (over 130 or 215 g, for radio and acoustic tags, respectively). The small battery size further reduces the reception range and/or tag endurance, and and imposes additional restrictions for studies using internally tagged eels. Endurance could be substantially improved by using transmitters operating on duty cycles, i.e. from dusk to dawn as eels essentially are nocturnal fish (LaBar et al., 1987; McGovern & McCarthy, 1992; Baras et al., 1998). A possible alternative to programmable transmitters would consist in adapting the shape of the transmitter to the narrow body cavity and swimming of eels, e.g. by designing a flexible long transmitter with narrow diameter, of which the batteries and electronic circuitry would be linked by flexible connectors.

## Acknowledgements

This work is part of the FIRST research programme granted to E.B. by the Ministry of New Technologies (DGTRE) of the Walloon Government.

## References

Baras, E., 1991. A bibliography on underwater telemetry. Can. Rep. Fish. aquat. Sci. 1819: 1–55.

Baras, E., 1992. Etude des stratégies d'occupation du temps et de l'espace chez le barbeau fluviatile, *Barbus barbus* (L.). Cah. Ethol. 12: 125–442.

Baras, E. & J. P. Lagardère, 1995. Fish telemetry in aquaculture: review and perspectives. Aquacult. Int. 3: 1–26.

Baras, E. & J. C. Philippart (eds), 1996. Underwater biotelemetry. University of Liège, Belgium: vi + 257 pp.

Baras, E., C. Birtles, L. Westerloppe, X. Thoreau, M. Ovidio, D. Jeandrain & J. C. Philippart, 1998. A critical review of surgery techniques for implanting telemetry devices into the body cavity of fish. In Y. Le Maho (ed.), Proceedings of the 5th European Conference on Wildlife Telemetry. Strasbourg, France, 26–30 August 1996: 10 pp, in press.

Baras, E., D. Jeandrain, B. Serouge & J. C. Philippart, 1998. Seasonal variations in time and space utilization by radio-tagged yellow eels *Anguilla anguilla* (L.) in a small stream. Hydrobiologia 371/372: 187–198.

Baras, E., L. Westerloppe & J. C. Philippart, 1996. Adéquation de procédures d'implantation chirurgicale pour le PIT tagging de juvéniles du tilapia *Oreochromis niloticus*. Rapports de la Station Aquacole de Tihange 96: 1–14.

Carmichael, G. J., 1991. Recovery of channel catfish from abdominal surgery. Progress. Fish Cult. 53: 193–195.

Ernande, B., 1995. Etude du rythme d'activité de l'anguille européenne *Anguilla anguilla* L. en marais maritime par télémétrie acoustique. M.S. Thesis, CNRS Ifremer, 41 pp. + ann. CREMA l'Houmeau, Place du Séminaire, BP 5, F-17137 L'Houmeau, France.

Hart, L. G. & R. C. Summerfelt, 1975. Surgical procedures for implanting ultrasonic transmitters into flathead catfish (*Pylodictis olivaris*). Trans. am. Fish. Soc. 104: 56–59.

Helfman, G. S., D. L. Stoneburner, E. L. Boseman, P. A. Christian & R. Whalen, 1983. Ultrasonic telemetry of American eel movements in a tidal creek. Trans. am. Fish. Soc. 112: 105–110.

Knights, B. C. & B. A. Lasee, 1996. Effects of implanted transmitters on adult bluegills at two temperatures. Trans. am. Fish. Soc. 125: 440–449.

LaBar, G. W., J. A. Hernando Casal & C. Fernandez Delgado, 1987. Local movements and population size of European eels, *Anguilla anguilla*, in a small lake in south-western Spain. Envir. Biol. Fishes 19: 111–117.

MacFarland, W. N. & G. W. Klontz, 1969. Anesthesia in fishes. Federal Proceedings 28: 1535–1540.

Marty, G. D. & R. C. Summerfelt, 1990. Wound healing in channel catfish by epithelialization and contraction of granulation tissue. Trans. am. Fish. Soc. 119: 145–150.

McGovern, P. & T. K. McCarthy, 1992. Local movements of freshwater eels (*Anguilla anguilla* L.) in western Ireland. In I. G. Priede & S. M. Swift (eds), Wildlife Telemetry: Remote Sensing and Monitoring of Animals. Ellis Horwood, Chichester, UK: 319–327.

Nemetz, T. G. & J. R. MacMillan, 1988. Wound healing of incisions closed with a cyanoacrylate adhesive. Trans. am. Fish. Soc. 117: 190–195.

Nielsen, J., 1988. Marking and tagging methods applied to eel, *Anguilla anguilla* (L.). EIFAC FAO occasional paper 21.

Petering, R. W. & D. L. Johnson, 1991. Suitability of cyanoacrilate adhesive to close incisions in black crappies used in telemetry studies. Trans. am. Fish. Soc. 120: 535–537.

Prentice E. F., T. A. Flagg & R. S. McCutcheon, 1990. Feasibility of using implantable passive integrated transponder (PIT) tags in salmonids. Am. Fish. Soc. Symp. 7: 317–322.

Priede, I. G. & S. M. Swift (eds), 1992. Wildlife Telemetry: Remote Sensing and Monitoring of Animals. Ellis Horwood Ltd., Chichester, UK, 708 pp.

Roberts, R. J., A. MacQueen, W. M. Shearer & H. Young, 1973. The histopathology of salmon tagging. I. The tagging lesion in newly tagged parr. J. Fish Biol. 5: 497–503.

Stasko, A. B. & S. A. Rommel Jr., 1974. Swimming depth of adult American eels (*Anguilla rostrata*) in a saltwater bay as determined by ultrasonic tracking. J. Fish. Res. Bd Can. 21: 1148–1150.

Summerfelt, R. C. & L. S. Smith, 1990. Anesthaesia, surgery, and related techniques. In C. B. Schreck & P. B. Moyle (eds), Methods for Fish Biology. Am. Fish. Soc., Bethesda, Maryland: 213–272.

Tesch, F. W., 1974. Speed and direction of silver and yellow eels, *Anguilla anguilla*, released and tracked in the open North Sea. Ber. dt. wiss Kommn. Meereforsch. 23: 181–197.

Tesch, F. W., 1989. Changes in swimming depth and direction of silver eels (*Anguilla anguilla* L.) from the continental shelf to the deep sea. Aquat. Living Resour. 2: 9–20.

Thoreau, X. & E. Baras, 1996. Anaesthesia and surgery procedures for implanting telemetry transmitters into the body cavity of tilapias *Oreochromis aureus*. In E. Baras & J. C. Philippart (eds), Underwater Biotelemetry. University of Liège, Belgium: 13–22.

Winter, J. D., 1983. Underwater biotelemetry. In L. A. Nielsen & J. D. Johnsen (eds), Fisheries Techniques. Am. Fish. Soc., Bethesda, Maryland: 371–395.

*Hydrobiologia* **371/372:** 113–116, 1998.
*J.-P. Lagardère, M.-L. Bégout Anras & G. Claireaux (eds), Advances in Invertebrates and Fish Telemetry.*



# A method for the external attachment of acoustic tags on roundfish

Guy Claireaux & Christel Lefrançois
*Centre de Recherche en Écologie Marine et Aquaculture Place du Séminaire, B.P. 5, F-17137 L'Houmeau, France (e-mail: gclairea@ifremer.fr)*

*Key words:* fish, tagging, telemetry, heart rate

## Abstract

The present paper proposes a new method for the external attachment of an acoustic tag on roundfish. The tag is placed in anterio-ventral position and attached with two wires passed around the pelvic girdle. This tagging method was tested on Atlantic cod and European sea bass using acoustic heart rate transmitters. Preliminary results concerning metabolic demand (oxygen consumption), heart rate and behaviour of free swimming individuals are presented and discussed.

## Introduction

The development of acoustic telemetry techniques from simple tracking devices to more sophisticated, interactive and energy demanding systems has often been synonymous with increased load to be carried by animals. This evolution has also resulted to greater difficulties in the implementation of the most widely used tag-implantation techniques, i.e. in the stomach or the coelom. External attachment had to be considered and various techniques were proposed and experimented (for review see Barras & Lagardère 1995). Carr & Chaney (1977) attached ultrasonic transmitters to the caudal peduncle of red drum (*Sciaenops ocellatus* L) while Holland et al. (1985) used nylon pull-ties inserted through the dorsal musculature of tuna. Mellas & Haynes (1985), using rainbow trout (*Oncorhynchus mykiss* W) and white perch (*Morone americana* G), and Matthews et al. (1990) using rockfish (*Sebastes spp.*), tested an attachment method which consisted in securing the transmitter onto a side-saddle harness. The study of heart rate requires that the tag be placed as close as possible to the heart in order to reduce the probability of rupture of the wires connecting the ECG electrodes to the transmitter. Priede & Young (1977) working on brown trout (*Salmo trutta* L), Armstrong et al. (1989) studying pike (*Esox lucius* L) and Sureau & Lagardère (1991) using European sole (*Solea solea* L) and European sea bass (*Dicentrarchus labrax* L) sutured heart rate transmitters dorsaly at the vertical of the heart. However, if this methods was adequate when using relatively small acoustic tags, its application with larger tags may be difficult for reasons related to fish's balance or anatomy. The objective of the present study was to examine an alternative method consisting in attaching a transmitter in anterio-ventral position and closer to the fish's centre of gravity. Preliminary results concerning the energetic expenditure, the cardiac activity and the behaviour of tagged European sea bass and Atlantic cod (*Gadus morhua* L) are presented.

## Surgical procedure

Fish (Atlantic cod and European sea bass) were anaesthetised (0.2 ml $l^{-1}$ of 2-phenoxyethanol or 100 mg $l^{-1}$ of ethyl-m-aminobenzoate) and placed belly up on an operating table. During the tagging procedure, gills were continuously perfused with oxygenated water containing a small dose of anaesthetic. Using a curved surgery needle (Arkansas n°6, triangular section), two attachment wires were inserted through the skin posterior to the pelvic fins and passed around the pelvic girdle (Figure 1). In some instances, a third wires was passed through the Maeckel cartilage between the

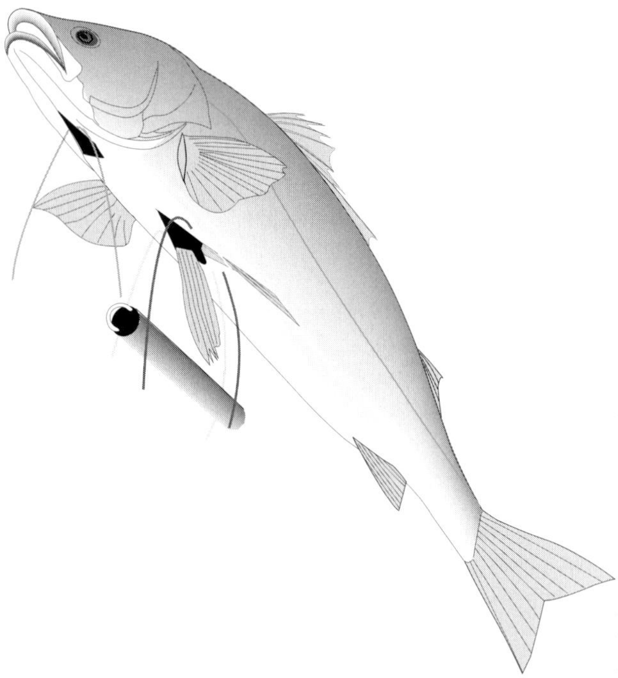

*Figure 1.* Technique for attaching an acoustic tag in anterio-ventral position.

opercula. These wires were made of a stainless steel core sheathed with kevlar (tensile strength 4.5 kg). The transmitter was then placed between the pelvic fins and secured. A drop of cyanoacrylic glue was put on the knots and a penicillin/streptomycin mixture was applied on the skin around the punctures. The total operation was performed in less than 15 min. Fish revived within 5 min after being ventilated with anaesthetic free water and they were then returned to their acclimation tank.

## Preliminary validation of the method

### *Protocol*

Four Atlantic cod and five sea bass weighing 0.9 to 2.4 kg, and acclimated to the experimental temperature (bass: 25 °C; cod: 5 °C), were tagged with a heart rate transmitter (Vemco V16H) according to the method described above. The tag length and diameter were 10 cm and 1.6 cm respectively and it weighed 37.5 g in air (i.e. 1.7 to 4.2% of the fish's body weight). ECG electrodes (stainless steel wires insulated to within 1 mm of the tip) were then inserted through the skin and muscles near the pericardial cavity and sutured. Figure 2 is an underwater photograph of a free swimming Atlantic cod carrying a heart rate transmitter. After a 24 h post-surgery recovery period fish were transferred to a 240 l respirometer. A complete description of the experimental set-up is given in Lefrançois et al. (1998). Following another 3–4 day acclimation period, the pump controlling the water supply to the respirometer was shut off at regular time intervals and the water oxygenation level was recorded for the next 15 minutes using a polarographic oxygen electrode connected to an oxygen meter (Orbisphere laboratories). Fish oxygen consumption being equal to the rate of decrease of the water oxygen content, routine metabolic rate (RMR in $mgO_2\ kg^{-1}\ h^{-1}$) was calculated using the following formula:

$$\mathrm{RMR} = \Delta C_\mathrm{w}\mathrm{O}_2 \times \Delta T^{-1} \times \mathrm{VOL}_\mathrm{resp} \times m^{-1}$$

where $\Delta C_w O_2$ is the change in water oxygen concentration ($mgO_2\ l^{-1}$), $\Delta T$ is the measuring time (h), $VOL_{resp}$ is the volume of the respirometer minus the volume of the fish (l), and $m$ is the mass of the fish (kg).

While in the respirometer, fish's heart rate was continuously monitored by recording the pulse rate emitted by the transmitter. The acoustic pulses were detected by an hydrophone (Vemco VH65) connected to a receiver (Vemco VR20) interfaced with a computer. Cardiac frequency was calculated from the time period between two consecutive pulses.

The impact of tagging on the behaviour of fish was roughly estimated during feeding periods through the visual observation of tagged sea bass in a group of five untagged animals.

## Results and discussion

The behaviour of animals was not altered by the tagging procedure and tagged individuals from both species resumed feeding within 24 h after surgery. Although fish carried the tag in ventral position, we observed no obvious changes in their ability to grab food spread at the bottom of the experimental tank. Buoyancy was also not noticeably modified as tagged individuals were able to remain motionless and steady in the water column. During feeding periods, tagged dominants were still aggressive and continued to chase subordinates around the tank, suggesting that their social rank was not altered.

In the present study, as fish were swimming freely in the respirometer we were unable to determine the

*Figure 2.* Underwater photograph of an Atlantic cod (1.8 kg) tagged with an heart rate acoustic transmitter.

proportion of the total oxygen demand attributable to the additional load represented by the transmitter. However, based on data reported by Schurmann & Steffensen (1997) on Atlantic cod and by Claireaux & Lagardère (submitted) on sea bass, we compared the routine metabolic rates of tagged fish to the standard metabolic rate (SMR) of untagged, unfed and resting animals (Table 1). The mean RMR of free swimming and tagged 25 °C-acclimated sea bass only reached 1.4 × SMR, while it represented 1.8 × SMR in 5 °C-acclimated cod. For comparison, it has been shown that the post-prandial oxygen consumption (apparent specific dynamic action) alone could reach up to 5 times SMR in juvenile Atlantic cod fed *ad libitum* (Soofiani & Hawkins, 1982). In bass the full metabolic scope corresponds to 3.8 × SMR at 25 °C (Claireaux & Lagardère, submitted). Whilst the impact of transmitter size on the hydrodynamics of externally-tagged fish cannot be disputed, these low levels of routine metabolic rate in tagged fish calls for a careful revaluation of the energetic significance of the 'tag-weight < 2% body-weight' criterion (Winter, 1983). In this regard it could be added that many top carnivores species can ingest up to 10% their body-weight in one meal (Beamish, 1974; Jobling, 1981; Lucas et al., 1991; Lucas & Priede, 1992; Dos Santos & Jobling, 1995). The low energetic impact of tag carrying is also corroborated by the analysis of the cardiac activity of these animals. When oxygen demand of sea bass or Atlantic cod is experimentally raised to approach the animal's maximum aerobic metabolic capacity, heart rate also increases tending towards a temperature-dependent maximum value. This uppermost heart beat frequency has been estimated at 100–105 beats $min^{-1}$ in 25 °C-acclimated sea bass (Lefrançois, Claireaux & Lagardère unpublished results) and at 35–40 beats $min^{-1}$ in 5 °C-acclimated Atlantic cod (Claireaux, Webber, Lagardère & Kerr, unpublished results). In both species these beating frequencies are well above the values observed in the present study 24 h post-surgery (Figure 3).

*Table 1.* Mean routine metabolic rate (RMR, $mgO_2\ h^{-1}\ kg^{-1}$) and ratio RMR/SMR in cod and sea bass: a: from Schurmann & Steffensen (1997) and corrected to a 1 kg body weight exponent of 0.82. b: from Claireaux & Lagardère (in press).

| | Atlantic cod | European sea bass |
|---|---|---|
| Tagged fish: | | |
| RMR ± sem | 43.1 ± 4.7 | 131.2 ± 3.3 |
| (*n*) | (4) | (5) |
| Untagged fish: | | |
| SMR ± sem | 24.4 ± 4.8[a] | 91.2 ± 5.1[b] |
| (*n*) | (10) | (16) |
| RMR/SMR | 1.8 | 1.4 |
| Temperature | 5 °C | 25 °C |

After carrying the tag for 3 weeks (the battery life span), fish were inspected and signs of abrasion were observed. These injuries were successfully reduced by coating the tag with medical-graded liquid silicone rubber (Dow Corning; Silastic Q7-4840) or by using a thin sheet of neoprene between the fish and the tag as recommended by Matthews et al. (1990).

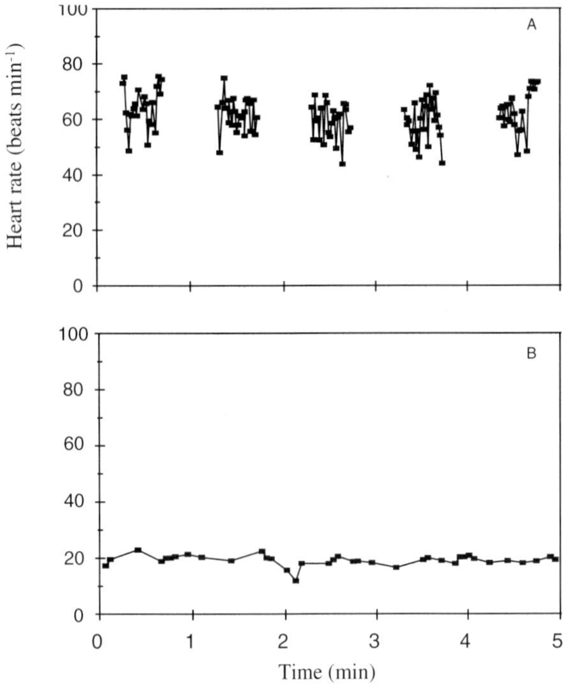


*Figure 3.* Examples of heart rate recording in unrestrained sea bass at 25 °C (A) and Atlantic cod at 5 °C (B).

## Conclusion

The tagging method tested was found satisfactory. No obvious alteration in the animal's behaviour or energy demand was observed, even if the ratio tag weight/fish weight reached in some instances 4.2%. The procedure described was easy to implement and adequate for behavioural or physiological studies. The impact of tagging on the swimming abilities of the two species tested was not evaluated, but it may be presumed that drag effect will increase with increasing swimming velocities. Thus, like other external attachment techniques (Mellas & Haynes, 1985) the proposed method should be applied only if fish swimming speed and endurance are not critical, and if entanglement-related risks are minimised. In the present study, fish only carried the transmitter during a relatively short period of time and an extensive validation of the technique is still required particularly with regard to the mid and long term effects of the tag

## Acknowledgments

Authors are indebted to M. Guillaut and M. Prineau for their skilful technical help and to Y. Descatoire for his patience while finalising the graphic illustrations.

## References


Armstrong, J. D., M. C. Lucas, I. G. Priede & L. De Vera, 1989. An acoustic telemetry system for monitoring the heart rate of pike, *Esox lucius*, and other fish in their natural environment. J. exp. Biol. 143: 549–552.

Baras, E. & J. P. Lagardère, 1995. Fish telemetry in aquaculture: review and perspectives. Aquat. Int. 3: 77–102.

Beamish, F. W. H., 1974. Apparent specific dynamic action of large-mouth bass, *Micropterus salmoides*. J. Fish. Res. Bd Can. 31: 1763–176.

Carr, W. E. S. & T. B. Chaney, 1977. Harness for an attachment of an ultrasonic transmitter to the red drum, *Sciaenops ocellata*. U.S. National Marine Fisheries Service, Fish. bull. 74: 998–1000.

Claireaux, G. & J. P. Lagardère. Influence of temperature, oxygen and salinity on the metabolism of the European sea bass (*Dicentrarchus labrax*). J. Fish. Biol., submitted.

Dos Santos, J. & M. Jobling, 1995. Test of a food consumption model for the Atlantic cod. ICES J. Mar. Sci. 52: 209–219.

Holland, K., R. Brill, S. Fergusson, R. Chang & R. Yost, 1985. A small vessel technique for tracking pelagic fish. U.S. Nat. Mar. Fish. Serv. Mar. Fish. rev. 47: 26–32.

Jobling, M., 1981. The influence of feeding on the metabolic rate of fishes: a short review. J. Fish Biol. 18: 385–400.

Lefrançois, C., G. Claireaux & J.-P. Lagardère, 1998. Heart rate telemetry to study environmental influences on fish metabolic expenditure. Hydrobiologia 371/372: 215-224.

Lucas, M. C. & I. G. Priede, 1992. Utilization of metabolic scope in relation to feeding and activity by individual and grouped zebrafish, *Brachydanio rerio*. J. Fish Biol. 41: 175–190.

Lucas, M. C., I. G. Priede, J. D. Armstrong, A. N. Z. Gindy & L. De Vera, 1991. Direct measurements of metabolism, activity and feeding of pike, *Esox lucius* L., in the wild, by the use of heart rate telemetry. J. Fish Biol. 39: 325–345.

Matthews, K. R., T. P. Quinn & B. S. Miller, 1990. Use of ultrasonic transmitters to track demersal rockfish movement on shallow rocky reefs. Am. Fish. Soc. Symp. 7: 375–379.

Mellas, E. J. & J. M. Haynes, 1985. Swimming performance and behavior of rainbow trout (*Salmo gairdneri*) and white perch (*Morone americana*): effect of attaching telemetry transmitters. Can. J. Fish. aquat. Sci. 42: 488–493.

Priede, I. G. & A. H. Young, 1977. The ultrasonic telemetry of cardiac rhythms of wild brown trout (*Salmo trutta* L.) as an indicator of bioenergetics and behaviour. J. Fish Biol. 10: 299–318.

Schurmann, H. & J. F. Steffensen, 1997. Effects of temperature, hypoxia and activity on the metabolism of juvenile Atlantic cod. J. Fish Biol. 50: 1166–1180.

Soofiani, N. M. & A. D. Hawkins, 1982. Energetic costs at different levels of feeding in the juvenile cod, *Gadus morhua*. J. Fish Biol. 21: 577–592.

Sureau, D. & J. P. Lagardère, 1991. Coupling of heart rate and locomotor activity in sole *Solea solea* (L.), and bass, *Dicentrarchus labrax* (L.), in their natural environment by using ultrasonic telemetry. J. Fish Biol. 38: 399–405.

Winter, J. D., 1983. Underwater biotelemetry. In Fisheries techniques. L. A. Nielsen & D. L. Johnsen (eds). Am. Fish. Soc., Bethesda: 371–395.

*Hydrobiologia* **371/372:** 117–121, 1998.
*J.-P. Lagardère, M.-L. Bégout Anras & G. Claireaux (eds), Advances in Invertebrates and Fish Telemetry.*



# Regurgitation rates of intragastric transmitters by adult Atlantic salmon (*Salmo salar* L.) during riverine migration.

G. W. Smith[1], R. N. B. Campbell[2] & J. S. MacLaine[2]
[1]*Freshwater Fisheries Laboratory Field Station, 16 River Street, Montrose DD10 8DL, Scotland*
*(e-mail: g.w.smith@marlab.ac.uk)*
[2]*The Tweed Foundation, Drygrange Steading, Melrose TD6 9DJ, Scotland*

*Key words: Salmo salar*, radio telemetry, intragastric tagging, regurgitation

## Abstract

Regurgitation rates of tagged salmon in the wild are difficult to estimate as few fish are recovered and thus the fate of the majority of tags is difficult to assess unambiguously. During a study on the River Tweed between 1994 and 1996, 281 returning adult salmon were caught by net and coble, and radio tagged. Salmon recaptured by rod and line were used as a sample of the radio-tagged population from which the frequency of tag regurgitation was determined. A total of 27 tagged fish were recaptured by anglers over the 3 year study period, of which 4 (14.8% ± 1.2% standard error) did not have radio tags in their stomach. Regurgitation rates were consistent between years; 16.7% in 1994 (2 out of 12), 12.5% in 1995 (1 out of 8) and 14.3% in 1996 (1 out of 7).

Regurgitation was not associated with the immediate aftermath of tagging. However, too few salmon were recaptured in the present study to identify other factors which may be important in determining transmitter regurgitation rates in radio tracking studies. No statistically significant relationship could be demonstrated between the lengths of the fish regurgitating their tags and those of the fish which retained them (Mann Whitney test: $N = 27$, $U = 33.5$, $P = 0.392$). Fish which regurgitated tags were tagged over a wide range of months (between April and September) and the subsequent positions of regurgitated tags were widely scattered within the river system.

## Introduction

In recent years, fish tracking techniques have been successfully used to study a number of aspects of the riverine migration of returning adult Atlantic salmon (*Salmo salar* L.). These include estuarine migration (Potter, 1988; Clarke et al., 1991; Smith, 1992) and river entry (Smith et al., 1994), riverine migration patterns (Hawkins and Smith, 1986; Webb, 1989; Milner, 1990; Laughton, 1991), salmon movements in relation to dams (Webb, 1990) and aspects of spawning behaviour (Webb & Hawkins, 1989; Laughton & Smith, 1992). A fundamental assumption underlying all tracking studies is that the behaviour of tagged individuals is not compromised by fish capture and handling nor by the transmitter attachment techniques. Although the precise effects of different attachment techniques may vary between species (Mellas & Haynes, 1985), intragastric tagging of adult salmon has generally been adopted as the standard method (Hawkins & Urquhart, 1983; Varallo, 1988) since it requires no invasive surgery and relatively little handling and recovery time.

However, intragastrically-tagged fish are free to regurgitate the transmitter (Mellas & Haynes, 1985), and fish which are actively feeding in the wild may regurgitate their tags relatively soon after tagging (Glass et al., 1992). Although adult salmon generally do not feed in fresh water, they are also capable of regurgitating their tags (McLeave et al., 1978). The retention of tags by actively feeding gadoid fish has been the subject of laboratory study (Lucas & Johnstone, 1990; Armstrong et al., 1992), but regurgitation rates of intragastrically-tagged salmon in rivers have not been

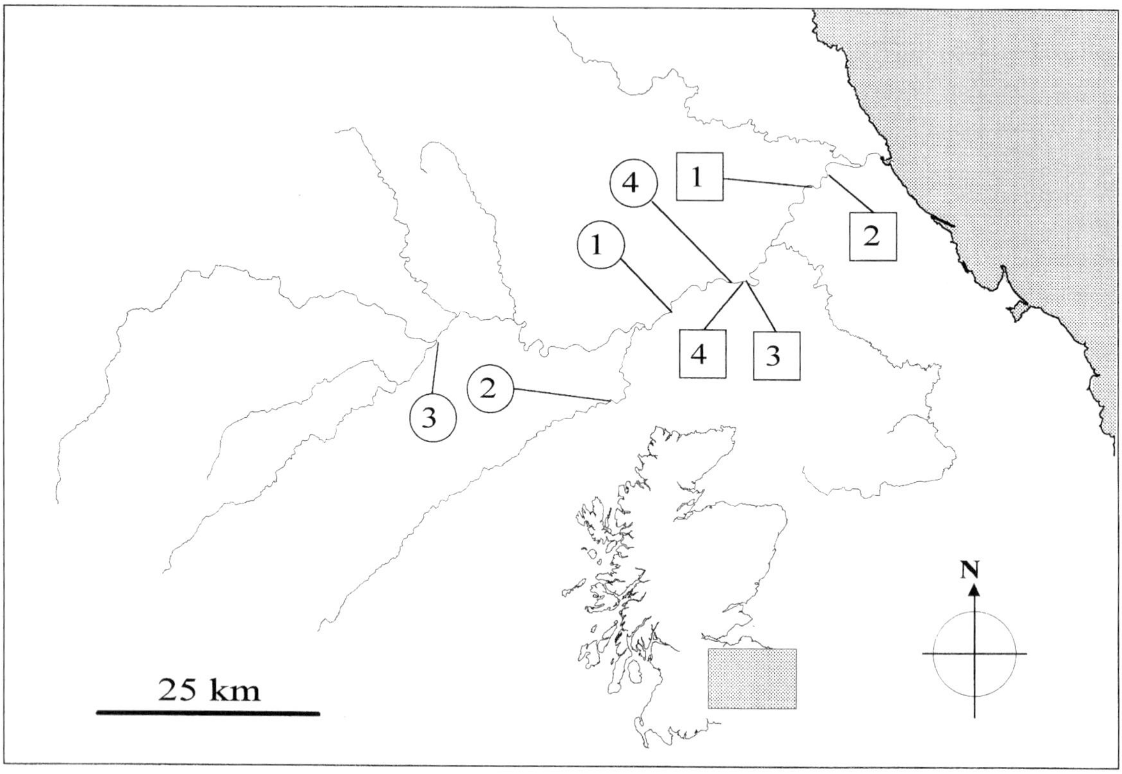


*Figure 1.* Map of the Tweed river system showing where each fish found to have regurgitated its tag was caught (circle) together with the position of the regurgitated tag (square). The labels 1 to 4 are used to indicate the pairs of positions for each of the 4 salmon which regurgitated tags. The location of the Tweed is also indicated on a map of Scotland.

examined. The present study aims to estimate the regurgitation rates of returning adult salmon in the wild by considering tagged salmon recaptured by anglers on the River Tweed.

## Materials and methods

As part of a large scale study on the spawning distributions of salmon in the River Tweed (Smith, 1994), returning adult Atlantic salmon were caught by a commercial net and coble fishery (Shearer, 1992) in the lower 2 km of the river (Figure 1). Fresh-run fish were removed from the net before it was beached and were then anaesthetised in approximately 1:10,000 solution of MS 222 (Sandoz). Fork-length was measured, a scale sample taken for later age determination and an individually numbered Floy tag inserted at the base of the dorsal fin. A transmitting tag was gently pushed into the salmon's stomach via a perspex tube placed in the mouth, and the fish was allowed to recover for up to 10 minutes before being returned to the estuary close to the point of capture. The radio tags used in the present study were developed by the MAFF Directorate of Fisheries Research (DFR), Lowestoft. They are cylindrical in shape, weigh approximately 8 g (between 0.1 and 0.4% of the estimated body weight of the study salmon) in air and 3 g in water and are 47 mm long with a diameter of 15 mm. Individual tags were identified by their unique combination of the radio frequency of the tag and an associated pulse repetition rate.

Posters were distributed throughout the Tweed catchment publicising the rewards which were available on return of both the floy tag (£ 3) and the radio tag (£ 15) from captured study fish. In each instance when floy tags were returned without radio tags, the location of the radio tag within the river system was determined.

*Table 1.* The numbers of fish tagged, caught by anglers, and those found to have regurgitated their radio transmitter in each year of the study.

| Year | Number of salmon radio tagged | Salmon reported caught by anglers | |
|---|---|---|---|
| | | Number caught | Number having regurgated transmitters |
| 1994 | 102 | 12 | 2 |
| 1995 | 86 | 8 | 1 |
| 1996 | 93 | 7 | 1 |

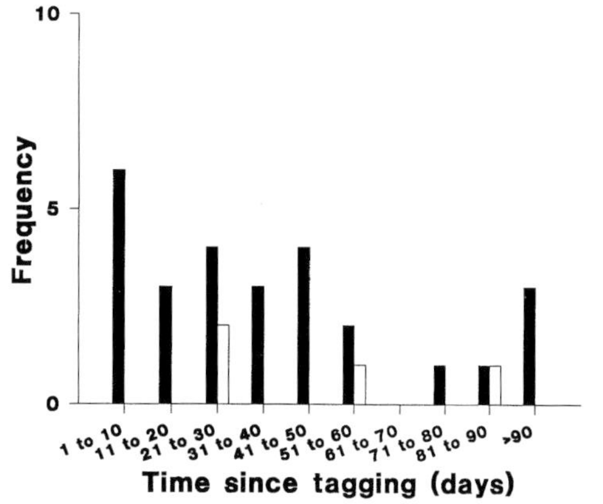


*Figure 2.* Frequency distribution of the number of salmon taken by anglers at successive 10 day periods from the time the fish were tagged. Salmon which were found to have retained the radio transmitters in their stomachs (black bars) are shown separately from those which regurgitated their tags (white bars).

## Results

A total 281 returning adult salmon was radio tagged between 1994 and 1996 (Table 1). 27 (9.6% ± 1.2% standard error) of these fish were caught by anglers, of which total of 4 (14.8% ± 1.2% standard error) were found to have regurgitated their transmitters. Regurgitation rates were consistent between years; 16.7% in 1994 (2 out of 12), 12.5% in 1995 (1 out of 8) and 14.3% in 1996 (1 out of 7) (Table 1).

Figure 1 shows the sites where fish which were known to have regurgitated their tags were taken by anglers and also the sites where the tags were subsequently located. Although all situated in the lower half of the river system, regurgitated tags were not located in the immediate vicinity of the tagging site, nor were they generally found in the area where the tagged fish were caught.

There was no indication from the analysis that the proportion of fish which regurgitated their tags increased with time after tagging. Approximately 50% of tagged fish were taken by anglers within 30 days of tagging, and 2 of the 4 fish which regurgitated their tags were also taken within this time (Figure 2). Regurgitation of transmitters was also not associated with tagging in any particular month of the year. Fish tagged in each of the months April, May, June and September were subsequently found to have regurgitated their tag.

Tagged fish ranged in length from 59 cm to 84 cm (Figure 3). There was no significant difference between the lengths of fish which had regurgitated their tags and those which did not (Mann Whitney test: $N = 27$, $U = 33.5$, $P = 0.392$).

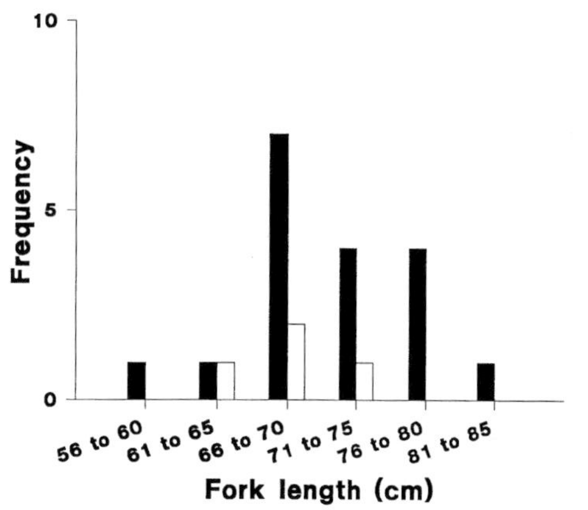


*Figure 3.* Frequency distribution of the fork lengths of salmon caught by anglers. Salmon which were found to have retained the radio transmitters in their stomachs (black bars) are shown separately from those which regurgitated their tags (white bars).

## Discussion

During their riverine migration, returning adult salmon may remain in particular areas for weeks or even months before resuming their upstream migration (Milner, 1990). When a signal is located in a restricted area for several months, however, particularly during the period leading up to spawning time and beyond, regurgitation of the tag is one possible explanation, others may be the death of the animal due to disease or natural predators. In practice, it is generally difficult to recover tags from rivers, and therefore confirmation of the fate of particular fish is rarely possible. Thus, estimates of regurgitation rates by radio tagged salmon in the wild are important for the interpretation of tracking data sets.

In this study, angler-caught salmon were used as a sample of the radio-tagged population from which a clear indication of whether each fish had regurgitated its tag could be determined. The data indicated that the sampling method was not biassed by salmon regurgitating the tag in response to capture by the angler. Indeed, 85% of salmon which were caught had retained their tags, while those which had regurgitated the tag generally did so some distance downstream of the capture site (Figure 1). It is unlikely that salmon regurgitated tags at the capture site which were then subsequently washed downstream. In an experiment specifically designed to test the extent to which radio transmitters may be moved downstream, Webb (1990) reported that of 7 tags identical to those used in the current study, released below Pitlochry dam, none moved more than 10 m. downstream.

In common with other studies (Shearer, 1992), however, the angler exploitation rate on tagged salmon was relatively small, 9.6%, and thus only 4 fish which had regurgitated their tags were identified. Although this represented only a very small proportion of the total number of radio tagged salmon, the finding that calculated regurgitation rates were similar, between 12.5% and 16.7%, in each of the three years of the study lends support to the proposition that they are accurate estimates of the regurgitation rates in the population as a whole.

Quantitative analysis of the factors associated with tag regurgitation in salmon during riverine migration is, however, severely compromised both by the small number of fish recovered and the low regurgitation rate. It is not clear, for example, whether the lack of relationship between regurgitation rate and either fish length or month when the fish was tagged was due to the very small sample size or indeed reflects the lack of a significant effect. Regurgitation was not associated with the immediate aftermath of tagging (Figures 1 and 2), although a more precise understanding of how the proportion of fish which have regurgitated their tags varies as time since tagging increases will require greater numbers of fish to be recovered in any future studies.

The current study indicates that returning adult Atlantic salmon intercepted in the lower reaches of the River Tweed and intragastrically tagged with radio transmitters experience an overall regurgitation rate of around 15%. Similar analyses are required from other studies where large numbers of salmon have been radio tagged before it can be concluded with confidence that similar rates may be expected in other river systems or under other handling and tagging procedures. In addition, it would be instructive to investigate other methods of recapturing the fish. In-river netting and trapping may both increase the recapture rates of tagged fish while also testing whether salmon taken by anglers were representative of the population of tagged salmon as a whole.

## Acknowledgments

We are grateful to the netsmen on the River Tweed who provided the salmon to tag and to the anglers who subsequently recaptured them; and to Tony Hawkins, Dick Shelton, Alan Youngson and Alastair Johnstone who commented on an earlier draft of this paper.

## References

Armstrong, J. D., A. D. F. Johnstone & M. C. Lucas, 1992. Retention of intragastric transmitters after voluntary ingestion by captive cod, *Gadus morhua* L. J. Fish. Biol. 42: 135–137.

Clarke, D., W. K. Purvis & D. Mee, 1991. Use of telemetric tracking studies to examine environmental influence on catch/effort indices. A case study of Atlantic salmon (*Salmo salar* L.) in the River Tywi, S. Wales. In I. Cowx (ed.), Catch Effort Sampling Strategies. Fishing News Books, Oxford: 33-48.

Glass, C. W., A. D. F. Johnstone, G. W. Smith & W. R. Mojsiewicz, 1992. The movements of saithe (*Pollachius virens*, L.) in the vicinity of an underwater reef. In I. G. Priede & S. M. Swift (eds), Wildlife Telemetry: Remote Monitoring and Tracking of Animals. Ellis Horwood Ltd, Chichister: 328–341.

Hawkins A. D. & G. W. Smith, 1986. Radio-tracking observations on Atlantic salmon ascending the Aberdeenshire Dee. Scott. Fish. Res. Rep. 36: 24 pp.

Hawkins, A. D. & G. G. Urquhart, 1983.Tracking fish in the sea. In A. D. Macdonald & I. G. Priede (eds), Experimental Biology at Sea. Academic Press, London: 103–166.

Laughton, R., 1991. The movements of adult Atlantic salmon (*Salmo salar* L.) in the River Spey as determined by radio telemetry during 1988 and 1989. Scott. Fish. Res. Rep. 50: 35 pp.

Laughton, R. & G. W. Smith, 1992. The relationship between date of river entry and the estimated spawning positions of adult Atlantic salmon (*Salmo salar* L.) in two major Scottish east coast rivers. In I. G. Priede & S. M. Swift (eds), Wildlife Telemetry: Remote Monitoring and Tracking of Animals. Ellis Horwood Ltd, Chichister: 423–433.

Lucas, M. C. & A. D. F. Johnstone, 1990. Observations on the retention of intragastric transmitters, and their effects on food consumption, in cod, *Gadus morhua* L. J. Fish Biol. 37: 647–649.

McCleave, J. D., J. H. Power & S. A. Rommel Jr., 1978. Use of radio telemetry for studying upriver migration of adult Atlantic salmon (*Salmo salar*). J. Fish Biol. 12: 549–558.

Mellas, E. J. & J. M. Haynes, 1985. Swimming performance and behaviour of Rainbow Trout (*Salmo gairdneri*) and White Perch

(*Morone americana*): effects of attaching telemetry transmitters. Can. J. Fish. aquat. Sci. 42: 488–493.

Milner, N. J. (ed.), 1990. Fish movement in relation to freshwater flow and quality. Atlantic Salmon Trust/Wessex Water Workshop Proceedings. Atlantic Salmon Trust, Pitlochry.

Potter, E. C. E., 1988. Movements of Atlantic salmon in an estuary in S.W. England. J. Fish Biol. 33: 153–159.

Shearer, W. M., 1992. The Atlantic salmon. Natural history, exploitation and future management. John Wiley and Sons, New York: 244 pp.

Smith, G. W., 1992. Salmon movements in relation to river flow: Estuarine net catches of adult Atlantic salmon (*Salmo salar* L.) and tracking observations. Proceedings of the Institute of Fisheries Management 22nd Annual Study Conference, University of Aberdeen, 10th to 12th September 1991: 37-46.

Smith, G. W., 1994. Keeping track of Tweed salmon. Salmon Net 25: 33–37.

Smith, G. W., I. P. Smith & S. M. Armstrong, 1994. The relationship between river flow and entry to the Aberdeenshire Dee by returning adult Atlantic salmon (*Salmo salar* L.). J. Fish Biol. 45: 953–960.

Varallo, P. V., 1988. Report on M.A.F.F/W.A.A. Fisheries Technical Liaison Committee Workshop on Acoustic and Radio Tracking of Adult Migratory Salmonids. Report No. 88/8.

Webb, J., 1989. The movements of adult Atlantic salmon in the River Tay. Scott. Fish. Res. Rep. 44: 32 pp.

Webb, J., 1990. The behaviour of adult Atlantic salmon ascending the Rivers Tay and Tummel to Pitlochry dam. Scott. Fish. Res. Rep. 48: 27 pp.

Webb, J. & A. D. Hawkins, 1989. The movements and spawning behaviour of adult salmon in the Girnock Burn, a tributary of the Aberdeenshire Dee. Scott. Fish. Res. Rep. 40: 42 pp

*Hydrobiologia* **371/372:** 123–132, 1998.
*J.-P. Lagardère, M.-L. Bégout Anras & G. Claireaux (eds), Advances in Invertebrates and Fish Telemetry.*



# New approaches to the behavioural ecology of decapod crustaceans using telemetry and electronic tags

Juan Freire & Eduardo González-Gurriarán
*Departamento de Bioloxía Animal, Bioloxía Vexetal e Ecoloxía, Universidade da Coruña, Campus da Zapateira s/n, E-15071 A Coruña, Spain, e-mail: jfreire@udc.es*

*Key words:* behavioural ecology, crustaceans, decapods, electronic tags, migrations, telemetry

## Abstract

Decapod crustaceans have complex life histories and behaviour in aspects such as foraging, mating and reproduction, moulting and growth, habitat selection and migration. New technologies have enabled us to use an individual, field-based approach to analyze these problems, although they have been less developed in decapods than in marine vertebrates. These new possibilities are discussed here mainly from a biological point of view. There is a brief review of previous applications of telemetry to analyze habitat selection, foraging behaviour, energetics, moulting site selection and migrations in decapods, and two case studies are discussed in more detail. The first one refers to the study of differences in habitat use and movement patterns in juveniles and adults of coastal species that show ontogenetic habitat shifts, related to differences in selective pressures affecting both life history stages (predation risk, and growth and reproduction optimization). The second case study is dedicated to the migratory patterns in spider crabs combining telemetry and electronic tags. Operational limitations in tracking make it impossible to get detailed information on movement patterns during migration, which in turn involve an important bathymetric gradient and a change in the oceanographic environment (mainly temperature). Monitoring depth and temperature in the immediate habitat of the animals, using electronic data storage tags recovered by the fishery, allow for movement patterns to be modeled using supplementary information on the topography and hydrography of the study area. This approach is being tested using both telemetry and electronic tags simultaneously.

## Introduction

The application of telemetry techniques to marine organisms has made it possible to track and determine their location on a continuous and individual basis in the field (Stasko & Pincock, 1978; Hawkins & Urquhart, 1983; Priede & Swift, 1992; Urquhart & Stewart, 1993) as well as to monitor physiological and behavioural parameters (O'Dor et al., 1988; Butler, 1989; Wolcott & Hines, 1989a; Kasello et al., 1992). With this information on an individual level, it is possible to analyze the relationship between physiological and behavioural changes and the environmental factors which determine them. Decapod crustaceans have complex life histories and behaviour in aspects such as foraging, mating and reproduction, moulting and growth, habitat selection and migration. Some of these biological characteristics make these marine invertebrates, along with the cephalopods, appropriate for the application of telemetry following the techniques designed for fishes or specific new methods. Since many decapod species have a large body size and high mobility, and they occupy easily accessible benthic and coastal habitats, it is possible to track them telemetrically.

Decapod crustaceans have certain peculiarities in their life histories which should be taken into consideration when establishing the methodology of study. The discontinuous growth through moults brings about the loss of the exoskeleton and consequently of external tags, although some species do undergo a terminal moult. However the presence of an exoskeleton offers some advantages for attaching external tags and

*Table 1.* Studies on the behavioural ecology of marine decapod crustaceans that have used biotelemetry techniques

| Species | Author | Tracking system | Carrier | Telemetered variables | Behaviour | Observations |
|---|---|---|---|---|---|---|
| BRACHYURAN CRABS | | | | | | |
| CANCRIDS | | | | | | |
| *Cancer magister* | Smith & Jamieson (1991) | D | U | Location | Movement, Dispersal | |
| *Cancer pagurus* | Bottoms & Marlow (1979) | – | U | Heart rate | | Laboratory testing |
| | Hall et al. (1991) | AC | U | Location | Movement | Sediment disturbance |
| | Skaaja et al. (1998) | AC | U | Location | Movement | |
| PORTUNIDS | | | | | | |
| *Callinectes sapidus* | Wolcott & Hines (1990), Shirley & Wolcott (1991) | D | U | Location, Ecdysis | Habitat use | |
| | Wolcott & Hines (1989a, b), Nye (1989) | C | U | Location, Madibular muscle activity | Foraging | |
| | Clark, Hines & Wolcott (pers. comm.), Wolcott (1995) | AC,C | U | Location, Mandibular muscle activity, Cheliped display | Foraging, Agonistic behaviour | |
| | Hines et al. (1995), Hines, Wolcott, Freire & Ruiz (unpubl. data) | D | U | Location | Movement, Habitat use | |
| *Scylla serrata* | Hill (1978) | C | U | Location | Movement | |
| MAJIDS | | | | | | |
| *Chionoecetes opilio* | Maynard & Webber (1987), Maynard (1991) | AC,D | U | Location, Branchial pressure | Movement, Energetics, Habitat use | Energetics in laboratory |
| *Maja squinado* | González-Gurriarán & Freire (1994), Hines et al. (1995) | D | U | Location, Temperature | Movement, Migration, Habitat use | |
| OCYPODIDS | | | | | | |
| *Ocypode quadrata* | Wolcott (1980a, b) | D | R | Location, Heart rate, Temperature | Movement, Energetics, Habitat use | Intertidal |
| ANOMURAN CRABS | | | | | | |
| *Paralithodes camtschaticus* | Monan & Thorne (1973), Stone et al. (1992, 1993) | D | U | Location | Movement, Migration, Habitat use | |
| LOBSTERS | | | | | | |
| *Homarus americanus* | O'Dor & Webber (1991) | AC | U | Location, Temperature | Movement | |
| *Homarus gammarus* | Collins & Jensen (1992) | D | U | Location | Movement | |
| | Collins et al. (1993), Smith et al. (1998) | AC | E | Location | Movement, Habitat use | |
| *Nephrops norvegicus* | Chapman et al. (1975) | AC | U | Location | Movement | |
| | Bjordal et al. (1993) | AC | U | Location | Movement | |
| *Panulirus cygnus* | Phillips et al. (1984), Jernakoff (1987a, b), Jernakoff et al. (1987) | AC | E | Location | Movement, Habitat use | |

Tracking system: D, discontinuous, C, continuous (non automated); AC: automated continuous (stationary array of omnidirectional hydrophones or aerials)
Carrier: U, ultrasonic; R, radio; E, electromagnetic
Telemetered variables: environmental variables (temperature) are only included if they were measured using telemetry (variables measured using other systems after positioning are not included)
Behaviour: Movement, includes speed and distance of displacements, orientation and timing. Habitat use, characterization of the habitat used in relation to its environmental characteristics (depth, temperature, salinity, vegetation)

locating internal structures. Decapods are mobile organisms and the large-sized walking species show major differences in the energetic costs of movement respect to swimming fishes. Although, as a rule, the range of movements in decapods is smaller than in fishes (see references cited in Table 1), some species cover large distances during migration.

## Methods for the study of behaviour in decapod crustaceans

The primary source of information on the behaviour of decapods comes from laboratory studies which entail certain limitations, particularly in large-sized species, and they do not allow for an analysis of certain aspects such as migrations. Urquhart & Stewart (1993) have reviewed the methods for the field observation of the behaviour of fishes which are basically the same as those used on decapods.

Direct observation by divers or video provides information on habitat use, movements and biotic interactions (e.g. Karnofsky et al., 1989 a,b), but it is not possible to record the physiological activity and this type of observations are restricted to shallow waters with adequate light (diving) and/or small areas (both methods).

The mark-recapture experiments (see review by Emery & Wydoski, 1987) have used magnetic tags (Hurley et al., 1990) or biological tags (Shelton & Chapman, 1987) involving internal implantation, or external tags (for example T-bar tags, Hurley et al., 1990). In all cases tagging presents different problems related to handling the animal, mortality induced, or tag shedding. Despite these limitations, the mark-recapture experiments have provided data of great interest on population dynamics and large scale migration patterns in crustaceans. However, from a behavioural point of view, their application is limited since only sporadic data on the location and biological state of the tagged animals can be recorded. Moreover, the fact that the animal must be recovered means that these methods apply mostly to commercial species where the fishing effort allows an acceptable recapture rate.

As an alternative to these tags, which we may call conventional, systems have been developed based on passive integrated transponders (PIT tags), which, because of their small size, make it possible to study juvenile individuals or small-sized species (Van Montfrans et al., 1986, 1991; Fitz & Wiegert, 1991). Despite the progress made with this new system, the problem of recapture persists. In this case, given the technical characteristics of this system, the samplings are generally realized for this purpose, which makes it possible to apply an experimental design, more appropriate than the commercial fisheries, however the size of the study area is limited.

Telemetry is a step forward in the study of behaviour in the field as compared to the methods discussed earlier. Ultrasonic telemetry techniques have led to detailed studies of the behaviour and energetics of marine organisms in the field (O'Dor et al., 1988, 1993, 1994; Butler, 1989; Wolcott & Hines, 1989b, 1990; Shirley & Wolcott, 1991; Kasello et al., 1992). The technological advancements in the construction of ultrasonic transmitters have led to the use of much smaller components and an increase in battery life, making it possible to apply them to invertebrates such as decapod crustaceans (see below) or cephalopods (e.g. O'Dor et al., 1993, 1994).

The recent introduction of electronic data storage tags or archival tags, allows for the continuous monitoring of the immediate environment, the microenvironment, of the animal, although they do not eliminate the necessity for recapture. To date the use of these tags has been much more restricted than the application of telemetry, not counting the great number of studies done on marine mammals and birds, but they have begun to provide important information on the migration of several species of fish (Metcalfe et al., 1994; Gunn et al., 1994, Sturlaugsson, 1995; Thorsteinsson, 1995).

## A brief review of the application of telemetry in the study of the behavioural ecology of decapod crustaceans

Table 1 presents a review of the studies that have applied telemetry techniques to marine crabs and lobsters. The most commonly used systems have been the ultrasonic ones, except in specific cases such as in intertidal areas or structurally complex habitats (e.g. reefs) where, on occasion, radiotelemetry or electromagnetic systems have been used respectively. Wolcott (1995) has recently reviewed the application of telemetry techniques in the study of behavioural and physiological ecology of decapod crustaceans. This author analyzes the different options currently available, mainly from a technological standpoint. The conclusions we may draw from this review are that with the technology currently available it is possible to record information on virtually any variable related to the behaviour, physiology and microenvironment of these animals.

Despite the wealth of potential applications of telemetry, most of these studies have consisted only of tracking the animals tagged with pingers in order to get information on their patterns of movement, dispersal and migrations, and habitat use. In most cases the characteristics of the environment occupied by the tagged individuals have been recorded by means of samplings coinciding with the positioning of their location without the use of telemetry (with some exceptions as is the case of temperature). Although these types of studies have not developed the full range of potential that telemetry has to offer, in terms of internal and microenvironmental variables, they have implemented automated systems of continuous tracking which have provided detailed information on the movements of the tagged individuals in space and time.

However, the use of telemetry of physiological and behavioural parameters in the field has not gained widespread use in decapods. Despite the early experiments and calibrations carried out in the laboratory as far back as the 1970's (Bottoms & Marlow, 1979), only Wolcott (1980a, b) working with *Ocypode quadrata* and Wolcott, Hines and collaborators (see references in Table 1 and a brief description in Wolcott, 1995) with *Callinectes sapidus* have performed experiments in the field in which the energetic costs or different aspects of foraging and agonistic behaviour have been measured. Thus the wide range of telemetry systems presented by Wolcott (1995) as technologically feasible at the present time remain largely unexplored in terms of their potential application in the field.

## Two case studies of the use of telemetry and electronic tags in decapod crustaceans

This Section provides a more in-depth look at two examples of the application of telemetry and related techniques, such as the use of electronic data storage tags, for the study of the behavioural ecology of crustaceans. These case studies aim to illustrate the wide potential of telemetry, even with the simplest applications, for testing in the field different hypotheses on behavioural mechanisms and life histories of decapods. The application of new technologies available that will help overcome some of the limitations inherent to telemetry is discussed.

### *Habitat utilization and movement patterns in juvenile and adult crabs*

The results presented by several authors on different coastal species of crabs and lobsters (see Hines et al., 1995), using research techniques such as field samplings, mark-recapture or laboratory experiments have, in most cases, pointed to the existence of differences in the characteristics and range of movements and habitat selection between juveniles and adults, often involving migrations (directional long-distance movements that entail drastic changes in habitat) after the pubertal moult.

González-Gurriarán & Freire (1994), Hines et al. (1995) and Hines, Wolcott, Freire & Ruiz (unpublished data) analyzed ontogenetic changes (between juveniles approaching maturity and adults) in two species of crabs with great differences in their life histories and habitat (the coastal spider crab *Maja squinado* and the estuarine blue crab *Callinectes sapidus*, in the NE and NW Atlantic respectively). The two studies are an example of simple telemetry applications in behavioural ecology, using discontinuous tracking to estimate the variables of movement (speed, distance and orientation) and habitat use (depth, bottom type)

*Maja* and *Callinectes* have patterns of behaviour and habitat utilization similar to those described in other species. The telemetry experiments show the immediate behavioural mechanisms that determine these patterns, making it possible to obtain individualized information on free-ranging animals (Table 2). In both species, juveniles present restricted (in areas $<$ 500 m), non-directional movements in shallow, structurally complex habitats (rocky kelp beds or soft bottoms with wood debris). On the other hand adults have broader ranges of movement with a strong directionality (in zones $>$ 1 km and far greater distances during migration, at much faster speeds than the juveniles) in the deeper areas (up to over 100 m in *Maja*), where they use, to a greater extent, less complex soft bottoms. In *Maja*, once adults reach sexual maturity after the terminal pubertal moult, they exhibit two clearly differentiated phases in their movements and habitat (Table 2). In the first phase, prior to migration, their behaviour is similar to that of juveniles, although both depth and speed increase slightly. The second phase, which corresponds to beginning of the

*Table 2.* Movements patterns of juvenile and adult crabs tracked with ultrasonic telemetry. Experiments on *Maja squinado* were carried out in the Ría de Arousa (Galicia, NW Spain) from summer 1993 to winter 1994, and on *Callinectes sapidus* in the Rhode River (a subestuary of Chesapeake Bay, USA) in summer 1993. For each parameter, mean and standard error (in parentheses) are shown. Positioning was carried out with respect to landmarks and buoys (juveniles) or using GPS (adults) for *Maja* and using landmarks for *Callinectes*. Body size was measured as carapace length in *Maja* and carapace width in *Callinectes*. The orientation was measured using the statistic r which represents the directionality of the different displacements (between successive locations) of each individual (0: random orientation, 1: all displacements with the same bearing). See González-Gurriarán & Freire (1994) and Hines et al. (1995) for more details

| | *Maja squinado* | | | *Callinectes sapidus* | |
|---|---|---|---|---|---|
| | | Adults | | | |
| | Juveniles | Pre-migration | Start of migration | Juveniles | Adults |
| Number of crabs | 8 | 5 | 6 | 12 | 8 |
| Body size (mm) | 107.0 (5.6) | 145.0 ( 5.9) | 150.0 ( 6.6) | 100.2 ( 2.3) | 148.5 ( 3.2) |
| Tracking time (days) | 43.5 (6.1) | 52.9 (28.8) | 27.6 ( 5.5) | 12.8 ( 0.8) | 10.2 ( 1.9) |
| Speed (m day$^{-1}$) | 6.5 (1.2) | 22.1 ( 6.1) | 75.8 (16.2) | 156.7 (22.3) | 328.1 (43.9) |
| Depth (m) | 4.5 (0.7) | 7.3 ( 0.8) | 14.0 ( 2.5) | 1.3 ( 0.1) | 2.6 ( 0.2) |
| Orientation (r) | 0.14 (0.03) | 0.35 ( 0.11) | 0.81 ( 0.05) | 0.32 ( 0.05) | 0.61 ( 0.08) |

migration, is characterized by directional movements towards deeper zones at a greater speed.

These findings have led to a number of hypotheses on the adaptive value of the ontogenetic changes in behaviour and habitat utilization, which are based on the type of selective pressures affecting juveniles and adults, and which would be similar in the different decapods living in the coastal areas. Predator avoidance and growth optimization are the selective forces of the behavioural patterns observed in juveniles. The maximization of reproductive success, through a successful incubation process, largely dependent upon temperature and salinity, and the release of larvae in habitats adequate to survival, determines adult behaviour and brings about the postpubertal migrations. The results from the use of telemetry have made it possible to characterize the habitat and movements of juveniles and adults, and therefore, lay the foundations for future studies that will allow the proposed hypotheses to be tested. For example, Hines & Ruiz (1995) show that the nearshore shallows constitute a refuge for the juveniles of*Callinectes* against cannibalism.

### *Migrations in spider crabs*

Telemetry has a number of drawbacks stemming from the difficulty in the continuous tracking of animals at sea. In fact, in the case of the spider crab *Maja squinado*, these operational problems prevent the detailed tracking in the phase where the animals migrate to deep off-shore waters (González-Gurriarán & Freire, 1994).

In species where the recapture of tagged individuals is feasible, mainly due to the fact that they are exploited by fisheries, alternative (non-telemetry) techniques have been used which involve the attachment of electronic tags to store data on the microenvironment or physiological or behavioural variables of the animal. These data can be downloaded after recapture takes place. The miniaturization of these tags has allowed them to be used in the study of fish migrations (Metcalfe et al., 1994; Gunn et al., 1994, Sturlaugsson, 1995; Thorsteinsson, 1995).

This application of this methodology is based on the fact that during their migrations the tagged animals move across important environmental gradients, some of which are used as an orientation mechanism. The continuous monitoring of the environmental parameters involved (such as depth, temperature, salinity or light) along with the baseline data on habitat characteristics (topography, oceanography) make it possible to reconstruct the tracks of the animals, estimating the temporality, directionality, and movements carried out throughout the migration. Arnold & Holford (1995) have presented a simulation model based on tidal streams, which when linked to previous knowledge on behaviour will enable us to reconstruct the movements of different fish species from temperature and depth data assimilated continuously by electronic tags.

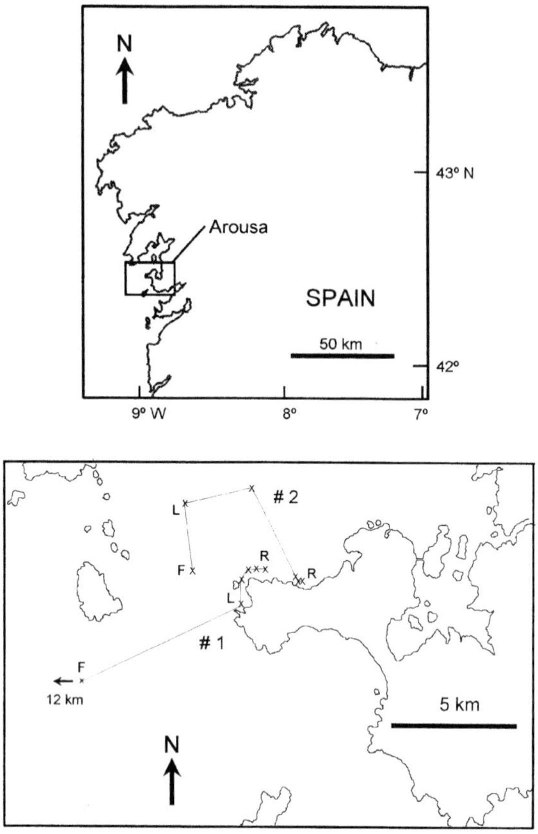


*Figure 1.* Migrations of the spider crab *Maja squinado* in the Ría de Arousa and adjacent coastal area (Galicia NW Spain) using telemetry and electronic data storage tags. The tracks of two adult females, tracked approximately at weekly intervals using telemetry in 1997, are shown (release point R, last telemetered location L and point of recapture by the commercial fishery F are shown).

Based on the hypothesis that the migrations of the spider crab *Maja* involve important bathymetric changes and movements between water masses having distinct oceanographic characteristics (temperature and, to a lesser extent, salinity), a experiment have been started for the purpose of calibration and use of electronic tags as a tool to study migrations. Moreover, earlier data point to a high fishing mortality rate (tangle-nets and glass-box gears) on the SW coast of Galicia (NW Spain). Previous telemetry experiments had over 70% recapture rates. However, *Maja* undergoes a terminal moult, which means that adults, which is the life stage involved in the migrations do not have additional ecdyses (González-Gurriarán et al., 1995) and that they may be tagged with devices attached to the carapace.

In the summer of 1996 an experiment was begun in the outer area of the Ría de Arousa (Figure 1) in order to compare the behavioural patterns of primiparous and multiparous adult females (in their first annual reproductive cycle or consecutive cycles, respectively; see González-Gurriarán et al., 1993, 1998) by means of a combination of telemetry and electronic tags. Each individual was tagged with an ultrasonic transmitter (V16 of VEMCO, Canada and CHP-87-S-M of SONOTRONICS, United States) and an electronic tag for temperature and depth (DST-100 of STAR ODDI, Iceland and MINILOG-TDX of VEMCO), which comprised < 5% of the body weight. We carried out tracking discontinuously, at intervals of approximately 1 week, in order to locate the animals with telemetry.

Our results to date have shown a recapture rate > 60% (11 recaptures from 18 releases of tagged females). The information provided by the electronic tags point to autumn migrations defined by important bathymetric changes during short periods of time (Figure 2). During this migration, females moved from an average depth of 12 m (range: 7–17 m) to 63 m (range: 36–96 m). The mean duration of these movements to deep waters was of 10 days (range 4–16 days) and occurred in early November (mean dates of start and end of migratory movements: 7 and 16 November), although extended from September to January for the different crabs. During these movements the tagged animals travel through clearly differentiated thermal environments characterized by different substrate types. The available data on bathymetry in the study area and its oceanography (based on a network of oceanographic stations sampled weekly and continuous records of temperature in the shallow areas where the animals were released) allow us to reconstruct the tracks of the tagged individuals. These estimates may be analyzed in relation to the positionings provided by telemetry. The preliminary analyses of the available data suggest that these two methods of estimating migration routes are basically in agreement. Moreover, this information enables us to define the behavioural rules that animals follow to orient themselves throughout the migration, based primarily on the depth gradient. This migration is characterized by a shift towards the deeper zones in autumn, when the temperature and salinity in the shallow waters drop, and a return to these areas at the end of winter and spring, which coincides with the incubation period.

These preliminary results point to the possibility of reconstructing the tracks of animals equipped with

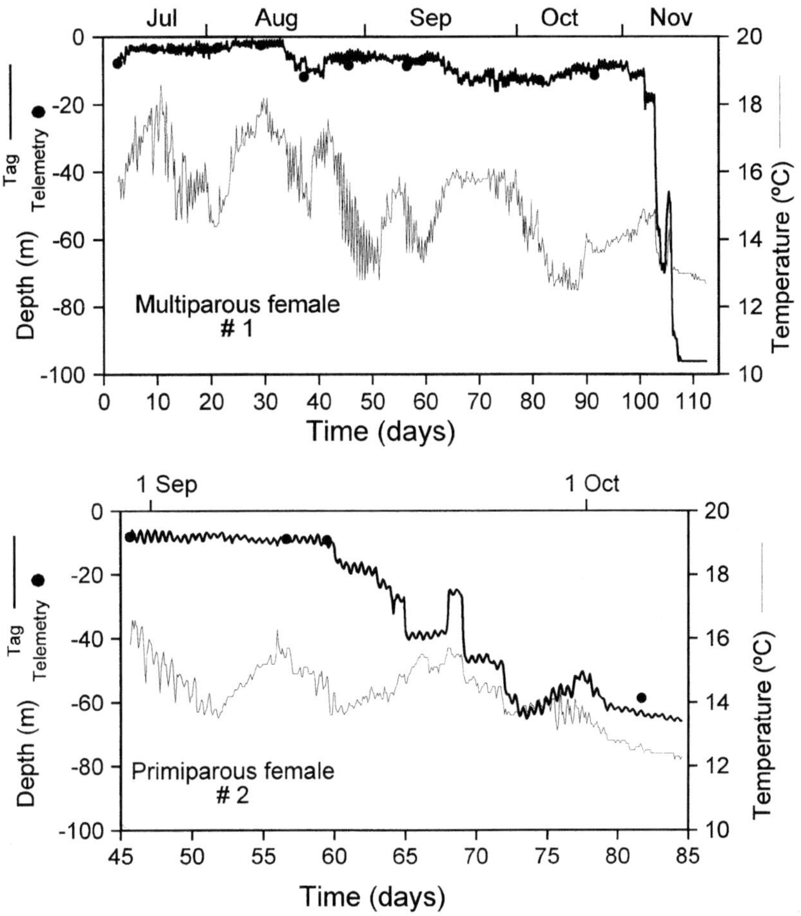


*Figure 2.* Migrations of the spider crab *Maja squinado* in the Ría de Arousa and adjacent coastal area (Galicia NW Spain) using telemetry and electronic data storage tags. Temporal evolution of the habitat depth and temperature of two females measured every two hours by means of eletronic tags and the depth at which they were located using telemetry. Depth was not corrected by the effect of the tide (maximum variation of aprox. ±3 m). The data series of the electronic tags covers the period from the release of the females to their recapture by the commercial fishery (day 0 = 15 July 1996). The tracks of both females obtained by telemetry are showed in Figure 1.

electronic tags based on information from their microenvironment provided by these devices and on the understanding of their orientation mechanisms. The second part of the experiment will attempt the large-scale use of electronic tags in order to determine the spatial variability (between nursery grounds) and between males and females in migration characteristics.

## Future directions in the telemetry of decapods and alternative methods

The future development of telemetry and related techniques in the study of the behavioural ecology of decapod crustaceans appears to be feasible using the technology currently available. Three broad areas of interest may be defined in which the technology which is currently available or soon to be developed, can help to provide a more in-depth knowledge of the biological problems posed by the decapods.

The automated systems of ultrasonic telemetric tracking have already been used both in fishes (Urquhart & Smith, 1992; Bjordal et al., 1993) as well as in invertebrates (Maynard & Webber, 1987; O'Dor et al., 1988; Wolcott, 1995). The use of these telemetric systems for physiological and behavioural parameters would provide detailed information on the temporal and spatial patterns of individual behaviour and on biotic interactions (mating, foraging and social behaviour) and with the physical environment (O'Dor et al., 1988; Wolcott, 1995). These systems may be applied to animals that carry out their activity in restricted areas. However, with animals that move over great distances, telemetry has its limitations due to the fact that it depends on direct work in the sea. The current problem with electronic tags is the recapture of the tagged animal and the bias created by the spatial distribution of the fishing effort. The development of recovery systems, whether they involve the tag (which may fall off the animal at any time, and emitting a signal that will allow it to be recovered) or information (through interrogation at permanent stations and activation of an ultrasonic system for transmitting information) will enable us to solve the problems facing us at the present time.

The telemetry of physiological parameters (Butler, 1989) has revolutionized our understanding of bioenergetics in fishes (Lucas et al., 1993; Claireaux et al., 1995a, b) and cephalopods (O'Dor et al., 1993, 1994). The applications developed for crustaceans are presently restricted to the calibration in the laboratory of the relationship between the recorded variable and energy consumption (Maynard, 1991), and to studies in the field that relate foraging to muscular activity (Nye, 1989; Wolcott & Hines, 1989a, b; Wolcott, 1995). Nowadays it is possible to monitor different physiological processes (such as heart rate and respiratory activity by recording biopotentials or pressure variations in the branchial chamber), which are closely linked to energy consumption. Thus, in the near future it will be possible to use to determine the costs of the different activities carried out by crustaceans in the field. Another drawback in this sense is the need to record data continuously, which will be mitigated by the development of automated systems discussed earlier.

Lastly, all of these possible future developments, in addition to the techniques currently in use, will benefit unquestionably from the miniaturization of the transmitters and electronic tags used. This has expedited the transfer of the technology used in large-sized animals such as birds and mammals to fishes, and subsequently to invertebrates, and it has served to broaden the range of species and life stages to include small-sized animals. These advancements may lead to the widespread use of these systems as methods of behavioural and physiological 'observation' in the marine environment.

## Acknowledgements

Part of the research presented in this paper was funded by the Consellería de Pesca, Marisqueo e Acuicultura of the Xunta de Galicia, the Dirección General de Investigación Científica y Técnica (DGICYT) of the Spanish Ministerio de Educación y Cultura, and the Fundación Caja de Madrid. We are grateful to Dr A. H. Hines for his insights in the use of telemetry with decapods and for the discussion of different aspects of our research projects.

## References

Arnold, G. P. & B. H. Holford, 1995. A computer simulation model for predicting rates and scales of movement of demersal fish on the European continental shelf. ICES J. mar. Sci. 52: 981–990.

Bjordal, A., A. Engas, A. V. Soldal & T. J. Ovredal, 1993. A new radio link telemetry positioning system. Experiences from tracking of fish and crustaceans. ICES, Fish Capture Comm., C.M. 1993/B:23: 1–6.

Bottoms, A. & J. Marlow, 1979. A new ultrasonic tag for the telemetry of physiological functions from aquatic animals. Mar. Biol. 50: 127–130.

Butler, P. J., 1989. Telemetric recording of physiological data from free-living animals. In P. J. Grubb & J. B. Whittaker (eds), Toward a More Exact Ecology, Blackwell Scientific Publications, Oxford: 63–84.

Chapman, C. J., A. D. Johnstone & A. L. Rice, 1975. The behaviour and ecology of the Norway lobster, *Nephrops norvegicus* (L.). In H. Barnes (ed.), Proceedings of the 9th European Marine Biology Symposium, Aberdeen University Press, Aberdeen: 59–74.

Claireaux, G., D. M. Webber, S. R. Kerr & R. G. Boutilier, 1995a. Physiology and behaviour of free-swimming Atlantic cod (*Gadus morhua*) facing fluctuating temperature conditions. J. exp. Biol. 198: 49–60.

Claireaux, G., D. M. Webber, S. R. Kerr & R. G. Boutilier, 1995b. Physiology and behaviour of free-swimming Atlantic cod (*Gadus morhua*) facing fluctuating salinity and oxygenation conditions. J. exp. Biol. 198: 61–69.

Collins, K. J. & A. C. Jensen, 1992. Acoustic tagging of lobsters on the Poole Bay artificial reef. In I. G. Priede & S. M. Swift (eds), Wildlife Telemetry. Remote Monitoring and Tracking of Animals, Ellis Horword, Chichester: 354–358.

Collins, K. J., E. K. Free, A. C. Jensen & S. Thompson, 1993. Analysis of Poole Bay, U.K. lobster data. ICES, Shellfish Committee C.M. 1993/K:48: 1–8.

Emery, L. & R. Wydoski, 1987. Marking and tagging of aquatic animals: an indexed bibliography. U.S. Fish Wildl. Serv., Resour. Publ. 165: 1–57.

Fitz, H. C. & R. G. Wiegert, 1991. Tagging juvenile blue crabs, *Callinectes sapidus*, with microwire tags: Retention, survival, and growth through multiple molts. J. Crust. Biol. 11: 229–235.

González-Gurriarán, E., L. Fernández, J. Freire, R. Muiño & J. Parapar, 1993. Reproduction of the spider crab *Maja squinado* (Brachyura: Majidae) in the southern Galician coast (NW Spain). ICES, Shell. Comm. C.M., 1993/K:19: 1–15.

González-Gurriarán, E., L. Fernández, J. Freire & R. Muiño, 1998. Mating and role of seminal receptacles in the reproductive biology of the spider crab, *Maja squinado* (Decapoda: Majidae). J. exp. mar. Biol. Ecol. 220: 269–285.

González-Gurriarán, E. & J. Freire, 1994. Movement patterns and habitat utilization in the spider crab *Maja squinado* (Herbst) (Decapoda, Majidae) measured by ultrasonic telemetry. J. exp. mar. Biol. Ecol. 184: 269–291.

González-Gurriarán, E., J. Freire, J. Parapar, M. P. Sampedro & M. Urcera, 1995. Growth at moult and moulting seasonality of the spider crab, *Maja squinado* (Herbst) (Decapoda: Majidae) in experimental conditions: implications for juvenile life history. J. exp. mar. Biol. Ecol. 189: 183–203.

Gunn, J., T. Polacheck, T. Davis, M. Sherlock & A. Betlehem, 1994. The development and use of archival tags for studying the migration, behaviour and physiology of southern bluefin tuna, with an assessment of the potential for transfer of technology to groudfish research. ICES C.M.1994/Mini:2.1: 1–23.

Hall, S. J., D. J. Basford, M. R. Robertson, D. G. Rafaelli & I. Tuck, 1991. Patterns of recolonisation and the importance of pit-digging by the crab *Cancer pagurus* in a subtidal sand habitat. Mar. Ecol. Prog. Ser. 72: 93–102.

Hawkins, A. D. & G. G. Urquhart, 1983. Tracking fish at sea. In A. G. MacDonald & I. G. Priede (eds), Experimental Biology at Sea, Academic Press, London: 103–166.

Hill, B. J., 1978. Activity, track and speed of movement of the crab *Scylla serrata* in an estuary. Mar. Biol. 47: 135–141.

Hines, A. H. & G. M. Ruiz, 1995. Temporal variation in juvenile blue crab mortality: nearshore shallows and cannibalism in Chesapeake Bay. Bull. Mar. Sci. 57: 884–901.

Hines, A. H., T. G. Wolcott, E. González-Gurriarán, J. L. González-Escalante & J. Freire, 1995. Movement patterns and migrations in crabs: telemetry of juvenile and adult behaviour in *Callinectes sapidus* and *Maja squinado*. J. mar. biol. Ass. U.K. 75: 27–42.

Hurley, G. V., R. W. Elner, D. M. Taylor & R. F. J. Bailey, 1990. Evaluation of snow crab tags retainable through molting. Am. Fish. Soc. Symp. 7: 84–93.

Jernakoff, P., 1987a. An electromagnetic tracking system for use in shallow water. J. exp. mar. Biol. Ecol. 113: 1–8.

Jernakoff, P., 1987b. Foraging patterns of juvenile rock lobsters *Panulirus cygnus* George. J. exp. mar. Biol. Ecol. 113: 125–144.

Jernakoff, P., B. F. Phillips & R. A. Maller, 1987. A quantitative study of the nocturnal foraging distances of the western rock lobster *Panulirus cygnus* George. J. exp. mar. Biol. Ecol. 113: 9–21.

Karnofsky, E. B., J. Atema & R. H. Elgin, 1989a. Field observations of social behavior, shelter use, and foraginig in the lobster, *Homarus americanus*. Biol. Bull. 176: 239–246.

Karnofsky, E. B., J. Atema & R. H. Elgin, 1989b. Natural dynamics of population structure and habitat use of the lobster, *Homarus americanus*, in a shallow cove. Biol. Bull. 176: 247–256.

Kasello, P. A., A. H. Weatherley, J. Lotimer & M. D. Farina, 1992. A biotelemetry system recording fish activity. J. Fish Biol. 40: 165–179.

Lucas, M. C., A. D. F. Johnstone & I. G. Priede, 1993. Use of physiological telemetry as a method of estimating metabolism in the natural environment. Trans. am. Fish. Soc. 122: 822–833.

Maynard, D. R. & D. M. Webber, 1987. Monitoring the movements of snow crab (*Chionoecetes opilio*) with ultrasonic telemetry. In Proceedings Oceans'87. The ocean – an international workplace. Volume 3: Marine Sciences, Fifth working symposium on oceanographic data systems and underwater work systems: 962–966.

Maynard, D. M., 1991. Biophysical ecology of snow crab *Chionoecetes opilio*; relating respiration rates and walking speeds from the laboratory to fields observations. Biology Department, Dalhouisie University, Halifax, Nova Scotia, 89 pp.

Metcalfe, J. D., G. P. Arnold & B. H. Holford, 1994. The migratory behaviour of plaice in the North Sea as revealed by data storage tags. ICES C.M.1994/Mini:1.1: 1–13.

Monan, G. E. & D. L. Thorne, 1973. Sonic tags attached to Alaska king crab. Mar. Fish. Rev. 35: 18–21.

Nye, L. A., 1989. Variation in feeding behavior of blue crabs (*Callinectes sapidus* Rathbun) measured by ultrasonic telemetry. Department of Marine, Earth and Atmospheric Sciences, University of North Carolina, Raleigh, 82 pp.

O'Dor, R. K., D. M. Webber & F. M. Voegeli, 1988. A multiple buoy acoustic-radio telemetry system for automated positioning and telemetry of physical and physiological data. In C. J. Amlaner (ed.), Biotelemetry X, Proc. 10th Int. Symp. Biotelemetry, The University of Arkansas Press: 444–452.

O'Dor, R. K. & D. M. Webber, 1991. Ultrasonic telemetry. The Lobster Newsletter 4: 1–2.

O'Dor, R. K., J. Forsythe, D. M. Webber, J. Wells & M. J. Wells, 1993. Activity levels of *Nautilus* in the wild. Nature 362: 626–628.

O'Dor, R. K., J. A. Hoar, D. M. Webber, F. G. Carey, S. Tanaka, H. R. Martins & F. M. Porteiro, 1994. Squid (*Loligo forbesi*) performance and metabolic rates in nature. Mar. Freshwat. Behav. Physiol. 25: 163–177.

Phillips, B. F., L. M. Joll & D. C. Ramm, 1984. An electromagnetic tracking system for studying the movements of rock (spiny) lobsters. J. exp. mar. Biol. Ecol. 79: 9–18.

Priede, I. G. & S. M. Swift (eds), 1992. Wildlife Telemetry. Remote Monitoring and Tracking of Animals. Ellis Horwood, Chichester.

Shelton, P. M. J. & C. J. Chapman, 1987. A living tag for recording moult histories in crustaceans. J. Cons. int. Explor. Mer 43: 209–215.

Shirley, M. A. & T. G. Wolcott, 1991. A telemetric study of microhabitat selection by premolt and molting blue crabs, *Callinectes sapidus* (Rathbun), within a subestuary of the Pamlico River, North Carolina. Mar. Behav. Physiol. 19: 133–148.

Skaaja, K., A. Fernö, S. Løkkeborg & E. K. Haugland, 1998. Basic movement pattern and chemo-oriented search towards baited pots in edible crab (*Cancer pagurus* L.). Hydrobiologia 371/372: 143–153.

Smith, B. D. & G. S. Jamieson, 1991. Movement, spatial distribution, and mortality of male and female Dungeness crab *Cancer magister* near Tofino, British Columbia. Fish. Bull. 89: 137–148.

Smith, I. P., K. J. Collins & A. C. Jensen, 1998. Electromagnetic telemetry of lobster (*Homarus gammarus* (L.)) movements and activity: preliminary results. Hydrobiologia 371/372: 133–141.

Stasko, A. B. & D. G. Pincock, 1978. Review of underwater biotelemetry, with emphasis on ultrasonic techniques. J. Fish. Res. Bd Can. 34: 1261–1285.

Stone, R. P., C. E. O'Clair & T. C. Shirley, 1992. Seasonal migration and distribution of female red king crabs in a southeast Alaskan estuary. J. Crust. Biol. 12: 546–560.

Stone, R. P., C. E. O'Clair & T. C. Shirley, 1993. Aggregating behavior of ovigerous female red king crab, *Paralithodes camtschaticus*, in Auke Bay, Alaska. Can. J. Fish. Aquat. Sci. 50: 750–758.

Sturlaugsson, J., 1995. Migration study of Atlantic salmon (*Salmo salar* L.) in coastal waters W-Iceland-Depth movements and se temperatures recorded at migration routes by data storage tags. ICES, Anad. Catad. Fish. Comm. C.M. 1995/M:17: 1–13.

Thorsteinsson, V., 1995. Tagging experiments using conventional tags and electronic data storage tags for the observations of migration, homing and habitat choice in the Icelandic spawning stock of cod. ICES, Fish. Capt. Comm. C.M. 1995/B:19: 1–16.

Urquhart, G. G. & G. W. Smith, 1992. Recent developments of a fixed hydrophone array system for monitoring movements of aquatic animals. In I. G. Priede & S. M. Swift (eds), Wildlife Telemetry. Remote Monitoring and Tracking of Animals, Ellis Horword, Chichester: 342–353.

Urquhart, G. G. & P. A. M. Stewart, 1993. A review of techniques for the observation of fish behaviour in the sea. ICES mar. Sci. Symp. 196: 135–139.

Van Montfrans, J., J. Capelli, R. J. Orth & C. H. Ryer, 1986. Use of microwire tags for tagging juvenile crabs (*Callinectes sapidus* Rathbun). J. Crust. Biol. 6: 370–376.

Van Montfrans, J., C. H. Ryer & R. J. Orth, 1991. Population dynamics of the blue crabs *Callinectes sapidus* Rathbun in a lower Chesapeake Bay tidal marsh creek. J. exp. mar. Biol. Ecol. 153: 1–14.

Wolcott, T. G., 1980a. Heart-rate telemetry using micropower integrated circuits. In C. J. Amlaner & D. W. MacDonald (eds), Handbook for Biomedical Telemetry and Radio Tracking, Pergamon Press, Oxford: 279–286.

Wolcott, T. G., 1980b. Optical and radio-optical techniques for tracking nocturnal animals. In C. J. Amlaner & D. W. MacDonald (eds), Handbook for Biomedical Telemetry and Radio Tracking, Pergamon Press, Oxford: 333–337.

Wolcott, T. G., 1995. New options in physiological and behavioural ecology through multichannel telemetry. J. exp. mar. Biol. Ecol. 193: 257–275.

Wolcott, T. G. & A. H. Hines, 1989a. Ultrasonic telemetry transmitters for behavioral studies on free-ranging blue crabs. In C. J. Amlaner (ed.), Biotelemetry X, Proc. 10th Int. Symp. Biotelemetry, The University of Arkansas Press, Fayetteville: 285–295.

Wolcott, T. G. & A. H. Hines, 1989b. Ultrasonic telemetry of muscle activity from free-ranging marine animals: A new method for studying foraging by blue crabs (*Callinectes sapidus*). Biol. Bull. 176: 50–56.

Wolcott, T. G. & A. H. Hines, 1990. Ultrasonic telemetry of small-scale movements and microhabitat selection by molting blue crabs (*Callinectes sapidus*). Bull. Mar. Sci. 46: 83–94.

*Hydrobiologia* **371/372:** 133–141, 1998.
*J. P. Lagardère, M.-L. Bégout Anras &G. Claireaux (eds), Advances in Invertebrates and Fish Telemetry.*



# Electromagnetic telemetry of lobster (*Homarus gammarus* (L.)) movements and activity: preliminary results

I. P. Smith, K. J. Collins & A. C. Jensen
*Department of Oceanography, University of Southampton, Southampton Oceanography Centre, European Way, Southampton SO14 3ZH, U.K.*
*E-mail: philip.smith@soc.soton.ac.uk*

*Key words:* lobster, *Homarus gammarus*, electromagnetic telemetry, movement, activity, artificial reef

## Abstract

Individual European lobsters were tracked on an artificial reef using an electromagnetic telemetry system, which detected movements between reef units and also recorded body movements (pitching and rolling) indicated by a tilt switch incorporated into the transmitting tag. Several environmental variables were recorded simultaneously by the telemetry system, which was self-contained on the seabed. To date, 26 lobsters (14 female, 12 male) have been tagged, up to seven have been monitored simultaneously and four individuals have been tracked for over 6 months. Movements between units of the artificial reef (over open seabed away from shelter) were predominantly nocturnal, although daytime movements occurred during periods of low light levels associated with increased turbidity resulting from wave action. A marked decline in the number of inter-reef movements from late summer to winter was most closely related to decreasing water temperature. Activity indicated by the tilt switch was greater at night for most lobsters, but there were moderate levels of activity during the day. As with inter-reef movements, activity declined as winter progressed and in addition the diel pattern diminished.

## Introduction

European lobsters (*Homarus gammarus* (L.)) are large, commercially important crustaceans, yet there is little published information about their behaviour in the wild. Detailed information about lobster movements and variation in activity in the field is valuable from ecological (Sutherland, 1996) and fishery management perspectives (Bannister, 1986). The paucity of information on lobster behaviour is probably largely a consequence of the difficulties in observing nocturnal marine animals. Mark-recapture studies have provided a broad indication of the mobility of *H. gammarus*: most individuals remain within small areas for long periods, but a few travel more than 10 km along the coast (Simpson, 1961; Hepper, 1978; Cooper & Uzmann, 1980; Jensen et al., 1994). There is no indication of the seasonal onshore/offshore migration evident in some populations of the closely related American lobster (*H. americanus* Milne Edwards) (Cooper & Uzmann, 1980). More detailed monitoring on a smaller scale with ultrasonic tracking (e.g. Collins & Jensen, 1992) is limited by reflection and attenuation of the signal by the irregular rocky substratum.

Electromagnetic tracking uses low frequency electromagnetic/inductive coupling between transmitting and receiving coils of wire (Ramm, 1980). The range of detection is low (a few metres) compared with ultrasound transmission, but the signal is not severely attenuated or reflected by rock. Collins et al. (1994) used an electromagnetic tracking system to monitor movements of *H. gammarus* on an artificial reef some 3 km from the shore. Collins et al. (in press) described an improved system that allowed more lobsters to be tracked simultaneously and which additionally conveyed independent information about lobster activity. The purpose of the present paper is to present preliminary results to illustrate the potential of the electromagnetic telemetry system. To preview the type of information available, data are presented from a small number

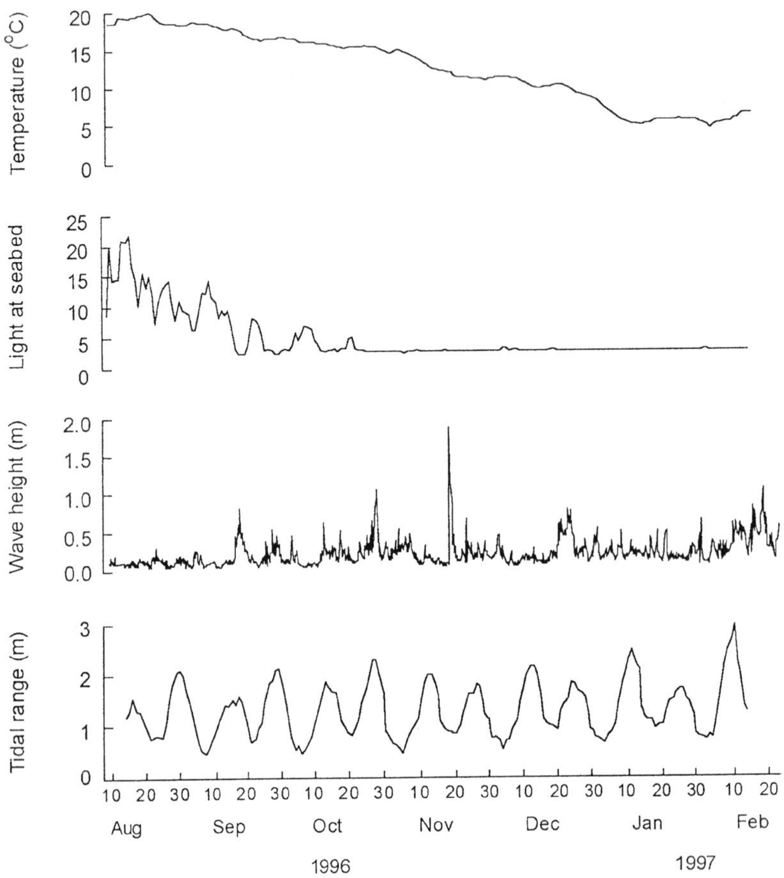


*Figure 1.* Environmental conditions during the study period. Light measurements are displayed in relative units from an analogue to digital converter.

of individual lobsters tracked for prolonged periods. A more comprehensive statistical analysis based on a greater number of individuals will be reported shortly.

## Materials and methods

### *Study site*

Lobsters were tracked on an artificial reef in Poole Bay on the central south coast of England. The reef consists of eight conical piles (1 m high, 4 m diameter) of blocks (0.4 × 0.2 × 0.2 m) made of concrete or cement-stabilised pulverised fuel ash. The piles, subsequently referred to as reef units, are arranged in two rows of four, aligned east–west, occupying an area of sedimentary seabed 15 m × 35 m at a depth of 10.5 m below chart datum (Collins et al., 1991).

*Table 1.* Derivation of activity level from the number of actuations of a tilt switch incorporated within the electromagnetic tag

| Activity level | Number of tilts (denary) | | Number of tilts (binary) | |
|---|---|---|---|---|
| | Min | Max | Min | Max |
| 0 | 0 | 0 | 00000000 | 00000000 |
| 1 | 1 | 1 | 00000001 | 00000001 |
| 2 | 2 | 3 | 00000010 | 00000011 |
| 3 | 4 | 7 | 00000100 | 00000111 |
| 4 | 8 | 15 | 00001000 | 00001111 |
| 5 | 16 | 31 | 00010000 | 00011111 |
| 6 | 32 | 63 | 00100000 | 00111111 |
| 7 | 64 | 127 | 01000000 | 01111111 |
| 8 | 128 | 255 | 10000000 | 11111111 |

### *Telemetry system*

The design of the transmitting tag and the receiving system has been described in detail by Collins et al.

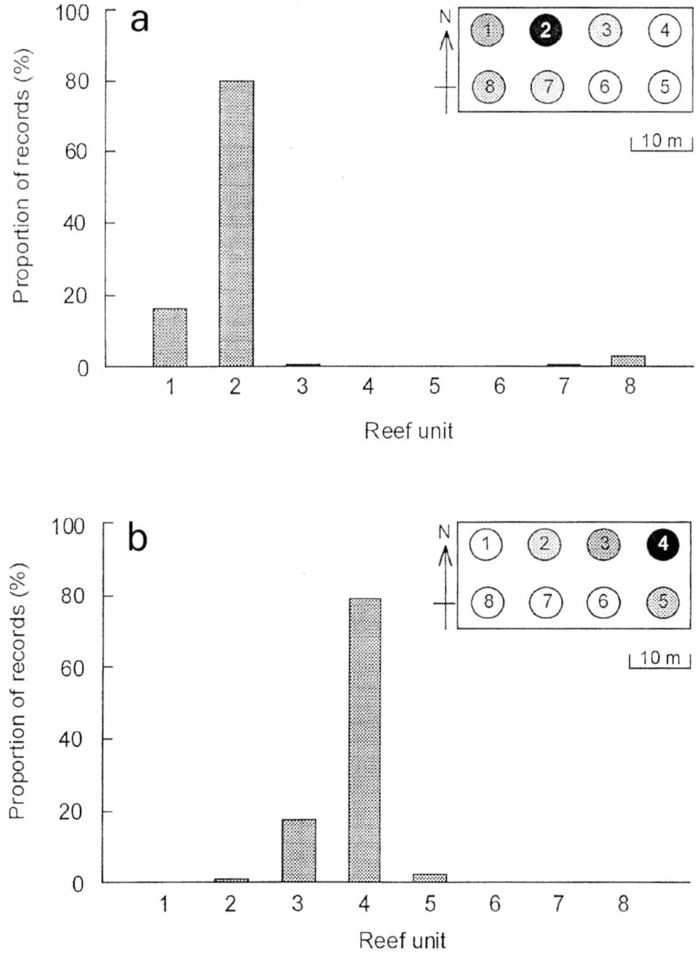


*Figure 2.* Examples of the proportion of records of individual lobsters on each reef unit. a. Female of carapace length 99 mm ($n = 8526$ records); b. female of carapace length 112 mm ($n = 8358$). These individuals were chosen to illustrate occupancy of different reef units. Schematic diagrams of the artificial reef show the spatial arrangement of reef units with degree of occupancy indicated by shading.

(in press). In summary, the tags were controlled by a microprocessor and transmitted a coded signal every 30 s at 32.7 kHz, which conveyed tag identity (potentially 1–99) and activity level (integer values 0–8). Activity level was derived from the number of tilts $> 20°$ from the horizontal in a 10 min period detected by a tilt switch incorporated into the electronic tag. At the end of each 10 min period, the activity level transmitted was set to the level of the most significant digit of the binary number of tilts and the counter within the microprocessor was reset to zero (Table 1). Tag signals were detected by 5 m diameter loop aerials laid on the seabed around each of the reef units connected, *via* a selector switch, to a receiver (Mariner Radar, Lowestoft). The receiver was adjusted for a range of detection of 2–3 m, to give a small overlap in reception area of adjacent aerials on the open seabed between reef units. Output from the receiver was passed to a computer (PC/104 core module, Ampro Computers Inc, Sunnyvale, California) contained in a waterproof housing with the receiver.

The computer was programmed to operate the system on a 10 min cycle: environmental sensors were interrogated, then the receiver was switched to each of the eight aerials in turn for 1 min. The identity and activity level of detected tag signals were decoded and recorded on magnetic disk (400 Mb 2.5″ hard drive) with the time and the aerial number. The following environmental variables were measured: temperature (National Semiconductor LM35 precision Centigrade

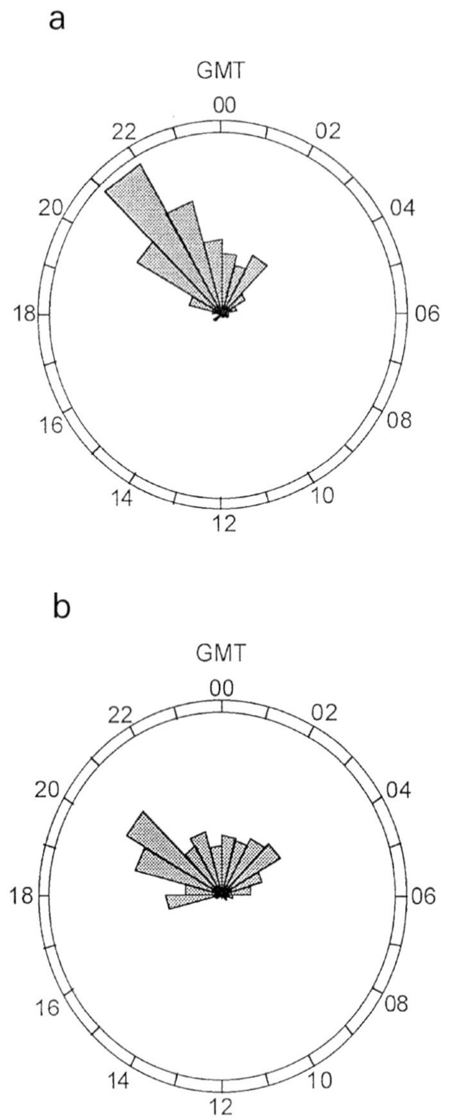


*Figure 3.* Number of movements between reef units by two lobsters in relation to time of day from August to October. The circular histograms have the same scale (0–60 movements). a. Female of carapace length 99 mm; b. male of carapace length 75 mm.

temperature sensor), light (R.S. Ltd 305–462 general purpose photodiode in a linear photometer FET op-amp circuit), hydrostatic pressure (Sensor Technics SSC3000 temperature compensated silicon stainless steel pressure sensor) and current speed (modified Braystoke current meter with impeller magnet/reed switch pulsing). In addition, surface wave height was estimated four times a day from 2050 readings taken from the pressure sensor at a sampling frequency of 4 Hz.

The system was powered by one or two sets of six 12 V sealed lead acid batteries (Yuasa NP15-12) in a separate housing. The system components were connected with underwater pluggable connectors (Subconn, MacArtney A/S, Esbjerg), enabling divers to replace the batteries and data logger at intervals of up to four weeks. Data were then downloaded to a desktop computer for analysis.

*Lobster capture and tagging*

From June to September 1996, parlour pots (Lovewell et al., 1988) placed adjacent to each of the eight reef units were baited and left for one or two days before divers removed lobsters and brought them to the surface in individual net bags. A few lobsters were also caught by hand. Lobsters were measured and the transmitting tag was attached to the dorsal surface of the cephalothorax with quick-setting epoxy resin after drying the carapace with propanone. The tag was attached with the integral tilt switch oriented so that pitch and roll would be detected equally. Divers then returned lobsters to the reef unit where they were caught. Lobsters were maintained in cool, dark, moist conditions during the tagging procedure, which involved aerial exposure of approximately 10 min.

The tags consisted of a battery and components contained within the transmitting coil and the whole encapsulated in epoxy resin, giving a disc of diameter 40 mm and depth 10 mm (Collins et al., in press). The mass of the tag (36 g) was approximately 7% of the body mass of the tagged lobsters.

To date, 26 lobsters (14 females, 12 male) of carapace length 75–139 mm have been tagged. Ten lobsters have been tracked for longer than 1 month, four (one male, three females) have been tracked for over 6 months and up to eight individuals have been monitored simultaneously.

## Results

*Environmental conditions*

Water temperature reached a maximum in late August and declined at a variable rate thereafter (Figure 1). Wave height increased sporadically in response to fluctuations in wind speed and was generally greater during late autumn and winter. Increased wave action caused

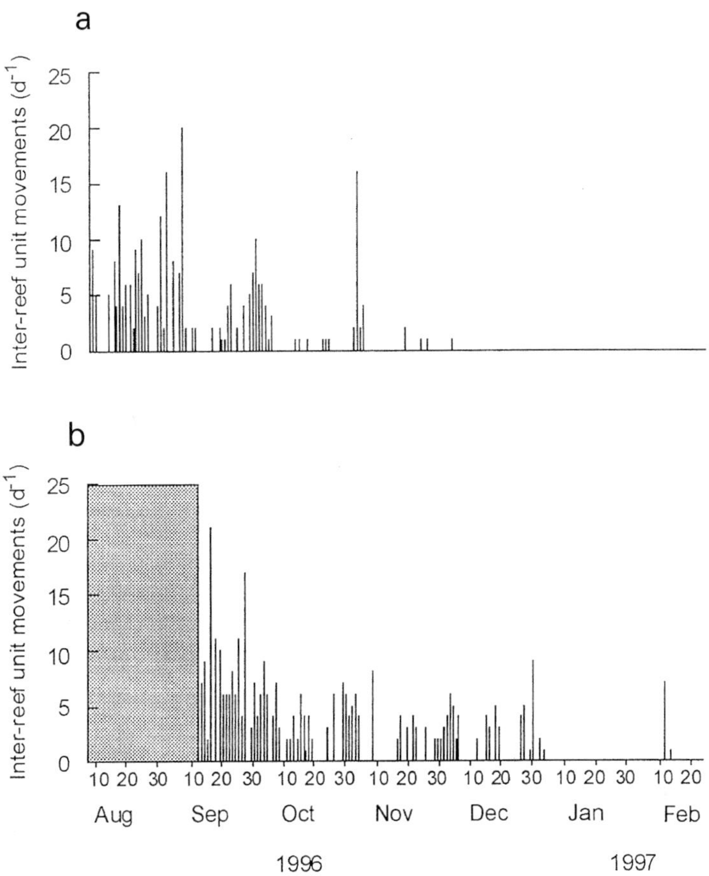


*Figure 4.* Daily number of movements between reef units by two lobsters in relation to time of year. a. Female of carapace length 99 mm, tagged 9/8/96; b. male of carapace length 75 mm, tagged 14/9/96 (grey box indicates period prior to tagging). In both cases, lobsters were detected within the study site until after the end of the period shown.

greater turbidity, thus reducing light level at the sea bed. The semilunar cycle of tidal range evident from measurements of hydrostatic pressure would be most likely to affect lobsters by the greater tidal current speeds (up to 0.6 m s$^{-1}$ at 1 m above the seabed) experienced during large ('spring') tides.

*Movements among reef units*

Most lobsters were detected predominantly on a particular reef unit with small numbers of records on other units (Figure 2). The majority of movements were between the mainly inhabited reef unit and an adjacent unit. Inter-reef movements were predominantly nocturnal, with a peak in the early part of the night for most lobsters, although there was a degree of individual variability in this pattern (Figure 3).

For each of the four lobsters tracked for over 6 months, there was a marked decline in the number of inter-reef unit movements from late summer to winter (Figure 4). For each individual, stepwise regression was used to test the relationship between daily number of movements (square root transformed) and water temperature, mid-day illumination (averaged over the period 11:00–13:00 UT+0), wave height, tidal range and movements up to five days previously (to account for autocorrelation). The daily number of inter-reef unit movements of each of the lobsters was significantly related to water temperature (partial regression coefficients for temperature, $b_t > 0$, $t = 3.658$–$6.115$, $P < 0.0005$), but not to the other environmental variables ($P > 0.05$). Overall regression results for individual lobsters: female, carapace length (CL) 99 mm, $F_{3,181} = 43.641$, $P < 0.0005$, $r^2 = 0.410$; female, 112 mm CL, $F_{2,178} = 27.144$, $P < 0.0005$,

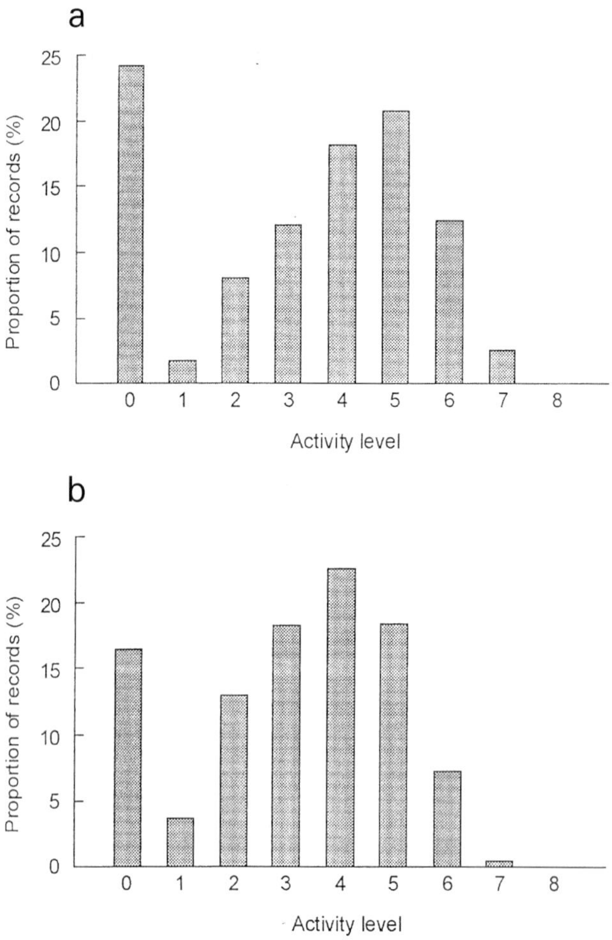


*Figure 5.* Frequency distribution of activity levels for two lobsters. a. Female of carapace length 99 mm ($n = 8526$ records); b. male of carapace length 75 mm ($n = 5167$).

$r^2 = 0.225$; female, 93 mm CL, $F_{2,154} = 48.659$, $P < 0.0005$, $r^2 = 0.379$; male, 75 mm CL; $F_{3,148} = 30.854$, $P < 0.0005$, $r^2 = 0.372$.

*Activity*

Most lobsters spent much time inactive or at intermediate levels of activity (Figure 5). Activity was generally greatest in the early part of the night, at around the same time as the peak in inter-reef movements (Figure 6a), although some individuals showed no clear pattern of activity in relation to time of day (Figure 6b). As with inter-reef movements, activity was significantly related to water temperature (stepwise multiple regression, $b_t > 0$, $t = 2.579–7.783$, $P < 0.0005$ for all four lobsters). In addition, the activity of two female lobsters was positively related to wave height ($t = 2.594$, $P = 0.01$, $t = 6.430$, $P < 0.05$, respectively) and the activity of another female was negatively related to mid-day illumination ($t = 2.219$, $P = 0.004$). Activity was not significantly related to tidal range in any of the four lobsters ($P > 0.05$). Overall regression results: female 99 mm CL, $F_{3,183} = 84.141$, $P < 0.0005$, $r^2 = 0.573$; female 112 mm CL, $F_{3,177} = 151.765$, $P < 0.0005$, $r^2 = 0.715$; female 93 mm CL, $F_{4,149} = 199.961$, $P < 0.0005$, $r^2 = 0.839$; male 75 mm CL, $F_{2,150} = 275.916$, $P < 0.0005$, $r^2 = 0.783$. In addition to the overall reduction in activity levels, the nocturnal pattern commonly observed during summer and autumn diminished in winter (Figure 6a).

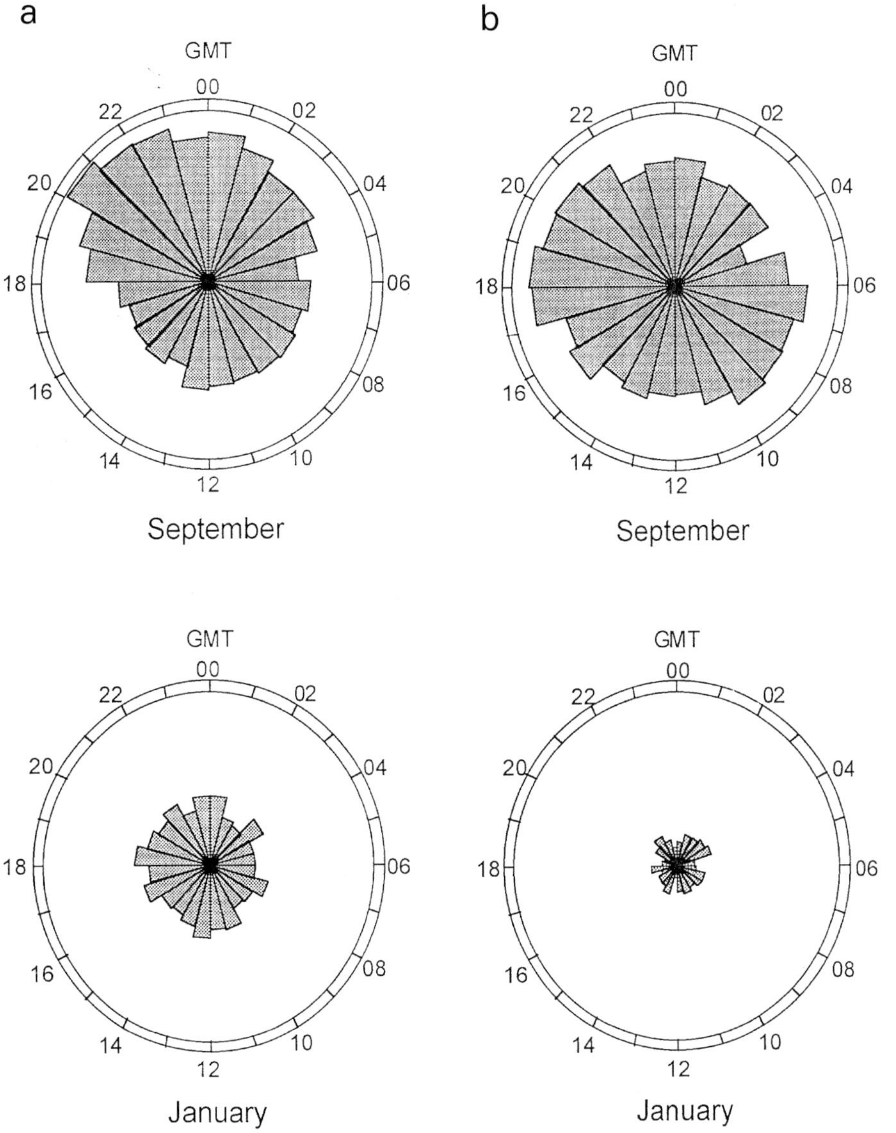


*Figure 6.* Mean activity level of two lobsters in relation to time of day in September and January. a. Female of carapace length 99 mm ($n = 3221$ and 2855 records in January and September, respectively); b. male of carapace length 75 mm ($n = 1832$ and 2709 in January and September, respectively). The circular bar charts have the same scale (mean activity 0.5–5.0).

## Discussion

Electromagnetic telemetry has proven to be an effective way of monitoring movements and activity of individual lobsters for long periods. Although lobster movements could only be monitored on a small scale, such short range movements are likely to be an important aspect of habitat use by lobsters, given that they apparently spend long periods within small areas (Simpson, 1961; Hepper, 1978; Cooper & Uzmann, 1980; Jensen et al., 1994). Small scale movements can also be sufficient to lead to capture in baited traps fished commercially or for stock surveys. Indeed, during the present study lobsters have been tracked during the period prior to capture in traps placed at the study site (unpublished observations). Larger areas could be monitored with a greater number of individual aerials, but this would involve either a longer operating cycle (to monitor each aerial for the same time as at present), which would reduce the temporal resolution of tracking, or shortening the listening period for each aerial in conjunction with increasing the tag transmission rate, which would reduce the tag battery life.

Preliminary results indicate that movements between reef units over open seabed are predominantly nocturnal, often occurring shortly after sunset. It is well known that, in general, clawed lobsters (Chapman et al., 1975; Cooper & Uzmann, 1980; Ennis, 1984; Karnofsky et al., 1989) and other marine decapods (Warner, 1977; Herrnkind, 1980; Barshaw & Spanier, 1994) are nocturnal and this probably reflects their vulnerability to predation by diurnally active fish and cephalopods (Cobb & Wahle, 1994). The diel activity patterns of lobsters may also reflect that of nocturnal prey, such as small crustaceans and polychaetes, although lobsters also feed on sessile organisms, such as molluscs and algae (Cooper & Uzmann, 1980), that would be equally available during the day or night. Activity indicated by the tilt switch was also greater at night, but this measure also suggested that lobsters were by no means completely inactive during the day. It is not possible to tell from the telemetry data whether lobsters were on the surface of the reef or inside cavities within it, but diver observations indicate that lobsters rarely make excursions away from shelter during the day (Jensen et al., 1994). It therefore seems as though a certain amount of body movement occurs within shelters, but the nature and function of this activity is unknown at present. Wickins et al. (1996) described a variety of behaviours exhibited by juvenile *H. gammarus* within mud burrows in aquaria, which were mostly related to feeding and construction, maintenance and ventilation of the burrow.

For the few lobsters tracked long enough to examine seasonal changes, the frequency of inter-reef movements and activity declined as winter progressed. These changes were most closely and consistently related to water temperature and were not unexpected for a poikilothermic animal. Indeed, studies using other methods have indicated that locomotor activity and feeding by lobsters diminishes as water temperature falls in winter (Cooper & Uzmann, 1980; Ennis, 1984; Karnofsky et al., 1989). However, the detail afforded by electromagnetic telemetry should eventually allow assessment of the importance of other environmental factors and of the extent to which males and females of various sizes differ in seasonal changes in activity. For example, movements by lobsters away from shelter may be reduced during periods of increased wave action, since waves create oscillatory water movement at the seabed and flume experiments have demonstrated that lobster mobility can be impaired by moving water (Howard & Nunny, 1983). There was some indication from the present data that body movements were positively related to wave height, possibly reflecting passive movement by wave-induced water movement at the seabed.

In addition to the decline in activity with falling water temperature, the diel pattern of activity expressed by most lobsters during summer and autumn diminished. This may simply reflect an overall depression of activity in low water temperatures. Alternatively, the loss of the diel pattern may indicate that lobster activity varies in response to light level at the seabed, since daytime light levels were considerably reduced during the winter months. There was limited evidence from one individual that, once the effect of temperature was accounted for, activity was greater at times of reduced illumination. Another possible explanation is that there are seasonal changes in endogenous rhythmicity, perhaps under the influence of water temperature, as has been found in other crustaceans (De Coursey, 1983). Further field data in conjunction with experiments in controlled conditions would be required to distinguish among, or assess the relative contribution of, these processes.

As longer time series and data from more lobsters are accumulated, a more complete description of seasonal and environmental variation in the movement and activity patterns of individuals of different sex, moult state and maturity will be possible. This information will contribute to a greater understanding of the habitat requirements of lobsters, their patterns of resource utilisation and their susceptibility to capture in commercial fisheries and stock surveys.

## Acknowledgements

This study was funded by the United Kingdom Ministry of Agriculture Fisheries and Food. We are grateful for the assistance given in the field by Jenny Mallinson, Frank Elson and volunteer divers. John French and Mike Wilkin helped us develop the telemetry system hardware and software, respectively. The contribution of Dr R. C. A. Bannister and Dr J. T. Addison (CEFAS Lowestoft) in establishing this project is gratefully acknowledged.

## References


Bannister, R. C. A., 1986. Assessment and population dynamics of commercially exploited shellfish in England and Wales. Can. spec. Publ. Fish. aquat. Sci. 92.

Barshaw, D. & E. Spanier, 1994. The undiscovered lobster: a first look at the social behaviour of the Mediterranean slipper lobster, *Scyllarides latus* (Decapoda, Scyllaridae). Crustaceana 67: 187–197.

Chapman, C. J., A. D. F. Johnstone & A. L. Rice, 1975. The behaviour and ecology of the Norway lobster, *Nephrops norvegicus* (L.). In H. Barnes (ed.), Proceedings of the 9th European Marine Biological Symposium. Aberdeen University Press, Aberdeen: 59–74.

Cobb, J. S. & R. A. Wahle, 1994. Early life history and recruitment processes of clawed lobsters. Crustaceana 67: 1–25.

Collins, K. J., J. French & A. C. Jensen, 1994. Electromagnetic tracking of lobsters on an artificial reef. In Electronics in Engineering, 19–21 July 1994, Cambridge. Conference Publication No. 394, Institution of Electrical Engineers, London: 1–5.

Collins, K. J. & A. C. Jensen, 1992. Acoustic tagging of lobsters on the Poole Bay artificial reef. In I. G. Priede & S. M. Swift (eds), Wildlife Telemetry. Remote Monitoring and Tracking of Animals. Ellis Horwood Ltd, London: 354–358.

Collins, K. J, A. C. Jensen & A. P. M. Lockwood, 1991. Artificial reefs: using coal-fired power station wastes constructively for fishery enhancement. Oceanol. Acta 11: 225–229.

Collins, K. J., I. P. Smith & A. C. Jensen, 1998. Lobster (*Homarus gammarus*) behaviour studies using electromagnetic telemetry. In Fifth European Conference on Wildlife Telemetry, Strasbourg 25–30 August 1996: Paper SVIII–5, in press.

Cooper, R. A. & J. R. Uzmann, 1980. Ecology of juvenile and adult *Homarus*. In J. S. Cobb & B. F. Phillips (eds), The Biology and Management of Lobsters. II. Ecology and Management. Academic Press, New York: 97–142.

De Coursey, P. J., 1983. Biological timing. In J. Vernberg & W. B. Vernberg (eds), Biology of Crustacea. 7. Behavior and Ecology. Academic Press, New York: 107–162.

Ennis, G. P., 1984. Territorial behavior of the American lobster *Homarus americanus*. Trans. am. Fish. Soc. 113: 330–335.

Hepper, B. T., 1978. Population dynamics of the lobster *Homarus gammarus* (L.) off the coasts of England. Fish. Res. tech. Rep. M.A.F.F. Dir. fish. Res., Lowestoft 41.

Herrnkind, W. F., 1980. Spiny lobsters: patterns of movement. In J. S. Cobb & B. F. Phillips (eds), The Biology and Management of Lobsters. I. Physiology and Behavior. Academic Press, New York: 349–407.

Howard, A. E. & R. S. Nunny, 1983. Effects of near-bed current speeds on the distribution and behaviour of the lobster, *Homarus gammarus* (L.). J. exp. mar. Biol. Ecol. 71: 27–42.

Jensen, A. C., K. J. Collins, E. K. Free & R. C. A. Bannister, 1994. Lobster (*Homarus gammarus*) movement on an artificial reef: the potential use of artificial reefs for stock enhancement. Crustaceana 67: 198–211.

Karnofsky, E. B., J. Atema & R. H. Elgin, 1989. Natural dynamics of population structure and habitat use of the lobster, *Homarus americanus*, in a shallow cove. Biol. Bull. 176: 247–256.

Lovewell, S. R., A. E. Howard & D. B. Bennet, 1988. The effectiveness of parlour pots for catching lobsters (*Homarus gammarus* (L.)) and crabs (*Cancer pagurus* L.). J. Cons. int. Explor. Mer 44: 247–252.

Ramm, D. C., 1980. Electromagnetic tracking of rock lobsters (*Jasus novaehollandiae*). Aust. J. mar. Freshwat. Res. 31: 263–269.

Simpson, A. C., 1961. A contribution to the bionomics of the lobster (*Homarus vulgaris* Edw.) on the coast of North Wales. Fishery Invest., Lond., Ser. 2 23(7): 28 pp.

Sutherland, W. J., 1996. From Individual Behaviour to Population Ecology. Oxford University Press, Oxford, 213 pp.

Warner, G. F., 1977. The Biology of Crabs. Elek Science, London, 202 pp.

Wickins, J. F., J. C. Roberts & M. S. Heasman, 1996. Within-burrow behaviour of juvenile European lobsters *Homarus gammarus* (L.). Mar. freshwat. Behav. Physiol. 28: 229–253.

*Hydrobiologia* **371/372:** 143–153, 1998.
*J. P. Lagardère, M.-L. Bégout Anras &G. Claireaux (eds), Advances in Invertebrates and Fish Telemetry.*



# Basic movement pattern and chemo-oriented search towards baited pots in edible crab (*Cancer pagurus* L.)

Katrine Skajaa, Anders Fernö, Svein Løkkeborg[1] & Eli K. Haugland[1]
*Department of Fishery and Marine Biology, University of Bergen, High Technology Centre, N-5020 Bergen, Norway (e-mail: katrine.skajaa@ifm.uib.no)*
[1] *Institute of Marine Research, C. Sundtsgt. 64, N-5024 Nordnes, Norway*

*Key words:* edible crab, *Cancer pagurus*, ultrasonic telemetry, movement pattern, chemoreception

## Abstract

The basic (natural) movement pattern and gear-induced behaviour of female edible crab *Cancer pagurus* L. were studied by means of a stationary positioning system. Nine crabs were tagged with ultrasonic transmitters and their positions were monitored approximately every third minute for 9–24 days. Because the crabs sometimes hid, the signals were often received improperly, resulting in inaccurate position fixes, and a computer program to distinguish between movement and inactivity was developed.

Edible crabs showed a nocturnal activity cycle. During the day crabs were in hiding as indicated by less accurate position fixes. There were variations in activity levels both between and within individuals. Three crabs hardly moved at all during the study, while the others were active during some, or all nights. The crabs did not return to the same place after movement, but stayed close to the hydrophone triangle (290 × 350 × 175 m) throughout the study. The activity and movement patterns are discussed in relation to optimal foraging behaviour and predator avoidance.

Six baited pots were set in the area on seven nights. Four of the crabs located baited pots during the study. Three crabs located pots more than once, resulting in a total of nine localisations. Three of the tagged crabs were caught. The localisations of baited pots were divided into four categories on the basis of the current direction at the start of the localisation and whether the crab was caught or not, in order to suggest the probability of the localisation being chemically stimulated. At least six of the localisations seemed to be induced by chemical attraction to the bait. Pots were always located during night at the time of high activity. The searching distances ranged from 12 to 48 m. Speed of movement during searching was higher than speed during basic movement.

## Introduction

Most studies of the movement of edible crab (*Cancer pagurus* L.) have used tagging-recapture methods and investigated migration or long-range movement (Williamson, 1900; Meek, 1913; Mason, 1965; Gundersen, 1977, 1979; Camus, 1982; Bennett & Brown, 1983; Latrouite & Le Foll, 1989). These studies have shown that crabs may move considerable distances and that females move further than males. The micro- and meso-scale movement of edible crab, however, is poorly known. Ultrasonic telemetry with a stationary positioning system gives us an opportunity to follow closely the movement of crabs in time and space.

The natural behaviour of a species sets limits for the operation and design of fishing gear. Some distance from a pot the crab will react to the bait as if it was a natural food item. In the vicinity of the pot, however, the behaviour is affected by the shape of the pot, though it is still a natural reaction in the sense that the animal reacts to the baited gear as it reacts to a novel item in its natural environment. In order to understand responses to fishing gear, and the variation in response, it is therefore important to gain knowledge about the basic

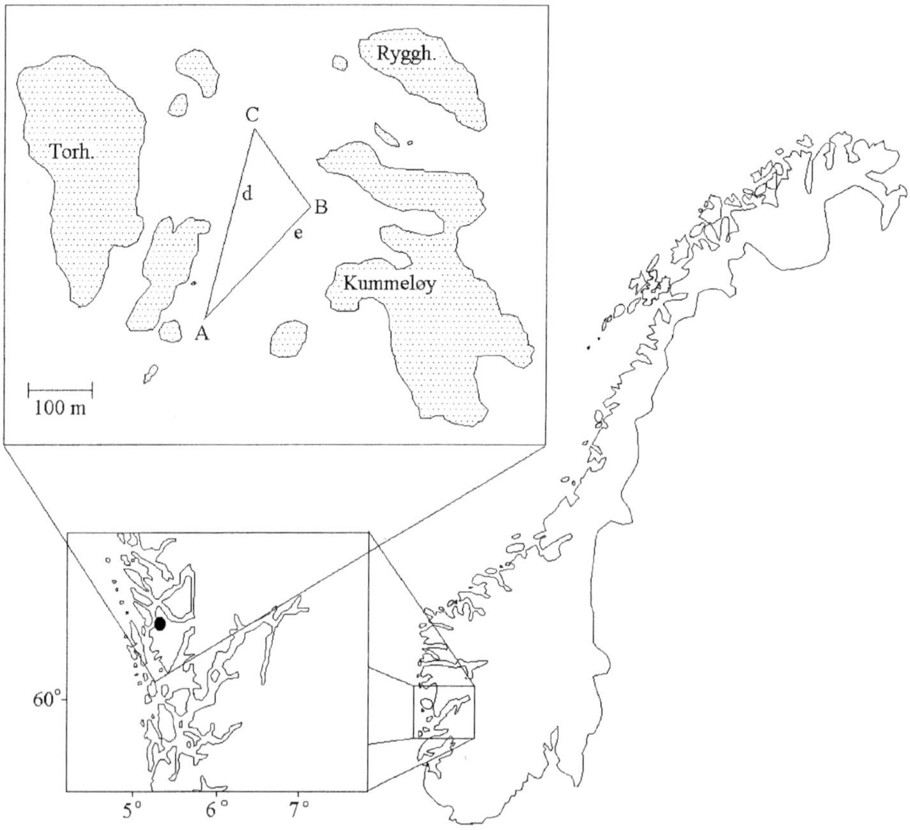


*Figure 1.* Study area for crabs relative to Bergen (•). The positions of the three hydrophones (A, B and C) and the two current meters (d and e) are indicated.

behaviour (natural behaviour in the absence of fishing gear) of the animal concerned (Fernö, 1993).

Until recently most studies of gear-induced behaviour have focused on behaviour close to fishing gears, and scuba diving, underwater TV cameras and remotely operated vehicles have all contributed to our understanding of such behaviour (Ben-Tuvia & Dickson, 1969; Wardle & Hollingworth, 1993; Zimmer-Faust et al., 1995). Baited fishing gears attract animals from a distance, and none of these methods is able to demonstrate either the mechanism that attracts the animals to the gear or the basic movements of the species. In this study a stationary positioning system was used to investigate the movement patterns of edible crab under natural conditions and in response to baited pots.

## Materials and methods

### *Study site*

The study was conducted in Austevoll, a group of islands close to Bergen on the west coast of Norway (Figure 1), in June and July 1994. The area consists of many small islets and, according to local fishermen, is a natural habitat for crabs. The depth of the sound in which the study was conducted ranged from 20 to 30 m, and scuba diving in the area revealed a seabed with shellsand that was relatively flat, but with some heights, valleys and stones. Current direction and current speed were recorded every 10 min with current meters (SD6000, Sensordata A/S, Bergen) suspended one metre above the seabed at two locations inside the experimental area. Tidal currents change with the lunar cycle, and in the beginning of the study the current was in one direction only, but during the experiments with baited gear the current turned regularly with the

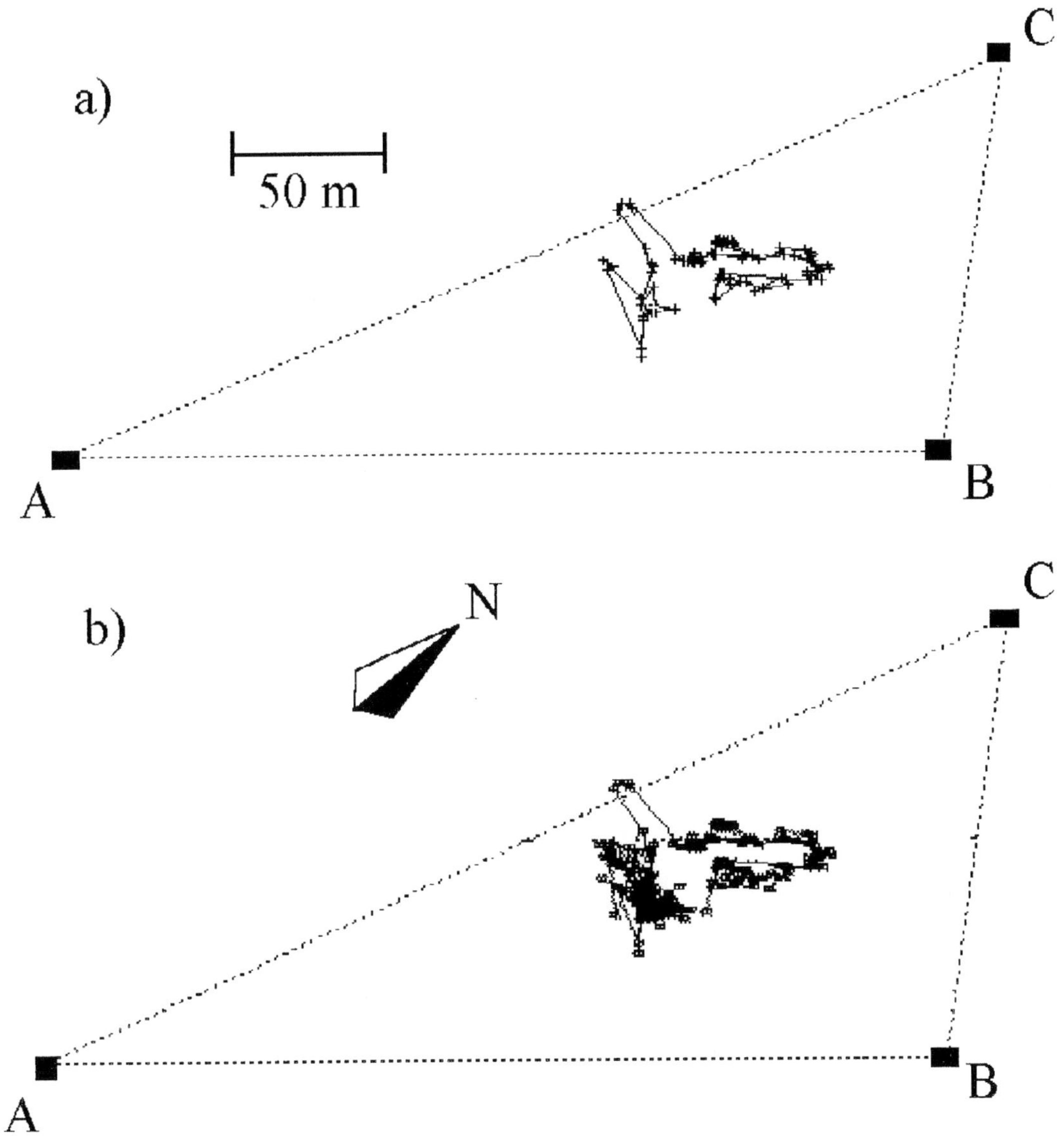


*Figure 2.* Movement of one crab through one day with data modified by the movement analysis computer program (a) and with unmodified data (b).

tide. The current velocity was usually below 2 cm $s^{-1}$ (threshold of the current meter).

*Telemetry equipment*

The positioning system (VR-20, Vemco Ltd., Halifax, Canada) had a fixed array of three hydrophone buoys anchored to the seabed with radio transmission to a base station. The buoys received acoustic pulses from transmitters attached to the crabs, and on the basis of the arrival time of the pulses, the base station calculated the position of the acoustic transmitters relative to the three buoys. A number of received pulse-times (3 to 15 depending on how many pulses that were received successfully in 15 sec) were aligned and an average position (*X* and *Y* position) was calculated together with a deviation value ($DEV = ((\bar{X} - X_i)^2 + (\bar{Y} - Y_i)^2)$). Crystal-controlled transmitters with a stable frequency which produced one pulse per second were used. The cylindrical transmitters (V16-1H, length 48 mm, diameter 16 mm, weight in water 9 g) had electronics and batteries in epoxy casing. Under optimal acoustic

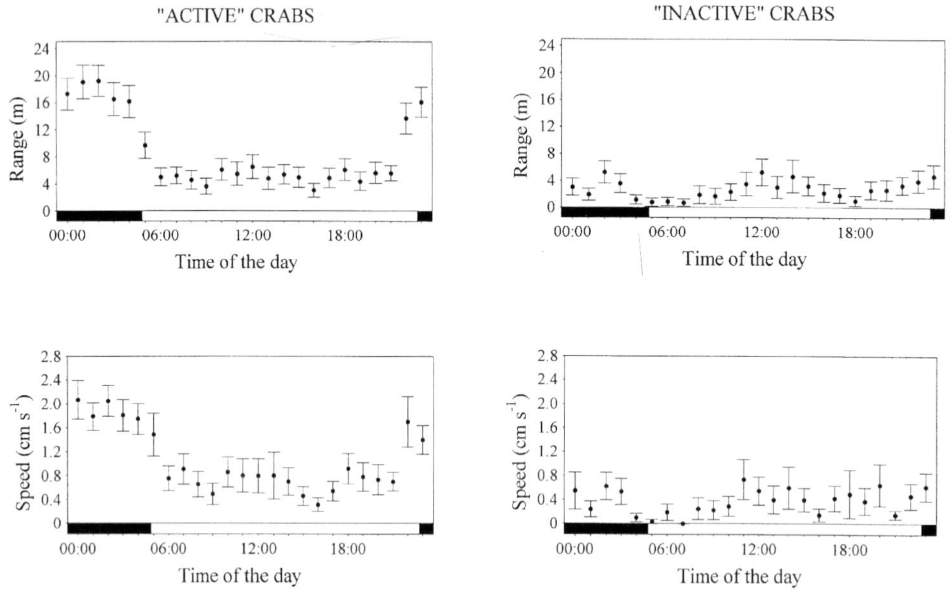


*Figure 3.* The mean speed during the 24 h (+2 h GMT) cycle and the mean range within which the crab moved within a period of one hour for active and inactive crabs. Vertical bars denote standard error. The dark and light sections indicate night and day in the middle of the study period.

conditions a transmitter within the hydrophone array could be positioned to an accuracy of about one metre. When the transmitter was outside the buoy triangle the accuracy decreased. Under poor acoustic conditions echoes and reverberations can interfere with the signal and cause the pulse to be received late or missed altogether. Each crab was tracked approximately every third minute except during the early part of the study when they were tracked more often because fewer individuals were tagged. A more detailed description of the positioning system is provided by Engås et al. (1996).

*Study animals and experimental procedure*

Nine female crabs with carapace widths ranging from 125 to 170 mm (mean = 142 mm) were tagged. Telemetry studies of this kind allow a limited number of animals to be tracked, and because male and female crabs may behave differently (Robles et al., 1989, *C. productus* L.) only female crabs were used in this study. The results therefore apply only to females. The crabs were caught in pots set close to the experimental area at depths of 5 to 30 m. The pots (nephrops pots, modified for crab fishing by enlarging the entrances), which were also used in the experiment, measured 50 × 50 × 25 cm and consisted of a frame of stainless steel covered with a net with openings on two sides (Bjordal, 1979). The bait was fresh saithe (*Pollachius virens* L.) cut into pieces and placed in a mesh bag inside the pot (about 100 g saithe per pot).

Crabs were tagged immediately after the pots were retrieved. Laboratory and field observations indicated that transmitters fastened only with quick epoxy glue were easily lost. The transmitters were therefore attached to the crabs with wires through 1 mm holes in each side of the carapace, and for the first six crabs with quick epoxy also. The dorsal part of the carapace was dried with acetone and scrubbed carefully with sandpaper before the transmitter was fastened. The tagging procedure took about one hour, after which the crab was released into the experimental area. Tracking began 10–15 min after release. Only medium sized female crabs with no lost appendages were used, except for one crab that lacked two legs. None of the crabs was ovigerous.

After tracking basic behaviour for 8–10 days, five or six baited pots were set in the area every day between 14:30 and 00:00 (GMT + 2 h) and retrieved the following day. After seven nights with pots in the area, studying both basic and gear-induced behaviour, basic behaviour was again studied for seven days. Since the current direction turned almost 180° with the tide, pots were set both up-stream and down-stream of the crabs. The distance between the crabs and the nearest pot var-

*Table 1.* Categories of localisations of baited pots based upon the statement in the header being true (+) or false (−)

| Localisation Category | Current directed towards the crab at the start of the localisation | Crab caught |
|---|---|---|
| 1 | + | + |
| 2 | + | − |
| 3 | − | + |
| 4 | − | − |

*Table 2.* Summary on tracking information of individual crab. Size is measured as the carapace width and rounded off to the nearest 5 mm. Total range is the range the crabs occupied throughout the study. The first six crabs were categorised as 'Active crabs' and showed distinct movement in contrast to the 'Inactive crabs' (the last three)

| Crab no. | Size (mm) | Total range (m) | Tracking time (days) |
|---|---|---|---|
| 1 | 125 | 195 | 13 |
| 2 | 150 | 329 | 11 |
| 3 | 140 | 134 | 13 |
| 6 | 170 | 79 | 10 |
| 7 | 135 | 402 | 24 |
| 8 | 130 | 88 | 24 |
| 4 | 140 | 44 | 8 |
| 5 | 150 | 44 | 14 |
| 9 | 140 | 67 | 21 |

ied from 1 to 59 m. It turned out to be difficult to set the pots exactly at the target spot so the direction from the pot to the crab relative to the current direction varied. During most nights all the crabs had at least one pot in their vicinity.

Except for the first two days, the positions of the pots were recorded by a transmitter attached to the pot and retrieved with a rope after the position was recorded. On the first two days the positions were estimated relative to the buoys and marks in the landscape. One of the pots carried a transmitter throughout each night to give a reference point.

The catch of each pot was recorded and size and sex of the crabs were noted. The transmitters recovered from tagged crabs that were caught were attached to other crabs caught on the same day.

### *Statistical validation of movement*

Ultrasonic telemetry gives inaccurate positions when tracking animals that hide in a shelter or a burrow (Phillips et al., 1984). Due to the inaccurate position fixes the estimated movement speed and distance would be highly overestimated, if the data are analysed in the recorded form. In order to differentiate between movement and inactivity of the crab, a computer program (movement analysis program) was developed to modify the position data before the data were analysed statistically. The program replaces the positions (*X* and *Y*) in a time interval that contains 20 positions, with the mean *X* and *Y* values for the same interval when two criteria are satisfied. The criteria used to determine whether the crab had moved or was inactive were based on the deviation values monitored for each position fix, indicating if the crab was hiding, and the regression values found for the *X* and *Y* position values relative to time, indicating systematic movement of the crab.

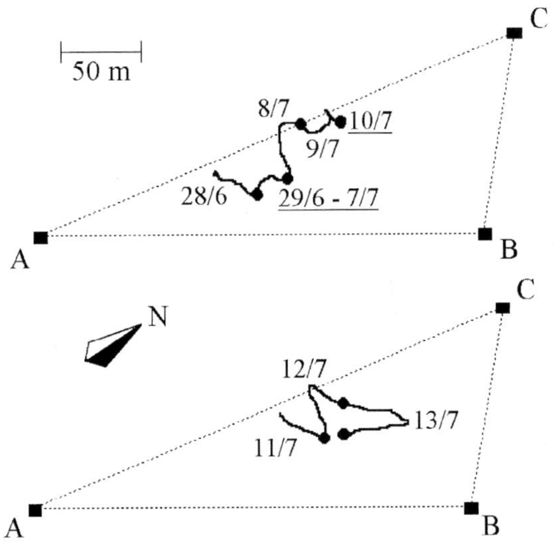


*Figure 4.* Movement of two crabs during a period of 3 and 13 days. Dots on the lines indicate the position during inactivity. Underlined dates indicate periods with no movement.

The program replaced the *X* and *Y* positions with the mean *X* and *Y* values for all succeeding intervals that satisfied the two requirements, i.e. where at least ten of the positions had a high deviation value (> 5) and where none of the regression values (*X* or *Y* relative to time) was significantly different from zero. Figure 2 shows the movement of one crab based on modified and unmodified data. The movement analysis program is described in detail in Skajaa (1997).

By swapping the positions with the mean values for the X and Y positions for a given interval it was indicat-

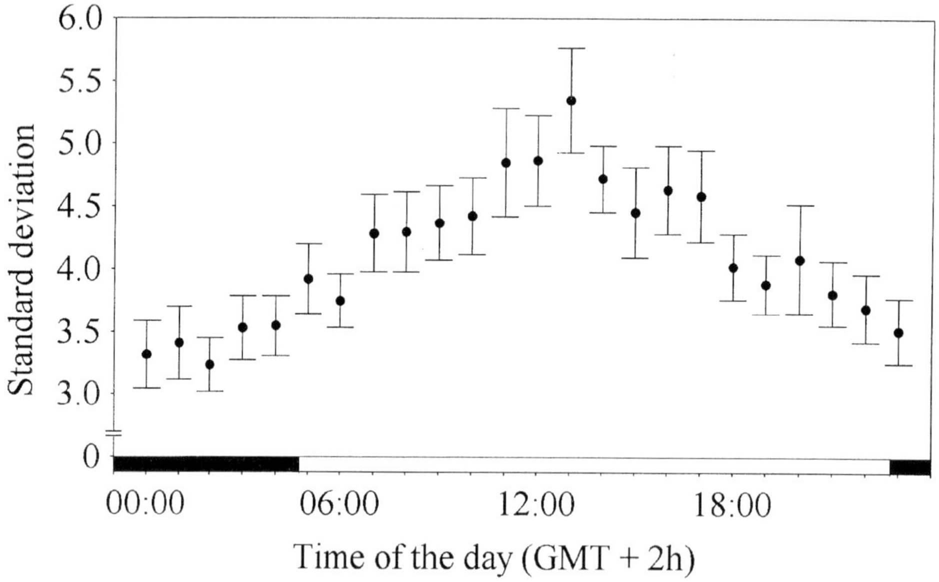


*Figure 5.* 'Degree of hiding' as indicated by mean SD for the position fixes per hour (mean ± SE). All crabs were included. The dark and light section indicate night and day in the middle of the study period.

ed that the crab was stationary during this interval. This could underestimate the activity, since the crabs might have moved within a restricted area around a burrow. However, during a period with exchanged values, the crab was either inactive or moving within a restricted area, and the main movement pattern when the crab moved from one place to another will still be shown by the data. Modification by the movement analysis program was done conservatively, i.e. if the crab obviously moved the positions were not exchanged even if this meant that the positions were not always exchanged when the crab might have been inactive. This will cause an overestimation of the movement speed during inactivity, which may counteract the previously mentioned underestimation.

Also when the crab was moving, most of the positions had a relatively high deviation value. The reason for this may be background noise, echoes or that the crab moved in the vicinity of stones. The calculated movement speed during activity could thus be somewhat overestimated.

### *Localisation of baited pot*

All the occasions on which the crab moved towards and came into contact with the pot were regarded as a localisation of the pot. The localisation was defined as starting (search started) when the crab began to move after having been inactive for a period of time (at least three position fixes), except in three of the reactions where the start of the localisation was clearly recognised by a significant change in movement direction ($> 90°$) followed by relatively straight movement towards the gear. The localisations were divided into four categories depending on the current direction relative to the crab at the start of the localisation, and whether the crab was caught or not (Table 1). The current was defined to be towards the crab when the deviation between the current direction and movement direction was $180° \pm 60°$.

An additional requirement for the localisations in categories 1 and 2 was that the period of time from when the gear was set to when the crab responded had to be long enough for the odour to reach the crab under the prevailing current velocity. Those localisations where the current was heading for the crab when its search started (categories 1 and 2) were regarded as chemically stimulated reactions, but it can not be ruled out that localisations in categories 3 and 4 might also have been reactions to chemical attractants (see discussion).

### *Data analysis*

Speed, distance and direction of movement between position fixes were calculated for all positions where

*Table 3*. Mean carapace width (mm) of female (F) and male (M) crabs caught during the experiment. There was no catch on 9. July and size was not measured on 10. July

| Date | Sex | Number | Mean size |
|---|---|---|---|
| 10. July | F | 2 | |
| | M | 3 | |
| 11. July | F | 10 | 146 |
| | M | 9 | 139 |
| 12. July | F | 10 | 143 |
| | M | 5 | 146 |
| 13. July | F | 9 | 140 |
| | M | 3 | 140 |
| 14. July | F | 3 | 145 |
| | M | 4 | 150 |
| 15. July | F | 8 | 151 |
| | M | 2 | 133 |
| Sum/average | | 68 | 144 |

the time since the last position fix did not exceed 10 min. Mean speed and range (diameter of the area the crab was occupying) per hour and for each three-hour period were also calculated. The 'degree of hiding' was indicated by the mean standard deviation (SD) per 3 h period for the position fixes. The SD of each position fix was calculated by the equation:

$$SD = \sqrt{\mathrm{DEV} / pulses}$$

where

SD = standard deviation
DEV = deviation value calculated for each position fix
pulses = number of pulses received for each position fix

## Results

### *Basic behaviour*

Data from the first 24 h of tracking were excluded from the analysis because three of the nine crabs showed a significantly higher degree of activity (i.e. a lower number of positions were exchanged by the movement analysis program) during the first tracking day than in each of the next six tracking days (*chi-square* test, $p < 0.01$, df = 1). This indicates that the tagging procedure influenced the behaviour of the crabs. Throughout the experimental period all the crabs remained within the range of the positioning system, except two that were out of range for a short period of time.

The nine crabs showed great individual variations in activity and total range (Table 2). Six of the crabs showed distinct movements regularly or during part of the study period, whereas three crabs were almost inactive. When the data were analysed, the crabs were therefore categorised as 'active' and 'inactive'.

'Active crabs' were most active at night between 22:00 and 04:00 (GMT + 2 h, Figure 3), with significant differences within the 24h cycle for mean movement speed (ANOVA $\chi^2 = 37.24$, $p < 0.001$, df = 7) and range (ANOVA $\chi^2 = 44.74$, $p < 0.001$, df = 7, the 24 h cycle was divided into 8 × 3 h periods and tested with a Friedman ANOVA). Among inactive crabs there were no significant differences, but they appeared to be least active in the morning and evening. The active crabs did not move every night. Some crabs were inactive for several nights and days followed by an active period of several nights (Figure 4). The crabs did not return to the same location after movement. There were significant differences in 'degree of hiding' during the 24h cycle in both active and inactive crabs (ANOVA for dependent groups, $F = 20.86$, $p < 0.001$, df = 7, 420), with hiding being most frequent during the day (Figure 5).

The current direction did not seem to influence the movement direction of the crabs (tested with ANOVA, dividing current direction and movement direction into eight groups of 45°), neither when all data where the crab moved were included (unmodified data, F = 1.64, $p > 0.1$, df = 7, 5660), nor when only data where the crab moved faster than 4.8 cm $s^{-1}$ (mean speed during reactions to baited gear, $F = 1.02$, $p > 0.4$, df = 7, 1454) were included. Moving at a speed faster than the mean speed during reaction may indicate that the crab was reacting to an attractant.

### *Gear-induced behaviour*

The average carapace width of the female crabs caught during the experiments with pots in the area was 143 mm (ranging from 115 mm to 180 mm) compared to 142 mm for the nine tracked crabs (Table 3). Four of the nine tagged crabs located baited pots, and three of these were caught (Table 4). All nine localisations were observed at night between 22:45 and 01:30 (GMT + 2 h). On six occasions crabs located baited pots without being caught. One crab was caught the

*Table 4.* Localisations of baited pots. The three upper localisations resulted in capture. The reaction distance is the distance (m) between the crab and the gear at the time the search started and the distance moved is the total distance (m) moved during the reaction. The response direction is the direction of the crab relative to the current direction (the deviation between the current and movement direction). The current was towards the crab when the deviation between the current direction and movement directionwas 180° ± 60°. Category 2: The current was towards the crab when the search started, the crab was not caught. Category 3: The current was not directed towards the crab, the crab was caught. The pots were set between 17:20 and 00:00 (GMT + 2 h)

| Crab no. | Date | Search started | Searching time (min) | Response distance | Distance moved | Response direction | Current velocity (cm $s^{-1}$) | Category |
|---|---|---|---|---|---|---|---|---|
| 5 | 10–11. | 00:15 | 55 | 25 | 75 | 60° | < 2 | 3 |
| 1 | 09–10. | 22:40 | 70 | 18 | 72 | 86° | < 2 | 3 |
| 3 | 12–13. | 01:07 | 142 | 48 | 176 | 0° | < 2 | 3 |
| 3 | 09–10. | 01:30 | 15 | 12 | 18 | 160° | < 2 | 2 |
| 3 | 10–11. | 01:00 | 75 | 45 | 147 | 167° | 8 | 2 |
| 1 | 08–09. | 01:25 | 20 | 22 | 57 | 123° | < 2 | 2 |
| 3 | 12–13. | 23:10 | 50 | 23 | 183 | 130° | < 2 | 2 |
| 2 | 12–13. | 00:10 | 20 | 14 | 45 | 186° | < 2 | 2 |
| 2 | 13–14. | 00:15 | 105 | 41 | 280 | 148° | 5 | 2 |

first time it located the baited pots, while the other two had 1 and 3 localisations before being caught. In six of the nine localisations the current was running in the direction of the crab when the search started (category 2), but the recorded current was not in the direction of the crabs in the three localisations that led to capture (category 3).

The time from when the gear was set until the crab started to search varied from 60 to 407 min (mean = 177 min). The distance between crab and pot at the time when the search started ranged from 12 to 48 m (mean = 28 m). The time between either setting the pot or from when the current turned towards the crab, until the crab initiated its search, varied from 10 min to 3 h 50 min (mean = 2 h). There was no correlation between this time and the response distance of the crab ($r = -0.37$, $p > 0.4$), but a positive correlation between the response distance and the time taken by the crab to locate the gear was found ($r = 0.86$, $p < 0.05$).

When the crab located a pot, it seldom moved towards it in a straight line. It often seemed to start the search in a slightly different direction than towards the pot, but after a while it turned and headed more or less in the direction of the pot. The ratio of 'total distance moved: response distance' (frequency of turning) may indicate how easily the crab located the pot. There was no correlation between frequency of turning and response distance ($r = 0.27$, $p > 0.4$), nor between frequency of turning and time to locate gear ($r = 0.36$, $p > 0.3$).

The four crabs that located pots showed significantly higher speeds during localisations (4.8 cm $s^{-1}$) than during basic movement (3.6 cm $s^{-1}$) in the corresponding time period (between 22:00 and 03:00, Mann-Whitney $U$ test, $U = 40617.5$, $p < 0.001$). This analysis was based exclusively on data where crabs were moving (not exchanged positions).

## Discussion

### *Basic behaviour*

Female edible crabs showed a nocturnal rhythm of activity. Ansell (1973) found up to five times as high oxygen consumption in edible crab at night than during the day, suggesting a higher rate of nocturnal activity. Several other decapod crustaceans have also been shown to exhibit a nocturnal activity rhythm (Fielder, 1965; Atkinson, 1974; Atkinson & Naylor, 1976; Hill, 1976, 1978; Rebach, 1985; Bjordal, 1986; Jernakoff, 1987; Karnofsky et al., 1989a; Chatterton & Williams, 1994). Biological rhythms are often controlled by an endogenous physiological pacemaker coupled to an environmental cycle that adjusts the internal clock to local time (DeCoursey, 1983). In this study the activity of edible crabs was associated with the light-dark cycle.

It is not possible to determine from the present data whether other external factors influenced the rhythm, or whether the rhythm was endogenously controlled.

Generally, the cause of rhythmicity is believed to be an adaptive response to variation in resources or risk of predation that are cyclical themselves (DeCoursey, 1983; Bell, 1990). Wolffish, seals, cod and birds may prey upon adult crabs (Rae, 1967, 1968; Rae & Shelton, 1982; Gundersen, pers. com.), and owing to limited visibility during nighttime a nocturnal rhythm of activity should be favoured by the crab to reduce the risk from visual predators. Availability of prey may also partly explain the nocturnal rhythm. Most published reports have shown that cancrid crabs are generalists preying mainly on molluscs and slow-moving crustaceans (Shelton et al., 1979; Drummond-Davis et al., 1982; Elner et al., 1985; Cerda & Wolff, 1993; Stehlik, 1993; Karlsson & Cristiansen, 1996). These prey should be easy to catch during the day and are unlikely to account for the nocturnal activity pattern. However, edible crabs also successfully subdue a wide size range of decapod crustaceans, some of which are relatively fast-moving (Lawton, 1989). Further, Gottshall (1977) found that the stomach of the closely related Dungeness crab (*Cancer magister* Dana) contained 40 different food items, with clams, fish, isopods and amphipods being the most frequent (see also Butler, 1954; Stevens et al., 1982). A nocturnal activity cycle might widen the food spectrum of the crab because of increased pursuit success at night due to its inconspicuousness and because many crustacean prey species also show a nocturnal activity cycle which may increase the rate of prey encounter (e.g. *Carcinus maenas* L., Naylor, 1958, 1960).

Edible crab seemed to spend more time hiding during the day than at night. Hall et al. (1991) observed that edible crab remained in pits in the sand when inactive. Several other decapod crustaceans have also been shown to spend time within a shelter (Edwards, 1958; Chapman & Rice, 1971; Hockett & Kritzler, 1972; Atkinson, 1974; Hazlett & Rittschof, 1975; Hill, 1976; Karnofsky et al., 1989b). Staying in a burrow provides protection from predators, but it also means spending less time searching for food. On the other hand, the sheltering behaviour may be a 'sit and wait' strategy, and edible crab provided with a shelter has been observed to adopt an ambush predation technique (Lawton, 1989).

The crabs categorised as inactive seemed to occupy one shelter during the observation period, while the active ones occupied different shelters each day. Some of the active crabs were inactive for several days, indicating that one individual could be active or inactive at different times. Whether or not to stay in shelter, whether to use only one shelter and how much time spent hiding may depend on feeding motivation, amount of food available, predation pressure and presence of other shelters. These factors influence the animal continuously and may cause behavioural variations within an individual (Atkinson & Naylor, 1973; Hill, 1976; Jernakoff, 1987; Karnofsky et al., 1989b; Eggeleston & Lipcius, 1992). The observation that active crabs occupied different shelters indicate that edible crab does not use a core area (Burt, 1943) as observed in other species (Reese, 1975; Hawkins et al., 1980; Fish & Savitz, 1983; Clark & Green, 1990; Morrissey & Gruber, 1993; Vannini & Cannicci, 1995; Engås et al., 1996)

During the observation period all the crabs remained within the range of the hydrophones. There was thus no long-range movement by female crabs in this study that was conducted during summer. Female edible crabs can move over considerable distances, but such long-range movements seem to take place during winter in connection with spawning (Williamson, 1900; Meek, 1913; Mason, 1965; Gundersen, 1979; Bennett & Brown, 1983; Woll, 1981).

*Gear-induced behaviour*

Crabs were observed locating baited gear only between 22:45 and 01:30 even though the pots were set at different times. This indicates that crabs do not react to chemical attractants unless 'the time is right', which seemed to be the time when the crabs were usually active. This is in agreement with the suggestion that crustaceans are more likely to be attracted to chemical attractants when they are in an internally aroused state (Zimmer-Faust, 1989; Zimmer-Faust et al., 1996). There were large variations in activity level between individual crabs. This variation in activity and presumably internal state should be reflected in the probability to enter an odour plume and the tendency to react to stimulation. Even though the number of crabs locating pots was low, the fact that only one of nine localisations was by an inactive crab indicates a connection between activity state and localisation of baited fishing gear.

Four of the nine crabs located baited pots, and three of these reacted more than once. All localisations of gear, whether the current was heading towards the crab or not, were included. The current direction measured by the two current meters differed which indicated that current conditions in the area were heterogeneous. It was thus difficult to exclude any localisations as definitely not being chemically stimulated. Localisations in categories 1 and 2 were more likely to have been chemically stimulated than those in categories 3 and 4. Besides showing chemically stimulated reactions to the pot, the crab may also locate the pots by random movement or visual attraction. The visual capacity of edible crabs is not known, but some species of crab appear to see quite well at night (Warner, 1977). A stationary pot with no resemblance to food is unlikely to stimulate a visually induced reaction to food, but the crab may be attracted to the pot as a shelter.

Crabs occasionally did not react to pots although the current seemed to be towards the crab. The reason may have been low feeding motivation, access to food at their current location or difficulty to localize the stimulus source because of low current velocity or turbulent current conditions caused by the seabed (Weissburg & Zimmer-Faust, 1993).

An interesting difference was found between speed of movement during localisation and during basic movement at night, with faster movement occurring during the localisations. High speed during localisations increases the chance of being the first to find the source of the attractant and could thus improve the competitive ability of an individual crab.

## Acknowledgements

We wish to thank Austevoll Aquaculture Station for room and hospitality during the field work and also Jan Erik Fosseidengen for help and good advice.

## References

Ansell, A. D., 1973. Changes in oxygen consumption, heart rate and ventilation accompanying starvation in the decapod crustacean *Cancer pagurus*. Neth. J. Sea Res. 7: 455–475.

Atkinson, R. J. A., 1974. The activity rhythm of *Goneplax rhomboides* (L.). Mar. Behav. Physiol. 2: 325–335.

Atkinson, R. J. A. & E. Naylor, 1973. Activity rhythms in some burrowing decapods. Helolander wiss. Meeresunters. 24: 192–201.

Atkinson, R. J. A. & E. Naylor, 1976. An endogenous activity rhythm and the rhythmicity of catches of *Nephrops norvegicus* (L.). J. exp. mar. Biol. Ecol. 25: 95–108.

Bell, W. J., 1990. Searching behaviour. The behavioural ecology of finding resources. Chapman and Hall, London, 309 pp.

Bennett, D. B. & C. G. Brown, 1983. Crab (*Cancer pagurus*) migrations in the English Channel. J. mar. biol. Ass. U.K. 63: 371–398.

BenTuvia, A. & W. Dickson, 1969. Proceedings of the FAO conference on fish behaviour in relation to fishing techniques. Bergen, Norway, 19–27 October 1967. FAO Fish. Rep., 1, 2, 3, 884 pp.

Bjordal, Å., 1979. Rapport fra forsøksfiske etter sjøkreps med teiner på vestlandet. Fiskets gang 25: 779–783.

Bjordal, Å., 1986. The behaviour of Norway lobster towards baited creels and size selectivity of creels and trawls. FiskDir. Skr. HavUnders. 18: 131–137.

Burt, W. H., 1943. Territoriality and home range concepts as applied to mammals. J. Mammal. 24: 346–352.

Butler, T. H., 1954. Food of the commercial crab in the Queen Charlotte Islands region, Canada. Fish. Res. Bd Can., Prog. Rep. Pac. Coast Stn. 99: 3–5.

Camus, P., 1982. Marquage d'une population côtière de tourteaux (*Cancer pagurus* L.) en baie d'Audierne (Bretagne sud) – Premiers résultats. ICES, Shellfish Committee, C.M. 1982/K:18, 13 pp.

Cerda, G. & M. Wolff, 1993. Feeding ecology of the crab *Cancer polyodon* in La Herradura Bay, Northern Chile. II. Food spectrum and prey consumption. Mar. Ecol. Prog. Ser. 100: 119–125.

Chapman, C. J. & A. L. Rice, 1971. Some direct observations on the ecology of the Norway lobster *Nephrops norvegicus*. Mar. Biol. 10: 321–329.

Chatterton, T. D. & B. G. Williams, 1994. Activity patterns of the New Zealand cancrid crab *Cancer novaezelandiae* (Jacquirot) in the field and laboratory. J. exp. mar. Biol. Ecol. 178: 261–274.

Clark, D. S. & J. M. Green, 1990. Activity and movement patterns of juvenile Atlantic cod, Gadus morhua, in Conception Bay, Newfoundland, as determined by sonic telemetry. Can. J. Zool. 68: 1434–1442.

DeCoursey, P. J., 1983. Biological timing. In F. J. Vernberg & W. B. Vernberg (eds), The Biology of Crustacea – Behaviour and Ecology. Academic Press, New York: 107–150.

Drummon-Davis, N. C., K. H. Mann & R. A. Pottle, 1982. Some Estimates of Population Density and Feeding Habits of the Rock Crab, *Cancer irrotatus*, in a Kelp Bed in Nova Scotia. Can. J. Fish. Aquat. Sci. 39: 636–639.

Edwards, R. L., 1958. Movements of individual members in a population of the shore crab, *Carcinus maenas* L., in the littoral zone. J. anim. Ecol. 27: 37–45.

Eggeleston, D. B. & R. N. Lipcius, 1992. Shelter seclection by spiny lobster under variable predation risk, social conditions, and shelter size. Ecology 73: 992–1011.

Elner, R. W., P. G. Beninger, L. E. Linkletter & S. Lanteigne, 1985. Guide to indication fragments of principal prey taxa in the stomachs of two common Atlantic crab species: *Cancer borealis*, Stimpson (1859) and *Cancer irrotatus*, Say (1817). Can. tech. Rep. Fish. Aquat. Sci. 1403, 19 pp.

Engås, A., S. Løkkeborg & J. T. Øvredal, 1996. Natural and fishing-gear-induced behaviour of tagged fish, studied by means of a stationary positioning system. In E. Baras & J. C. Philippart (eds), Underwater Telemetry: 203–211. Proceedings of the First Conference and Workshop on Fish Telemetry in Europe. University of Liege, Belgium.

Fernö, A., 1993. Advances in understanding of basic behaviour: consequences for fish capture studies. ICES mar. Sci. Symp. 196: 5–11.

Fielder, D. R., 1965. The spiny lobster, *Jasus lalandei* (H. Milne-Edwards), in South Australia. Aust. J. mar. Freshwat. Res. 16: 351–367.

Fish, P. A. & J. Savitz, 1983. Variations in Home Ranges of Largemouth Bass, Yellow Perch, Bluegills, and Pumpkinseeds in an Illinois Lake. Trans. am. Fish. Soc. 112: 147–153.

Gottshall, D. W., 1977. Stomach contents of northern California Dungeness crabs, Cancer magister. Calif. Fish. Game 63: 43–51.

Gundersen, K. R., 1977. Migration of crab in Norwegian waters. ICES C.M. Shellfish Comitee, Doc (mimeo) 39: 1–8.

Gundersen, K. R., 1979. Some results of tagging experiments on the edible crab (*Cancer pagurus*) in Norwegian waters. Rapp. P.-v. Réun. Cons. int. Explor. Mer. 175: 222–224.

Hall, S. J., D. J. Basford, M. R. Robertson, D. G. Raffaelli & I. Tuck, 1991. Patterns of recolonisation and the importance of pit-digging by the crab *Cancer pagurus* in a subtidal sand habitat. Mar. Ecol. Prog. Ser. 72: 93–102.

Hawkins, A. D., G. G. Urquhart & G. W. Smith, 1980. Ultrasonic Tracking of Juvenile Cod by Means of a Large Spaced Hydrophone Array. In C. J. Amlaner, Jr. & D. W. MacDonald (eds), A Handbook on Biotelemetry and Radio Tracking. Pergamon Press, Oxford: 461–470.

Hazlett, B. & D. Rittschof, 1975. Daily Movements and Home Range in *Mitrax spinosissimus* (Majidae, Decapoda). Mar. Behav. Physiol. 3: 101–118.

Hill, B. J., 1976. Natural food, foregut clearance-rate and activity of the crab *Scylla serrata*. Mar. Biol. 34: 109–116.

Hill, B. J., 1978. Activity, Track and Speed of Movement of the Crab *Scylla serrata* in an Estuary. Mar. Biol. 47: 135–141.

Hockett, J. C. & H. Kritzler, 1972. Capture-recapture methods with *Uca*. Biol. Bull. 142: 49–56.

Jernakoff, P., 1987. Foraging patterns of juvenile western rock lobster *Panulirus cygnus* George. J. exp. mar. Biol. Ecol. 113: 125–144.

Karlsson, K. & M. E. Cristiansen, 1996. Occurrence and population composition of the edible crab (*Cancer pagurus*) on rocky shores of an islet on the south coast of Norway. Sarsia 81: 307–314.

Karnofsky, E. B., J. Atema & R. H. Elgin, 1989a. Natural Dynamics of Population structure and Habitat Use of the Lobster, *Homarus americanus*, in a shallow Cove. Biol. Bull. 176: 247–256.

Karnofsky, E. B., J. Atema & R. H. Elgin, 1989b. Field Observations of Social Behavior, Shelter Use, and Foraging in the Lobster, *Homarus americanus*. Biol. Bull. 176: 239–246.

Latrouite, D. & D. Le Foll, 1989. Donnes sur les migrations des crabes torteau *Cancer pagurus* et les araignees de mer *Maja squinado*. Océanis 15: 133–142.

Lawton, P., 1989. Predatory interaction between the brachyuran crab *Cancer pagurus* and decapod crustacean prey. Mar. Ecol. Prog. Ser. 52: 169–179.

Mason, J., 1965. The Scottish crab-tagging experiments, 1960–61. Rapp. P.-v. Reun. Cons. int. Explor. Mer. 156: 71–80.

Meek, A., 1913. The migrations of crabs. Report of the Dove Marine Laboratory, Cullercoats, Northumberland 2: 13–20.

Morrissey, J. F. & S. H. Gruber, 1993. Home Range of Juvenile Lemon Sharks, Negaprion brevirostris. Copeia 1993: 425–434.

Naylor, E., 1958. Spontaneous tidal and diurnal rhythms of locomotor activity in *Carcinus maenas* (L.). J. exp. Biol. 35: 602–610.

Naylor, E., 1960. Locomotor rhythms in a population of *Carcinus maenas* (L.) from non-tidal conditions. J. exp. Biol. 37: 481–488.

Phillips, B. F., L. M. Joll & D. C. Ramm, 1984. An electromagnetic tracking system for studying the movements of rock (spiny) lobsters. J. exp. mar. Biol. Ecol. 79: 9–18.

Rae, B. B., 1967. The food of cod in the North Sea and on West of Scotland fishing grounds. Mar. Res. 1967, 68 pp.

Rae, B. B., 1968. The food of seals in the Scottish waters. Mar. Res. 1969, 23 pp.

Rae, B. B. & R. G. J. Shelton, 1982. Notes on the food of nine species of elasmobranch (Part I) and nine species of demersal teleost (Part II) fishes from Scottish waters. ICES C.M. 1982/G: 56, 5 pp.

Rebach, S., 1985. Rhythmicity under constant conditions in the rock crab, Cancer irrotatus. Bull. mar. Sci. 36: 454–466.

Reese, E. S., 1975. A comparative field study of the social behavior and related ecology of reef fishes of the family Chaetodontidae. Z. Tierpsychol. 37: 37–61.

Robles, C., D. A. Sweetnam & D. Dittman, 1989. Diel variation of intertidal foraging by Cancer productus L. in British Columbia. J. Nat. Hist. 23: 1041–1049.

Shelton, R. G. J., J. A. M. Kinnear & K. Livingstone, 1979. A preliminary account of the feeding habits of the edible crab, *Cancer pagurus* L. of N. W. Scotland. Int. Counc. Explor. Sea Comm. Meet. (Shellfish and Benthos Comm.) K: 35: 1–6.

Skajaa, K., 1997. Basic movement pattern and chemo-oriented search toward baited gears in demersal species; a field study of ling and edible crab. *Cand. scient.* thesis, Department of Fisheries and Marine Biology, University of Bergen, Norway, 138 pp.

Stehlik, L. L., 1993. Diets of the brachyuran crabs Cancer irrotatus, C. Borealis, and Ovalipes ocellatus in the New York Bight. J. Crust. Biol. 13: 723–735.

Stevens, B. G., D. A. Armstrong & R. Cusimano, 1982. Feeding habits of the Dungeness crab *Cancer magister* as determined by the index of relative importance. Mar. Biol. 72: 135–145.

Vannini, M. & S. Cannicci, 1995. Homing behaviorur and possible cognitive maps in crustacean decapods. J. exp. mar. Biol. Ecol. 193: 67–91.

Wardle, C. S. & C. E. Hollingworth, 1993. Fish behaviour in relation to fishing operations. ICES mar. Sci. Symp. 196, 216 pp.

Warner, G. F., 1977. The biology of crabs. Elek Science London, London, 201 pp.

Weissburg, M. J. & R. K. Zimmer-Faust, 1993. Life and death in moving fluids: hydrodynamic effects on chemosensory-mediated predation. Ecology 74: 1428–1443.

Williamson, H. C., 1900. Contributions to the life-history of the edible crab (*Cancer pagurus*, Linn.). Rep. Fish. Bd Scot., Ann. Rep. 18: 77–143.

Woll, A. K., 1981. Krabbens (*Cancer pagurus*) biologi i midtre og ytre Romsdal. *Cand. scient.* thesis. University of Trondheim, Norway, 75 pp (in Norwegian).

Zimmer-Faust, R. K., 1989. The relationship between chemoreception and foraging behavior in crustaceans. Limnol. Oceanogr. 34: 1367–1374.

Zimmer-Faust, R. K., C. M. Finelli, N. D. Pentcheff & D. S. Wethey, 1995. Odor plumes and animal navigation in turbulent water flow: A Field Study. Biol. Bull. 188: 111–116.

Zimmer-Faust, R. K., P. B. O'Neill & D. W. Schar, 1996. The relationship between predator activity state and sensitivity to prey odor. Biol. Bull. 190: 82–87.

*Hydrobiologia* **371/372:** 155–171, 1998.
*J. P. Lagardère, M.-L. Bégout Anras &G. Claireaux (eds), Advances in Invertebrates and Fish Telemetry.*



# Telemetry applied to behaviour analysis of yellowfin tuna (*Thunnus albacares*, Bonnaterre, 1788) movements in a network of fish aggregating devices

Francis Marsac[1] & Patrice Cayré[2]
[1] *ORSTOM-HEA, B.P. 5045, F-34032 Montpellier, France; e-mail: marsac@orstom.fr*
[2] *ORSTOM, 213 rue Lafayette, F-75480 Paris Cedex 10, France*

*Key words:* tuna, *Thunnus albacares*, behaviour, FAD, feeding, Indian Ocean

**Abstract**

This paper analyses acoustic tagging of 8 yellowfin tunas (*Thunnus albacares*, Bonnaterre, 1788) undertaken around Fish Aggregating Devices (FADs) in La Reunion island (Indian Ocean). Emphasis is laid on the horizontal movements and thus complete previous studies on vertical movements around the same FADs. The first result of the present study deals with the relative dwelling time of yellowfin over the distance to FAD where tagged, in 0.9 km intervals. Distributions of dwelling time are different between day and night. During daytime, the fishes remain in the close vicinity of the FAD (mostly within a 1.8 km radius), whilst a drastic disassociation to the FAD occurs at night. During the day, the attractive influence of the FAD disappears 9.3 km away; this finding leads to the suggestion that a minimum distance of 18 km between neighbouring FADs should be applied in La Reunion to avoid overlapping radii of influence. The second group of results points out the potential use of the vertical and total swimming speeds as indicators of the foraging activity of the fish and of the type of movements (behaviour of tight association to FAD, transit among FADs or offshore migration, away from the area of FADs). The day/night change affects the vertical activity, with large magnitude of vertical movements exhibited at night. The total speed during the offshore movements is estimated about 1.2 m $s^{-1}$: the shift from a phase of tight association to FAD toward a phase of transit movement is characterised by an increase of the total speed. A typology of the relationships between swimming speed and feeding activity is proposed and discussed.

## Introduction

Following previous ultrasonic tagging experiments in the Pacific Ocean (Yuen, 1970; Carey & Olson, 1982; Cayré & Chabanne, 1986; Holland et al., 1990), acoustic tracking of tuna was initiated in the Indian Ocean in 1989, within the framework of a Regional Tuna Project undertaken by the 'Commission de l'Océan Indien'. The experiments aim to improve the knowledge of tuna behaviour that are associated with Fish Aggregating Devices (FADs) and thus explore more efficient ways for the exploitation of these migrating resources. FADs appeared in the early 80's as 'miracle solutions' for making populations of tunas available to small scale fisheries of island developing countries (Preston, 1982; Buckley, 1986; Marsac & Stequert, 1987; Willmann, 1990; Sims, 1992). Following some success in the Pacific islands, the technique rapidly expanded to other tropical oceans, the south west Indian Ocean in particular. The dynamics of the aggregation needed to be explained, and the high resolution in time and space provided by the acoustic telemetry makes it one of the most powerful tools dealing with this issue.

The first experiments in the Indian Ocean were made in Comoros Islands (Cayré, 1991); then other tracking cruises were carried out in La Reunion and Mauritius around FADs. For long, the analysis of the surveys remained fairly descriptive but, as a greater number of fishes was tagged, advanced studies became possible, focusing on vertical distribution and the effects of the nearby physical environment of the fish (Cayré & Marsac, 1993).

In this paper, we focus more on the horizontal movements of the yellowfin tunas (*Thunnus albacares*, Bonnaterre 1788) tagged at FADs, but still keeping some reference to the vertical activity. The issues addressed are dual: (1) is there a typical horizontal distribution of the yellowfin around FADs, and any day/night effect affecting this distribution, and (2) can quantitative (swimming speed) and qualitative (type of movement) variables express behavioural changes exhibited by the tunas while at the FAD or departing from it.

## Materials and methods

From February 1994 until December 1996, four tagging cruises were carried out in the vicinity of FADs anchored off La Reunion island, in the south-west tropical Indian Ocean (Figure 1). This island is influenced by trade winds almost all the year round, and the experiments were made on the leeward side (the West coast) where better working conditions at sea can be met.

### *Capture and tagging technique*

Fishing, tagging and tracking were conducted from a small chartered vessel (10 m long, big game fishing type). The trips were scheduled during the peak tuna fishing season which stretches barely from December to April in La Reunion. The fish were all caught at FADs, using a trolling technique. When hooked, the tuna was rapidly hauled on board and laid on the tagging cradle. A constant flow of sea water was supplied in the mouth of the animal to keep the gills well oxygenated and humid. The transmitter was attached at the back of the second dorsal fin using two nylon 'tie-wraps' through the dorsal pterygiophores and the muscle (Holland et al., op. cit.; Figure 1). The tagging operation lasted approximately one minute, after which the fish was gently released into the water.

### *Tracking equipment and strategy*

The tracking equipment was manufactured by Vemco$^{TM}$ (Halifax, Nova Scotia, Canada). It comprised the following items:

- pressure sensitive ultrasonic transmitters: those used for tuna were 74 mm long, 16 mm diameter and have a life of nearly 5 d according to the depth reached. The signal can be detected at a range of 600 to 800 m. The weight of the tag in the water represents 1 to 2‰ of the fish weight.
- a directional hydrophone was mounted on a V-fin depressor towed by the boat.
- a receiver (VR-60) was used for decoding the signals and recording depth data every 15 s.

The survey was a continuous search of the strongest signal intensity, by frequent changes of course. When the fish was rather stationary, the vessel kept moving to maintain proper gliding of the V-fin depressor and optimal detection. The latitudes and longitudes given by GPS were noted in a logbook every 5 min with the corresponding depth of the fish. It was assumed the consecutive positions of the tracking vessel were similar to the positions and horizontal movements of the fish.

### *Data preparation*

The series of depth records were downloaded from the receiver memory and processed to produce swimming depth values by one minute bins. A simple average was calculated when several measurements were available for a given bin. In other cases (loss of signal for a few moments), the non-sampled bins were filled in by linear interpolation between the closest values. When the signal was lost for more than ten minutes, no interpolation was attempted and the bins remained blank. A similar procedure was applied to position data, i.e. an interpolation on a one minute time basis for both latitudes and longitudes.

### *Estimation of swimming speed*

The tracking technique does not allow accurate estimation of the horizontal small scale movements (Cayré & Chabanne, 1986). In the situation of a stationary fish, the movements of the boat tend to magnify the actual movements of the fish, and therefore its swimming speed would be overestimated (case A). Conversely, we should also consider the case of a fish exhibiting quick movements on a very fine scale, with signals being detected properly by an almost motionless vessel. For example, along a five minute straight section made by the vessel at a speed of 7 km h$^{-1}$ (600 m long), the fish can return to the same point after a diversion of up to 500 m away from the vessel's course (a dis-

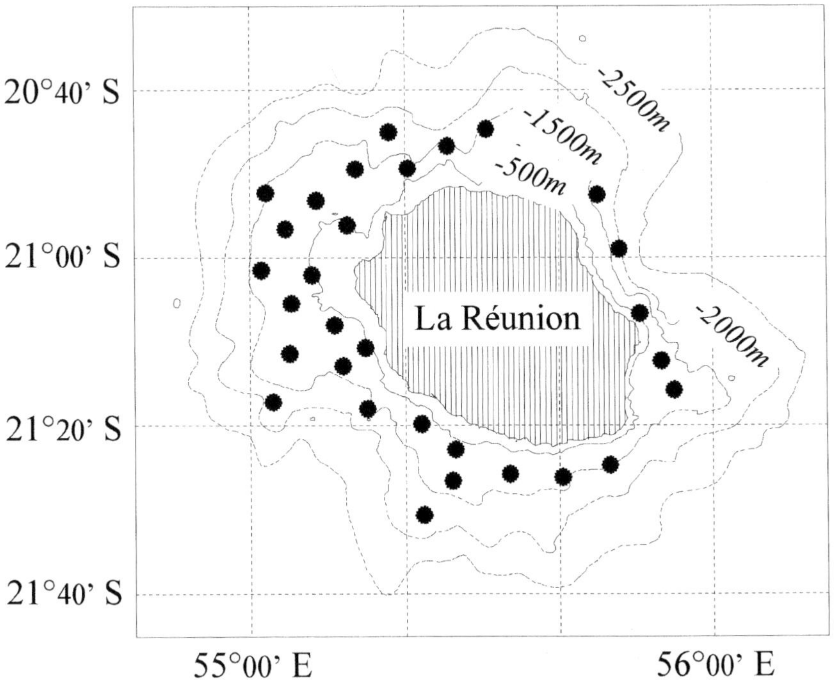


*Figure 1.* Bathymetric map of the West coast of La Reunion with location of FADs.

tance where the signal is still audible): in such a case, the distance covered by the fish would be twice longer than the boat's one and the swimming speed will be underestimated (case B). Considering more complex movements (sinusoids about the main course), the factor would be greater than 3 (case C). Therefore, caution should be paid in the analysis of very small scale movements, which is particularly the case around FADs. In order to provide a rough order of magnitude of the swimming speed, we used a raising factor of 2 estimated for the intermediate situation (case B) to correct the raw horizontal speed during the phase of association to the FAD.

On the other hand, we can consider that vertical speed (VS) to be properly assessed and can be used without any correction for a comparative analysis. The VS is axis-oriented: negative/positive values corresponding to a downward/upward motion.

A resultant speed (RS) is calculated from the combination (by Pythagorus) of the vertical and raw horizontal speeds. The corrected RS combining the vertical and the corrected horizontal speeds will be applied on FAD sections only.

*Behaviour typology*

Eight yellowfin tuna tracked during a significant amount of time (> 10 h) are herein considered (Table 1). All the fishes were tagged in the close vicinity of a FAD (300 m radius). Consequently, the FAD of tagging will be considered as the spatial reference from which the movements of every fish are analysed. Therefore, the observations can be classified in 2 phases:

- a phase of tight association to the original FAD of tagging
- a transit phase when the fish leaves the FAD, either to patrol within the area where the FAD network is deployed (FAD-associated behaviour), or to swim offshore, far away from the FADs network area (Offshore-moving behaviour).

All the fishes went through the first phase, at least during the hours following tagging. Later, the tunas tracked diverged in two groups:

- fishes remaining in the coastal area, exhibiting the FAD-associated behaviour (fish no. 1, 2, 3, 5, 7; Table 1);

*Table 1.* Characteristics of the 8 yellowfin tuna tracking operations.

| Fish no. | Fork length (cm) | FAD name | FAD no. | Date | Tracking duration (h) | Moon phase | Type of behaviour |
|---|---|---|---|---|---|---|---|
| 1 | 110 | Le Port | 16 | Feb 19, 1994 | 44 | Full moon | FAD |
| 2 | 49 | Le Port | 16 | Feb 23, 1994 | 21 | Full moon | FAD |
| 3 | 104 | St Leu | 4 | Mar 27, 1995 | 19 | Dark night | FAD |
| 4 | 95 | Stella | 5 | Mar 29, 1995 | 24 | Dark night | OFFSHORE |
| 5 | 58 | Stella | 5 | Apr 1, 1995 | 30 | Dark night | FAD |
| 6 | 70 | St Gilles 5M | 3 | Dec 18, 1995 | 21 | Last quarter | OFFSHORE |
| 7 | 84 | St Gilles 5M | 3 | Dec 20, 1995 | 11 | Last quarter | FAD |
| 8 | 76 | Etang Salé | 25 | Dec 3, 1996 | 30 | Dark night | OFFSHORE |

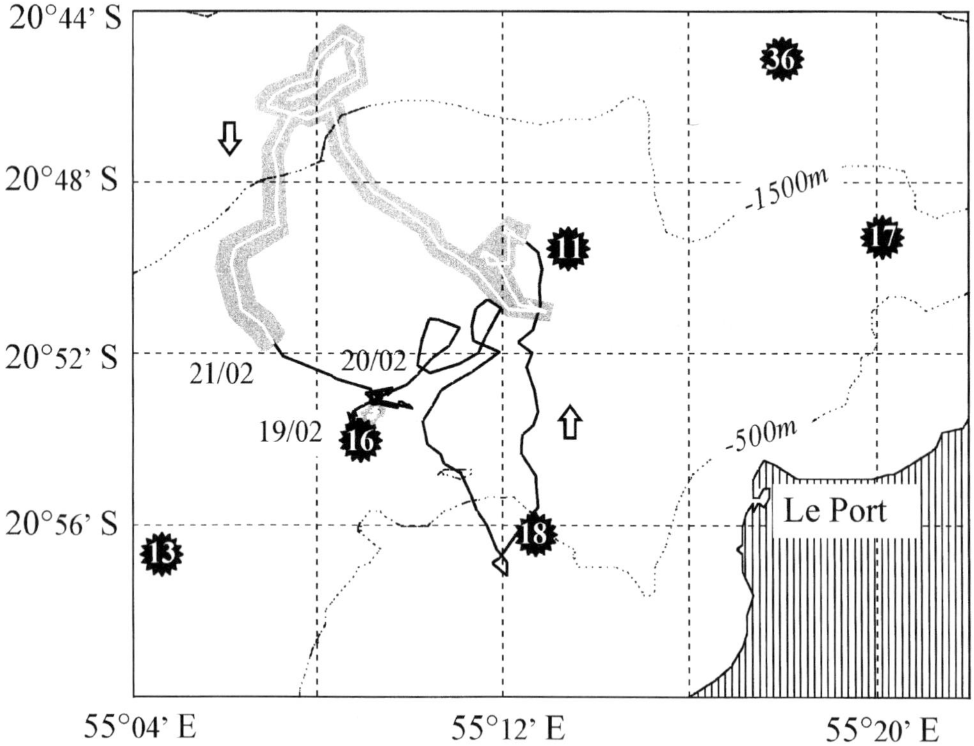


*Figure 2.* Course of the fish no. 1 (FL 110 cm), Feb. 19 1994, 11:00 to Feb. 21, 6:30.

– fishes escaping from the attraction spheres of the FADs, exhibiting the offshore-moving behaviour (fish no.4, 6, 8; Table 1).

Finally, the moon phase was also considered: in February 1994, the tracks were conducted during full moon nights whilst the other cruises took place in half or new moon nights. This leads to another level of data clustering, according to 3 periods: daytime, moonlight and night time.

*Horizontal distribution around FADs*

The horizontal distribution of the fish around the FAD is analysed on a linear and unilateral scale, starting at the position of the FAD of tagging. However, a FAD moves about its anchoring location, because of the current and the fact that the mooring line is always longer than the actual depth (the surplus of length versus depth is around 20 to 30% in La Reunion). With a mooring depth of 500 to 1000 m, the radius of maximal shift can range from 415 to 830 m. As a comparison, an accurate survey of a FAD in Hawaii (Holland et al., op. cit.) shows a 600 m shift radius within 7 h, which is of the same magnitude as our estimate. Cayré & Chabanne (op. cit.) observed a 1700 m shift radius in a depth twice as great as in La Reunion (2000 m), i.e. a similar ratio between actual depth and FAD's shifting area. In view of this uncertainty, we decided to use 926 m (0.5 nautical mile) increments to classify the fish positions, analysing separately daytime (6:00 to 18:00) and night time (19:00 to 5:00) periods. Twilight periods will be considered as well: they denote the transitions between sunset and night and between night and sunrise.

## Results

### *FAD-associated behaviour*

- Fish no. 1 (Figure 2). The 44 h track of this tuna is a typical example of strong link with FADs. The fish remained in the close vicinity of the FAD where tagging took place (at 11:00) until midnight. Then, shortly after the moon set (1:00), the animal made a 5 h overnight excursion before moving back toward the FAD immediately after sunrise. Three hours later, the fish left and swam in different directions during 6 h before swimming 6.5 km straightway to another FAD (no. 18) where it stayed for 1 h. Then, it moved almost straightway toward FAD no. 11, crossing the 13 km distance in less than 2 h. The fish visited the FAD very briefly and left at sunset. During the first part of the night, the tuna kept an offshore-oriented course then moved back to the tagging FAD by sunrise, where the track was ended.
- Fish no. 2 (Figure 3). This juvenile tuna was tagged on the same FAD (no. 16) at 14:00. The fish exhibited a similar FAD-associated behaviour during daytime, then swam 7.4 km away during the night. It was lost at dawn then found later (9:00) in the vicinity of the initial FAD where the track ended.
- Fish no. 3 (Figure 4). This tuna was tagged at 11:00 at FAD no. 4, and remained extremely close to the FAD until sunset. The FAD was left at night and the fish took a course to the North, following the 500 m isobath parallel to the coast. Just after dawn, it passed within a few hundred meters of FAD no. 16, but kept straight on, and increased its swimming speed when it was lost.
- Fish no. 5 (Figure 5). Tagged in the morning at the FAD no. 5, this yellowfin tuna remained stuck to it at daytime. The fish finally left the buoy just before sunset for an erratic overnight excursion, at a distance never exceeding 7.4 km from the initial FAD, and passing at less than 3.7 km from another FAD (no. 4). At dawn and during the following hours, the tuna kept a coastward course ending at less than 3.7 km from the shore. At 15:00, when the track was stopped, the fish had reached a third FAD (no. 34).
- Fish no. 7 (Figure 6). The fish was tagged at 18:00 at the FAD no. 3 and departed 4 h later for a night time excursion, patrolling parallel to the coastline above the 300-400 m isobath. The track ended at the FAD no. 4 at dawn the following day.

### *Offshore moving behaviour*

- Fish no. 4 (Figure 7). The fish was tagged at the FAD no. 5. As previously observed in the FAD-associated behaviour tunas, it stayed close to the FAD during daytime, then left the buoy for a straight course to the South, out of the coastal area where FADs are anchored. The track was stopped when the fish was 41 km offshore.
- Fish no. 6 (Figure 8). This tuna was tagged at 17:00 at the FAD no. 3 which was left at sunset while patrolling within a 5.5 km radius from the FAD. At moonrise (quarter moon), the fish departed, kept swimming along the 500 m isobath and passed in the vicinity of another FAD (no. 19). After sunrise, it took a westward course which took it 46.3 km offshore where the track was stopped. It must be noted that, during this daytime excursion, the course was suddenly diverted toward an undersea relief elevation (300 m above the seabed) located 2380 m below the surface (21°06′ S–55°10′ E).
- Fish no. 8 (Figure 9). The tagging took place at the FAD no. 25, at 17:30 and the fish remained nearby until 21:00. Afterwards, it departed and followed a general offshore movement to the Northwest, without any diversion to FADs at sunrise. However, this track must be analysed with caution because of an anomalous swimming behaviour that could be explained by a lingering death: at 22:15, while patrolling close to the surface, the fish suddenly started to swim deeper, and the depth increased continuously with time until 7:00 when it reached 270 m. According to the time spent in cold waters (more than 2 h below 20 °C without upwards move-

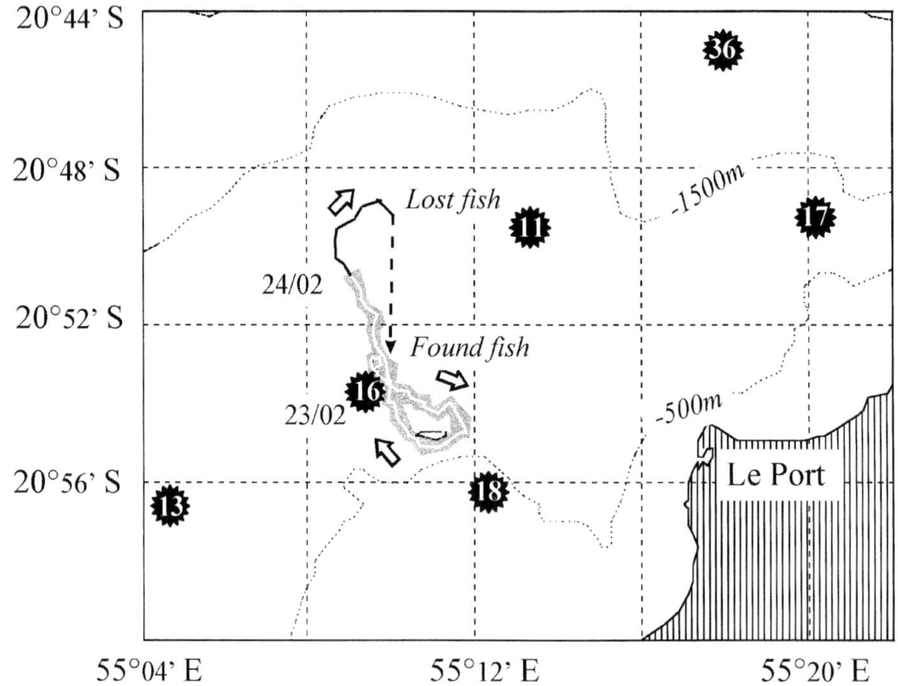


*Figure 3*. Course of the fish no. 2 (FL 49 cm), Feb. 23 1994, 13:30 to Feb. 24, 10:00.

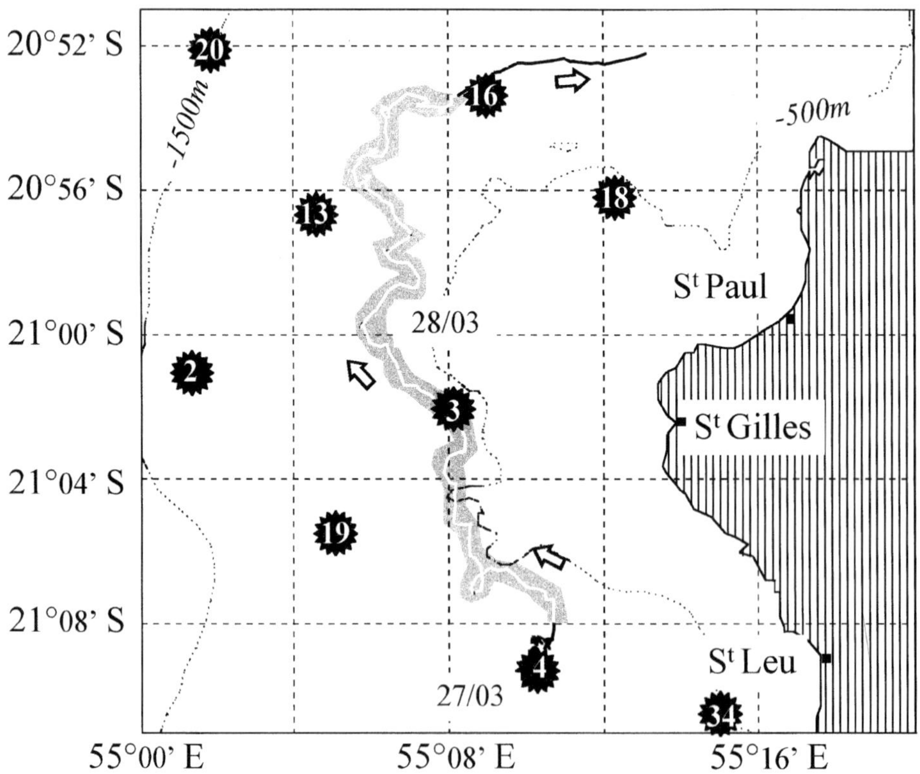


*Figure 4*. Course of the fish no. 3 (FL 104 cm), Mar. 27 1995, 11:00 to Mar. 28, 7:30.

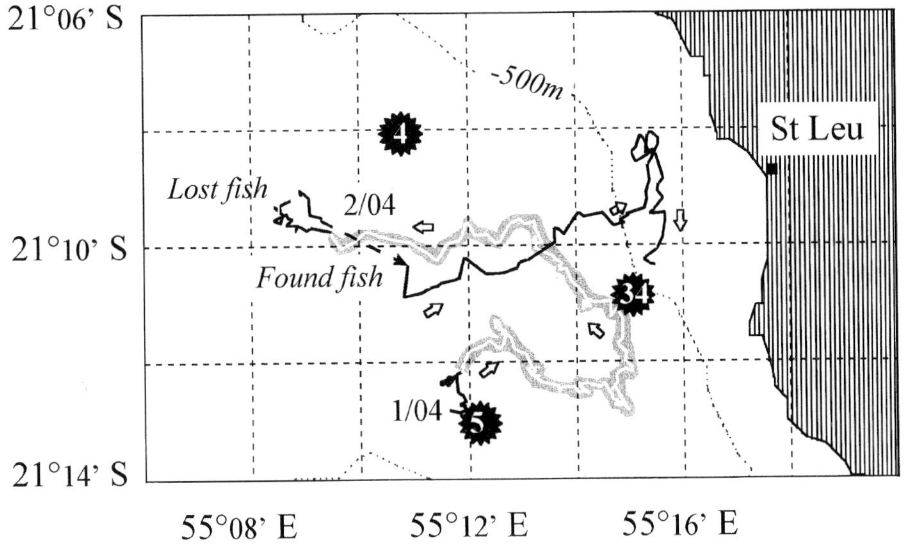


*Figure 5.* Course of the fish no. 5 (FL 58 cm), Apr. 1 1995, 8:50 to Apr. 2, 14:50.

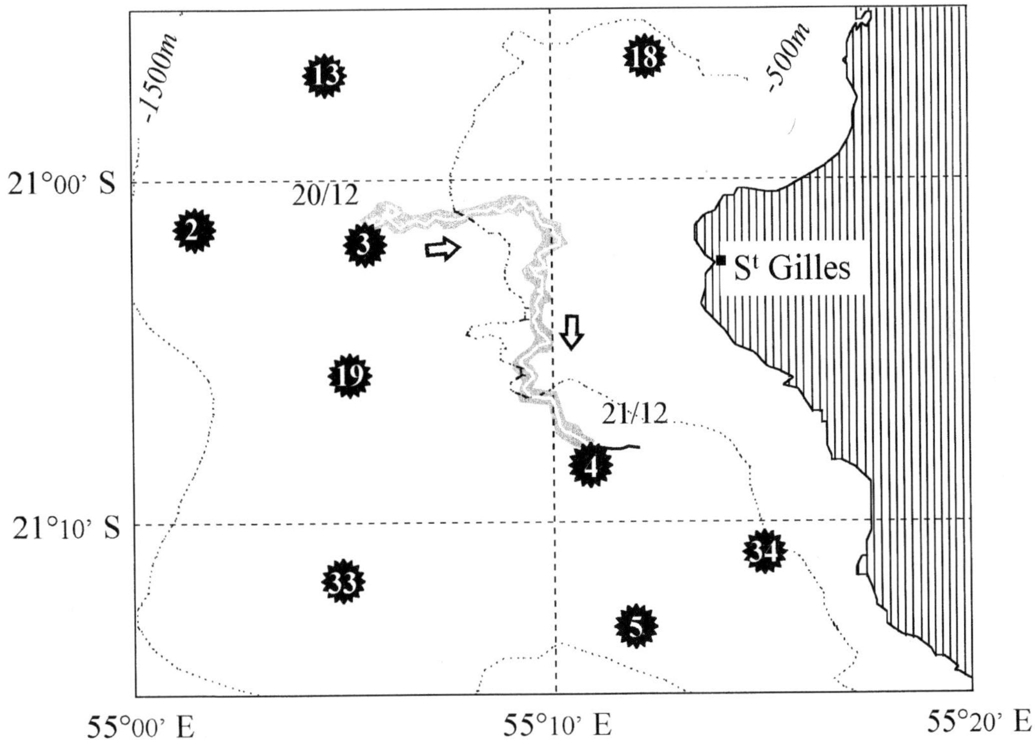


*Figure 6.* Course of the fish no. 7 (FL 84 cm), Dec. 20 1995, 18:00 to Dec. 21, 5:40.

ment to regain heat in the mixed layer), it is likely that the fish was injured (attack from a predator?) and slowly dying.

*Horizontal distribution pattern around FADs*

Most of the tuna were caught upstream from the FAD. This was also reported by Holland et al. op. cit., and

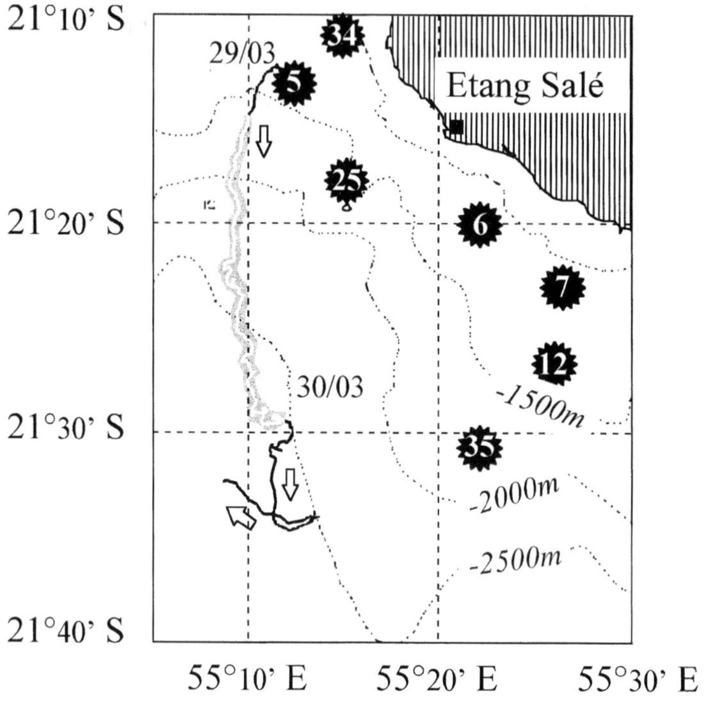


*Figure 7.* Course of the fish no. 4 (FL 95 cm), Mar. 29 1995, 13:00 to Mar. 30, 13:15.

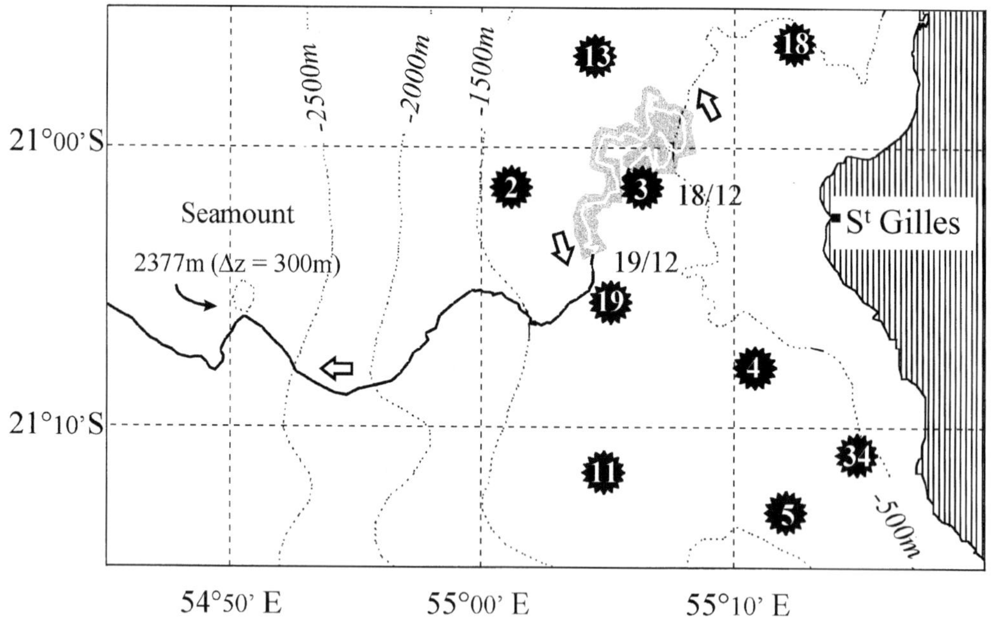


*Figure 8.* Courses of the fish no. 6 (FL 70 cm), Dec. 18 1995, 17:10 to Dec. 19, 14:30.

even in technical handbooks for fishermen in the South Pacific. The up-current orientation of the predators can be explained by a more efficient way to intercept incoming preys joining the FAD for shelter. This position can also match with the distribution of prey staying at a slight distance from the structure, like baitfish

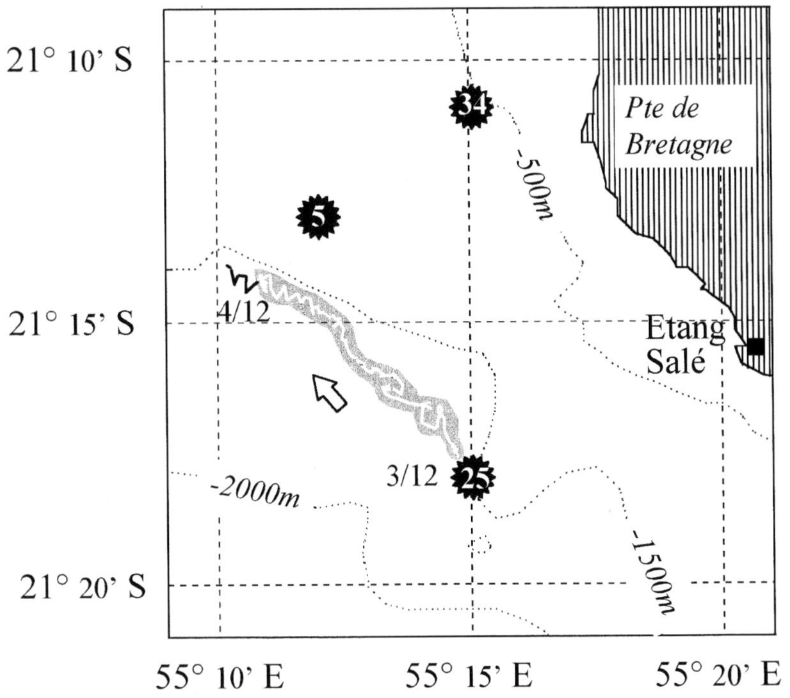


*Figure 9.* Course of the fish no. 8 (FL 76 cm), Dec. 3 1996, 17:30 to Dec. 4, 19:40.

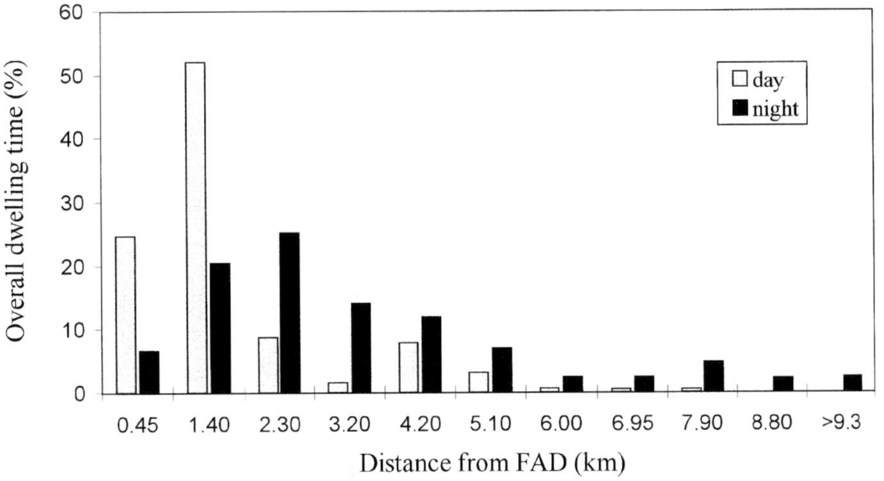


*Figure 10.* Overall dwelling time versus radial distance to FAD, by 0.90 km (0.5 nautical mile) interval.

groups maintaining an up-current position as observed in Florida (Klima & Wickam, 1971). However, this might not be an absolute rule, since downstream position has also been reported around FADs in Vanuatu (Cillauren, 1994).

All fish positions to a maximum distance of 18 km from the FAD were pooled to provide a general distribution (Figure 10). The median of the daytime distribution is at 0.9 to 1.8 km whereas at night it is 1.8 to 2.7 km, with greater skewness. Most of the dwelling time of a yellowfin during its FAD-associated phase is spent within 9.3 km from the FAD (100% at day, 97.5% at night).

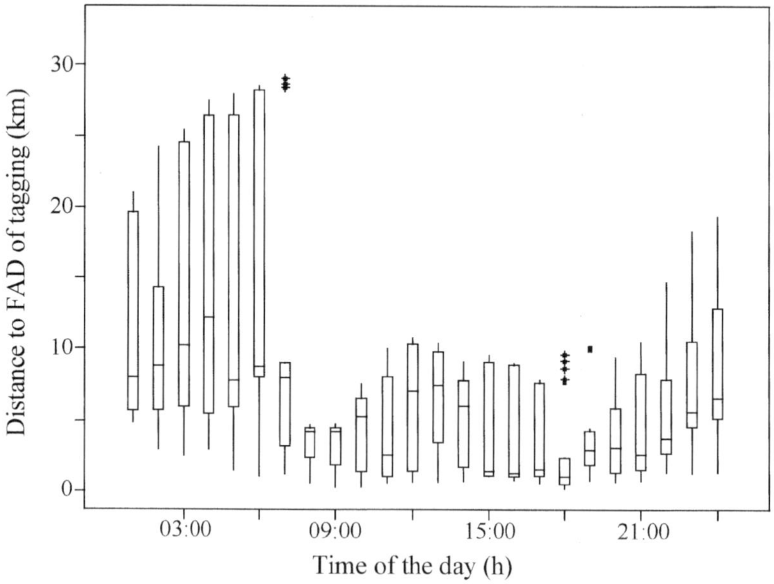


*Figure 11.* Box and whisker plot of the distance to FAD of tagging over the time of the day.

Considering the fish with a FAD-associated behaviour, the hour effect was tested with a non parametric ANOVA procedure, the Kruskall-Wallis one way analysis by ranks. The high value of the test statistic indicates a significant effect ($\alpha < 0.01$) of the hour. The FAD-dwelling phase is well defined from 7:00 to 19:00, and departure from the FAD appears progressively from 20:00 onwards (Figure 11).

*Swimming speed*

The swimming pattern appears to be affected when the fish is shifting from one behavioural phase to another, and between different periods. Some examples are given by the tracks of fish no. 3, 4 and 5. Swimming depths, VS and RS are presented in Figures 12 to 14. During the first hour of tracking, the vertical movements often showed a greater amplitude, a consequence of the stress following the tagging operation. This initial period being excluded, the FAD phase at daytime was characterised by a slow VS (0.08 to 0.12 m $s^{-1}$) and an RS within the range 0.22 to 0.32 m $s^{-1}$. According to the assumed underestimation of the horizontal speed by a factor of 2, the corrected RS would then be in the range 0.35 to 0.52 m $s^{-1}$. At dusk (18:00 to 19:00), while leaving the FAD and entering a transit phase, the fish drastically changed its swimming behaviour, moving actively upwards and downwards, increasing both VS (respectively 0.22, 0.33 and 0.15 m $s^{-1}$) and RS (1.40, 1.20 and 0.90 m $s^{-1}$). At dawn, still in a transit phase, the vertical activity slowed down (0.08 to 0.11 m $s^{-1}$) whilst the average RS was maintained at the same level as during the night (1.10 m $s^{-1}$). Consequently, the day/night change affects the vertical activity, whilst the shift from a FAD phase to a transit phase controls the RS level. Overall, the tunas moving offshore have a median RS of 1.24 m $s^{-1}$. In the close surroundings of the FAD of tagging (0 to 5.5 km), the corrected RS is estimated at 0.87 m $s^{-1}$.

– Fish size effect?

Average RS and its variance were calculated for the transit phase only, for the three periods already defined, i.e. day, moonlight and night (Figure 15). The swimming speeds (RS) and their variances do not show any trend with the size of the fish (as already observed in Hawaii, Holland et al., op. cit.) and is comprised between 0.70 and 1.50 m $s^{-1}$ whatever the periods considered.

– Hour of the day effect?

The median VS (0.05 m $s^{-1}$) corresponding to a FAD-associated behaviour within the interval 0 to 5.5 km, does not vary much over the hour. The largest skewness is observed from 15:00 to 22:00 (Figure 16a).

A very different pattern characterises the offshore-moving behaviour at distances greater than 9.3 km from the FAD, with a diurnal variability of the vertical activity (Figure 16b). The statistical tests (Kruskall-Wallis) discriminate three different phases: 19:00 to

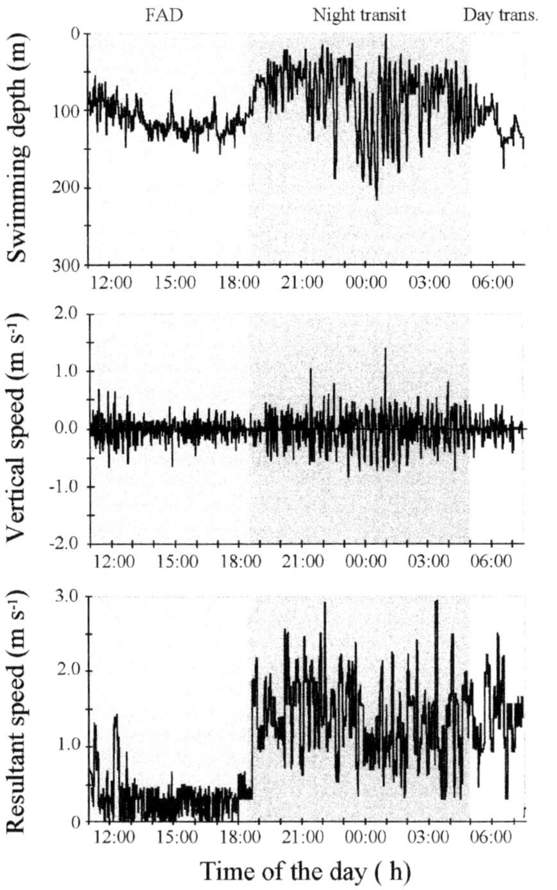


*Figure 12*. Swimming depth, vertical and resultant speed of the yellowfin no. 3.

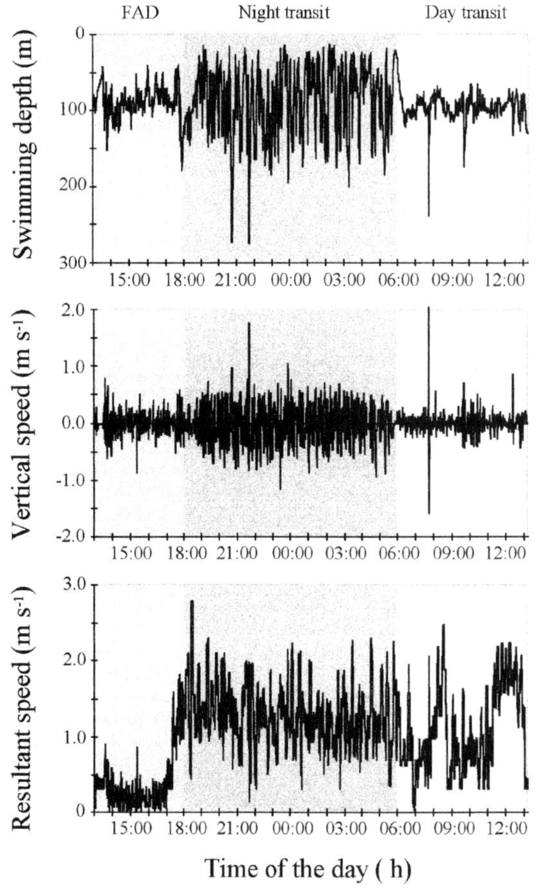


*Figure 13*. Swimming depth, vertical and resultant speed of the yellowfin no. 4.

3:00 (night) with high vertical speeds (0.35 m $s^{-1}$), 6:00 to 13:00 (day) with low activity (0.02 m $s^{-1}$) and 4:00 to 6:00 (morning twilight) as a transition situation between night and day patterns.

To a lesser extent, the RS in an offshore-moving behaviour exhibits a diurnal variability as well, with top speed in the middle of the day (1.8 to 1.9 m $s^{-1}$). There is no hour effect from 20:00 to 4:00, with a median speed of 1.20 m $s^{-1}$, but a significant hour effect for the rest of the day (Figure 17).

## Discussion

### *Horizontal distribution of dwelling time*

The tracked yellowfin tunas stayed closer to the FAD during the day than at night. This is likely to be the result of an attraction effect (vision and other factors) exerted by the vertical anchor line which provides a spatial reference in a 3D environment. The device also aggregates other small species preyed on by tuna; this might be a significant stimulus to retain predators in the vicinity of the structure. Observations on stomach contents of yellowfin caught around FADs in La Reunion (Conand, pers. com.) indicate that foraging activity starts in the twilight period preceding the sunrise, and increases during the day. Though it is agreed that a FAD cannot supply a prey turnover supporting the feeding requirements of a whole school of tunas, opportunistic foraging activity does occur around FADs at daytime.

During daytime, the fishes remain close to the FAD (mostly in the interval 0.9–1.8 km). The bulk of the horizontal movements, during the FAD phase, are within a 1.8 km radius (77% of the total time). Conversely, the less pronounced mode at night (1.8 to 2.7 km) and the skewness of the distribution (as far as 9.3 km; Figure 10) denotes a disassociation, followed by overnight

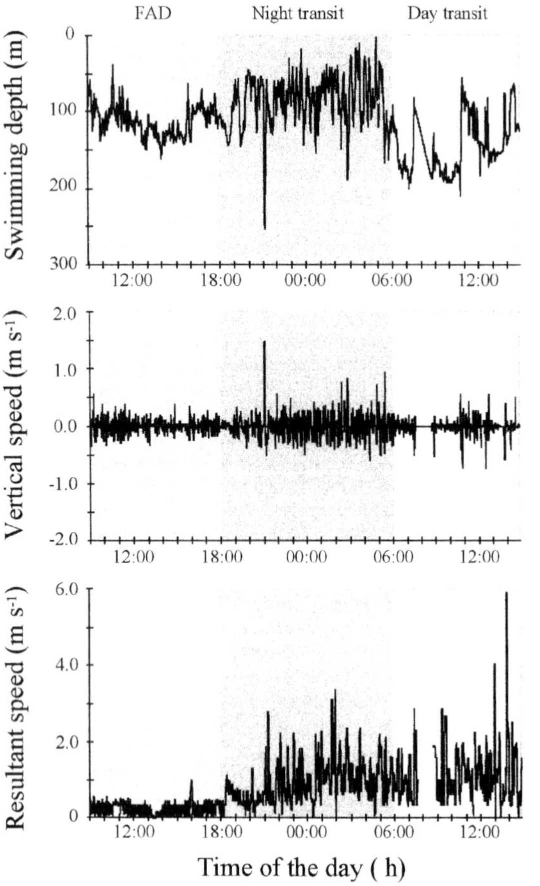


*Figure 14.* Swimming depths, vertical and resultant speed of the yellowfin no. 5.

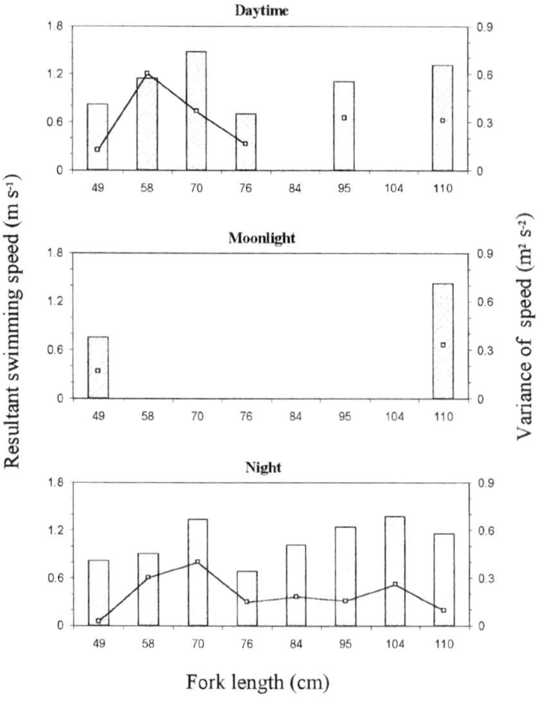


*Figure 15.* Resultant swimming speed (bars) and variance (line) by period of the day (daytime, moonlight and night) and size of the fish (fork length).

excursions away from the FAD of tagging, a behavioural pattern reported by various authors in different regions. Buckley & Miller (1994) suggest that keeping in the immediate vicinity of stationary objects, such as FADs, could make yellowfin tuna (especially small individuals) more vulnerable to attacks by nocturnal predators, like swordfish. If so, it might bring a piece of explanation to the night departures of tunas off the FAD.

The overnight maximum excursion radius is 13 km whilst movements are confined within 9.3 km during the day. These values are in agreement with previous studies: 11 km in Polynesia (Cayré & Chabanne, 1986); 8 km calculated by model (Hilborn & Medley, 1989); 9.3 km in Hawaii (Holland et al., 1990), 13 km in Comoros (Cayré, 1991). Consequently, in La Reunion, the FADs should not be set at distances less than 18 km to avoid overlapping the respective radii of influence.

### *Composite distribution in a vertical plane*

The vertical distribution of the fish is constrained by the surrounding oceanic environment features. The combination of temperature gradients and oxygen concentration can explain a great part of the swimming behaviour in depth (Cayré & Marsac, 1993). In La Reunion, the oxygen is not a limiting factor (3.5 ml $l^{-1}$ at 200 m) and the thermocline is not strongly marked (0.5 to 1.4 °C 10 $m^{-1}$). Thus, the model is not properly applicable. However, relative swimming depths published by Marsac et al. (1996) for FAD can be used to depict the composite depth distribution of the FAD-associated yellowfin in La Reunion. The overall distribution has a prominent mode around 125 m and a secondary one at about 25 m depth (Figure 18).

The vertical plane is defined by the radial distance to FAD (X axis) and the depth (Y axis). Vertical distributions by 10 m steps, from 0 to 250 m, are available at each 900 m radial increment from the FAD, for both daytime and night time periods. The sum of the probabilities at depths for a given interval being equal to 1, it is then possible to merge the horizontal and vertical distributions and generate a matrix of probabilities by

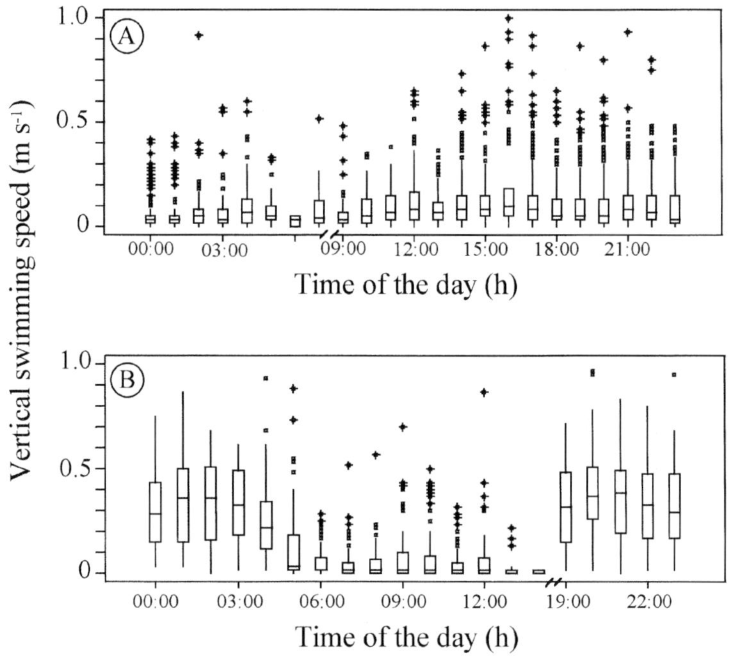


*Figure 16*. Box and whisker plot of the vertical speed of (A) yellowfin within a 5.5 km radius around the FAD of tagging and (B) yellowfin with an offshore-moving behaviour at distances greater than 9.3 km from the FAD of tagging.

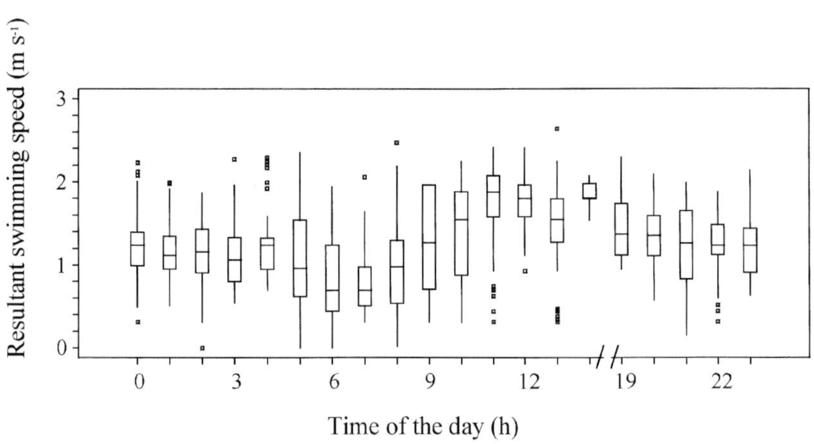


*Figure 17*. Resultant speed of offshore moving yellowfin at distances greater than 9.3 km from the FAD of tagging.

multiplying the values of the two distributions for each couple 'distance to FAD, depth'. When applied to a given biomass of tuna, the matrix of probability provides a set of values (in weight) which clearly indicates the places where the fish are likely to congregate. The distribution of the biomass in the vertical plane is then calculated is as follows:

$$B_{x,y,t} = B_{\text{FAD}} \cdot f(x,t) \cdot g(y,t),$$

where $B_{\text{FAD}}$ = total biomass associated to a FAD; $t$ = period factor (day or night); $f(x,t)$ = horizontal distribution at period $t$; $g(y,t)$ = vertical distribution at period $t$.

As an example, we assumed that a 50 $10^3$ kg is associated to a FAD in La Reunion at daytime. The resultant matrix (Figure 18) shows a lesser biomass in the surface layer but a slow decreasing rate over the distance to FAD. Yellowfin congregate at 100 m depth

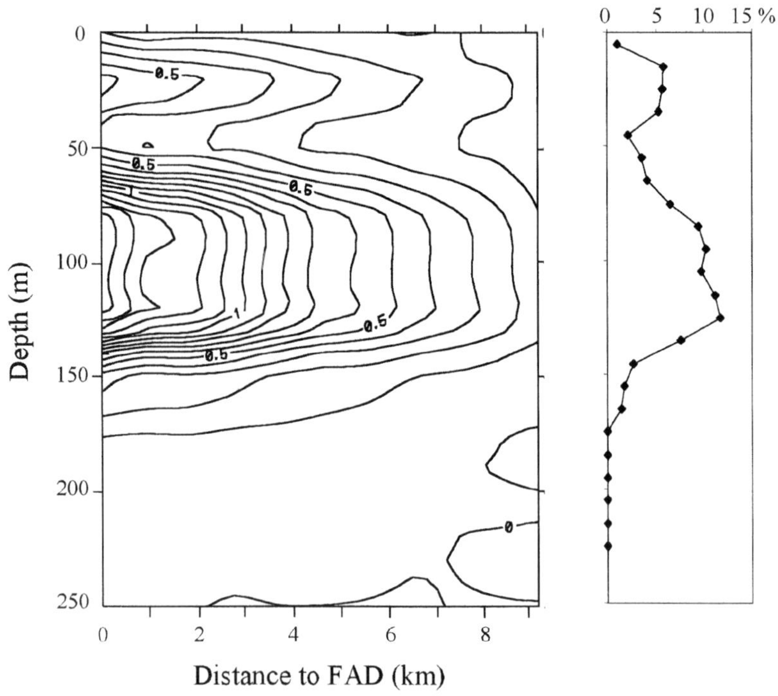


*Figure 18.* Spatial distribution (in $10^3$ kg) in a vertical plane of a yellowfin biomass of 50 $10^3$kg associated to a FAD at daytime (left) and relative dwelling time (in %) in depth used in the model (right).

with more than 1000 kg expected within a 3 km radius. If regular monitoring of the vertical distributions can be made around some FADs (using echosounding techniques for example, with relative scale of abundance in depth) this kind of figures can provide valuable information for the local fishermen, who can adjust the hooks depth of their fishing gears.

*Swimming speed and behavioural inferences*

From our observations, the median swimming speed (RS) of the yellowfin patrolling offshore is 1.20 m $s^{-1}$ at night and ranges from 0.70 to 1.90 m $s^{-1}$ during the day. These estimates match quite well with those calculated in Hawaii (0.89 to 1.81 m $s^{-1}$, Holland et al., op. cit.) and in the Eastern Pacific (0.67 to 2.17 m $s^{-1}$, Carey & Olson, 1982).

The swimming speed on FAD is estimated to be about 0.87 m $s^{-1}$. Those estimates are consistent with the Magnuson's model (1978) that predicts a sustained speed of 0.50 to 0.60 m $s^{-1}$ as the threshold preventing a yellowfin from sinking. Cayré & Chabanne (1986) found much lower values (minimum speed of 0.20 m $s^{-1}$, and 0.60 to 0.80 m $s^{-1}$ in periods of strong activity). Recent experiments in French Polynesia gave estimates of the same magnitude (Josse, pers. com.). This points out the uncertainty in the calculation of the swimming speed when a fish exhibits very limited horizontal movements. In this paper, a correcting factor of 2 was proposed but it remains purely a working assumption. A more detailed analysis (by simulation) of different movements patterns of the fish within the audible sphere of the hydrophone, when the boats goes along straight transects at different time intervals appears necessary to provide a more objective estimate of the horizontal swimming speed. Another method would be a direct measurement of the speed using a mini-flowmeter set on the tag (like the model designed by VEMCO™) or a rotor speedometer (Block et al., 1992). When comparing calculated speeds with the theoretical threshold proposed by Magnuson, we should also consider the flow of the current which provides naturally a gliding potential for the fish. In such a case, even a slow ground speed against the stream can prevent the fish from sinking.

The information provided by the analysis of VS and RS can be interpreted together with swimming behaviour, foraging activity and energy management.

*a) Phase of tight association to FAD*
The moderate swimming activity of yellowfin in the close vicinity of FADs could denote an optimal behaviour enabling hunting (opportunistic foraging) while saving energy by limited movements. The FAD would be perceived as an anomaly in the field of the natural topography of the nearby island and would enhance the probability of meeting conspecifics (meeting point hypothesis, Soria & Dagorn 1992, Dagorn 1994, Fréon & Misund, in press).

*b) Transit phase*

- general:
  - most of the night time transit sections are characterised by a high vertical activity, with repetitive and symetrical movements upwards and downwards about the mean swimming depth. Periodograms of the vertical movements for one of the fish (fish no. 3) indicate a dominant period at 13 min. This behaviour can be interpreted in different ways: (a) resting or energy saving movement, with alternate sinking (passive gliding) and climbing (active swimming), or (b) systematic sampling of the water column in order to locate patches of food in the dark.
  - the moonlit night transit sections present an RS similar to the daytime transit sections and a relatively high vertical activity. These features might denote a feeding behaviour in shallow depth. among FADs:
  - yellowfin tuna exhibit a higher VS and sudden changes of RS (rushes) during daytime, which may indicate foraging activity in the lower part of the thermocline;
  - the tunas transit from one FAD to another: though they do not necessarily stop at the intermediate FADs, this behaviour points out the ability for tunas to memorise specific locations, possibly in relation to different clues. Auditive detection of the FAD's anchor line (low frequency sounds) might be possible; in the same way, olfactive detection from the object itself or from conspecifics (excretion matters) could be considered, but none of these assumptions have been experimentally tested. The underlying stimuli of the attraction to a floating object remain uncertain, but it is likely that tuna find in their surrounding environment a combination of clues and use them to navigate as optimally as possible.
- offshore, away from the FADs:

A striking feature is the low vertical speed and high resultant speed at daytime. This might indicate a clear oriented movement towards a target (for instance, a seamount as observed with fish no. 6) at a stable depth.

Tunas are essentially day-feeders and the feeding activity is presumed to be limited at night (Kobayashi & Yamaguchi, 1971; Legand et al., 1972; Roger & Grandperrin, 1976). However, active hunting behaviour during the night was observed around the artificial source of light supplied by a vessel. Tunas have special adaptations to benefit from their habitat and maximise opportunistic foraging. In particular, the head of the pineal body (a light-sensitive organ) is large and lies below a cartilaginous window in the frontal bone (Rivers, 1953). This organ is activated during twilight and can play as a detector of shadows (Hanyu, 1978). Moreover, advanced predatory species benefit from a more efficient visual system in low ambient light (a larger proportion of retinal cones) than their prey (Munz & McFarland, 1973). Therefore, the success rate of attacks is increased during the short twilight periods. Bright moonlit nights would also allow tunas to chase and prey upon the migrating mesopelagic fauna coming up from deep daytime depths, as it is known that species of this fauna (crustaceans, bathypelagic fishes, cephalopods) constitute a prominent component of yellowfin diet (Brock, 1985; Roger, 1994a, 1994b). The diel variability in the dynamics of baitfish schools can also be taken into account to explain a continuing feeding activity during crepuscular and moonlight hours. At dusk, the schools start dispersing, but to a lesser extent than expected, especially for big schools which remain easily detected by echo-sounder (Fréon et al., 1996), thus providing easy targets for predators.

## Conclusions

This analysis points out particularities in the swimming behaviour of yellowfin patrolling in an area where FADs are anchored. Behavioural changes can be noticed when fish shift from a tight association phase with a FAD to a transit phase, either to swim among other FADs or to escape from a network of coastal FADs. The diel variability need to be considered as well. Overall, it appears that the day/night change affects the vertical activity, whilst the shift from a FAD phase to a transit phase controls the resultant swimming speed.

The fish remain tightly associated to the FAD during the day, at a short distance from the mooring line (less than 2 km), and the distribution ends at a maximum distance of about 9 km. The attraction potential of a FAD will be optimised if there is no overlap with the neighbouring ones. Therefore, we can consider that in La Reunion, the distance between FADs should never be less than 18 km. However, it must be kept in mind that the attraction potential also depends on the available biomass in the area. FAD will not generate tuna but alter the distribution and vulnerability of the resource that is present. The seasonal variation in abundance is even more marked when the area where FADs are anchored is at the boundary of the geographical distribution of the species.

Swimming speed appears to be an indicator of foraging activity and nature of movements. However, the technique used to estimate swimming speed is only reliable in relatively long and straight tracks and remains questionable in situations of small horizontal movements (the FAD situation, in particular). Direct measurement of the flow by a specific sensor set on/inside the tag would enable a better estimate of this parameter. The average speed calculated while yellowfin are moving offshore (1.24 m $s^{-1}$) is reliable and appear very consistent within those calculated in other regions for this species (Hawaii, Eastern Pacific).

## Acknowledgements

We are grateful to Dr François Conand, who participated actively in all the cruises, for having organised and monitored the tracking experiments in La Reunion. His help in some topics covered by this paper is also greatly appreciated.

## References

Block, B. A., D. Booth & F. G. Carey, 1992. Direct measurement of swimming speeds and depth of blue marlin. J. exp. Biol. 166: 267–284.

Brock, R., 1985. Preliminary study of the feeding habits of pelagic fish around Hawaiian FADs or can FADs enhance local fisheries productivity. Bull. mar. Sci. 37: 40–49.

Buckley, R., 1986. Fish aggregating device (FAD) enhancement of offshore fisheries in American Samoa. South Pacific Commission, SPC Fish. Newsletter 37: 37–41.

Buckley, R. & B. S. Miller, 1994. Feeding habits of yellowfin tuna associated with fish aggregation devices in American Samoa. Fifth International Conference on Aquatic Habitat Enhancement, Long beach, CA (USA), 3–7 Nov. 1991. Bull. mar. Sci. 55: 445–459.

Carey, F. G. & R. J. Olson, 1982. Sonic tracking experiments with tunas. Collective Volume of Scientific Papers ICCAT, 17: 458–466.

Cayré, P. & F. Marsac, 1993. Modelling the yellowfin tuna (*Thunnus albacares*) vertical distribution using sonic tagging results and local environmental parameters. Aquat. Living Resour. 6: 1–14.

Cayré, P. & J. Chabanne, 1986. Marquage acoustique et comportement de thons tropicaux (albacore: *Thunnus albacares* et listao: *Katsuwonus pelamis*) au voisinage d'un dispositif concentrateur de poissons. Océanogr. Trop. 21: 167–183.

Cayré, P., 1991. Behaviour of yellowfin tuna (*Thunnus albacares*) and skipjack tuna (*Katsuwonus pelamis*) around fish aggregating devices (FADs) in the Comoros Islands as determined by ultrasonic tagging. Aquat. Living Resour. 4: 1–12.

Cillauren, E., 1994. Daily fluctuations in the presence of *Thunnus albacares* and *Katsuwonus pelamis* around fish aggregating devices anchored in Vanuatu, Oceania. Fifth International Conference on Aquatic Habitat Enhancement, Long Beach, CA (USA), 3-7 Nov. 1991. Bull. mar. Sci. 55: 581–591.

Dagorn, L., 1994. Le comportement des thons tropicaux modélisé selon les principes de la vie artificielle. PhD Thesis, ENSA de Rennes, France: 250 p.

Fréon, P., F. Gerlotto & M. Soria, 1996. Diel variability of school structure with special reference to transition periods. ICES J. mar. Sci. 53: 459–464.

Fréon, P. & O. A. Misund, 1998. Dynamics of pelagic fish distribution and behaviour: effects on fisheries and stock assessment. Fishing News Book, Blackwell (in press).

Hanyu, I., 1978. Salient features in photosensory function of teleostean pineal organ. J. comp. Biochem. Physiol. 61A: 1240–1242.

Hilborn, R. & P. Medley, 1989. Tuna purse seine fishing with fish aggregating devices (FAD): models of tuna FAD interactions. Can. J. Fish. aquat. Sci. 46: 28–32.

Holland, K. N., R. W. Brill & R. K. C. Chang, 1990. Horizontal and vertical movements of yellowfin and bigeye tuna associated with fish aggregating devices. Fish. Bull. 88: 493–507.

Klima, E. F. & D. A. Wickham, 1971. Attraction of coastal pelagic fishes with artificial structures. Trans. Am. Fish. Soc. 100: 86–99.

Kobayashi, H. & Y. Yamaguchi, 1971. Feeding ecology and hooking tendency of tunas and marlins in the eastern equatorial Pacific. Bull. Jap. Soc. Sci. Fish. 27: 83–89.

Legand, M., P. Bourret, P. Fourmanoir, R. Grandperrin, J. A. Gueredrat, A. Michel, P. Rancurel, R. Repelin & C. Roger, 1982. Relations trophiques et distributions verticales en milieu pélagique dans l'océan Pacifique intertropical. Cahiers ORSTOM, Sér. Océanogr. 10: 303–393.

Magnuson, J. J, 1978. Locomotion by scombrid fishes: hydrodynamica, morphology and behaviour. In Hoar H. S. & D. J. Randall (eds), Fish Physiol. 7: 239–312.

Marsac, F. & B. Stequert, 1987. La pêche des thons autour d'épaves ancrées dans l'océan Indien. La Pêche Maritime 66: 439–446.

Marsac, F., P. Cayré & F. Conand, 1996. Analysis of small scale movements of yellowfin tuna around fish aggregating devices (FADs) using sonic tagging. In Anganuzzi, A. A, K. A. Stobberup & N. J. Webb (eds), Proceedings of the Expert Consultation on Indian Ocean tunas, $6^{th}$ session, Colombo, Sri Lanka, 25–29 Sept. 1995. Indo-Pacific Tuna Project Collective Volume 9: 151–159.

Munz, F. W. & W. N. McFarland, 1973. The significance of spectral position in the rhodopsins of tropical marine fishes. Vision Res. 13: 1829–1874.

Preston, G., 1982. The Fijian experience in the utilisation of fish agregating devices. South Pacific Commission, SPC/Fisheries 14/WP25: 61 pp.

Rivers, R., 1953. The pineal apparatus of tunas in relation to phototactic movement. Bull. mar. Sci. Gulf Caribb. 2: 168–180.

Roger, C. & R. Grandperrin, 1976. Pelagic food webs in the tropical Pacific. Limnol. Oceanogr. 21: 731–735.

Roger, C., 1994a. The plankton of the tropical western Indian ocean as a biomass indirectly supporting surface tunas (yellowfin, *Thunnus albacares* and skipjack, *Katsuwonus pelamis*). Envir. Biol. Fishes 39: 161–172.

Roger, C., 1994b. Relationships among yellowfin and skipjack tuna, their prey-fish and plankton in the tropical western Indian Ocean. Fish. Oceanogr. 3: 133–131.

Sims, N., 1992. A cost-benefit analysis of fish aggregation devices (FADs) in the artisanal tuna fishery in Rarotonga, Cook Islands. South Pacific Commission, Inshore Fisheries Research Project. Techn. Doc. 1: 10 pp.

Soria, M. & L. Dagorn, 1992. Rappels sur le comportement grégaire. In Stretta, J. M. (ed.), Action Incitative Comportement Agrégatif (AICA), Montpellier, June 1992. Document Interne Centre ORSTOM Montpellier: 5–9.

Willmann, R., 1990. Economic and social aspects of artificial reefs and fish aggregating devices. Symposium on artificial reefs and fish aggregating devices as resource enhancement and fisheries management tools. Indo-Pacific Fishery Commission, Colombo, Sri Lanka, 14–17 May 1990, IPFC/Symp 90/28:11 pp.

Yuen, H.S, 1970. Behaviour of skipjack tuna (*Katsuwonus pelamis*) as determined by tracking with ultrasonic devices. Can. J. Fish aquat. Sci. 27: 2071–2079.

*Hydrobiologia* **371/372:** 173–180, 1998.
*J.-P. Lagardère, M.-L. Bégout Anras & G. Claireaux (eds), Advances in Invertebrates and Fish Telemetry.*



# Spawning migration of radio tagged landlocked Arctic Charr, *Salvelinus alpinus* L. in Ennerdale Lake, the English Lake District

D. J. F. McCubbing[1], B. D. Bayliss[2] & V. M. Locke[2]
[1] *302-2170 West 3rd Av. Vancouver. BC, Canada V6K 1L1 (e-mail: don.mcc@intergate.bc.ca)*
[2] *Environment Agency, Chertsey Hill, London Rd, Carlisle, CA1 2QX, England*

*Key words:* Arctic charr, radio telemetry, spawning, migration


## Abstract

A sample of twenty Arctic charr, *Salvelinus alpinus* L. from Ennerdale lake were externally tagged, using a small radio transmitter, during their annual spawning migration. Ten fish (5 male, 5 female) were captured from each of the two spawning tributaries using electrofishing techniques. Following release, migration behaviour was observed remotely through a network of automatic radio listening stations placed at the mouths of the two spawning streams. Over 106 detections were made of tagged fish entering the two streams, during the observed spawning period of 15th November to 4th December 1996.

Time of entry and stream residence duration were examined. On average male and female charr showed no difference in initial entry time to the spawning grounds each night although charr entered Smithy Beck earlier than the River Liza. Male charr stayed resident on the spawning grounds for slightly longer than female charr although residence duration's varied throughout the experimental period, with some fish remaining resident on the spawning grounds during daylight hours at the peak of spawning activity. Repeated migrations from the lake into one or other of the spawning tributaries was observed for 9 female and all 10 male charr. Significant differences occurred between the frequency of these repeated spawning migrations with females making on average 2.7 migration runs, whilst males made on average 6.1 migrations.

The selectivity of spawning stream was also observed with 7 of the 20 fish showing migration runs of greater than 10 minutes, into both rivers during the course of the spawning period. Male fish originally captured in the River Liza appeared to 'stray' most, whilst male fish caught in Smithy beck showed the least straying behaviour.


## Introduction

The Environment Agency has a statutory duty for the protection and conservation of fisheries in England and Wales. To achieve this aim, it is first necessary to understand the current status of stocks. This is of particular importance where stocks are of a nationally rare species. Whilst the arctic charr, *Salvelinus alpinus* L., is locally common within the English Lake District, occurring in at least seven of the major lakes (Frost, 1965), it is generally a rare species in the UK. The Lake District populations are relics of the Ice Age at which time the lakes are thought to have been populated with anadromous charr. The anadromous behaviour has since been lost in all the Lake District populations (although access to the sea is still possible) resulting in stocks that have formed very individual ecological and morphological characteristics (Frost, 1965). Maitland & Lyle (1991) highlighted the vulnerability of the Lake District charr populations, along with two populations of white fish, the schelly (*Coregonus lavaretus* L.) and the vendace (*C. albula* L.).

Farooqi (1992) identified the arctic charr population in Ennerdale as warranting detailed investigation, as this population of arctic charr is the only Lake District population that is exclusively river spawning (Partington & Mills, 1988). Further to this work, a recent investigation on the only other river spawning population of charr in England, (on the River Brathay,

Winderemere), suggested only a remnant number of fish remain (J. Foster, pers. com.).

In 1850, Dr Lietch of Keswick noted in his notebook on the 25th of November "... Went over to Ennerdale to see the spawning charr of that lake, which I was told run up in Martinmas Week and remain a fortnight, and deposit their spawn in 'The Char Dub' ". He further noted observations of thousands of fish weighing up to 250g. Although as recently as the 1970's rod exploitation of the charr stock was common practice, visual observations in the mid 1980's indicated a significant reduction in population since this historical note was made.

The key aim of this study was to remotely gather detailed information on the migratory behaviour of the charr population during its annual spawning migration. Concern over the effects of netting studies on fish behaviour (D. Smith, pers.com.) and spawning success had been raised. Whilst a video monitoring study of migration runs was effective (Chiri & Nicholson, 1995), it was restrictive in the detail of data it could collect. Information from remote monitoring with radio telemetry will assist in the understanding of population structure and ecology of these fish.

## Study area

Ennerdale Lake is situated to the West of the Lake District, in the North West of England. It acts as a catchment basin for the River Ehen, which flows from the lake west towards its confluence with the Irish Sea at Seascale (Figure 1).

It is known from previous studies (Farooqi et al., 1992), that the arctic charr in Ennerdale only spawn in two tributaries of the lake, namely the River Liza and Smithy Beck, (Figure 1). Spawning only takes place in the River Liza from 200 m upstream of the river lake confluence to downstream of an Irish pipe bridge (around 1000 m river length), whilst on Smithy Beck it is restricted to a reach, above the lake but downstream of a natural waterfall (around 300 m in total). Suitable spawning areas are typically less than 1 m deep with a fine pea sized gravel. Redds the size of a tennis ball are constructed by a spawning pair of charr and the eggs and milt are fertilised and buried in a similar way to other salmonids.

The need to understand the behaviour and stock levels of the charr population is tempered with the wish to avoid any interference with their normal spawning behaviour. Reasons for declines in charr numbers are postulated to include the acidification of the River Liza (Prigg, 1986), the construction and subsequent effects of an Irish pipe bridge (Reed, 1995; Palmer, 1996) and the availability of suitable spawning gravel and spawning access (Reed, 1995). Loss of spawning gravel through straightening the river channel for drainage purposes has also been observed in Smithy Beck, (D. Pearson, pers. com.).

Concerns over the ability of the charr population to maintain itself, has resulted in recent channel modifications and gravel seeding being undertaken on Smithy Beck with a significant increase in the area of spawning gravel already achieved. Proposed modifications to the Irish Bridge are been addressed and a design has been drawn up for a replacement (Palmer, 1996). In addition to these physical changes, invertebrate monitoring and the re-occurrence of salmon within the Liza suggest improving water quality, due to reduced acidic flushes, may also assist future recruitment of the population (McCubbing, 1997).

## Materials and methods

Twenty charr were captured for monitoring by electrofishing techniques (after sunset, 16:00 GMT) using a backpack battery powered control box with an output voltage of between 150–400 V., smoothed DC. Fish were detected using torchlight (within the spawning area of each sample stream) and then stunned temporarily to aid netting. Following capture the fish were held in a light proof tank to avoid undue stress from light, until ready for tagging.

Before tagging was undertaken, the fish were anaesthetised using a dilute solution of phenoxyethanol (concentration dependant on water temperature) to reduce post tagging trauma (E. Pottinger, pers. com.). When the fish were fully anaesthetised, showing no reaction to external stimuli (particularly light), they were removed individually from the light proof tank. Each fish was visually assigned a sex from morphological features, measured (fork length to the nearest mm) and notes on the fishes spawning condition were made.

The fish were tagged with small radio tags (Biotrak SS-2); externally mounted on the fish with temporary sutures. A tag cradle was developed from methods used by Batt et al. (1995) and Solomon & Storeton-West (1983). It consists of a small piece of neoprene rubber, 30 mm by 15 mm (dry suit cuff material), backed by nylon weave material (used in dry suits), that fits over the back of the fish. To one side of this

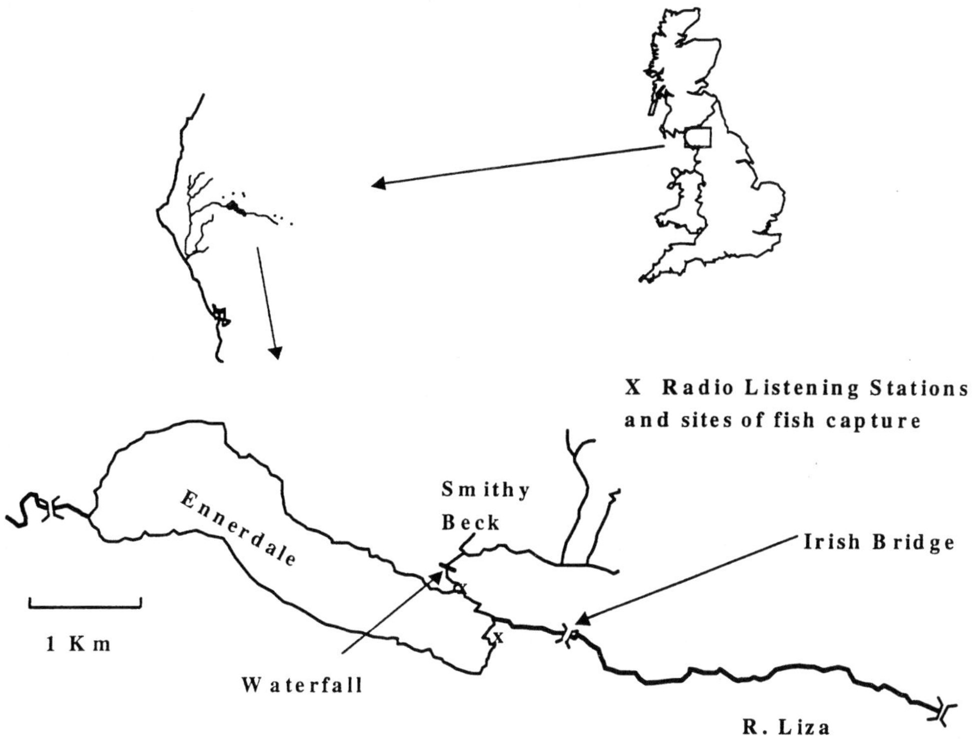


*Figure 1.* Location of study site, showing spawning areas and radio listening station deployment.

cradle is mounted the radio tag (using Aquasure adhesive) complete with internal battery and external aerial. Tags for use on male fish were painted with yellow fluorescent paint, whilst those for use on female fish were painted with orange fluorescent paint (Acrylic based water-resistant poster colour; Rowney). Fluorescent paints aided visual identification of tagged fish within the spawning population during stream migration with colour differences increasing the chance of visually identifying individual tagged fishes spawning behaviour.

The radio tag cradle with tag attached was sutured to the fish through the musculature just forward of the dorsal fin, being careful to avoid the fin ray and spinal cord. A single dissolving suture thread (Ethicon Coated Vicryl size 3/0 with 60 mm fine straight cutting needles) was used as this would allow for the complete loss of a tag if either the front or rear section of the attachment thread was broken. This avoided the risk of partial loss of the tag cradle, which might cause fish damage. Following tag attachment the fish was allowed to recover for at least 15 min in a second light proofed tank containing clean fresh water.

A previous pilot study undertaken to assess tag loss rates, showed charr released back into their river of capture, close to where they were caught showed a high incidence of tag loss. It later became apparent that these tag losses were probably due to mechanical removal by rubbing against boulders under which they had taken refuge. Further trials with fish released back into the lake, to allow for a period of adjustment to carrying a tag, showed significantly better tag retention. Tagged fish in this study were thus returned directly to the lake environment following tagging.

The radio tags on the fish were set to emit similar pulse rates but on distinct frequencies, from 716 to 899 kHz, to allow individual detection of each tagged fish. Monitoring of fish position was undertaken with remote radio listening stations, incorporating a Yaesu FT 290R receiver with ten pre set frequency channels and an omni directional aerial. Two stations were placed at the confluence of each of the spawning streams and the lake, in a position that ensured that tags could not be detected within the body of the lake (Figure 1) but would be detected when fish entered the river spawning area. Most of the spawning ground was within the detection range of the stations, so periods of

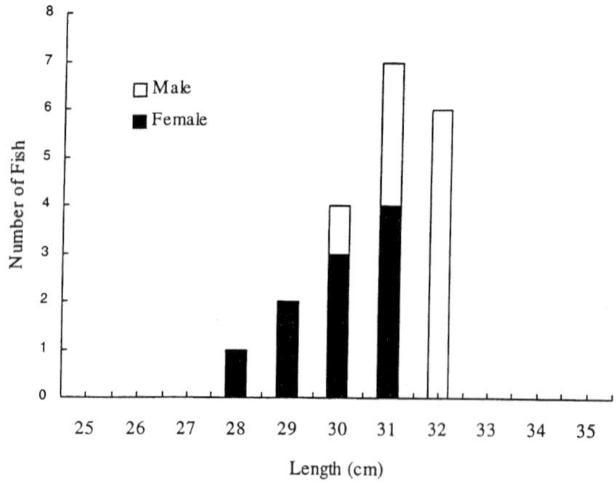


*Figure 2.* Length frequency of tagged charr.

residence could be established. In many cases manual recording of tagged fish using a portable Yaesu FT 290R receiver from the riverbank backed up the presence of tagged fish. Following release of the first tagged fish, monitoring was undertaken in both spawning streams every 5 minutes by remote station and nightly by hand held receivers.

## Results

Details of tagged fish length, sex, tag frequency and river of origin were recorded for each fish (as shown in Table 1), along with the number of spawning migrations observed and the average duration of these migrations. Fish lengths for male and female charr were found to be very similar to previous studies of this population (Farooqi et al., 1994). Male fish (mean length = 31.6cm, S.D. = 0.57) being of slightly greater mean length than female fish (mean length = 30.2 cm, S.D. = 1.35, $t = -3.48$, $P < 0.05$, two tailed $t$-test) (Figure 2).

A total of 106 discrete observations of tagged fish were made by the remote listening stations, (many confirmed by portable receiver) in the two week period after tagging. Many more repeat observations were also made as tagged fish remained within detection range of a listening station on the spawning grounds, after initial detection. Visual observations of tagged fish on the spawning areas were made on several occasions during the hours of darkness. I one case this included two male tagged fish and one female tagged fish in a shoal of around seven individuals. Five fish were assessed to have lost their tags during the spawning run; four at the confluence of the two tributaries with the lake (2 on the Liza and 2 on Smithy Beck), and one within Smithy Beck. These losses occurred some days after tagging and were assessed by the stationary source of signal over a period of a number of days or tag recovery. Tag losses probably resulted from spawning activity through mechanical damage to the suture threads during redd cutting, and generally occurred in male fish.

From the 106 detection's it was ascertained that nine of the female fish and all ten of the male fish were detected on at least one occasion. Male fish undertook on average 6.1 spawning migrations (S.D. = 3.14, range 1–10) whilst female fish made on average only 2.7 migrations (S.D. = 1.16, range 1–5). This difference is statistically significant ($t = 3.21$, $P < 0.01$, two tailed $t$-test). Male charr originally captured in the Liza made most spawning migrations (average 8.8, S.D. = 1.3, range 7–10), whilst males from Smithy Beck made least migrations (average 2.7, S.D. = 0.7, range 1–3). No relationship between fish length and number of migration runs was detected in either male or female fish. Occasions when individual tagged fish were recorded entering Smithy Beck and/or the River Liza more than once in the same night were also recorded. Five fish exhibited this behaviour, four males and one female. Four of these dual migrations were a combination of periods of residence in both rivers, whilst one fish was recorded to have spent two periods in Smithy Beck in the same night having left the area between each migration.

The time of first detection was noted each day and recorded as the time of entry to the spawning grounds if detection was for 10 min or longer. This eliminates single detections when fish made short exploratory migrations into the lower reaches of the river but did not stay resident long enough to reach the spawning areas. No statistically significant differences in the time of entry to the spawning grounds were detected for male compared to female fish, although differences did occur for fish entering Smithy Beck compared to those entering the River Liza (Figure 3). Charr entering Smithy Beck entered the spawning area significantly earlier than those entering the Liza ($p < 0.001$, Kolmogorov Smirnov two sample test). Over the study period, 95% of tagged fish recorded entered Smithy Beck before 18:00 GMT, whilst in comparison only 40% of tagged fish recorded in the River Liza had been detected by this time. Individual charr making more than one entry to the spawning grounds on different nights during the study showed no pattern in time of entry, when successive migrations are compared.

*Table 1.* Fish and tag data from radio tracking study

| Freq. (kHz) | Sex | Length (cm) | Capture site | Observed spawning migrations (nos.) | | Mean residence duration (minutes | |
|---|---|---|---|---|---|---|---|
| | | | | Smithy | Liza | Smithy | Liza |
| 800.2 | Female | 30.0 | Liza | 0 | 3 | – | 220 |
| 818.8 | Female | 31.5 | Liza | 0 | 5 | – | 381 |
| 830 | Female | 31.0 | Liza | 1 | 2 | 130 | 535 |
| 862.6 | Female | 31.0 | Liza | 3 | 1 | 443 | 945 |
| 881.7 | Female | 30.0 | Liza | 2 | 0 | 1250 | – |
| 716.4 | Female | 29.5 | Smithy | 3 | 0 | 408 | – |
| 755.4 | Female | 31.5 | Smithy | 0 | 1 | – | 200 |
| 775.3 | Female | 28.0 | Smithy | 0 | 2 | – | 170 |
| 843.6 | Female | 29.0 | Smithy | 2 | 0 | 845 | – |
| 884.7 | Female | 30.5 | Smithy | 2 | 0 | 173 | – |
| 767.7 | Male | 31.0 | Liza | 1 | 8 | 365 | 677 |
| 786.3 | Male | 32.0 | Liza | 3 | 4 | 360 | 274 |
| 808.6 | Male | 31.5 | Liza | 1 | 9 | 130 | 466 |
| 836.1 | Male | 32.0 | Liza | 3 | 5 | 373 | 286 |
| 810.2 | Male | 32.0 | Liza | 6 | 4 | 967 | 315 |
| 733.9 | Male | 32.0 | Smithy | 3 | 0 | 470 | – |
| 827.7 | Male | 30.5 | Smithy | 0 | 1 | – | 165 |
| 857 | Male | 32.0 | Smithy | 4 | 0 | 626 | – |
| 873.1 | Male | 31.0 | Smithy | 5 | 0 | 1013 | – |
| 898.5 | Male | 32.0 | Smithy | 4 | 0 | 513 | – |

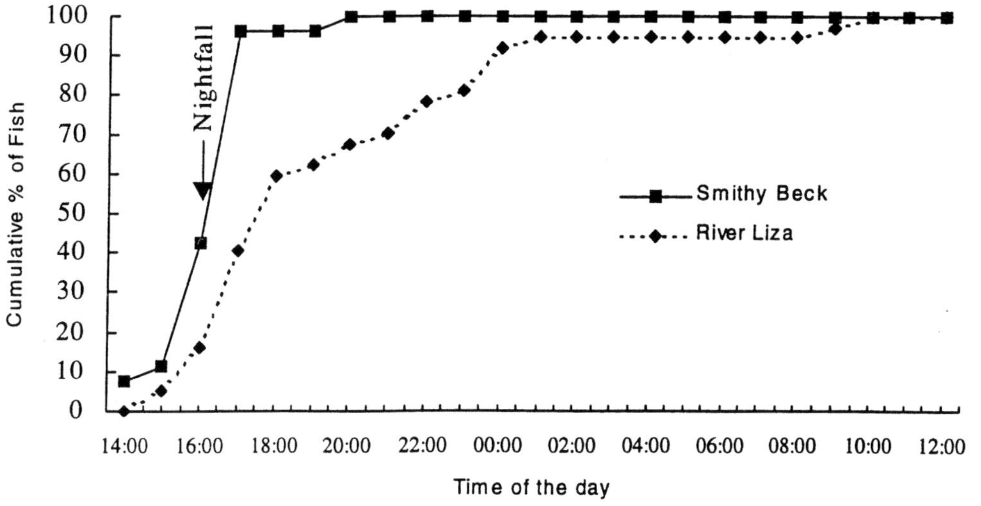


*Figure 3.* Cumulative time of entry to spawning tributaries, of migrating charr, through the entire spawning run.

Stream specificity was assessed, using the site of original capture as a home river and comparing where fish were detected in relation to this designation. Of the 20 fish tagged, 7 fish were subsequently recorded present for at least 10 min on both spawning areas, whilst a further 6 fish made shorter exploratory migrations into the lower reaches of each river. A difference was found in behaviour between male and female charr, with only 2 of the female charr but 5 of the male charr recorded present in both rivers. These results statistically significant due to the small sample size. There were also differences in the selectivity of fish initially captured in the River Liza compared to those originating from Smithy Beck, with detection of fish from the Liza significantly more likely in Smithy beck than the other way round ($z = 2.414$, $p = 0.016$ Wilcoxen signed rank test). This appeared to because of fish straying from the Liza.

Residence durations were calculated for both male and female fish, (ignoring spawning site) and for fish detected on each spawning area (Table 2). A mean period of residence of 543 min was detected for male charr, whilst female charr stayed resident for on average only 411 min (ignoring site of detection). This was not a significant difference due to the large variations in residence time in individual migrations (two-tailed Mann Whitney test, $W = 17.7$, $P = 0.77$). A similar situation was detected in residence duration (ignoring sex differences) if the site of migration was compared. Fish detected resident in the River Liza spent on average 399 min in the river, whilst fish detected in Smithy Beck spent an average of 599 min in the stream. These results again were not significant due to individual variances in migration residence (two-tailed Mann Whitney, $W = 23.79$, $P = 0.36$). Both male and female captured in Smithy beck stayed only briefly in the Liza during the few such migrations detected, however the low number of migrations precludes the use of statistical analysis.

## Discussion

Male and female charr captured for tagging were found to be of a similar fork length and probably age to charr sampled in the previous studies (see Farooqi et al., 1994). Thus the sample collected was thought to be representative of the spawning population as a whole. Only one fish showed any sign of having spawned before capture, thus the tagging date was prior to major spawning activity occurring.

During the study a large number of observations were made of tagged fish by the remote listening stations, compared to other published data (Solomon & Storten-West, 1983), partly due to the small study area and partly due to the concentration of fish in the catchment during the spawning period. As a result significant population and behavioural information was obtained from the recorded data. For example, it is widely understood that the handling and tagging of fish is a stressful event (Gray & Haynes, 1979) and that initial behaviour following such an intrusive event is unlikely to be natural. However in the case of this study, over 70% of the tagged fish returned to the spawning grounds just hours after the release, suggesting the urge to spawn may have outweighed the stress of the tagging event. Many studies detail a loss of contact with tagged fish during this initial post-tagging period (Bayliss & McCubbing, 1996).

Radio tracking data shows that male charr in this study made on average nearly twice the amount of spawning migrations that female fish exhibited. Such behaviour has been detailed in other studies (Fabricios & Gustafson, 1954; Moore, 1975) and may be the result of male charr spawning with several females (McCart, 1981), in an attempt to spread their genetic material through a wider section of the population. Such behaviour patterns in male and female fish were previously observed by Farooqi et al. (1993, 1994) in the Ennerdale system, when it was found that on any given night, 16%–57% of fish spawning were recaptures from earlier migrations in the same spawning year. Male fish dominated these recaptures, but no evidence for an imbalance in sex ratio can be determined from mark recapture data on this population (Farooqi et al., 1994).

The majority of tagged fish entered the two rivers after dusk, with the greater proportion waiting until complete darkness. The reasons for this behaviour pattern are probably attributed to an avoidance of light by the fish in an effort to reduce predation risks in the shallow spawning areas. Observed delays in migration into the river Liza after sunset may occur because of its clear water and wide exposed nature. Many nights during the spawning period were affected by bright moonlight from sunset until the moon passed behind local mountains around midnight. In comparison overhanging vegetation and narrower nature of Smithy beck may afford more shading and cover to spawning fish. Further investigations involving light intensity and charr behaviour may be appropriate.

*Table 2.* Summary means and standard deviations for spawning migration in arctic char, Ennerdale, 1996

| | Nos of fish | Mean duration of residence of spawning migration (min) Smithy Beck | River Liza | Combined |
|---|---|---|---|---|
| Smithy caught females | 5 | 428 (S.D. = 334) | 180 (S.D. = 126) | 380 (S.D. = 276) |
| Liza caught females | 5 | 479 (S.D. = 526) | 416 (S.D. = 296) | 433 (S.D. = 350) |
| All females | 10 | 470 (S.D. = 362) | 365 (S.D. = 283) | 411 (S.D. = 317) |
| Smithy caught males | 5 | 645 (S.D. = 724) | 165 (S.D. = 0) | 645 (S.D. = 649) |
| Liza caught males | 5 | 658 (S.D. = 582) | 460 (S.D. = 536) | 546 (S.D. = 676) |
| All males | 10 | 652 (S.D. = 640) | 449 (S.D. = 429) | 543 (S.D. = 486) |
| All fish | 20 | 580 (S.D. = 571) | 399 (S.D. = 463) | 482 (S.D. = 520) |

Typically male charr from previous studies have been thought to arrive on the spawning grounds earlier than female fish (Farooqi et al., 1993, 1994; Chiri & Nicholson, 1995). This behaviour was not found in the present study based on radio tracking data. It is proposed that what previous studies actually observed was the greater proportion of male fish entering the spawning grounds on any night, due to the higher level of repeat migrations in male fish. This would give the impression that male fish entered the rivers earlier than female fish, but subsequent analysis of the sex ratio of later entering fish remained constant in this study, refuting this idea.

The occurrence of charr on a spawning ground, other than their site of initial capture was noted in 35% of the charr tagged in 1996, with a further 35% making short (less than 10 min) migrations into the lower reaches of both rivers. A greater proportion of males exhibited this behaviour, which is probably due to their observed higher number of migrations in general. This is possibly a natural occurrence with fish seeking out a suitable mate and spawning site in the area available to them or it may be a reaction to factors affecting the spawning areas, particularly on the river Liza (Reed, 1996). Migrations by males into both spawning areas will reduce the risk of environmental perturbations on one river adversely affecting the genetic pool of the whole population in any year, although this may or may not be an adaptive response to such changes. Female charr on the other hand, make less migrations and may not adapt as well to changes in their spawning environment, with less 'straying events' recorded. The possibility that these patterns are occurring in a single homogenous population would seem unlikely, although cannot be ruled out, as the data is from only one year of study. However, previous mark/recapture experiments also found straying of male and female fish between the two rivers, with 11% of tagged 'Liza' charr recaptured in the Smithy Beck in 1992 (Farooqi et al., 1994). Recaptures of Smithy Beck fish in the River Liza as in that study were also less common. The reasons for this are unclear but may relate to the lower population levels estimated for Smithy Beck (around 243 fish, Chiri & Nicholson, 1995) and the better availability of suitable spawning substrate found in this beck, compared to the River Liza (population estimate of 1076, Farooqi et al., 1994).

The reasons for differences in residence duration during spawning migrations, between fish captured from Smithy beck and those captured from the Liza are difficult to explain. Males and females caught on the Liza spent on average around the same period of time on the spawning areas of both rivers during any single spawning run, whilst fish of both sexes captured on Smithy beck spent on average much shorter periods in the Liza. A possible explanation may be due to the degraded spawning substrate in the Liza. Should the two rivers have historically developed subpopulations, with separate preferred spawning sites, there is currently little advantage for fish originating from Smithy beck to spawn in the river Liza, where conditions are deteriorating. There would be advantages for fish from the Liza to spread their genetic material to Smithy beck. Investigations following removal of the Irish pipe bridge, subsequent improvements in spawning gravel, and continuing reductions in acidic episodes may answer some of these questions although design of such experiments is complex (Walters & Collie, 1989). During the peak of spawning activity several tagged and many untagged male charr were observed resident during daylight hours on the spawning grounds. Similar behaviour has been noted in sea trout populations, (Evans, 1994) and some charr populations (Johnson, 1994). As charr in this population normally avoid bright light (authors observations), this unusual behaviour may be an attempt to increase the likelihood of finding a suitable spawning partner or site.

## Conclusions

The radio tracking of charr using telemetry techniques was successful in meeting the studies objectives. Data were collected remotely on charr behaviour with the minimal of disturbance to spawning fish. Similarities to the results observed in previous mark/recapture and video analysis studies indicate the robustness of this technique. The need to understand fish behaviour in determining a suitable method for assessing population structure and size, has been addressed and whilst radio tracking does not offer a way of determining population size, it would allow in conjunction with alternative sampling strategies a better population estimates. The current and historical studies have highlighted the risks involved in visual observations being used to assess this spawning stock of charr, where assumptions on repeat spawning, stream residency and stream specificity may be made incorrectly. The data now available on this river spawning charr population should allow the population changes following improvements in habitat and water quality to be quantitatively assessed without concerns over the effect of such monitoring work on the behaviour and success of spawning fish.

## Acknowledgements

The assistance and support of the South West Cumbrian bailiff team is gratefully acknowledged in the collection of the data used in this study as was the co-operation of Forest Enterprise in allowing access to the site of study.

## References

Batt, P. I., E. Campbell-Lendrum & S. N. Young, 1995. Field trials with the new SS-2 button tag on sea trout using radio telemetry. In the Proceedings of the Institute of Fisheries Management 25th Annual Study Course, Lancaster 1994.

Bayliss, B. D. & D. J. F. McCubbing, 1996. Preliminary results of the Leven Estuary Fisheries Study. Environment Agency Technical Report. Northwest Region.

Brown, R. S & W. C. Mackay, 1995. Spawning ecology of cutthroat trout Oncorhynchus clarki in the Ram River, Alberta. Can. J. Fish. Aquat. Sci. 52: 983–992.

Chiri, F. & S. A. Nicholson, 1995. An assessment of the Spawning Stock of Arctic Charr (*Salvelinus alpinus* L.) using InfraRed TV Surveillance in Smithy Beck, Ennerdale Water NRA Northwest Region Report nos. NRA/NW/FTR/95/9.

Evans, D. M, 1994. Observations on the spawning behaviour of male and female adult sea trout *Salmo trutta* L., using radio telemetry. Fish. Mgmt Ecol. 1: 91–105.

Fabricius, E. & K. J. Gustafson, 1954. Further aquarium observations on the spawning behaviour of the char, *Salmo alpinus*, L. Rept. Inst. Freshwat. Res. 35: 58–104.

Farooqi, M. A., S. A. Nicholson, M. W. Aprahaminan & S. R. Douglas, 1992. Arctic charr from Ennerdale Water: A pilot Study NRA Northwest Region Report nos. NRA/NW/FTR/92/2.

Farooqi, M. A., S. A. Nicholson, M. W. Aprahaminan & S. R. Douglas, 1994. Arctic charr from Ennerdale Water: Relics of the Ice Age NRA North West Region Report nos. NRA/NW/FTR/94/2.

Frost, W. E., 1965. Breeding habits of Windermere charr, *Salvelinus willughbii* (Gunther), and their bearing on the speciation of these fish. Proc. r. Soc. biol. 163: 232–284.

Gray, R. H & J. M. Haynes, 1979. Spawning migration of adult chinook salmon (Oncorynchus tshawytscha) carrying external and internal radio transmitters. J. Fish. Res. Bd Can. 36: 1060–1064.

Johnson, L., 1994. Systems for Survival: of ecology, morals and Arctic charr. Trondheim: Keynote address to the Arctic Charr Symposium, June 1994.

Maitland, P. S. & A. A. Lyle, 1991. Conservation of freshwater fish in the British Isles: the current status and biology of threatened species. Aquat. Cons.: Mar. Freshwat. Ecosys. 1: 25–55.

Moore, J. P., 1975. Distribution, movements and mortality of anadromous Arctic char, *Salvelinus alpinus*, L. in the Cumberland sound area of Baffin Island. J. Fish. Biol. 7: 339–348.

McCubbing, D. J. F., 1997. The River Ehen Salmonid Fishery – Current Status and a Summary of Fisheries Work 1993–1996. Environment Agency Report. EA/NW/FTR/97/1.

Palmer, T., 1996. River Liza, Ennerdale Modelling of Irish Bridge, Environment Agency Report, November 1996.

Partington, J. D. & C. A. Mills, 1988. An electrophoretic and biometric study of Arctic charr, *Salvelinus alpinus* (L.), from ten British lakes. J. Fish Biol. 33: 791–814.

Prigg, R. F., 1986. Preliminary note on the present fish population and potential of Ennerdale Water tributaries, with special reference to salmon. NorthWest Water Rivers Division, Scientists Department Biology (North). Technical Note NC306 (2/86).

Reed, S., 1995. Irishmans Bridge: River Liza, Cumbria, A Fluvial Audit. National Rivers Authority Report November 1995.

Solomon, D. J. & T. J. Storeton-West, 1983. Radio tracking of migratory salmonids in rivers: development of an effective system. Fisheries Research Technical Report, Ministry of Agriculture, Fisheries and Food, Directorate of Fisheries Research, Lowestoft, No. 75.

Walters, C. J. & J. S. Collie, 1989. Experimental designs for the estimating transient response to habitat alterations: Is it practical to control for environmental interactions. Workshop on Effects of Habitat Alteration on Salmonid Stocks. Can. Spec. Pub. Fish aquat. Sci. 105: 13–20.

*Hydrobiologia* **371/372:** 181–186, 1998.
*J.-P. Lagardère, M.-L. Bégout Anras & G. Claireaux (eds), Advances in Invertebrates and Fish Telemetry.*



# Acoustic tracking of the dispersal of organic matter by scavenging fishes in the deep-sea

M. A. Collins, I. G. Priede, S. Addison, A. Smith & P. M. Bagley
*Department of Zoology, University of Aberdeen, Tillydrone Avenue, Aberdeen AB24 2TZ, United Kingdom.*
*E-mail: m.a.collins@aberdeen.ac.uk*

*Key words:* deep sea, *Coryphaenoides (Nematonurus) armatus*, transponder, food-fall, behaviour, teleost

## Abstract

Movements of the scavenging deep-sea fish *Coryphaenoides (Nematonurus) armatus* were tracked at depths of 2500–4800 m on the Porcupine Abyssal Plain and Porcupine Seabight in the NE Atlantic using the Aberdeen University Deep Ocean Submersible (AUDOS). The AUDOS was deployed with mackerel bait and three coded acoustic transponders (CATs) covered in mackerel flesh and each enclosed in a nylon bag. Time lapse photographs allowed identification of fish ingesting CATs and estimation of fish arrival times and staying times at the bait. The radial swimming speeds recorded were much slower than inferred in previous studies using older designs of tracking equipment.

It was found that *C.(N.) armatus* moved away from the bait source at an average rate of 1.05 m min$^{-1}$ and separated from one another at an average rate of 1.32 m min$^{-1}$ indicating solitary independent movement with no evidence of shoaling aggregation. The average time delay before food falls were intercepted by *C.(N.) armatus* in the study area was 26 min, the mean fish staying time was 115 min and the swimming time required to disperse the bait to background organic carbon density in the benthos was 46 min. Thus total time for food-fall dispersal after arrival on the sea floor was approximately 3 h.

## Introduction

In the decade since Priede & Smith (1986) first demonstrated tracking of deep sea fish using ingestible acoustic transmitters considerable technical progress has been made (Priede, 1996). Simple pingers were used in early work by Priede et al. (1990) and Armstrong et al. (1992a) and it was not possible to discriminate individual fish or their direction of movement. Never-the-less it was evident that at depths greater than 2500 m in both the North Pacific and North Atlantic Oceans the grenadier fish *Coryphaenoides (Nematonurus) armatus* (Hector) rapidly intercepted acoustically-tagged artificial food falls and transported them out of range of the tracking system within a time scale of less than one day. Priede et al. (1991) and Armstrong et al. (1992b) reviewed the data from these early studies. They applied an algorithm based on the attenuation of the pinger sound signal with range and estimated the mean rate of movement of fish away from the food source to be 6.96 m min$^{-1}$. This was taken to represent the rate at which organic matter arriving on the sea floor in the form of animal carcasses is dispersed by these scavenging fishes. Life in the deep sea is fueled largely by input of particulate organic carbon from the euphotic zone. In the temperate North Atlantic Ocean much of the input occurs in a seasonal pulse of detritus, usually 1–2 months after the surface spring plankton bloom (Lampitt, 1985). However larger food-falls, such as dead fish, squid, or marine mammals may play an important role in supplementing the food supply to the deep sea and enhancing local production (Isaacs & Scwartzlose, 1975; Hinga et al., 1979; Stockton & Delaca, 1982). The early acoustic tracking experiments provided the first insight into possible rates of transport of such material across the sea floor.

In subsequent work, instead of a pinger, a code activated transponder (CAT) and a more sophisticated tracking system with a scanning directional sonar were introduced (Bagley, 1992) which enabled several individually identified fish to be tracked simultaneously. It was discovered that *C.(N.) armatus* follow convoluted tracks across the sea floor swimming independently of bottom currents at speeds of up to 12 m min$^{-1}$ (Smith et al., 1997). In the Pacific Ocean there was evidence of a seasonal change in swimming speed (Priede et al., 1994b) and in the NE Atlantic speed increased with depth from a mean of 3 m min$^{-1}$ at 2500 m to 7.2 m min$^{-1}$ at 4000 m depth (Bagley et al., 1994). These more recent measurements however were speeds along the winding track of the fish and did not represent the mean rate at which matter was transported radially away from the food source. Smith et al. (1997) commented that, despite recording relatively high along-track speeds of *C.(N.) armatus* in spring on the Porcupine Abyssal Plain (NE Atlantic Ocean), the resultant radial speeds were the slowest ever recorded.

In this paper three aspects of movements of *C.(N.) armatus* are considered in the light of the recent measurements obtained using code activated transponders (Priede et al., 1996). Firstly the measurement of radial departure velocity as discussed by Priede et al. (1991) is re-examined. Secondly, taking advantage of the ability to track more than one fish simultaneously, the rate of increase in distance between individuals is examined. Finally a protocol is proposed for evaluating the time taken for a food fall to be totally dispersed.

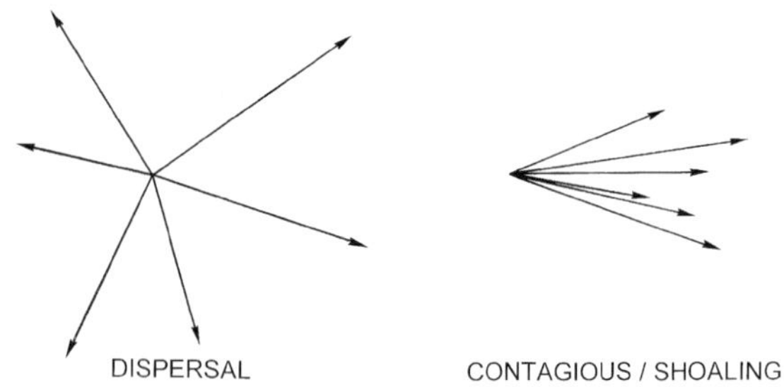


*Figure 1.* Schematic comparison of dispersal from the food source, at the origin of the arrows, resulting from different social behaviour patterns but with the same average radial swimming velocities represented by the lengths of the arrows. On the left F/R (see text) is approximately 1.3 whereas on the right F/R is approximately 0.5.

## Materials and methods

The Aberdeen University Deep Ocean Submersible (AUDOS) is a free-fall submersible vehicle that carries a photographic camera and flash unit (Camera Alive Ltd. C1800) together with the TRATEX (Transponding Acoustic Tracking Experiment) sonar system, an electromagnetic current meter and a compass (Bagley & Priede, 1997). Code activated transponders are suspended in baits attached to ballast below the vehicle. Each CAT responds to a unique two-pulse signal from the TRATEX system by transmitting a single return pulse. Valid returns received by the TRATEX are timed and stored in memory.

The AUDOS system was deployed on two cruises in the North East Atlantic. Challenger 94, in 1992 worked in the Porcupine Seabight 49°29′ N – 50°11′ N, 13°02′ W – 14°48′ W (Priede et al., 1994a) and Challenger 111 in 1994 worked on the Porcupine Abyssal Plain in the vicinity of 48°50′ N 16°30′ W (Smith et al., 1997) (Table 1). On all but one deployment, bait consisted of a single whole mackerel (*Scomber scombrus* L.), ca. 500 g wet weight, which was attached to a graduated cross-piece mounted directly above the ballast close to the sea floor. On deployment 5 of Challenger 111 a single 500 g squid (*Loligo forbesi* Steenstrup) was used as bait. Three CATs were used on each deployment, each with a different activation code. The CATs were enclosed in bait, mounted in nylon bags and hung from the ballast. Fish taking CATs were identified from photographs taken at one minute intervals, commencing shortly before the AUDOS reached the sea floor. On the sea floor the hydrophone on the TRATEX system initiated a scanning sequence every 15 min. in 52 steps to cover the full 360°.

The TRATEX system gave a time delay and bearing for each CAT return signal. Times were converted to ranges, assuming the velocity of sound in water to be 1534 m s$^{-1}$ (Mackenzie, 1981). The range and direction of the fish from AUDOS were converted to x,y co-ordinates using simple trigonometry. Return signals were not obtained for each CAT during every scanning cycle, but estimates of fish positions, at a given time (t), were obtained by linear interpolation. The intercept of a time-range regression was used to estimate the departure time of each fish. All the data were then pooled with departure time = 0 for analysis of overall mean radial speed.

The same radial swimming speed of fish however can result in very different dispersal patterns depending on whether fish are shoaling or solitary in their

*Table 1.* Details of deployments of AUDOS during cruises of the *RRS Challenger* C94 and C111 (Priede et al., 1994; Smith et al., 1997)

| Cruise | Deployment | Date | Depth (m) | *Coryphaenoides (Nematonurus) armatus* Number tracked | Arrival time (min) | Staying time (min) |
|---|---|---|---|---|---|---|
| *Challenger* 94 | 1 | 01-08-92 | 4050 | 2 | 28 | 90 |
| | 5 | 08-08-92 | 2897 | 3 | 13 | 86 |
| | 6 | 10-08-92 | 3500 | 3 | 16 | 60 |
| | 7 | 12-08-92 | 3988 | 2 | 40 | 103 |
| | 8 | 14-08-92 | 2501 | 2 | 23 | 185 |
| *Challenger* 111 | 1 | 07-04-94 | 4808 | 2 | 26 | 245 |
| | 2 | 11-04-94 | 4810 | 1 | 63 | 133 |
| | 3 | 13-04-94 | 4810 | 0 | 32 | 90 |
| | 4 | 15-04-94 | 4810 | 0 | 10 | 82 |
| | 5 | 19-04-94 | 4811 | 1 | 7 | 75 |
| | 6 | 29-04-94 | 4812 | 3 | 31 | 110 |
| Mean values | | | | | 26.3 | 114.5 |

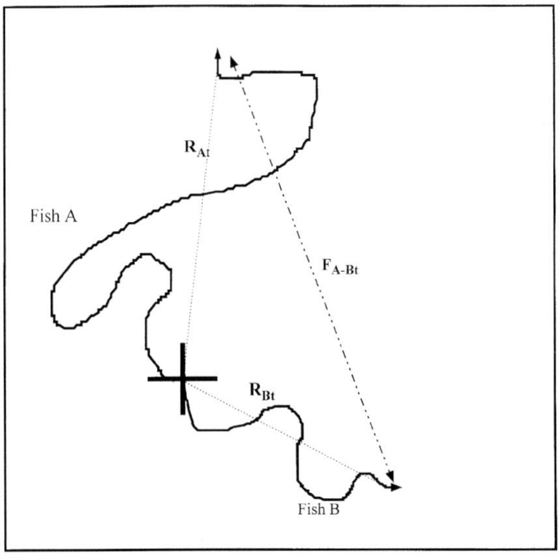


*Figure 2.* Diagram showing the measurements for two fish, A and B of radial range at time ($t$), $R_{At}$ and $R_{Bt}$ and the distance between them $F_{AB,t}$. The large cross represents the position of the food source beneath AUDOS and the wavy lines are the tracks of the two fish.

behaviour (Figure 1). To examine the behavioural characteristics of *C.(N.) armatus*, for all occasions when more than one fish was tracked, the distance between each pair of fish, the inter-fish distance ($F$) and the corresponding mean range ($R$) were calculated for each time increment of tracking (Figure 2). The value of $F$ can range between a value of 2, the case of two fish moving away in opposite directions and zero for fish moving away together as a tight shoal. For a large number of fish dispersing evenly with no formation of groups or shoals $F$ will equate to the mean length of all chords of a circle and have a value of $1.278R$.

## Results

*Coryphaenoides (Nematonurus) armatus* were tracked during 5 deployments of AUDOS during Challenger 94 (August) and 4 deployments during Challenger 111 (April) (Table 1). In 7 of these deployments more than one fish was tracked. Also recorded were the times of arrival of the first fish, (number of minutes after the baits reached the sea floor) and the mean staying time of fish as observed from the AUDOS camera (Priede et al., 1994; Smith et al., 1997).

Mean radial velocity of *C.(N.) armatus* departing from the bait source was estimated by fitting a regression line forced through the origin of the range-time plot (Figure 3) since all fish departed at $t = 0$. The mean value for the month of April (C111) was 1.03 m $min^{-1}$ and for the month of August (C94) 1.26 m $min^{-1}$. For

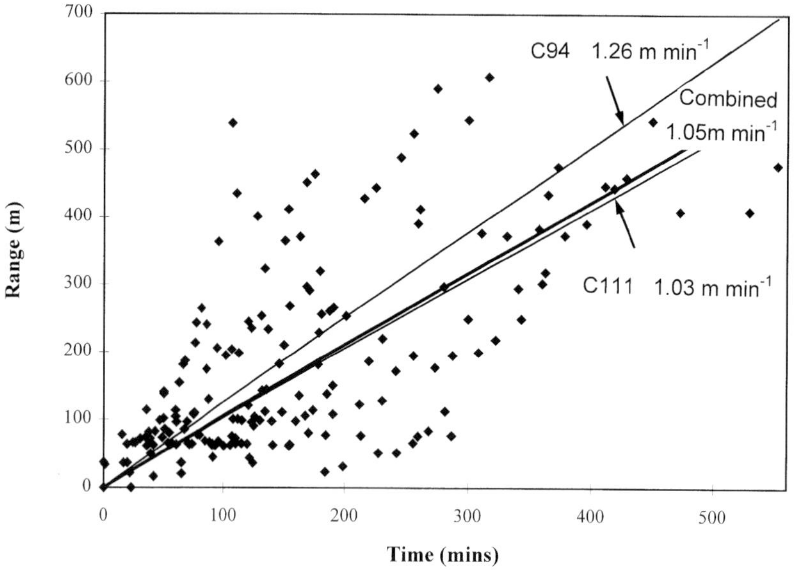


*Figure 3.* Distance from the food source as a function of time after departure of the grenadier *Coryphaenoides (Nematonurus) armatus* tracked using acoustic transponders. Regression lines were fitted to the Challenger 94 data from August 1992 ($R^2 = 0.477$) and the Challenger 111 data from April 1994 ($R^2 = 0.476$) as well as a combined regression for the entire data set represented by the thick line ($R^2 = 0.473$). The slopes represent the mean radial velocity values indicated.

deployment 5 on C111, the tracked fish departed the area and then returned, so calculations are based on the initial departure rate. Combining all the data for both cruises gives a radial velocity of 1.05 m min$^{-1}$.

The relationship between $F$ and $R$ was determined for all possible pairs of fish, *i.e.* three tracked fish produced three pairs. A linear regression was fitted to the data and the slope of $1.32R$, was close to the theoretical value of $1.278R$ if fish are evenly dispersed (Figure 4).

## Discussion

It is evident from this analysis, using data from different areas in the NE Atlantic, in two different years and in two different seasons, that the early estimates of radial dispersion rate of food falls by *C.(N.) armatus* published by Priede et al. (1991) are in urgent need of revision. The earlier technique yielded a value of 6.96 m min$^{-1}$ whereas using code activated transponders the present analysis gives an overall mean value of 1.05 m min$^{-1}$.

To explain this discrepancy it is instructive to re-examine the method of estimation of range using a pinger. In the early work, range was estimated by measuring the decline in signal strength detected at the receiver as the pinger moved away from the AUDOS vehicle (Armstrong et al., 1992b). Acoustic transmission loss (TL, dB) increases with range in accordance with the equation:

$$\mathrm{TL} = 20 \log_{10} r + \alpha r\, 10^{-3}, \tag{1}$$

where $r$ is range in metres and $\alpha$ is absorption coefficient in dB per km. Solutions of this equation were set up in a look-up table in the analysis software to calculate range from voltages recorded in the receiver system. In view of the logarithmic relationships involved small errors in measurement (or the assumed reference voltage) result in large errors at range values exceeding 100 m. For example, an error of 3% of full scale, gives a ten fold range error at 250 m. It was assumed that averaging over a large number of data points would overcome this problem. Back calculation indicates that there was an overall bias of 1–2% resulting in a apparent 7 fold over-estimate in dispersal velocity. The new findings analysed here illustrate the importance of development of the code activate transponder in order to gain a proper understanding of fish behaviour in the deep sea. The maximum likely error in range estimation using transponders, owing to changes in velocity of sound in sea water is less than 1% (Bagley, 1992). The use of signal attenuation as an indicator of range was justifiable in the context of unusual difficulties in getting any measurements at

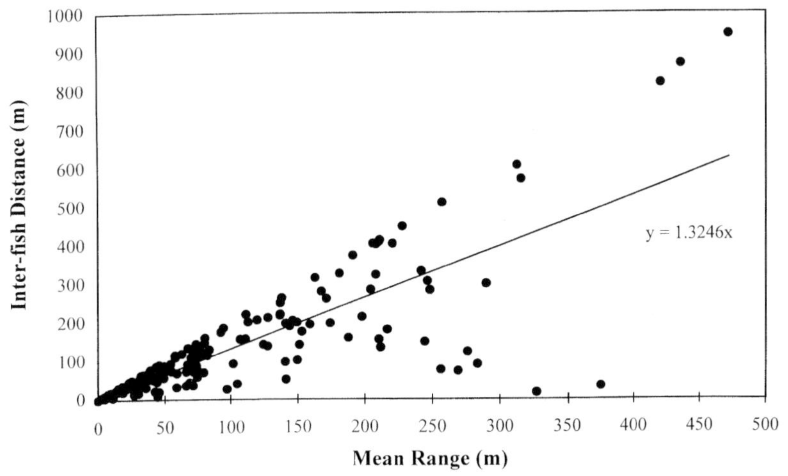


*Figure 4.* The relationship between inter-fish distance (*F*) and mean radial range (*R*) for all possible pairs of grenadiers tracked in the NE Atlantic during Challenger cruises 94 and 111. The maximum possible slope of this relationship is 2 and values close to zero would indicate shoaling behaviour.

all, in the deep-sea, but this method should not be recommended for wider application.

Smith et al. (1997) observed that *C.(N.) armatus* do not all depart from the food source at the same time. Furthermore the value of $F = 1.32R$ calculated here indicates that departing *C.(N.) armatus*, appear to space out over the sea floor as uniformly as possible. Thus apart from forming aggregations at food falls these fish show no evidence of group social behaviour. This analysis does not exclude the possibility of some bias in their departure directions, such as against the current (Priede et al., 1994b), or up the continental slope (Bagley et al., 1994). The *F/R* ratio cannot discriminate between dispersal around a circle and dispersal around a semicircle.

The meaning of these results in terms of dispersal of food-falls reaching the sea floor can interpreted in various ways. One view might be that once the food has been ingested then it is has been dispersed, since it is no longer available to other members of the benthic community. However some species of deep-sea fish (Jones et al., 1998) remain at the site of a large food-fall. In such circumstances the fish, with full stomachs, excreting, defecating, respiring and storing organic matter (as growth) represent a concentration of organic carbon. We propose that the food fall can be considered to be completely dispersed when the area A ($m^2$) enclosing the loci of the fish is such, that the organic carbon content $C_f$ (mg C) of the food fall, divided by that area is equal to the background organic carbon density $C_b$ (mg C $m^{-2}$).

$$C_f/A = C_b \tag{2}$$

Assuming an organic carbon content of approximately 7%, a 500 g mackerel, as used in this study would contain approximately 35 g of organic carbon. Given an approximate average organic carbon input to the deep sea of 4.8 mg $m^{-2}$ $d^{-1}$ (Lampitt et al., 1995) the mackerel bait is equivalent to the daily input into an area of 7292 $m^2$. Assuming that *C.(N.) armatus* depart independently in all directions at a mean radial velocity (v), the area of dispersal will correspond to a circle of area:

$$A = \pi\ (v\ .\ t)^2 \tag{3}$$

Solving area equation 3 with a radial velocity of 1.05 m $min^{-1}$ and an area of 7291 $m^2$ gives an estimate of t = 46 min, which is the time delay from departure of the fishes to complete dispersal of the food fall.

The process of dispersal of food falls in the NE Atlantic therefore can be summarised (Table 2) suggesting a total dispersal time of approximately 3 hours. This measure of dispersal time is to some extent arbitrary. In this case food-fall organic carbon in the fish doubles the background value. Various definitions can be considered with different frames of reference. For example, dispersal to 1% above the background value, or defined with reference to total benthic standing stock of organic matter rather than the daily particulate flux. There is no obvious single measure that can be used as an absolute measure of 'dispersal rate'. Work in this field will require the adoption of some kind of

*Table 2.* Mean dispersal time of food falls by *Coryphaenoides (Nematonurus) armatus* in the NE Atlantic

| | |
|---|---|
| First Fish arrival time | 26 min |
| Mean fish Staying time | 115 min |
| Time to dispersal at 1.05 m $min^{-1}$ | 46 min |
| TOTAL time | 187 min |

standardisation to allow comparison between different studies. The radial velocity *per se* is a useful unequivocal first measurement but interpretation in terms of a useful value that can be fitted into bio-geochemical models may be problematic. Finally despite the significant revision of estimated fish radial swimming speed the original conclusion by Priede et al. (1991) that 'organic food falls reaching sea floor are rapidly dispersed at abyssal depths' is not contradicted by this new analysis.

## Acknowledgements

We thank the ship's company of the *RRS Challenger* for their help in gathering data cited in this paper. The work was supported by NERC Grant No GR3/10162 and EU MAST contracts CT92-003 and CT95-0010.

## References

Armstrong, J. D., P. M. Bagley & I. G. Priede, 1992a. Photographic and acoustic tracking observations of the behaviour of the grenadier, *Coryphaenoides (Nematonurus) armatus*, the eel, *Synaphobranchus bathybius*, and other abyssal demersal fish in the North Atlantic Ocean. Mar. Biol. 112: 535–544.

Armstrong, J. D., P. M. Bagley & I. G. Priede, 1992b. Tracking deep-sea fish using ingestible transmitters and an autonomous sea-floor instrument package. In I. G. Priede & S. M. Swift (eds), Wildlife Telemetry: Remote Monitoring and Tracking of Animals. Ellis Horwood. Chichester: 376–386.

Bagley, P. M., 1992. A code-activated transponder for the individual indentification and tracking of deep sea fish. In I. G. Priede & S. M. Swift (eds), Wildlife Telemetry: Remote Monitoring and Tracking of Animals. Ellis Horwood. Chichester: 111–119.

Bagley, P. M. & I. G. Priede, 1997. An autonomous free-fall acoustic tracking system for investigation of fish behaviour at abyssal depths. Aquat. Living Res. 10: 67–74.

Bagley, P. M., A. Smith & I. G. Priede, 1994. Tracking movements of deep demersal fishes, *Coryphaenoides (Nematonurus) armatus*, *Antimora rostrata* and *Centroscymnus coelolepis* in the Porcupine Seabight, N.E. Atlantic Ocean. J. Mar. Biol. Assoc. U.K. 74: 473–480.

Hinga, K. R. J., J. McN. Sieberth & G. R. Heath, 1979. The supply and use of organic material at the deep-sea floor. J. mar. Res. 37: 557–579.

Isaacs, J. D. & R. A. Schwartzlose, 1975. Active animals of the deep-sea floor. Sci. Am. 233: 85–91.

Jones, E. G., M. A. Collins, P. M. Bagley, S. Addison & I. G. Priede, 1998. The tate of cetacean carcasses in the deep sea: observations on consumption rates and succession of scavenging species in the abyssal north-east Atlantic Ocean. Proceedings of the Royal Society, London B, 265: 1–9.

Lampitt, R. S., 1985. Evidence for the seasonal deposition of detritus to the deep-sea floor and its subsequent resuspension. Deep Sea Res. 32: 885–897.

Lampitt, R. S., R. T. C. Raine, D. S. M. Billett & A. L. Rice, 1995. Material supply to the European continental slope: a budget based on benthc oxygen demand and organic supply. Deep Sea Res. 42: 1865–1880.

Mackenzie, K. V., 1981. Nine-term equations for sound speed in the oceans. J. acoust. Soc. Am. 70: 807–812.

Priede, I. G., P. M. Bagley, S. Addison, S. Bradley & M. A. Collins, 1996. New Methods for Tracking Deep Sea Fishes. Proceedings of the 5th European Conference of Wildlife Telemetry. Strasbourg.

Priede, I. G., P. Bagley, J. D. Armstrong, K. L. Smith & N. R. Merrett, 1991. Direct Measurement of Active dispersal of Food-falls by abyssal demersal fishes. Nature 351: 647–649.

Priede, I. G., P. M. Bagley, A. Smith, S. Creasey & N. R. Merrett, 1994a. Scavenging deep demersal fishes of the Porcupine Seabight, (NE Atlantic Ocean); observations by baited camera, trap and trawl. J. Mar. Biol. Assoc. U.K. 74: 481–498.

Priede, I. G., P. M. Bagley & K. L. Smith Jr., 1994b. Seasonal change in activity of abyssal demersal scavenging grenadiers *Coryphaenoides (Nematonurus) armatus* in the eastern North Pacific Ocean. Limnol. Oceanogr. 39: 279–285.

Priede, I. G. & K. L. Smith Jr., 1986. Behaviour of the abyssal grenadier, *Coryphaenoides yaquinae*, monitored using ingestible acoustic transmitters in the Pacific Ocean. J. Fish Biol. 29 (Suppl. A): 199–206.

Priede, I. G., K. L. Smith Jr. & J. D. Armstrong, 1990. Foraging behaviour of abyssal grenadier fish: inferences from acoustic tagging and tracking in the North Pacific Ocean. Deep-sea Res. 37: 81–101.

Smith, A., I. G. Priede, P. M. Bagley & S. W. Addison, 1997. Interception and dispersal of artificial food falls by scavenging fishes in the abyssal NE Atlantic: early season observations prior to annual deposition of phytodetritus. Mar. Biol. 128: 329–336.

Stockton, W. L. & T. E. DeLaca, 1982. Food falls in the deep-sea:occurrence, quality and significance. Deep-Sea Res. 29: 157–169.

*Hydrobiologia* **371/372:** 187–198, 1998.
*J.-P. Lagardère, M.-L. Bégout Anras & G. Claireaux (eds), Advances in Invertebrates and Fish Telemetry.*



# Seasonal variations in time and space utilization by radio-tagged yellow eels *Anguilla anguilla* (L.) in a small stream

E. Baras, D. Jeandrain, B. Serouge & J. C. Philippart
*University of Liège, Laboratory of Fish Demography and Aquaculture*
*Mailing address: Chemin de la Justice, 10 B-4500 Tihange, Belgium; E-mail: E.Baras@ulg.ac.be*

*Key words:* home range, diel activity, foraging, behaviour, eel, *Anguilla anguilla*

## Abstract

Seven yellow eels (572–643 mm, 318–592 g) *Anguilla anguilla* (L.) were tagged with surgically implanted radio transmitters (activity circuit, 1.6–1.7 g) and tracked in the Awirs stream, a small (width $< 5$ m, depth from 0.1 to 1.2 m), densely populated (ca. 250 kg of eel ha$^{-1}$) tributary of the Belgian River Meuse. The eels were positioned daily from late April to mid-August, and their diel activity was studied during twenty four 24-h cycles. During day-time, the eels were resting in rootwads or in crevices inside stone walls or in crevices in between rocks. They became more active in the late afternoon but generally did not leave their residence before sunset, except under overcast weather. Activity peaked during the first part of the night then progressively vanished, and always ended before sunrise. The area exploited during night-time never extended over more than 40 m$^2$, except when the eel changed its residence. The intensity and timing of nocturnal activity and the extent of the daily activity area were dependent on water temperature (respectively $P < 0.0001$, $P < 0.05$ and $P < 0.0005$), with eels showing little or no activity when the diurnal temperature did not exceed 13 °C. Eels showed higher agitation under full moon and maintained their activity later in the night ($P < 0.05$). The eels showed restricted mobility, and occupied small stream areas (from 0.01 to 0.10 ha) in a non sequential mode, except for two fish which were displaced to the River Meuse by a spate in early June and were never recovered. The length and frequency of net daily journeys were higher ($P = 0.005$) at water temperatures above 16 °C in late May and June, which also corresponded to the period of immigration of eels from the River Meuse. This study thus shows that large yellow eels may adopt a highly sedentary lifestyle in a continental, fast flowing and densely populated environment, even at periods of the year when these stages usually show upstream migrations.

## Introduction

Due to its amphibiotic catadromous life cycle and huge migration distances in sea and fresh water, the European eel *Anguilla anguilla* (L.) has intrigued scientists for centuries (Sinha & Jones, 1975; Tesch, 1977; Deelder, 1984; McDowall, 1988). The recent decline of eel stocks revealed by the glass and silver eel fisheries (Brusle, 1989; Moriarty, 1990; Knights et al., 1996) further urges the need for providing more accurate information on their ecology. Eels are discrete animals, which essentially move at night and exhibit cryptic habits during daytime. These traits make them difficult to capture and observe in the wild, except at obstacles in estuaries and rivers, where traps at fish passes are frequently used to document the seasonal periodicity and intensity of their migration (Moriarty, 1986; Legault, 1987; Baras et al., 1996). Outside of this particular context, the study of time and space utilization by eels can hardly be achieved by methods other than biotelemetry (Winter, 1983; Priede & Swift, 1992; Baras & Philippart, 1996).

Biotelemetry techniques have been used to study the movements of silver eels *A. anguilla* or *A. rostrata* (Lesueur) in estuaries and coastal waters (Stasko & Rommel, 1974; Nyman, 1975; Helfman et al., 1983; Parker, 1995), in the open sea (Tesch, 1974) and their migration from the continental shelf to the deep sea (Tesch, 1978, 1979). Paradoxically, whilst the fresh

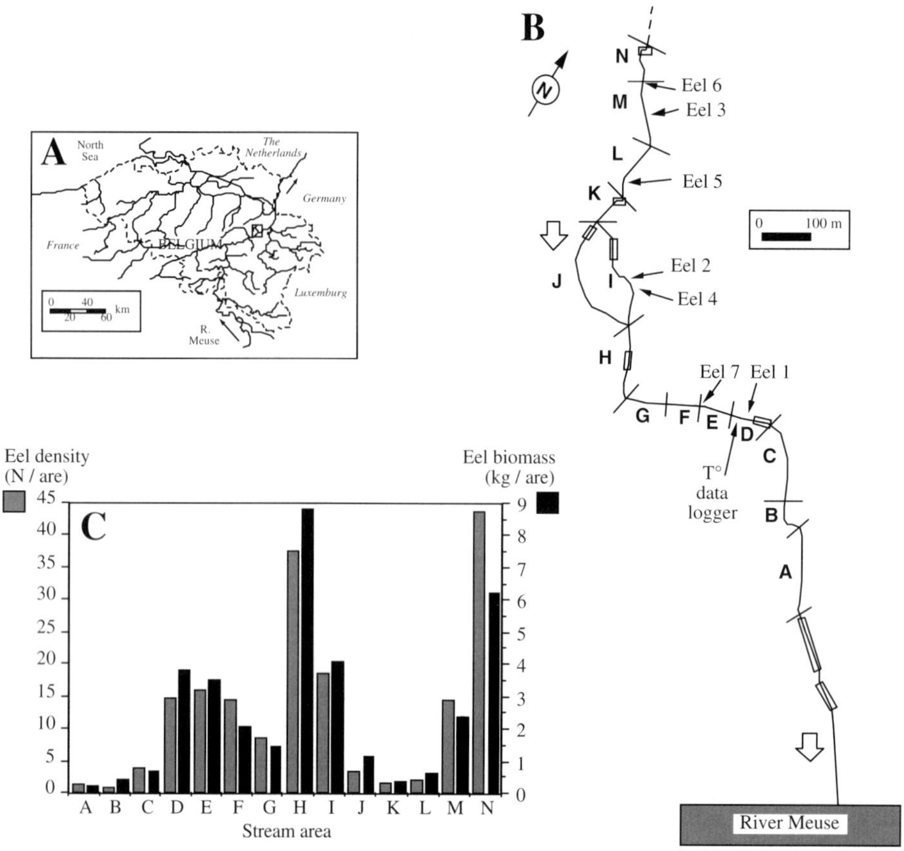


*Figure 1. A.* Location of the Awirs stream in Southern Belgium. *B.* Morphology of the study area, upstream of the confluence with the River Meuse in Engis; open squares correspond to tunnels under road bridges; the arrows indicate the sites of release of radio-tagged eels (1 to 7). *C.* Population density of eels in May 1996, in sectors A to N, delimited by transverse bars on map B, as determined by electric fishing and mark recapture experiments.

water stages represent more than 80% of the life cycle of eels, few telemetry studies have documented their ecology in fresh water ecosystems. Lake studies were conducted on *A. anguilla* in Sweden (Westerberg, 1982) and Spain (LaBar et al., 1987) and on *A. rostrata* in Vermont (USA; LaBar & Facey, 1983). The only detailed studies in riverine ecosystem were these by Dutil et al. (1988) on *A. rostrata* in a tributary of the Saint-Laurent River (Canada), and by McGovern & McCarthy (1992) on *A. anguilla* in River Clare (Ireland). Both sites were only a few kilometres away from the sea.

This study aimed to investigate the behavioural ecology (home range, activity rhythm patterns, effect of environmental factors on mobility) of yellow eels *A. anguilla* in a typical riverine fast flowing ecosystem, beyond any possible influence by lacustrine or marine ecosystems.

### *Study area*

The study was conducted in Southern Belgium, in the lower reaches of the Awirs stream, from 0.1 to 1.2 km upstream of the confluence with the River Meuse in Engis (225 km from the estuary in the Netherlands). In this area, which is typical of the trout zone (Huet, 1949), the stream width and depth range from 1.75 to 4.50 m, and from 0.10 to 1.20 m, respectively. Fast flowing ($\geq 50$ cm s$^{-1}$) riffles and rapids alternate with pools and glides, with an average distance of 30 m between consecutive morphodynamic units. The limits between some morphodynamic units have been artificially enhanced by the construction of small weirs, with less of 40 cm of difference in height. Cobble and stone (10–30 cm) are the most frequent substrata ($\geq 75\%$). The stream is lined by straight vertical banks (0.3 to 2.0 m high) which have been changed into stone walls over more than 60% of their length, as a preventive measure against flooding where it crosses the suburbs of Engis.

The fish community in the study area was analysed from electric fishing surveys before the tracking study and was found to contain 15 fish species, of which the most abundant were, in decreasing order of importance: the stone loach *Barbatula barbatula* (L.), the European eel, the gudgeon *Gobio gobio* (L.) and the roach *Rutilus rutilus* (L.). The size and weight of resident eels in spring 1996 averaged 477 mm (200–768 mm) and 215 g. Their density and biomass was estimated by mark recapture at $11.8 \pm 0.2$ eels $a^{-1}$, and $2.54 \pm 0.04$ kg $a^{-1}$. The variation of eel density and biomass within the study area is illustrated in Figure 1C.

## Material and methods

The eels used in this study were captured by DC electric fishing (EPMC, 2.5 kVA) in the aforementioned stretch of the Awirs stream during spring 1996. Experimental animals were selected based on several criteria. Eels from the upstream reaches were preferred to those of the downstream reaches (A–C, Figure 1B) which crossed urban areas. We always selected eels longer than 55 cm and weighing more than 275 g to make sure that the transmitter to body weight ratio in the air would not exceed 0.6% (Baras & Jeandrain, 1998, this volume). The transmitters were 40 MHz motion-sensitive units with internal coiled antenna (ATS Inc. model 377, current drain of 0.03 mA, minimum expected battery life of 20–25 d), with carrying frequencies separated by 10 kHz intervals. They measured 18 mm in length by 8 mm in diameter and weighed 1.6–1.7 g in air. The transmitters were tuned to transmit at a pulse rate of $40 \pm 1$ or $63 \pm 1$ pulses per minute, depending on the orientation of the fish (corresponding interpulse intervals of 1500 and 952 ms, respectively). Each fast movement or direction change of the fish caused the signal to shift from one pulse rate to the other. Transmitters were implanted via a mid ventral incision into the body cavity of fish anaesthetised with 2-phenoxy-ethanol (0.9 ml $l^{-1}$). The incision was closed with commercial grade cyanoacrilate adhesive, with a freshly cut fragment of eel's fin applied as a biological bandage on the drying adhesive, according to the procedure recently evaluated by Baras and Jeandrain (op. cit.).

Tagged eels were released at their capture site less than one hour after surgery to minimise the possible trauma and stress resulting from long term confinement (Parker, 1995; Baras et al., 1998). They were positioned at least once a day during daytime until the end of the transmitter battery life. Locations were made by triangulation from markers at 5–10 m intervals on the banks of the river, using a Fieldmaster receiver and diamond antenna (ATS Inc.) equipped with a compass. Due to the high conductivity of the Awirs stream (800–900 $\mu$S cm$^{-1}$), the low power of transmitter and the cryptic behaviour of eels, the reception range could be cut down to less than 5 m during daytime. This forced the observer to move very close to the fish but permitted accurate positioning. Water temperature was automatically recorded at 15 min intervals throughout the study with a datalogger (0.1 °C resolution; ControlOne) installed in the centre of the study area. Water level was measured at least once a day on a limnimetric scale installed at the temperature logger site.

The diel movement pattern of eel was analysed during twenty four 24-h cycles, from 3 May to 8 July, 1996. Fish were positioned at 10 min intervals and activity indices were measured in between, over 1 min intervals. Activity indices were deduced from the changes of pulse rates from the motion sensitive transmitters, which were all tuned at interpulse intervals of 952 and 1500 ms. Consequently, the maximum possible numbers of variations of pulse rate over one minute would be 49 (i.e. for a transmitter systematically changing from one pulse rate to the other over this period) and would be granted an activity index of 100%. Lower number of changes in the pulse rates were given proportionally lower indices, with a minimum of 0 for a transmitter constantly producing the same pulse rate (fast or slow) over one minute. An automatic listening station (Lotek SRX-400), connected to a fixed diamond antenna, has also been used to continuously record the variations of signal strength and pulse rate of eels over 24 h cycles. Due to the very weak power of transmitters, high water conductivity, intense jamming of radio signals by local industries, and additional loss of up to 12 dB when automatically recording pulse rate intervals, its use was restricted to extremely good conditions of signal propagation (five cycles in June and July). During 24 h cycles with the automatic listening station, the antenna was oriented all day round in the same direction, which permitted to discriminate, from signal strength variations, the times when the eel would leave its cryptic residence place and swim in open water. The observer addition-

*Table 1.* Characteristics of the radio-tagged eels in the Awirs stream. TBWR is the transmitter to body weight ratio in the air.

| Eel n° | Body Length (mm) | Weight (g) | TBWR (%) | Tracked in 1996 From | To | Days | Cause of tracking end | Recapture by electric fishing |
|---|---|---|---|---|---|---|---|---|
| 1 | 585 | 467 | 0.35 | 28 Apr | 08 Jun | 40 | Down to Meuse | None |
| 2 | 585 | 373 | 0.44 | 28 Apr | 29 May | 31 | Battery | 15 May |
| 3 | 564 | 385 | 0.43 | 28 Apr | 21 Jun | 54 | Battery | 05 Jun |
| 4 | 643 | 592 | 0.28 | 06 Jun | 08 Jun | 3 | Down to Meuse | None |
| 5 | 574 | 453 | 0.36 | 06 Jun | 18 Jul | 42 | Battery | 27 Jun, 18 Jul |
| 6 | 614 | 431 | 0.38 | 06 Jun | 17 Jul | 41 | Battery | 27 Jun, 18 Jul |
| 7 | 572 | 318 | 0.52 | 27 Jun | 03 Aug | 37 | End of study | 18 Jul |

ally located the fish at regular intervals to measure the extension of its daily activity area.

## Results

Seven eels have been tagged from 28 April to 3 August, 1996 (Table 1; Figure 1B). The transmitters lasted up to 54 d and permitted to measure 248 eel-days observations and 241 net daily journeys (n-1 observation, for each of the seven eels). All eels showed restricted movements during the tracking period (Figure 2). Their home range did not extend over more than 285 m (0.1 ha), except for eels n° 1 and 4, which descended into the River Meuse during a freshet on June 8th, when the water level rose by about 1 m following a summer rainstorm. Despite intensive research during the following weeks, both downstream and upstream of the study area, these two fish were never recovered.

Eels showed contrasted mobility and space utilization patterns. During the first tracking days, four of the seven eels (n° 1, 4, 6 and 7) consistently occupied sites at less than 5 m from their release site. By contrast, eels n° 2 and 3 were located 40–50 m downstream on the day following their release whilst eel n° 5 moved 100 m upstream then was displaced downstream during the freshet. Eels n° 1 and 2 occupied almost sequentially their home ranges and travelled over the longest distance on the same day (May 27). The journey of eel n° 2 was much longer and extended its home range to 285 m, presumably because the eel was crossing habitats with low suitability, as suggested by the very low estimates of population density within these sectors (J and K, Figure 1C). Eel n° 7 consistently remained in the site when it was released, except for two short excursions upstream, which were followed by homing movements. Eels n° 5 and 6, and to a lesser extent eel n° 3, utilised their home range in a radiative way, with upstream and downstream excursions followed by returns to residences previously occupied (Figure 2). The diurnal residences areas used by radio-tagged eels essentially corresponded to highly structured, cryptic habitats, such as burrows or cavities inside stone walls, boulder assemblages, rootwads or undercut banks (Figure 3). On one occasion only, an eel rested in an open habitat where it obviously burrowed into the deep silt substratum. The small regulation weirs in the stream apparently were frequently used as residences by eels but represented no obstacle to their mobility, as they cleared them at low water level (Figure 2).

Amongst the 241 daily journeys of eels, 128 (53.1%) were null and 76 others (31.5%) less than 10 m. The longest journeys corresponded to the two downstream migrations to the River Meuse under exceptionally high floods on June 8th. Except for this particular day, the water level varied very little in the Awirs stream, due to the regulation by the small weirs, and had no marked influence on the mobility of eel. The influence of thermal variables is illustrated in Figure 4. Eels showed higher mobility under higher temperatures, and proportionally higher when the water temperature was increasing or decreasing by more than 1 °C between consecutive days, than during periods with little variation.

Eels showed a marked nocturnal activity pattern throughout the study. During daytime, they were consistently located inside their cryptic residence, which strongly reduced the strength of radio signals (Figure 5). During the late afternoon, high activity indices

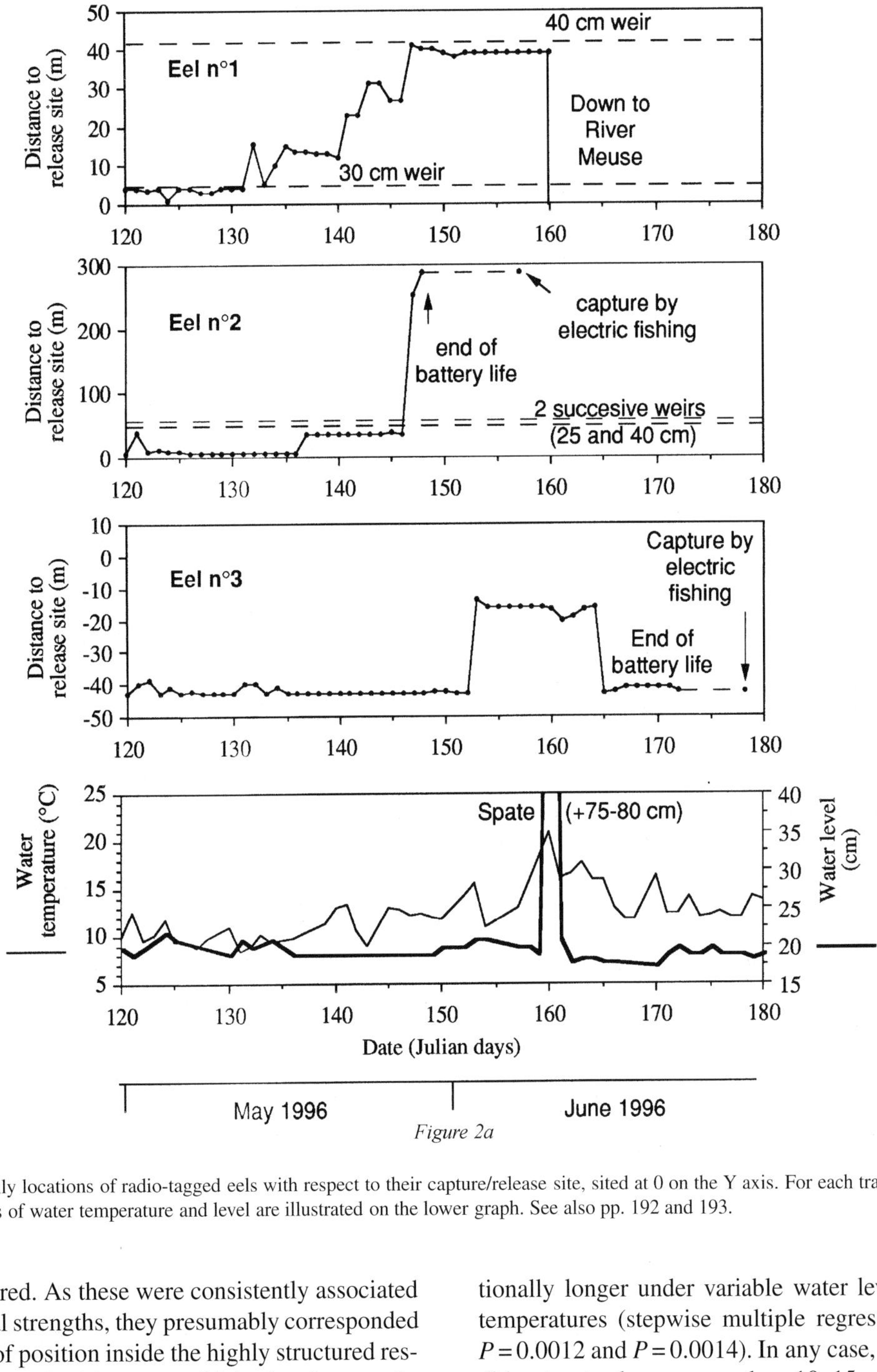


*Figure 2a*

*Figure 2a.* Daily locations of radio-tagged eels with respect to their capture/release site, sited at 0 on the Y axis. For each tracking period, the daily variations of water temperature and level are illustrated on the lower graph. See also pp. 192 and 193.

were measured. As these were consistently associated to low signal strengths, they presumably corresponded to changes of position inside the highly structured residence (Figure 3). Temporary high signal strengths were occasionally measured during this period, but the eels usually did not leave their residence before sunset, except on cloudy and/or rainy days, or when the water level and turbidity were increasing (Table 2). Eels exploited daily activity areas that were proportionally longer under variable water levels and high temperatures (stepwise multiple regression analysis, $P = 0.0012$ and $P = 0.0014$). In any case, the daily area did not extend over more than 10–15 m (ca. 40 $m^2$), except when the eel moved to another part of the stream and changed its residence.

During night-time, periods of high activity alternated with periods of complete rest, either inside or outside of the residence (e.g. Figure 5). The first activ-

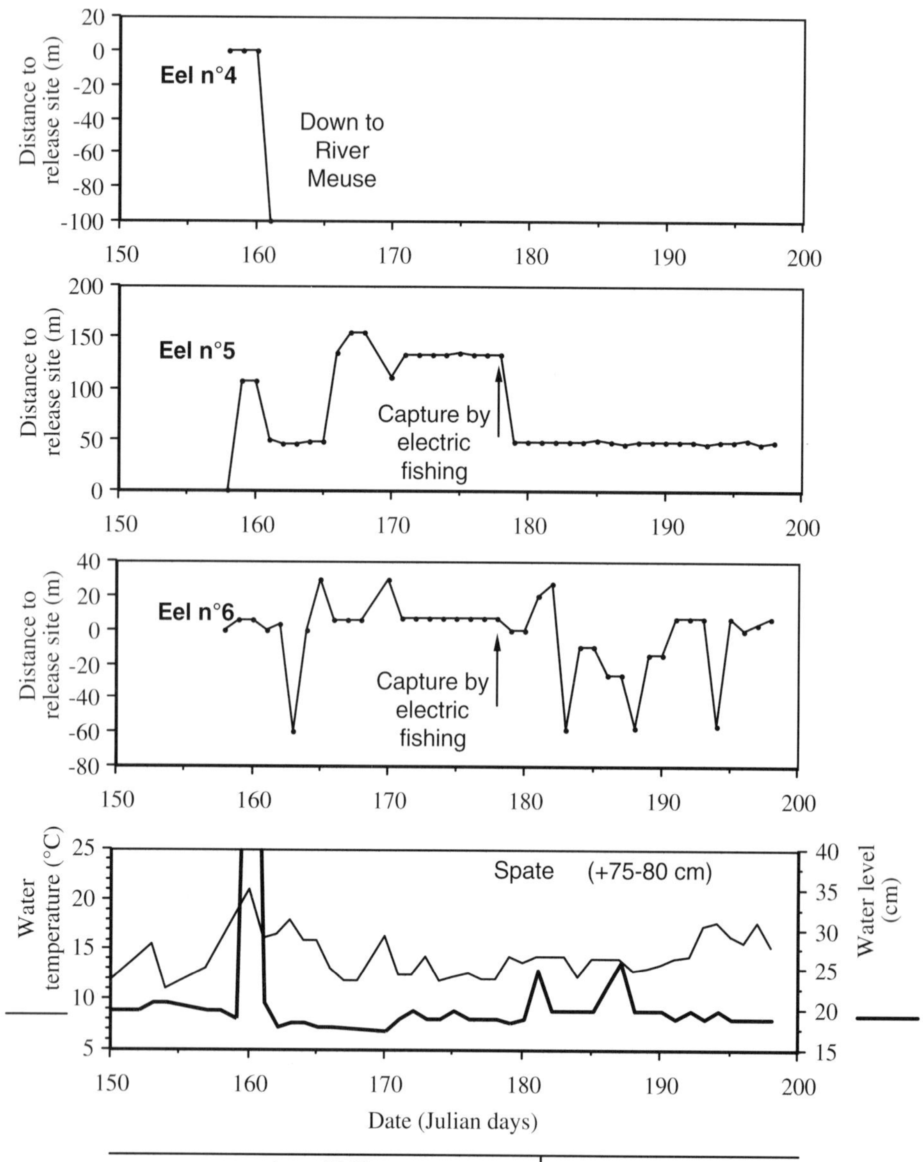


*Figure 2b.* See p. 191.

ity period took place 5 to 35 min after the eel left its diurnal residence, and almost always was the most intense of the night. Later on, the activity progressively vanished, and always ended before sunrise. Both the duration and intensity of activity varied considerably during the study, depending on environmental variables, including the lunar cycle (Table 2). Eels exposed at the light of full moon had higher overall activity indices than during other periods of the lunar cycle (Mann-Whitney $U$-test; $U=13$, $U'=67$, $P=0.034$). Their activity under moonlight consisted in short bursts, with most frequent changes of pulse rates but low variations of signal strength, that indicated a high degree of agitation, but over a most restricted area. This activity pattern contrasted with the more regular patterns observed during other nights (e.g. Figure 5). During these cycles, the eels maintained their activity later in the night ($U=4$, $U'=48$, $P=0.013$), after moonset. During nights with no influence of moonlight, the overall activity index was positively correlated with mean water temperature (Log–Log relationship, $F=53.4$, $R^2=0.750$, $P<0.0001$; $n=20$),

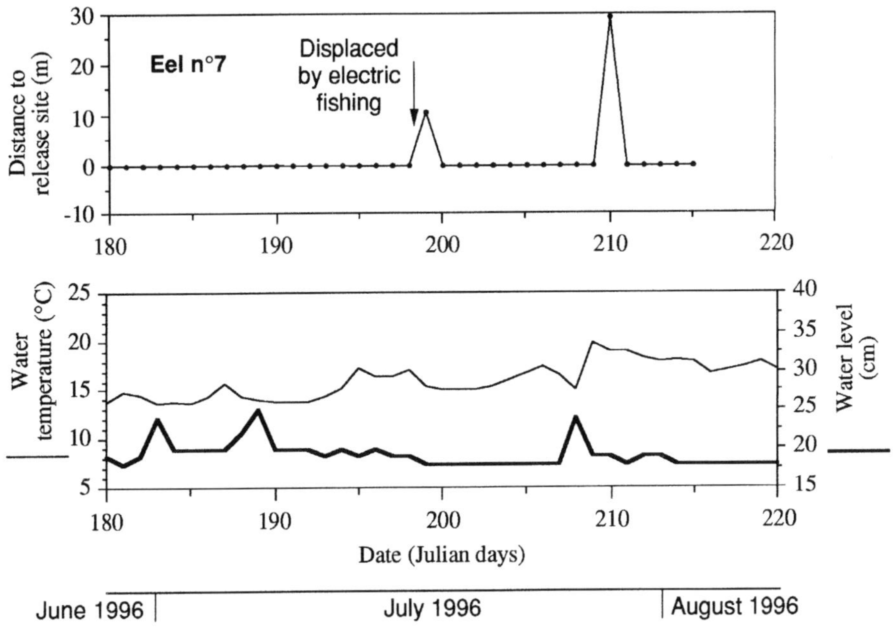


*Figure 2c.* See p. 191.

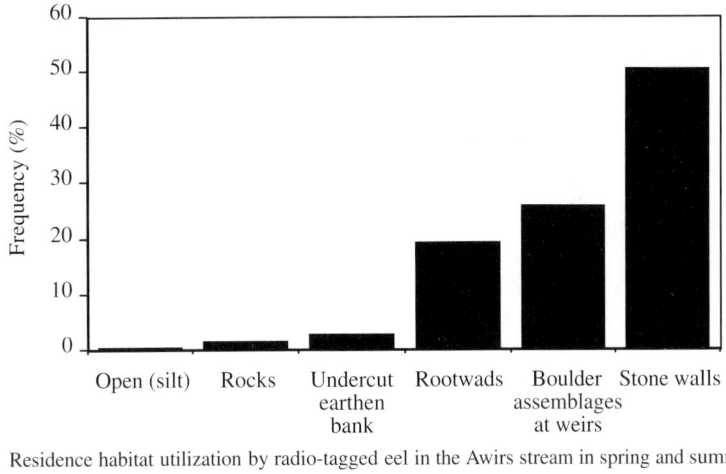


*Figure 3.* Residence habitat utilization by radio-tagged eel in the Awirs stream in spring and summer 1996.

with eels showing little or no activity when the diurnal temperature did not exceed 13 °C. Water temperature also influenced the timing of activity: the nocturnal activity of eels ended sooner under a steep nocturnal temperature decrease, and proportionally sooner at lower temperatures (stepwise multiple regression analysis, $P = 0.0096$ and $P = 0.0385$, for temperature gradient and temperature at sunrise, respectively).

## Discussion

All seven radio-tagged eels survived the surgical tagging procedure and consistently retained their transmitter throughout the study. The fish that were recaptured by electric fishing more than three weeks after tagging, had healed their abdominal incision. Their sacrification at the end of the study (eels n° 2, 3, 5 and 6) revealed no internal damage resulting from the presence of the transmitter inside the body cavity. The low mobility of eels n° 1, 2 and 3 during the two first tracking weeks (before the incision had healed) may

*Table 2.* Characteristics of 24 h cycles of activity of radio-tagged eels in the Awirs stream in spring-summer 1996. The duration of activity corresponds to the running sum of periods when eels showed activity (at least one change of pulse rate $\Delta R$ min$^{-1}$) and the mean activity index is calculated from the ratio (%) of active records over these periods. The overall activity index is the product of the two latter variables, and is expressed as the ratio (%) over a 24-h cycle. For eels which apparently remained at the same site during the whole tracking cycle, the extension of the daily activity area is set as the accuracy of the positioning at close range (0.5 m). The symbol * indicates that the activity area of the eel was illuminated by full moon light.

| Date (1996) | Water T° (°C) min–max | Water level (cm) | Eel n° | Start at (min before (−)/after (+) sunset) | End at (min before (−)/after (+) sunrise) | Duration activity (min) | Mean of activity index (%) | Overall activity index (%) | Extension of daily activity area (m) |
|---|---|---|---|---|---|---|---|---|---|
| 03–04 May | 10.1–13.6 | 27 ± 8.0 | 2 | 20:20 (− 47) | 05h40 (− 32) | 215 | 13.3 | 2.0* | 12.5 |
| 09–10 May | 8.2–12.3 | 19 ± 0.5 | 1 | – | – | 0 | 0.0 | 0.0 | 0.5 |
| 12–13 May | 10.7–11.0 | 21 ± 0.5 | 1 | – | – | 0 | 0.0 | 0.0 | 0.5 |
| 16–17 May | 9.8–11.5 | 19 ± 0.0 | 1 | 04:00 (+ 353) | 04:30 (− 82) | 30 | 8.2 | 0.2 | 3.0 |
| | | | 2 | – | – | 0 | 0.0 | 0.0 | 0.5 |
| | | | 3 | – | – | 0 | 0.0 | 0.0 | 0.5 |
| 20–21 May | 10.6–13.8 | 19 ± 0.0 | 1 | – | – | 0 | 0.0 | 0.0 | 0.5 |
| | | | 2 | 20:00 (− 93) | 23:55 (− 352) | 80 | 10.9 | 0.3 | 9.0 |
| | | | 3 | – | – | 0 | 0.0 | 0.0 | 0.5 |
| 26–27 May | 12.5–14.1 | 22 ± 6.0 | 1 | 19:30 (− 130) | 03:10 (− 149) | 200 | 8.7 | 1.2 | 10.0 |
| | | | 2 | 21:30 (− 10) | 05:00 (− 39) | 180 | 8.9 | 1.1 | 13.0 |
| 01–02 June | 12.4–16.2 | 21 ± 0.5 | 1 | 21:50 (+ 3) | 02:20 (− 195) | 150 | 14.7 | 1.5 | 2.0 |
| | | | 3 | 22:35 (+ 48) | 05:25 (− 10) | 185 | 32.2 | 4.1* | 7.0 |
| 11–12 June | 16.3–20.1 | 18 ± 0.5 | 3 | 22:45 (+ 50) | 01:52 (− 218) | 180 | 14.9 | 1.9 | 8.0 |
| | | | 5 | 22:16 (+ 21) | 05:17 (− 13) | 115 | 37.5 | 3.0 | 8.0 |
| | | | 6 | 22:30 (+35) | 04:10 (− 80) | 185 | 25.5 | 3.3 | 80.0 |
| 18–19 June | 13.2–17.5 | 18 ± 0.5 | 5 | 22:23 (+ 24) | 03:10 (− 139) | 135 | 35.4 | 3.3 | 10.0 |
| | | | 6 | 22:24 (− 25) | 23:35 (− 354) | 75 | 26.5 | 1.4 | 8.0 |
| 24–25 June | 12.4–13.1 | 19 ± 0.5 | 5 | – | – | 0 | 0.0 | 0.0 | 0.5 |
| | | | 6 | 22:05 (+ 5) | 04:55 (− 35) | 236 | 9.0 | 1.5 | 8.0 |
| 01–02 July | 12.9–14.6 | 20 ± 1.0 | 5 | 22:35 (+ 36) | 04:35 (− 59) | 110 | 16.0 | 1.2* | 6.0 |
| | | | 6 | 01:15 (+ 196) | 04:30 (− 64) | 110 | 24.1 | 1.8 | 6.0 |
| | | | 7 | 22:41 (+ 42) | 05:32 (− 2) | 244 | 50.2 | 8.5* | 8.0 |
| 08–09 July | 12.8–15.1 | 20 ± 0.5 | 7 | 22:18 (+22) | 04:00 (− 100) | 240 | 16.3 | 2.7 | 3.0 |

be regarded as a consequence of the tagging procedure. However, the comparison with eels n° 4, 5 and 6, which all moved over much longer distances within the first tracking days, suggests that this low mobility presumably was a consequence of low water temperature at that time of the year. These observations on wild fish in natural environments substantiate the adequacy of surgical tagging and incision closing with cyanoacrilate adhesives (Baras & Jeandrain, 1998, this volume).

This study documented the highly sedentary lifestyle of yellow eels in streams, both at the seasonal and daily levels. They occupied home ranges from 0.01 to 0.10 ha, which were much shorter than most estimates reported in mark-recapture or telemetry studies on eels. Mann (1965) reported home ranges as long as 30 km for *A. anguilla* in the River Elbe (Germany) and Parker (1995) recently documented a mean summer home range of 325 ± 64 ha for *A. rostrata* in a tidal estuary. Home ranges of *A. rostrata* and *A. anguilla* in lakes ranged from 2 to 65 ha (LaBar & Facey, 1983), and from 0.1 to 0.27 ha (LaBar et al., 1987), respectively. In salt marshes, *A. australis* (Richardson) and *A. rostrata* did not move over more than 400 m and 100 m, respectively (Beumer, 1979; Ford & Mercer, 1986). Studies in riverine ecosystems indicated home ranges of 0.5 to 2.0 ha for *A. rostrata* (Dutil et al., 1988), and of 0.1 to 0.6 ha for *A. anguilla* (McGovern & McCarthy, 1992). The difference between these studies and ours presumably

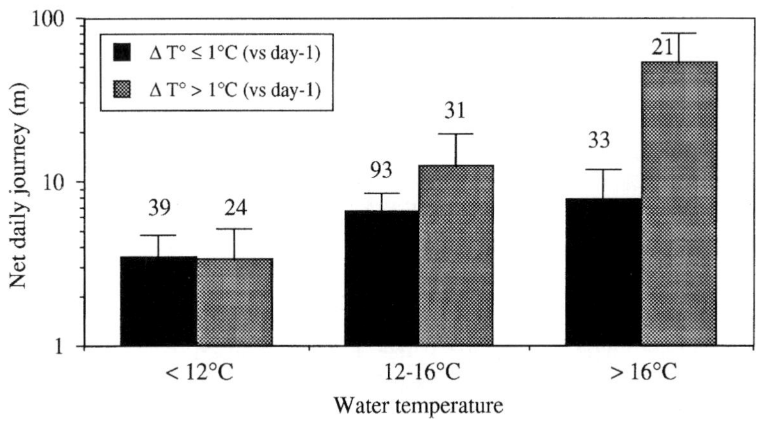


*Figure 4.* Mobility of radio-tagged eel in the Awirs stream depending on ambient water temperature ($T°$) and temperature variation between consecutive days ($\Delta T°$ d–d$^{-1}$). Values are means of net daily journeys ($N = 241$) of seven eels from late April to early August 1996 (2-way ANOVA; $T°$: $F = 6.39$, $P = 0.005$; $\Delta T°$ (d-d$^{-1}$): $F = 7.92$, $P = 0.002$; interaction: $F = 4.89$, $P = 0.008$).

originated from the small width and high complexity of the Awirs stream, which offered numerous refuges and did not force the eels to move over long distances to find alternative refuges. Chisnall & Kalish (1993), who studied the movements of *A. australis* and *A. dieffenbachii* (Gray) by mark recapture also concluded to small scale (< 140 m) movements in a small stream of New Zealand. Similarly, Gunning & Shoop (1962) observed that *A. rostrata* occupied home ranges of less than 50 m in small Louisiana streams but were using larger ranges in wider streams.

Beyond the relationship between stream width and home range size, the sedentary nature of yellow eels in this study may look surprising, as migration can take place at all stages of the life cycle and as eel populations are usually considered as open (Moriarty, 1983). However, several studies documented the fidelity of tagged eels to discrete refuges (Ford & Mercer, 1986; McGovern & McCarthy, 1992), as well as their precise homing abilities (*A. anguilla*, Rossi et al., 1987), even over distances exceeding 10 km (*A. rostrata*, Parker, 1995). These findings suggest some form of territoriality or at least attachment to precise loci, which can be interpreted within the context of optimal foraging strategies (Hart, 1986), i.e. eels exploiting more efficiently an environment which they have already experienced. The attachment to precise loci may proportionally be higher in densely populated environments, such as the Awirs stream, where the population density is sensibly higher than in most European waters (Tesch, 1977; Lobon-Cervia et al., 1990; Barak & Mason, 1992).

Variations of water levels were scarce in the Awirs stream and had no marked influence over daily journeys, except during the spate in early June when two eels were flushed down to the River Meuse. The reasons why these two eels did not home, whilst other eels (n° 3, 5, 6) displaced at shorter distances did, are unknown. Water temperature strongly influenced the movements of eels, which occupied restricted ranges and showed little or no diel activity when water temperature was below 12 °C. Several authors (Sinha & Jones, 1975; Tesch, 1977; Beumer, 1979; LaBar et al., 1987) already reported that eels had little or no activity at low temperatures. Comparatively, eels showed much more frequent and longer net daily journeys above than below 16 °C. This result is consistent with the observations by Sörensen (1950, in Deelder, 1984) and Moriarty (1986) who both found that yellow eels had no migratory activity below 14–15 °C. The higher mobility of 'sedentary eels' at higher temperatures may be regarded as a consequence of increasing energetic expenditures. The foraging activity of eels was longer, more intense and extended over a larger daily area when the water temperature was higher. This presumably resulted into a higher competition between resident eels, especially in a densely populated stream. Higher mobility of eels at high temperature may also reflect another form of population density pressure, due to the arrival of immigrants from the River Meuse, the ascent of which is also dependent on water temperature (Baras et al., 1996). Indeed, the periods when radio-tagged eels showed most frequent daily journeys (June 1996) precisely corresponded to the period of migration of yellow eels in the River

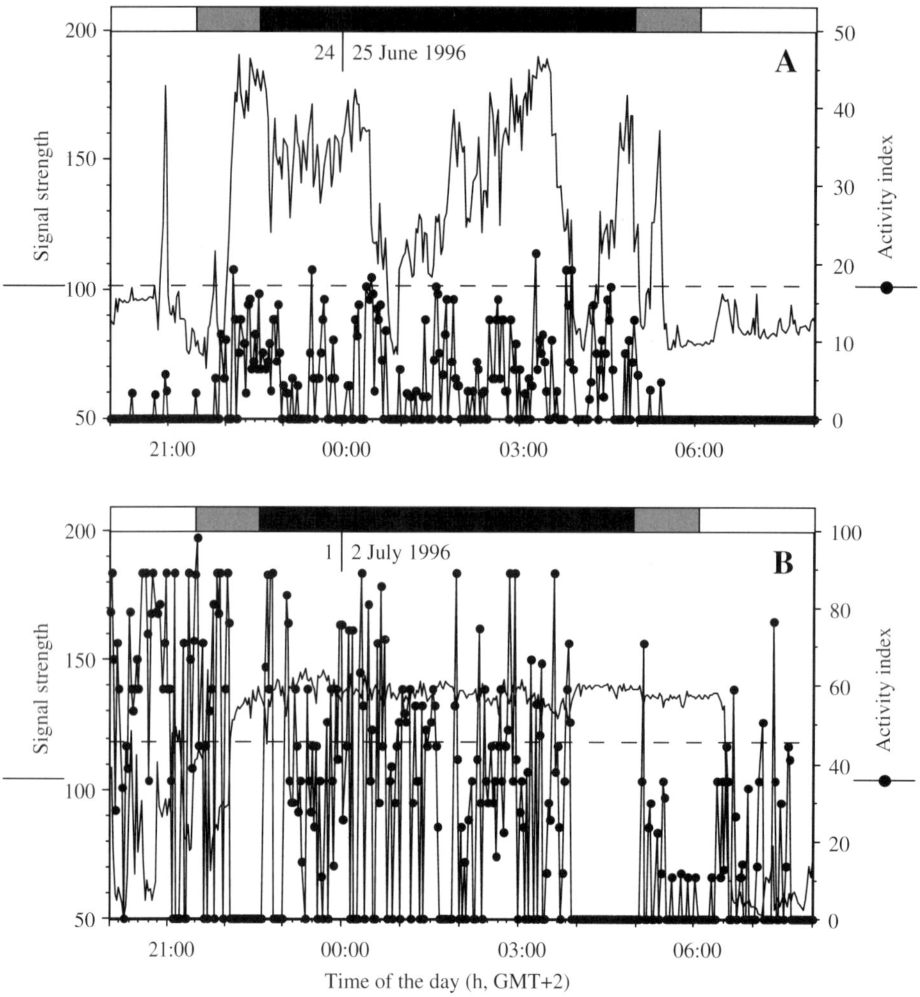


*Figure 5.* Illustration of diel activity of radio-tagged yellow eel in the Awirs stream in 1996, recorded by an automatic data-logging telemetry receiver. A. eel n° 6 on 24–25 June. B. Eel n° 7 on 1–2 July, under full moon. Signal strengths above the dotted line correspond to eels in open water, and below the dotted line to eels inside their cryptic residence (crevice in stone walls or in between rocks), which strongly reduces the signal strength. The activity index is proportional to the number of changes of pulse rates of activity transmitters over a 1 min interval (see methods). No activity index higher than 10% and no movement outside the residence between 08:00 and 20:00.

Meuse and to their immigration in the Awirs stream (Baras & Jeandrain, unpublished).

Consistent with the results of laboratory (Westin & Nyman, 1979), field trapping (Sorensen et al., 1986; Baras et al., 1994) or tracking studies (Helfman et al., 1983; LaBar & Facey, 1983; LaBar et al., 1987; McGovern & McCarthy, 1992; Parker, 1995), yellow eels in the Awirs stream almost exclusively moved at night and always ended their activity before sunrise. Some eels left their diurnal residence before sunset, but only under low light conditions (rain, dark cloudy sky). McGovern & McCarthy (1992) had already provided evidence that the diurnal swimming speeds of eels during overcast weather were similar to those during night-time. Ernande (1995) and Parker (1995) also reported the diurnal activity of eels under cloudy days. These results, as well as the selection of cryptic, highly structured refuges, corroborate the colloquial belief that eels are photophobic animals. As day light can hardly penetrate inside these refuges, eels may have little or no information on the light intensity in their daily activity area. This situation may account for the high activity indices measured during the late afternoon, when eels presumably made short movements at the limit of their residence, which were aborted until the light intensity decreased to a level similar to that at sunset. McGovern & McCarthy (1992) found that the swimming activity of eels was discontinuous, with sequences of continuous swimming lasting from 4 to 35 min, but never exceeding 6% of the tracking time. The activity of eels in the Awirs stream was consistent with this observation, except during the first part of the night, when eels could be active more than one hour in a row. This early nocturnal peak of activity during spring and summer may be interpreted within a motivational context, as the hunger stimulus presumably was very high for fish which had not been foraging since sunrise (ca. 16 h before).

The examination of faeces of fish captured by electric fishing revealed that eels were feeding on invertebrates, and essentially on small molluscs. This may partly explain the nature of the relationship between the lunar cycle and eel's activity in the Awirs stream. The moonlight at the full moon is known to depress the activity of benthic invertebrates (e.g. Neveu & Echaubard, 1975). This reduced prey availability may force foraging fish, the appetite of which is basically dependent on ambient water temperature, to be active for longer periods at full moon, even if these fish are photophobic (e.g. *Barbus barbus* (L.); Baras, 1992; 1995). However, the very short and intense bursts of activity over a restricted area under full moon, and the observation of eels showing additional activity in the late night, after moonset, both suggest that eels were not feeling comfortable under moonlight. This activity pattern may be compared with the observations of Ernande (1995) who also found that eels were travelling short distances under full moon and showed higher activity in the late night, and of McGovern & McCarthy (1992), who reported that eels were swimming three to four times faster under bright conditions than in darkness, presumably because they were feeling less comfortable under higher light intensity. The precise reason why eels chose to be active under uncomfortable conditions remains to be investigated. It may originate from a high hunger stimulus at sunset and relatively short time for foraging during summer nights, especially in a densely populated environment, where eels presumably compete for food. Further studies at different population densities and periods of the year could bring additional information to substantiate or invalidate this interpretation.

## Conclusions

This study provided evidence that large eels may adopt a highly sedentary lifestyle in a continental, fast flowing and densely populated environment, even at periods of the year (spring-summer) when these stages usually show upstream migrations. This finding, along with the home-site fidelity documented by displacement experiments (Parker, 1995), should be considered when establishing environmental management policies. The day-by-day positioning and minute-by-minute monitoring of activity patterns of eels by manual or automatic telemetry also permitted a better understanding of the influence of environmental variables on the way yellow eels utilise their environment. The finding that eels may be highly sedentary further increases the opportunity to automatically monitor the behaviour of radio-tagged fish over longer periods, and to analyse the adequacy of their time and space utilization strategies, i.e. testing for the consistency of activity patterns over consecutive 24 h cycles and investigating the growth rates of eels with different activity or home range patterns. With the continuous transmitting units used in this study, it was not possible to track eels over periods longer than 8 weeks. In the future, we intend to use programmable transmitters operating with duty cycles in order to document the mobility patterns of eels in the long run.

## Acknowledgements

The ‘eel’ research programme was supported by the Commission Provinciale de Liège du Fonds Piscicole’, coordinated by J. C. Philippart (FNRS). The authors wish to thank M. Ovidio, C. Birtles, L. Hill, G. Rimbaud, F. Giroux and G. Bataille who participated in the electric fishing and tracking surveys. We are indebted to the inhabitants of Engis who kindly permitted us to access their private properties when tracking eels.

## References

Barak, N. A. E. & C. F. Mason, 1992. Population density, growth and diet of eels, *Anguilla anguilla* L., in two rivers in eastern England. Aquacult. Fish. Mgmt. 23: 59–70.

Baras, E., 1992. Etudes des stratégies d’occupation du temps et de l’espace chez le barbeau fluviatile, *Barbus barbus* (L.). Cah. Ethol. 12: 125–442.

Baras, E., 1995. Seasonal activities of *Barbus barbus*: effect of temperature on time-budgeting. J. Fish Biol. 46: 806–818.

Baras, E. & D. Jeandrain, 1998. Evaluation of surgery procedures for tagging eel *Anguilla anguilla* with biotelemetry transmitters. Hydrobiologia 371/372: 107–111.

Baras, E. & J. C. Philippart (eds), 1996. Underwater Biotelemetry. University of Liège, Belgium, vi + 257 pp.

Baras, E., J. C. Philippart & B. Salmon, 1996. Estimation of migrant yellow eel stock in large rivers through the survey of fish passes: a preliminary investigation in the River Meuse (Belgium). In I.G.

Cowx (ed), Stock Assessment in Inland Fisheries. Fishing News Books, London: 83–92.

Baras E., B. Salmon & J. C. Philippart, 1994. Evaluation of a eel trap sampling method for the assessment of migrant yellow eels *Anguilla anguilla* (L) in the River Meuse. Bull. fr. Pêche Piscic. 335: 7–16.

Baras, E., C. Birtles, L. Westerloppe, X. Thoreau, M. Ovidio, D. Jeandrain & J. C. Philippart, 1998. A critical review of surgery techniques for implanting telemetry devices into the body cavity of fish. In Y. Le Maho (ed.), Proceedings of the 5th European Conference on Wildlife Telemetry. Strasbourg, France, 26–30 August 1996, in press.

Beumer, J. P., 1979. Feeding and movement of *Anguilla australis* and *A. reihnardtii* in Macleods Morass, Victoria, Australia. J. Fish Biol. 14: 573–592.

Brusle, J., 1989. L'anguille européenne (*Anguilla anguilla*), une espèce jugée commune jusqu'à la dernière décennie, mérite-t-elle d'acquérir aujourd'hui le statut d'espèce menacée? Bull. Soc. zool. France 114: 61–70.

Chisnall, B. L. & J. M. Kalish, 1993. Age validation and movement of freshwater eels (*Anguilla dieffenbachii* and *A. australis*) in a New Zealand pastoral stream. New Zealand J. mar. freshw. Res. 27: 333–338.

Deelder, C. L., 1984. Synopsis of biological data on the eel *Anguilla anguilla* (Linnaeus, 1758). FAO Fish. Synopsis 80: 73 pp.

Dutil, J. D., A. Giroux, A. Kemp, G. Lavoie & J. P. Dallaire, 1988. Tidal influence on movements and daily cycle of activity of American eels. Trans. am. Fish. Soc. 117: 488–494.

Ernande, B., 1995. Etude du rythme d'activité de l'anguille européenne *Anguilla anguilla* L. en marais maritime par télémétrie acoustique. M.S. Thesis, CNRS Ifremer, 41 pp. + ann. CREMA l'Houmeau, Place du Séminaire, BP 5, F-17137 L'Houmeau, France.

Ford, T. E. & E. Mercer, 1986. Density, size distribution and home range of American eels, *Anguilla rostrata*, in a Massachusetts salt marsh. Envir. Biol. Fishes 17: 309–314.

Gunning, G. E. & C. R. Shoop, 1962. Restricted movements of the American eel, *Anguilla rostrata* (LeSueur), in freshwater streams with comments on growth rate. Tulane Stud. Zool. 9: 265–272.

Hart, P. J. B., 1986. Foraging in teleost fishes. In T. J. Pitcher (ed.), The behaviour of teleost fishes. Croom Helm, London: 211–235.

Helfman, G. S., D. L. Stoneburner, E. L. Boseman, P. A. Christian & R. Whalen, 1983. Ultrasonic telemetry of American eel movements in a tidal creek. Trans. am. Fish. Soc. 112: 105–110.

Huet, M., 1949. Aperçu de la relation entre la pente et les populations piscicoles des eaux courantes. Schweiz Z. Hydrol. 11: 332–351.

Knights, B., E. White & L. A. Naismith, 1996. Stock assessment of European eel, *Anguilla anguilla* L. In I. G. Cowx (ed.), Stock Assessment in Inland Fisheries. Fishing News Books, London: 431–447.

LaBar, G. W. & D. E. Facey, 1983. Local movements and inshore population sizes of American eels in Lake Champlain, Vermont. Trans. am. Fish. Soc. 112: 111–116.

LaBar, G. W., J. A. Hernando Casal & C. Fernandez Delgado, 1987. Local movements and population size of European eels, *Anguilla anguilla*, in a small lake in south-western Spain. Envir. Biol. Fishes 19: 111–117.

Legault, A., 1987. L'anguille dans le bassin de la Sèvre niortaise. Publ. Dépt. Halieut. Ecole Nat. Sup. Agro. Halieut. 6: 295 pp., Nantes, France.

Lobon-Cervia, J., Y. Bernat & P. A. Rincon, 1990. Effects of eel (*Anguilla anguilla* L.) removals from selected sites of a stream on its subsequent densities. Hydrobiologia 206: 207–216.

Mann, H., 1965. Über das rückkehrvermogen verpflantzer fluss-aal. Arch. Fischereiwissenschaft 15: 177–185.

McDowall, R. M., 1988. Diadromy in fishes. Croom Helm, London, 308 pp.

McGovern, P. & T. K. McCarthy, 1992. Local movements of freshwater eels (*Anguilla anguilla* L.) in western Ireland. In I. G. Priede & S. M. Swift (eds), Wildlife Telemetry: Remote Sensing and Monitoring of Animals. Ellis Horwood, Chichester, UK: 319–327.

Moriarty, C., 1983. A population study of the eel *Anguilla anguilla* in Meelick Bay, Lough Derg. Fish. Bull. Dublin 7: 1–8.

Moriarty, C., 1986. Riverine migration of young eels *Anguilla anguilla* (L.). Fish. Res. 4: 43–58.

Moriarty, C., 1990. European catches of elver of 1928–1988. Int. Rev. Ges. Hydrobiol. 75: 701–706.

Neveu, A. & M. Echaubard, 1975. La dérive estivale des invertébrés aquatiques et terrestres dans un ruisseau du Massif Central: la Couze Pavin. Ann. Hydrobiol. 6: 1–26.

Nyman, L., 1975. Behaviour of fish influenced by hotwater effluents as observed by ultrasonic tracking. Rep. Inst. freshwat. Res. Drottningholm 54: 64–74.

Parker, S. J., 1995. Homing ability and home range of yellow-phase American eels in a tidally dominated estuary. J. mar. biol. Ass. U.K. 75: 127–140.

Priede, I. G. & S. M. Swift (eds), 1992. Wildlife Telemetry: Remote Sensing and Monitoring of Animals. Ellis Horwood Ltd., Chichester, UK, 708 pp.

Rossi, R., M. Bianchini, A. Carrieri & P. Franzoi, 1987. Observations on movements of yellow eels *Anguilla anguilla* after displacement from coastal waters to sea. J. Fish Biol. 31: 155–164.

Sinha, V. R. P. & J. W. Jones, 1975. The European freshwater eel. Liverpool University, Liverpool, 146 pp.

Sörensen, J., 1950. An investigation of some factors affecting the upstream migration of the eel. Rep. Inst. freshwat. Res. Drottningholm 32: 126–172.

Sorensen, P. W., M. L. Bianchini & H. E. Winn, 1986. Diel foraging activity of American eels, *Anguilla rostrata* (LeSueur), in a Rhode Island estuary. Fish. Bull. 84: 746–747.

Stasko, A. B. & S. A. Rommel Jr., 1974. Swimming depth of adult American eels (*Anguilla rostrata*) in a saltwater bay as determined by ultrasonic tracking. J. Fish. Res. Bd Can. 21, 1148–1150.

Tesch, F. W., 1974. Speed and direction of silver and yellow eels, *Anguilla anguilla*, released and tracked in the open North Sea. Ber. dt. wiss Kommn. Meereforsch. 23: 181–197.

Tesch, F. W., 1977. The Eel. Biology and Management of Anguillid Eels. Chapman & Hall, London.

Tesch, F. W., 1978. Telemetric observations on the spawning migration of the eel (*Anguilla anguilla*) west of the European continental shelf. Envir. Biol. Fishes 3: 203–209.

Tesch, F. W., 1979. Tracking of silver eels (*Anguilla anguilla*) in different shelf areas of the Northeast Atlantic. Rapp. P.V. Réun. Cons. Explor. Mer 174: 104–114.

Westerberg, H., 1982. Ultrasonic tracking of migrating eels in Lake Hjaelmaren. Rep. Inst. freshw. Res. Drottningholm 7, 1–22.

Westin, L. & L. Nyman, 1979. Activity, orientation and migration of Baltic eel (*Anguilla anguilla* L.). Rapp. Cons. Explor. Mer 174: 115–123.

Winter, J. D., 1983. Underwater biotelemetry. In L. A. Nielsen & J. D. Johnsen (eds), Fisheries Techniques. American Fisheries Society, Bethesda, Maryland: 371–395.

*Hydrobiologia* **371/372:** 199–206, 1998.
*J.-P. Lagardère, M.-L. Bégout Anras & G. Claireaux (eds), Advances in Invertebrates and Fish Telemetry.*



# Reactions of cod (*Gadus morhua* L.) in the pre-vessel zone to an approaching trawler under different light conditions

*Preliminary results*

Arill Engås, Eli Kyrkjebø Haugland & Jan Tore Øvredal
*Institute of Marine Research, P.O. Box 1870, N-5024 Bergen, Norway*
*e-mail: arill.engaas@imr.no*

*Key words:* cod, vessel sound, fish behaviour, acoustic tracking

## Abstract

The behaviour of cod (*Gadus morhua* L.) towards an approaching vessel during trawling was studied by means of a stationary positioning system. Three cod were tagged by allowing them to voluntarily swallow transmitters wrapped in bait. The behaviour was recorded before, during and after trawling, which was carried out during high and low light level.

Nine hauls were carried out, but only six hauls are included in the analysis, because of technical problems during three hauls. Only in two out of six hauls all three cod were observed.

Behaviour towards the approaching trawler differed from tow to tow. Only three out of thirteen cases were classified as reaction; all three occurred during daylight. In two of the three cases a change in the swimming direction away from the approaching vessel was observed, whereas a reduction in swimming speed was recorded in the third case. The reaction distance from the vessel, between 470 and 1470 m, corresponds to a sound level between 82 and 92 dB re 1 $\mu$Pa per Hz. There were indications of a lower level of sensory awareness during reduced light level.

## Introduction

Trawl surveys are today a critical component in fishery resource assessment world-wide. The main objective of a trawl survey is to obtain an indication of abundance (Pennington & Brown, 1981) and to reveal changes in population age structure and distribution over time. This objective is rarely met (Engås, 1994) because no sampling operation exhibits equal selectivity to all species and length groups within an area. One reason for this is the difference in behaviour of various species and sizes of fish as they encounter the vicinity of the vessel and trawl. Knowledge of the behaviour of fish throughout the whole catching process is thus vital to improve the reliability of stock assessment.

The use of scuba diving, remotely operated underwater vehicles and photographic techniques has contributed significantly to our understanding of fish behaviour patterns in the fishing zone of trawls, extending from the doors to the codend (Ben-Tuvia & Dickson, 1969; Walsh & Hickey, 1993; Wardle, 1993). However, there has been a lack of information on the behaviour of fish, especially of individual fish, in front of an approaching vessel. Echosounders and sonars have been used to study the behaviour of fish in front of an approaching vessel (Olsen et al., 1983; Ona & Godø, 1990; Misund et al., 1996), but have strong limitations in quantifying the behaviour of individual fish in this area, especially of fish close to the sea bed.

Because sound generated by a trawler falls within the hearing range of fish, fish may hear sounds from an approaching vessel and react at a long distance (Buerkle, 1977). Tracking individual acoustically tagged fish with sonar or hydrophones is a potential approach to a better understanding of the behaviour of individual fish in the pre-vessel zone. The experiment by Harden Jones et al. (1977), studying the behaviour of tagged plaice (*Pleuronectes platessa* L.)

using sector-scanning sonar gave valuable information on the behaviour of single fish towards the vessel and trawl gear, but unfortunately such experiments have not been followed up. Tracking of cod (*Gadus morhua* L.), acoustically tagged on deck, in front of a trawler, with a directional hydrophone mounted on board a rubber boat gave some information, but the methodology used had limitations (Engås et al., 1991). This paper presents the behaviour of three individual free-swimming cod, acoustically tagged close to bottom, and tracked with a stationary position system, in the pre-vessel zone during trawling at different light conditions.

## Materials and methods

### *Study locations, positioning system and tagging*

The experiment was conducted in Ramfjord (northern Norway), at 60–80 m depth, from 9–16 October, 1996. A stationary positioning system (VR-20, Vemco Ltd.) was used to track acoustically tagged cod. The system had a fixed array of three hydrophone buoys with radio transmitters to a base station (R/V 'Fjordfangst'). The hydrophone buoys, anchored to the sea bed in a triangular configuration, were 400 to 600 m apart.

Three cod were tagged *in situ* close to the sea bed with acoustic transmitters, by allowing the cod to voluntarily ingest transmitters wrapped in mackerel bait. Each transmitter was 48 mm long and 16 mm in diameter (weighing 9 g in water), and gave horizontal (two-dimensional) position fixes (x,y coordinates). An underwater camera mounted in a frame was used to identify the fish and to estimate the length of the tagged fish based on marks on the frame (10 cm apart). The three cod were all about 50 cm long. They are later referred to as fish A, B and C, respectively. One selected frequency (transmitter) was monitored every 28 s, i.e. the interval between positional fixes for each tag is multiples of 28 s. Normally several received pulses from each buoy are averaged to get one position. In this study, however, the tags were traced in fast mode, which means that every received position (time lag 1–5 s) was monitored. In the calculations of swimming speed and direction, however, only averaged values of the position fix were used. A more detailed description of the position system, transmitters and tagging procedures is given by Engås et al. (1996).

### *Experimental procedures*

The cod were tagged 20 days before the trawler, M/Tr 'Sara Karin' (15.5 m, 400 hp), arrived in the study area. During these 20 days, the natural behaviour of the cod and their response towards baited longline were studied. Trawling was conducted every day, at different light conditions. Nine trawl hauls were performed during the experiment, but three hauls were excluded from the analysis, because sound generated from the vessel disturbed the position system.

The shrimp trawl was shot with the codend open at a distance of about 3.5 km from the tagged fish. A transmitter was mounted on the starboard trawl door, and the warp length was 150 m. The horizontal distance between the vessel and the trawl doors was calculated to be approximately 130 m, based on a warp angle of 20 degrees. The vessel headed at a constant speed of 1 m $s^{-1}$ towards the centre of the triangle of hydrophones.

### *Sound monitoring*

The sound level and frequency spectrum of M/Tr 'Sara Karin' were measured during the fishing operation by a hydrophone (Bruel & Kjær, type 8104) suspended at a depth of 10 m. The measurements were made from R/V 'Fjordfangst', anchored in the investigation area with its engine turned off. Before and after the measurements, the ambient sound level was recorded. The sound were amplified, Bruel & Kjær, type 2635, and logged on a digital tape recorder (Sony Dat Pro II). The equipment was calibrated (Bruel & Kjær calibrator, type 4229) before and after the measurements. The sound was later analysed in 1/3-octave bands with a real-time analyser (Bruel & Kjær, type 2143).

### *Criteria for reaction*

Reaction was defined as an attempt by the fish to move to a lower sound intensity. Behavioural variation was expected due to genetic differences between individuals and variations in the individual trade off due to varying physical and physiological factors. Despite the variable behaviour, a criterion for reaction was defined. The behaviour of the fish had to satisfy the following criteria: before the trawl entered the study area, the fish must have shown consistency in movement by swimming in one direction for at least 30 min (at least 80% of a minimum of 10 direction observations had to be in a 135° sector). Reaction (R) to the approaching vessel was defined when the fish changed

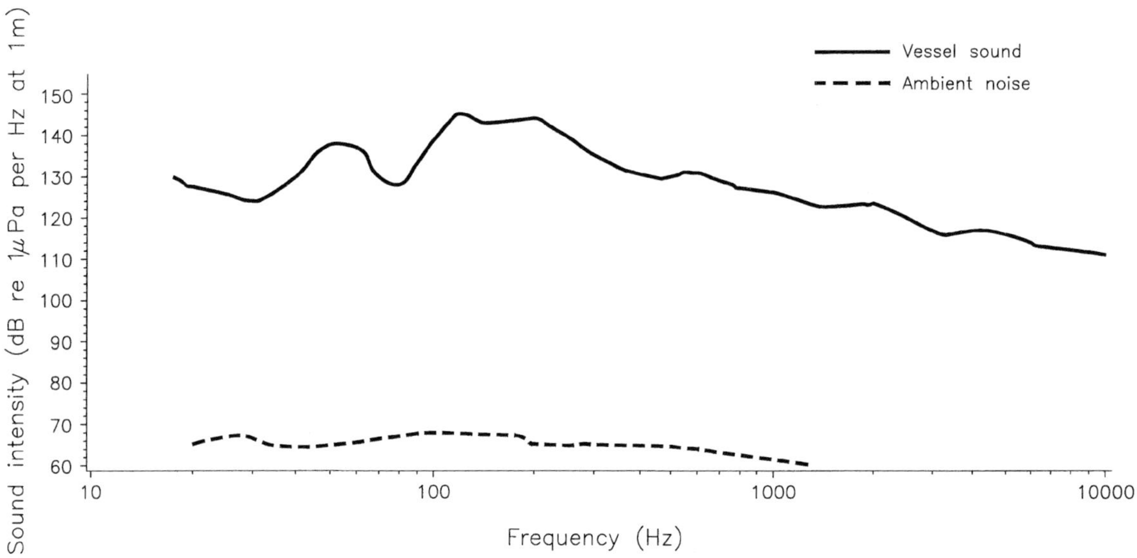


*Figure 1.* Ambient noise level at the experimental site and the spectral distribution of the sound of M/Tr 'Sara Karin' at the moment of passing the hydrophone. The vessel sound was recorded when the vessel was free-running at 1 m s$^{-1}$, measured on the starboard side 10 m below the sea level.

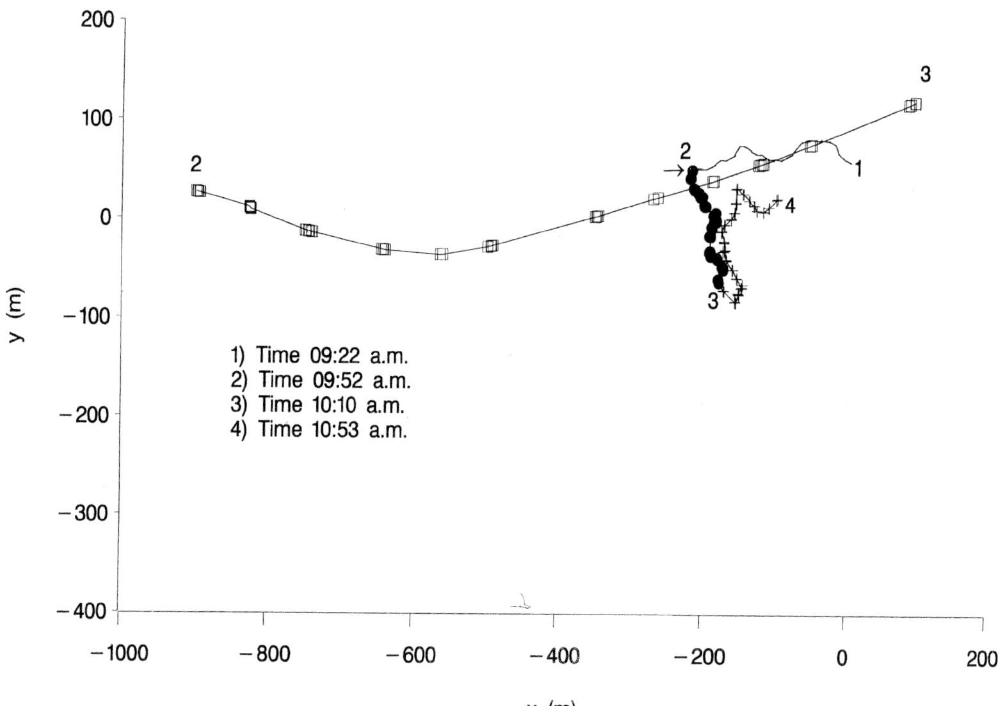


*Figure 2.* The true movement of cod (A) relative to the starboard trawl door. This cod (case xiii in Table 1) reacted (marked) when it changed swimming direction away from the approaching vessel. The movement of the cod is shown from time 1 to 4. The symbols show the positions of the starboard trawl door (open squares) relative to the fish before (plain line, time 1–2), during (dots, time 2–3) and after (crosses, time 3–4) trawling. The time of the position marked 2 and 3 of the cod correspond with the position 2 and 3 of the starboard trawl door. The axes represent horizontal coordinates (x, y).

*Table 1.* Summary of the different behavioural criteria of the cod and the following classification, reaction (R) or no reaction (N). If the criteria given to clarify the reaction was satisfied it is marked 'Y' = yes, else 'N' = not. When the behaviour of the fish satisfied the criterion 'before reaction' and one of the criteria 'during reaction', the possibility that cod reacted (R) is high. All other cases are characterised as no reaction (N). The time is set to the nearest hour. Because of large variation the description of each of the cases i–xiii) should be read together with the table.

| Case | Haul | Date Time | Fish | Before reaction: Stable direction | Reaction: Change direction | Reaction: Change speed | Classification | Reaction distance (m) fish-vessel | Noise (dB) at distance of reaction |
|---|---|---|---|---|---|---|---|---|---|
| i | 2 | 12 Oct | A | Y | N | Y | R | 1290 | 83 |
| ii | | 09 a.m. | B | Y | Y | N | R | 1470 | 82 |
| iii | | high light | C | N | N | Y | N | | |
| iv | 3 | 13 Oct | A | N | N | N | N | | |
| v | | 02 p.m. | B | N | N | N | N | | |
| | | high light | | | | | | | |
| vi | 4 | 13 Oct | A | Y | N | N | N | | |
| vii | | 10 p.m. | B | Y | N | N | N | | |
| | | low light | | | | | | | |
| viii | 6 | 14 Oct | A | N | N | N | N | | |
| ix | | 10 p.m. | B | N | N | N | N | | |
| | | low light | | | | | | | |
| x | 7 | 15 Oct | A | N | Y | N | N | | |
| xi | | 11 a.m. | B | N | N | N | N | | |
| xii | | high light | C | N | N | N | N | | |
| xiii | 9 | 16 Oct | A | Y | Y | N | R | 470 | 92 |
| | | 10 a.m. | | | | | | | |
| | | high light | | | | | | | |

i) The behaviour of the fish satisfy the criteria before the reaction. The fish slowed down and stayed in a restricted area when the vessel passed and the swimming speed was significantly (Duncan, $p < 0.05$) lower than before it slowed down. About 10 min after the trawl had passed the fish began to swim.
ii) The behaviour of the fish satisfy the criteria before the reaction. The fish changed (Duncan, $p < 0.05$) the direction of swimming when the vessel was approaching.
iii) In this case the fish seemed to swim along with the trawl door or trawl. The reaction distance to the trawl door was about 105 m and the fish swam along with the vessel or trawl for 9 min. The swimming speed during reaction was higher (Duncan, $p < 0.05$) than that before and after. (Due to the short period during which the fish swam along with the trawl, the calculation of the swimming speed during reaction was based on few observations.)
iv) No criterion was satisfied.
v) No criterion was satisfied.
vi) The fish swam towards the approaching vessel both before and during the trawling (Figure 3).
vii) The fish swam towards the approaching vessel. At a distance of about 612 m in front of the vessel the fish reduced its swimming speed (not significantly) and stayed in a restricted area during and after trawling.
viii) No criterion was satisfied.
ix) The fish reduced its swimming speed (not significantly) at a distance of 90 m in front of the vessel and stayed in a restricted area for at least one hour.
x) The fish seemed to react at a distance of 430 m in front of the vessel and the swimming direction before and after the believable reaction was different (Duncan, $p < 0.05$) but the behaviour did not satisfy the criteria before reaction (Figure 4).
xi) No criterion was satisfied.
xii) No criterion was satisfied.
xiii) The behaviour of the fish satisfy the criteria before the reaction. The swimming direction before and after reaction was different (Duncan, $p < 0.05$) (Figure 2).

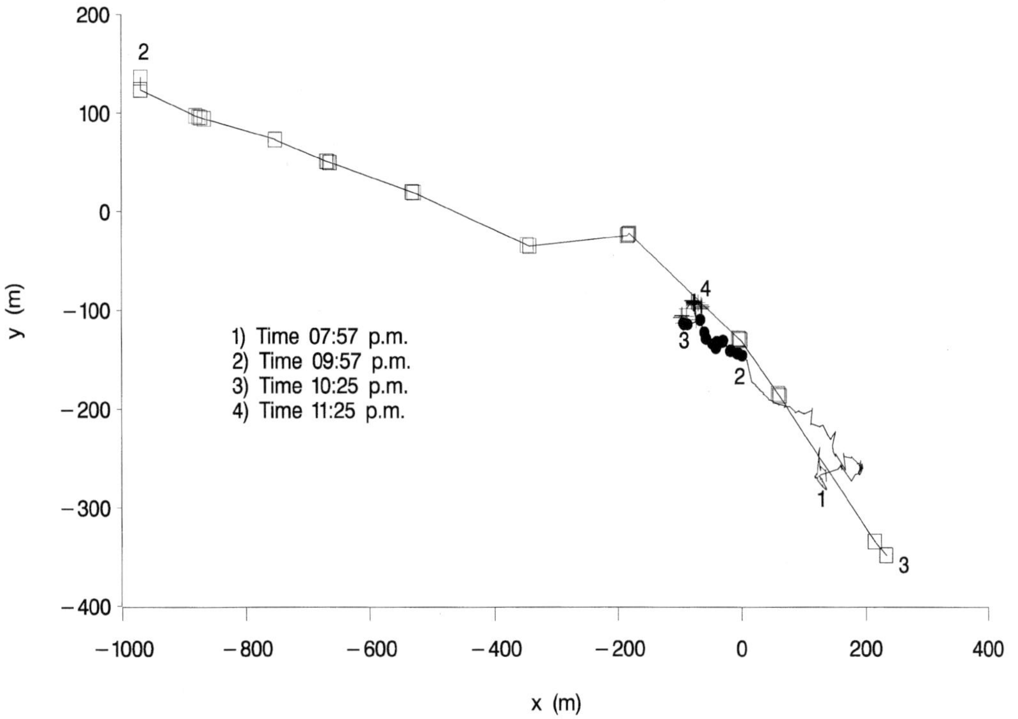


*Figure 3.* The true movement of cod (A) relative to the starboard trawl door (case vi). The symbols show the positions of the starboard trawl door (open squares) relative to the fish before (plain line, time 1–2), during (dots, time 2–3) and after (crosses, time 3–4) trawling. The movement of the cod is shown from time 1 to 4. The time of the position marked 2 and 3 of the cod correspond with the position 2 and 3 of starboard trawl door. The axes represent horizontal coordinates (x, y).

either its swimming direction (the swimming direction after reaction should be significantly different from the initial direction, ANOVA, Duncan's multiple range test, $p < 0.05$) (Zar, 1984) or its swimming speed (the swimming speed after the reaction should be significantly different from the initial swimming speed, ANOVA, Duncan's multiple range test, $p < 0.05$). If these criteria could not be satisfied, the case was characterised as no reaction (N).

To calculate the distance between the vessel and the fish, x–y positions of the vessel were back-calculated to the same time as the position of the fish. In all cases, it was assumed that the fish were close to the bottom.

## Results

### *Vessel sound analysis*

At the moment at which it passed R/V 'Fjordfangst', M/Tr 'Sara Karin' generated the highest sound intensities, between about 100 Hz and 250 Hz (Figure 1). The highest source level reached 145 dB re 1 $\mu$Pa per Hz at 130 Hz, i.e. approximately 80 dB re 1 $\mu$Pa per Hz above ambient sound level.

### *Swimming behaviour*

Thirteen cases are included in the study. Each is classified in one of the categories: reaction (R) or no reaction (N), depending on the behaviour of the fish relative to the criteria for reaction (Table 1). Each of the cases is unique and the classification of the behaviour should be read together with the notes in Table 1.

Reaction (R) to the approaching vessel was observed in three of thirteen cases, two cases for fish A and one case for fish B (Table 1). In the category reaction (R), fish seemed to show reaction at distances from 470 (case xiii, Figure 2) to 1470 m (case ii) in front of the vessel. In all three cases, the fish showed consistency in swimming behaviour for at least 30 min before the vessel arrived. When the vessel approached, the fish changed (Duncan, $p < 0.05$) their directions of swimming away from the approaching vessel in two

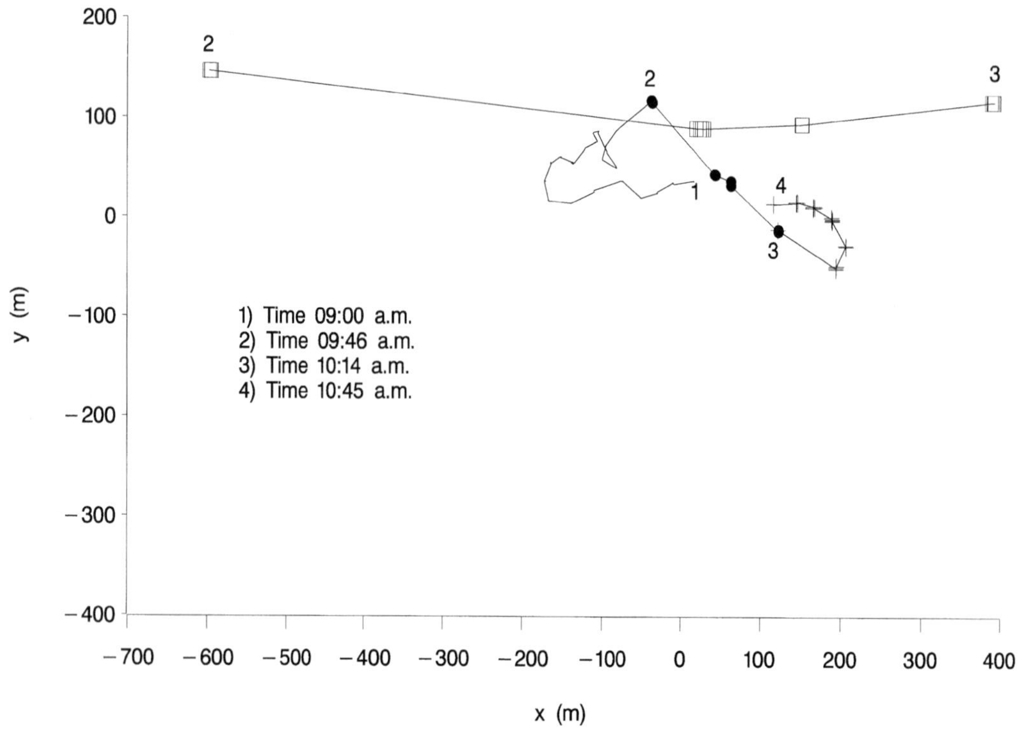


*Figure 4.* The true movement of cod (A) relative to the starboard trawl door (case x). The symbols show the positions of the starboard trawl door (open squares) relative to the fish before (plain line, time 1–2), during (dots, time 2–3) and after (crosses, time 3–4) trawling. The movement of the cod is shown from time 1 to 4. The time of the position marked 2 and 3 of the cod correspond with the position 2 and 3 of the starboard trawl door. The axes represent horizontal coordinates (x, y).

of the cases, but did not change the swimming speed. In case i, the fish reduced its swimming speed (Duncan, $p < 0.05$) from 0.13 $ms^{-1}$ to less than 0.07 $ms^{-1}$ (at a distance of approximately 1290 m from the vessel) and stayed in a limited area until the vessel had passed.

In the four cases where towing was conducted during low light levels, fish were observed to be closer to the vessel and trawl than during daylight, except on one occasion (case i). During low light levels, the vessel passed the fish at distances from about 20 to 50 m, whereas the distance at high light levels varied from about 40 to 300 m.

## Discussion

The observed behaviour of cod towards an approaching trawler varied from tow to tow and among the three cod within the same tow. A total of three of the thirteen recorded cases were classified as reaction to the approaching vessel. In two cases, the fish changed swimming direction as the vessel approached, but no change in swimming speed was observed. This observation is in accordance with the avoidance model proposed by Olsen et al. (1983), which assumes that fish start reacting to an approaching vessel simply by adjusting the swimming direction away from the vessel. Similar observations have been made for herring (Misund et al., 1996). The third case, where the fish reduced its swimming speed and stayed within a limited area until the vessel had passed, could be explained as pointed out by Wardle (1993), that vessel sound may arouse curiosity or enhance sensitivity such that a visual explanation is expected. This explanation is supported by the observation that when the trawl had passed maybe beyond the visual range, the fish continued with the same swimming direction and speed as before the trawler approached. Two out of three significant reactions were found during the first haul. Habituation to applied sound stimuli have been observed for fish when the sound stimuli have been repeated at very short time intervals of less than 10 minutes (Olsen, 1976). However, during this experiment trawling was conducted within a time interval of no less than six hours, a time interval during which no habituation were recorded to playbacks of vessel sound for cod in net pens (Engås et al., 1995).

As the reaction observed is well outside the visual range of the vessel, it is reasonable to believe that sound from the vessel and trawl is the determinant in triggering the reaction behaviour. The sound from the vessel had its highest intensity between about 100 and 250 Hz, which is within the range where cod have their highest sensitivity (Chapman & Hawkins, 1973). On the basis of a source level of 145 dB re 1 $\mu$Pa per Hz and a transmission loss of 20 log distance, the sound level at the reaction distance observed is from 82 to 92 dB re 1 $\mu$Pa per Hz. When extrapolating from Chapman & Hawkins (1973), this is about 2–12 dB respectively above the auditory threshold at a background sound level of about 65 dB re 1 $\mu$Pa per Hz, but 10–20 dB respectively below the reaction threshold usually reported (see Anon, 1994). A possible explanation for the low reaction threshold observed in the present study could be the low ambient sound in the study area. The threshold is found to increase in proportion to the ambient sea noise (Chapman & Hawkins, 1973). Sand and Karlsen (1986) showed that cod are also sensitive to infrasound (0,1–10 Hz). To what extent this part of the frequency spectrum affected the behaviour of the fish is unknown, due to limitations of the sound recording system in monitoring infrasound.

In five of the ten cases that were classified as no reaction (N), it is likely that a reaction did occur, despite the fact that the swimming behaviour did not satisfy the criteria for reaction. In four cases (vi (Figure 3), vii, viii and ix) where trawling was conducted during low light levels, the fish reduced swimming speed and stayed in a limited area during and after trawling. The distance between the fish and vessel, from the moment the fish stayed within a limited area, was about 610 and 90 m in cases vii and ix respectively. In one of the five cases (case x, Figure 4), the fish changed (Duncan, $p < 0.05$) the swimming direction away from the approaching vessel at a distance of 430 m in front of the vessel, but it did not satisfy the criteria of consistency in swimming behaviour before trawling.

The variations in behaviour from tow to tow may be due to variations in motivational state or in the level of background sound. It is interesting to note that in the four cases where towing was conducted in darkness, fish were observed to be closer to the vessel and trawl than in any other situation observed in daylight. Maybe this was due to a lower level of sensory awareness at low light levels, as questioned by Walsh & Hickey (1993), who found that cod and haddock rested on the bottom during low light level and seemed slow to react to the approaching trawl compared with the situation during high light levels. The level of ambient sea noise is expected to have changed during the experimental period, and if it was well above the level measured during calm sea conditions, that can explain the variations in behaviour observed. As pointed out earlier, variation in the thresholds is related to changes in the level of ambient sea noise.

## References


Anon, 1994. Report of the study group on research vessel noise measurement. ICES cm 1994/B:5, 81 pp.

Ben-Tuvia, A. & W. Dickson, 1969. Proceedings of the FAO Conference on fish behaviour in relation to fishing techniques. Bergen, Norway, 19-27 Oct 1967. FAO Fish. Rep. 1, 2, 3.

Buerkle, U., 1977. Detection of trawling noise by Atlantic cod (*Gadus morhua* L.) Mar. Behav. Physiol. 4: 233–242.

Chapman, C. J. & A. D. Hawkins, 1973. A field study of hearing in the cod, *Gadus morhua* L.. J. Comp. Physiol. 85: 147–167.

Engås, A., A. V. Soldal & J. T. Øvredal, 1991. Avoidance reactions of ultrasonic tagged cod during bottom trawling in shallow water. ICES C.M. 1991/B:41, 9 pp.

Engås, A., 1994. The effects of trawl performance and fish behaviour on the catching efficiency of demersal sampling trawl. In A. Fernö & S. Olsen (eds), Marine Fish Behaviour in Capture and Abundance Estimation. Fishing News Books, Oxford: 45–68.

Engås, A., O. A. Misund, A. V. Soldal, B. Horvei & A. Solstad, 1995. Reactions of penned herring and cod to playback of original, frequency-filtered and time-smoothed vessel sound. Fish. Res. 22: 243–254.

Harden Jones, F. R., A. R. Margetts, M. Greer Walker & G. P. Arnold, 1977. The efficiency of the Granton otter trawl determined by sector scanning sonar and acoustic transponding tags. Rapp. P.-v. Réun. Cons. int. Explor. Mer 170: 45–51.

Misund, O., J. T. Øvredal & M. T. Hafsteinsson, 1996. Reactions of herring schools to the sound field of a survey vessel. Aquat. Living Resour. 9: 5–11.

Olsen, K., 1976. Evidence for localization of sound by fish in schools. In A. Schuijf & A. D. Hawkins (eds), Sound Reception in Fish. Elsevier, Amsterdam: 257–270.

Olsen, K., J. Angel, F. Pettersen & A. Løvik, 1983. Observed fish reaction to a surveying vessel with special reference to herring, cod, capelin and polar cod. In O. Nakken & S. C. Venema, Symposium on Fisheries Acoustics. Selected papers of the ICES/FAO Symposium on Fisheries Acoustics. Bergen, Norway, 21–24 June 1982. FAO Fish. Rep. 300: 131–138.

Ona, E. & O. R. Godø, 1990. Fish reaction to trawling noise: the significance for trawl sampling. Rapp. P.-v. Réun. Cons. int. Explor. Mer 189: 159–166.

Pennington, M. & B. E. Brown, 1981. Abundance estimates based on stratified random trawl surveys. Can. Spec. Publ. Aquat. Sci. 58: 149–153.

Sand, O. & H. E. Karlsen, 1986. Detection of infrasound by the Atlantic cod. J. exp. Biol. 125: 197–204.

Walsh, S. J. & W. M. Hickey, 1993. Behavioural reactions of demersal fish to bottom trawls at various light conditions. ICES mar. Sci. Symp. 196: 68–76.

Wardle, C. S., 1993. Fish behaviour and fishing gear. In T. J. Pitcher (ed.), Behaviour in Teleost Fishes. Chapman & Hall, London: 609–643.

Zar, J. H., 1984. Biostatistical Analysis. Second edition. Prentice-Hall, New Jersey, 718 pp.

*Hydrobiologia* **371/372:** 207–213, 1998.
*J. P. Lagardère, M.-L. Bégout Anras &G. Claireaux (eds), Advances in Invertebrates and Fish Telemetry.*



# Changes in vertical distribution of sea bass (*Dicentrarchus labrax* L.) during a hypoxic episode

Henrik Schurmann, Guy Claireaux & Hervé Chartois
*Centre de Recherche en Ecologie Marine et Aquaculture de L'Houmeau, B.P.5, 17137 L'Houmeau, France; e-mail: schurmanpost8.tele.dk*

*Key words:* telemetry, salinity, oxygen, diurnal rhythm

## Abstract

The vertical distribution of free-swimming sea bass (*Dicentrarchus labrax* L.) was investigated in a 26 $m^3$ indoor tank by means of acoustic telemetry. During the experiment the bottom of the tank was made progressively hypoxic.

In normoxic conditions, sea bass displayed a clear diurnal rhythm during which they occupied the surface at night and swam deeper in the water column during the day. When a vertical oxygen gradient was established, this behaviour changed and sea bass reduced the amplitude of their daily migration to avoid being exposed to the hypoxic bottom water layer.

## Introduction

Owing to the energetic forcing resulting from tidal mixing, major currents and large scale physical processes, eutrophication was, until recently, supposed to be unlikely in most coastal ecosystems. However, the scale and scope of anthropogenic disturbances have increased extensively in the past decades, and it is now quite clear that inshore waters of the global oceans are severely affected by eutrophication (Kerr & Ryder, 1997). During that period, human activities have been concentrating along major rivers, in estuaries and on the coastlines, increasing the amount of nutrients, mainly nitrogen and phosphorous, been received by marine ecosystems. One of the many consequences of this drift has been a gradual rise in the oxygen consumption by these systems in relation with the mineralization of organic matter (Rydberg et al., 1990). Accordingly, over that period the number and seriousness of hypoxic episodes reported in coastal environments have been quickly augmenting (e.g., Baden et al., 1990).

Although shallow coastal waters play a major economical role as it supports most of the fisheries and related activities, the impact of temporary hypoxia on the productivity of these ecosystems has been very little studied. Fish that live in close association with the bottom, and particularly flatfishes with their poor locomotor ability, are vulnerable to hypoxic events. Petersen and Pihl (1995) reported that the growth rate of two flatfish species, the plaice (*Pleuronectes platessa* L.), and the dab (*Limanda limanda* L.), was reduced during hypoxic episodes in the sea between Denmark and Sweden (Kattegat). Furthermore, the total biomass of these species was also significantly reduced during these periods. For roundfish living in the water column, the situation may be expected to be different as their locomotor capacity enables them to seek better oxygenated surface layers (Pihl et al., 1991), an option which is not available to flatfishes. However, D'amours (1993) showed that temperature, but also ambient oxygenation, set the extent of the distribution of Atlantic cod (*Gadus morhua* L.) in the Gulf of Saint Lawrence (Canada). Similarly, Claireaux et al. (1995) showed that the presence of a deoxygenated water layer in the lowest part of the water column, profoundly modified the distribution of cod in a 10 m deep tank, altering their ability to feed on the bottom. The potential effects of organic enrichment and associated hypoxia on the trophic interactions in marine benthic communities were also emphasized by Pihl (1994).

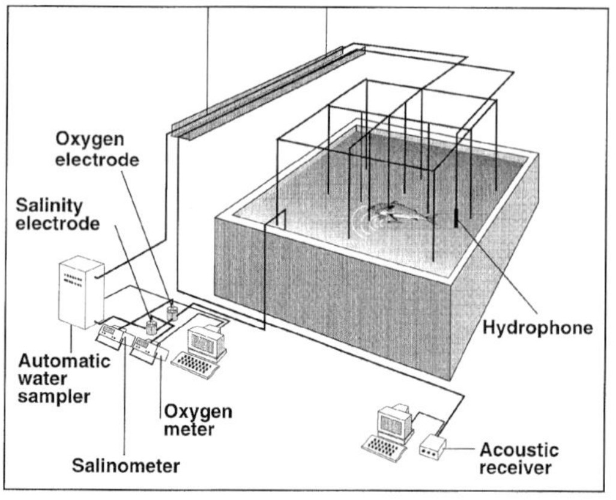


*Figure 1.* Schematic illustration of the experimental setup. The automatic water sampler with the 14 tubes (only 1 is shown from the sampler to the tank) collected water at different depths and positions in the tank. It then passed an oxygen, temperature and a salinity electrode, before it was send back to the tank. A computer controlled the water sampler and monitored oxygen, temperature and salinity data. A hydrophone was placed in the tank and the signal from this was received by a acoustic receiver and stored on a computer. See the text for further details.

In this context, the aim of the present study was to investigate the interactions between the vertical distribution of fish and water oxygenation conditions under controlled laboratory conditions. The use of acoustic telemetry enabled us to precisely follow the vertical movements of free-swimming sea bass (*Dicentrarchus labrax* L.) throughout a hypoxic episode at the bottom.

## Materials and methods

Experiments were carried out from March to April 1997. The results reported here cover 2 experiments with a total of 6 fish.

*The fish*

Sea bass weighing 800–1200 g were kept in natural light conditions in an indoor recirculated sea water system at a salinity of 29–30‰. The temperature of the system was not controlled and followed a natural cycle. At the time of these experiments the water temperature ranged from 17 to 20 °C. Fish were fed twice a week on dry pellets, and once a week on chopped oysters.

*The experimental tank*

Experiments were carried out in an indoor tank measuring $4.5 \times 4.5 \times 1.3$ m (26.3 $m^3$) (Figure 1). Artificial lighting followed the natural photoperiod. The temperature of the water was not controlled and varied between 19 °C and 21 °C during the experiments. The vertical temperature gradient in the tank never exceeded 0.25 °C.

Water temperature, salinity, and oxygen content were monitored using an automatic water sampler. In the tank, 14 sampling points were placed at different positions and depths. Via polyamide tubes, the automatic sampler sequentially pumped water from each of these points and passed it through oxygen, temperature and salinity electrodes (Orbisphere instruments and WTW LF 196), before it was send back to the tank. The water sampler pumped 125 ml $min^{-1}$ and alternated sampling point every 4th min. It was controlled by a computer which also collected oxygen, temperature and salinity data. In order to take into account the response time of the oxygen electrode, only records from the last 15 s of the 4 min sampling periods were considered in later analysis.

Manipulation of water salinity and oxygen saturation was carried out in two water gas equilibration and water mixing columns connected to the tank. The water flowing in the first column was pumped from the bottom of the tank and sent back to an inlet pointing downwards at the bottom. The water in the second column was pumped from the surface and sent back to an inlet pointing upwards, 60 cm below the surface.

*Ultrasonic transmitters*

For each experiment three fish were tagged with ultrasonic pressure transmitters (Vemco V16P-1LR). The transmitters were placed in an anterio-ventral position as described by Claireaux & Lefrançois (1998, this volume). The pulse rate of the transmitter was linearly related to the hydrostatic pressure (depth). Before each experiment, the batteries of the transmitters were changed allowing approximately 3 weeks of transmission. Tags where then calibrated at experimental temperatures and the accuracy of the depth measurements was ±5 cm. Each transmitter had a different acoustic frequency, and they were sequentially monitored for 5 measurements each (approximately 5 s) by a hydrophone plugged into a receiver (Vemco VR-20). The receiver was connected to a computer which stored the depth data for later analysis.

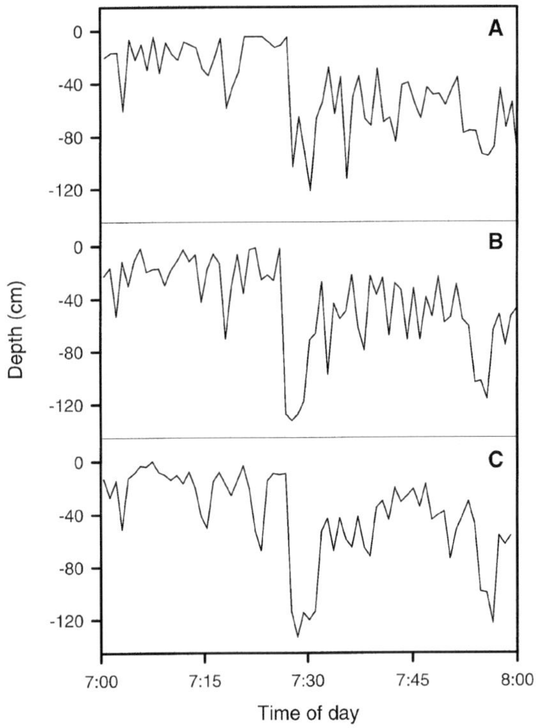


*Figure 2.* The occupied depth of three fish during 1 h of experiment, in a homogeneous environment (fish A = 998 g, fish B = 984 g, fish C = 861 g). The example shows that all three fish swam to the bottom when the light was turned on in the morning at 7:25, and after a short period they swam up to mid depth. Note the synchronization of the occupied depths.

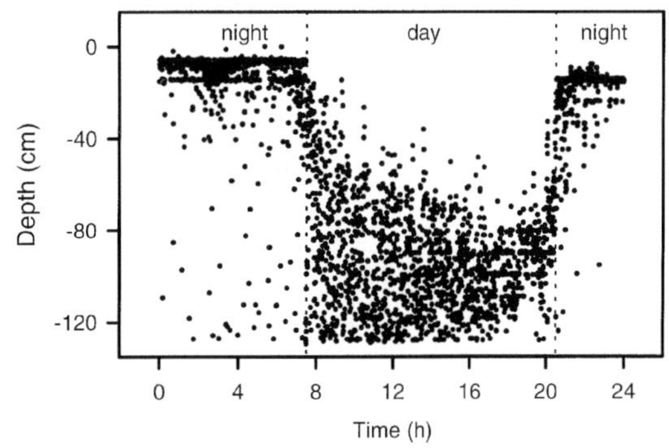


*Figure 3.* Example of the diurnal rhythm of a sea bass. During the night the fish stayed near the surface while it swam deeper during the day. The hatched line indicate light on and off.

*Experimental procedure*

After being tagged, fish were transferred to the experimental tank and allowed to recover and acclimate to the new surroundings for at least three days before any further manipulations. Following this habituation period, the salinity of the upper 60 cm of the water in the tank was decreased by adding pure fresh water in the column exchanging the surface layer. This was continued for 18 to 24 h until a salinity gradient of 5‰ between bottom and surface was obtained.

The halocline was created to minimize mixing of deoxygenated bottom water with oxygenated surface water. However, the pumps and the swimming movements of the fish induced some mixing of the water in the tank and some deoxygenation of the surface water could not be avoided. To maintain the salinity gradient, fresh water was occasionally added to the surface. Deoxygenation of the bottom layer was initiated two days after the salinity gradient had been established. It was carried out by bubbling nitrogen in the column exchanging the water at the bottom. The rate of deoxygenation was approximately 10% day$^{-1}$.

The duration of an experiment was between 2 and 3 weeks and during this period fish were fed daily on a maintenance ration.

## Results

*Swimming activity*

In general, the swimming pattern of the three sea bass which made up each experimental group were synchronized. From visual observations it was clear that fish often swam together, and the depth measurements confirmed these observations (Figure 2).

In homogeneous environmental conditions, sea bass displayed a clear diurnal rhythm in their vertical distribution. At night fish stayed at the surface, while they swam deeper in the water column during the day (Figure 3).

*Effect of salinity on the depth of the animals*

To separate the surface and the bottom water layers, a vertical salinity gradient was established. A comparison of the night and day distributions, before and after fresh water was added to the surface layer, showed that the salinity gradient had no significant influence on the diurnal rhythm of the bass (Mann-Whitney test) (Figure 4).

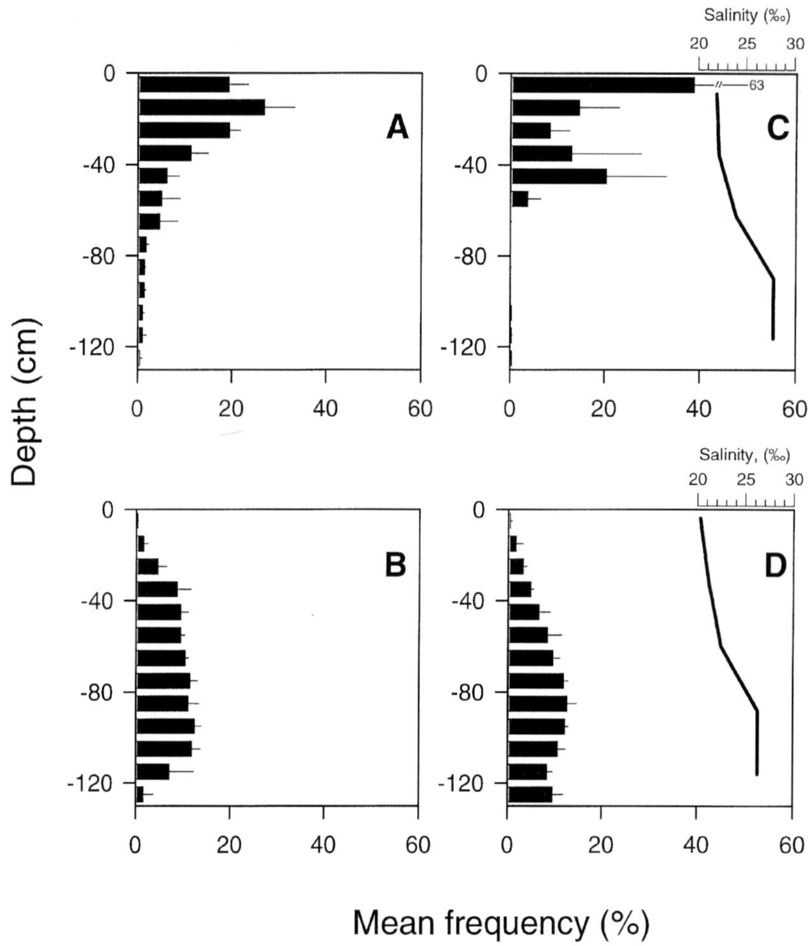


*Figure 4.* The depth distribution of three fish before and after establishment of a vertical salinity gradient. A and B are night and day distributions in a homogeneous environment, respectively. C and D are night and day distributions after the salinity gradient (indicated by the vertical line) was established. There was a significant difference between the night and day both before and after the gradient ($p < 0.01$, Man-Whitney test). A comparison of the night and the day distributions, showed that the salinity gradient had no effect on the diurnal rhythm. Values are mean and line on bars are SD.

### *Effect of oxygen*

The level of oxygen had a strong impact on the vertical distribution and behaviour of the sea bass. As hypoxia developed, an alteration in the daily vertical migration was observed. In contrast to the normoxic trial, fish occupied the oxygenated surface water during the daylight period rather than the deoxygenated bottom water. An example of this is shown in Figure 5. At the end of the experiment, the oxygen level at the bottom was increased again and sea bass returned to their normal vertical migratory behaviour (Figure 5).

## Discussion

The vertical migrations observed in the present investigation were related to photoperiod. Sea bass swam directly to the bottom just after the light was switched on in the morning (Figure 2) and stayed deeper in the water column during the light hours (Figure 3). Furthermore, the vertical swimming activity was higher during the day (Figure 3), which agrees with the result of Bégout Anras et al. (1997), who showed that sea bass in groups have the highest activity level during the day. On the other hand, we found no satisfactory explanation for their presence close to the surface during the night, although the same behaviour was

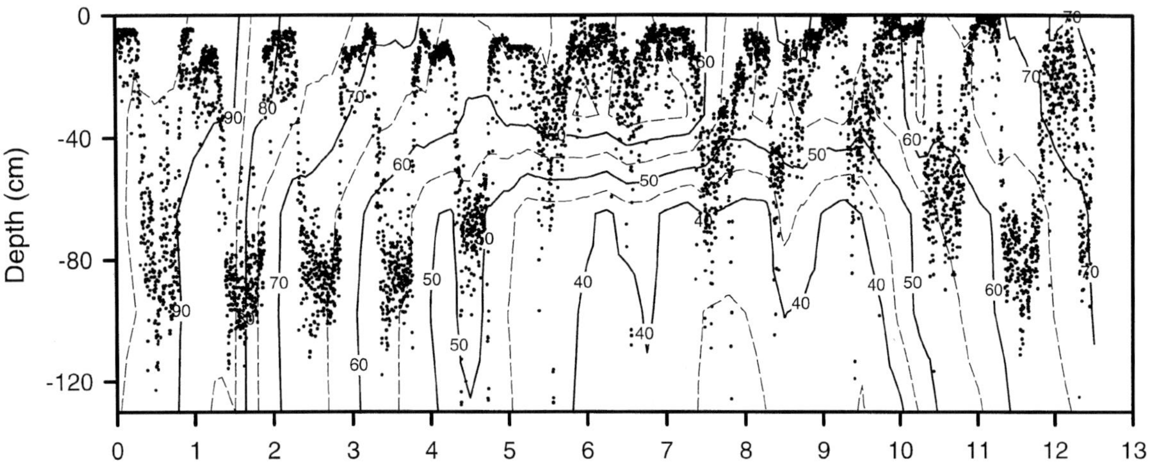


*Figure 5.* Example of the depth occupied by a sea bass during 13 days with an hypoxic episode at the bottom. The water temperature in this example was 20 °C. Establishment of the oxygen gradient was initiated the first day. The isobars indicate the oxygen saturation in %. From day 2 to 4 a gradient was slowly developing and no apparent change in the depth distribution had occurred. On day 5 oxygen saturation at the bottom reached 50%, and during the day the fish was not swimming as deep as the 4 preceding days. On day 6 and 7 the gradient became steeper, and the fish spend most of the time in the surface layer. On day 8, 9 and 10 the fish still avoided the hypoxic water at the bottom, but seemed to be swimming lower than the preceding 2 days. On day 11, 12 and 13 the water was reoxygenated and the fish swam lower during the day again. Note that the sea bass, almost always avoided the water with a oxygen saturation below 40% from day 6 to 10.

observed for the Atlantic cod (*Gadus morhua* L.) in similar conditions (Claireaux et al., 1995).

During these experiments we established a salinity gradient to prevent mixing between the hypoxic bottom and the oxygenated surface layers. Bass of all ages are euryhaline and do tolerate fresh water (Pickett & Pawson, 1994). Accordingly, the salinity gradient had no influence on the diurnal rhythm of the bass

There is a general agreement in the literature that fish avoid areas with low oxygen saturations. This can be concluded both from field studies (e.g. Coble, 1982; Pihl et al., 1991; Suthers & Gee, 1986), and laboratory experiments (e.g., Spoor 1990, Stott & Gross 1973). Disagreement persists, however, concerning how sensitive fish are to differences in the water oxygen saturation. Whitmore et al. (1960) found that fish displayed prompt avoidance reactions when deoxygenated water was encountered. These reactions were rapid and suggested an ability to detect differences in the oxygen saturation of the water. However, this result is rather uncommon, and other investigations have shown that the avoidance reactions are somewhat delayed (Höglund, 1961; Jones, 1952; Stott & Cross, 1973) this being interpreted as the fish do not react to hypoxic water until suffocation occurs. Similar experiments also showed that although fish avoid water with a very low oxygen saturation, they do not necessarily select the water with the highest oxygen availabili-

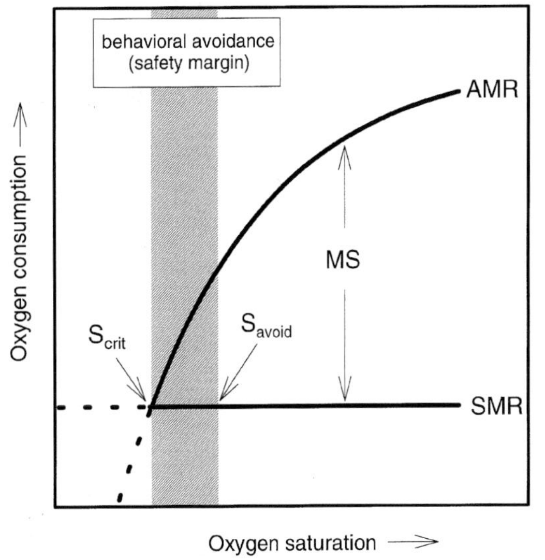


*Figure 6.* Schematic relationship between ambient oxygen saturation and standard (SMR) and active metabolic rate (AMR). The SMR can be maintained until the critical oxygen saturation ($S_{crit}$) while lower oxygen saturations are lethal. The AMR, and thus the metabolic scope (MS), is continuously decreasing with oxygen. In order to maintain some scope, the behavioural avoidance initiates at $S_{avoid}$. The shaded area represent a safety margin for the fish where it can still survive, but the room for metabolic manoeuvring is becoming so small that it avoids these oxygen saturations. At oxygen saturations above $S_{avoid}$, MS is high and the fish behaviour is not dominated by the oxygen saturation.

ty (Stott & Gross, 1973 Spoor, 1990; Claireaux et al., 1995). These latter observations are confirmed in the present investigation. In the presence of a vertical oxygen gradient, the amplitude of the daily vertical migration decreased and the fish stayed closer to the surface and almost always in water with an oxygen saturation above 40% (Figure 5). However, bass did not always select the highest available oxygen saturation (Figure 5). For example, on day 8 and 9 bass were swimming at a depth between 40 and 60 cm during the day, although the water at the surface had an oxygen saturation which was 10 to 20% higher. This suggests that the advantage of swimming higher in the water column was counteracted by the negative effect of light at the surface. A similar antagonism was reported by Suthers & Gee (1986) on perch (*Perca flavescens* L.). In their experiment perch could swim between two compartments, one was covered and contained hypoxic water, and the other, which offered no shelter, contained normoxic water. As long as the oxygen saturation in the hypoxic water remained above 37% perch preferentially occupied the covered, compartment, whereas they moved to the normoxic side at lower oxygen saturations. As mentioned above sea bass avoided water with an oxygen saturation of below 40% ($S_{avoid}$) (Figure 5). At 20 °C the minimum oxygen saturation required to maintain standard metabolic rate (SMR) is approximately 20% ($S_{crit}$) (Claireaux & Lagardère, submitted). While SMR can be maintained at oxygen saturations ranging from 20–100%, the highest possible oxygen consumption (AMR) decreases with oxygen within this range. The difference between AMR and SMR is called the metabolic scope (MS), and this decreases from its maximum in normoxia, to zero at 20% oxygen saturation. This relationship is schematically illustrated in Figure 6. In natural conditions, fish must be able to maintain a metabolic rate above SMR in order to perform basic activities such as feeding, swimming, maturation of gonads, growth and so on. It can be hypothesized that by avoiding water with an oxygen saturation below 40%, the bass ensures that the metabolic scope is sufficient to perform these essential activities. The difference between $S_{avoid}$ and $S_{crit}$ represent a safety margin for the fish. Within this margin the fish can still survive, but its performance is strongly restricted by the low oxygen saturation.

In conclusion the sea bass displayed a clear diurnal rhythm in the depth occurrence. During the night, bass occupied the surface while they swam deeper during the day. This behaviour remained unchanged after the establishment of a vertical salinity gradient. However, when the water at the bottom became hypoxic bass reduced the amplitude of their vertical migration and stayed closer to the surface during the day, presumably to optimize their aerobic metabolic capacity.

## Acknowledgements

This work was supported by a grant from the Danish Natural Science Research Council to HS. The authors wish to thank Yves Descatoire for the illustration of the tank, Marcel Guillaut for keeping the fish in good health and Michel Prineau for technical assistance.

## References

Baden, S. P., L.-O. Loo, L. Pihl & R. Rosenberg, 1990. Effects of eutrophication on the benthic communities including fish: Swedish west coast. Ambio 19: 113–122.

Bégout Anras M. L., J. P. Lagardère & J. Y Lafaye, 1997. Diel activity rhythm of seabass tracked in a natural environment: group effects on swimming patterns and amplitudes. Can. J. Fish. aquat. Sci. 54: 162–168.

Claireaux, G., D. M. Webber, S. R. Kerr & R. G. Boutilier, 1995. Physiology and behaviour of free-swimming Atlantic cod (*Gadus morhua*) facing fluctuating salinity and oxygenation conditions. J. exp. Biol. 198: 61–69.

Claireaux G. & J. P. Lagardère. Influence of temperature, oxygen and salinity on the metabolism of the European sea bass (*Dicentrarchus labrax*). J. Fish. Biol. (submitted)

Claireaux, G & C. Lefrançois, 1998. A method for the external attachment of acoustic tags on roundfish. Hydrobiologia 371/372: 113–116.

Coble D. W., 1982. Fish populations in relation to dissolved oxygen in the Wisconsin river. Trans. am. Fish. Soc. 111: 612–623.

Höglund L. B., 1961. The reactions of fish in concentrations gradients. Rep. Inst. Fresh. wat. Res. Drottningholm 43: 147 pp.

Jones J. R. E., 1952. The reactions of fish to water of low oxygen concentration. J. exp. Biol. 29: 403–415.

Ker S. R. & R. A. Ryder, 1997. The Laurentian Great Lakes experience: a prognosis for the fisheries of Atlantic Canada. Can. J. Fish. aquat. Sci. 54: 1190–1197.

Petersen J. K. & L. Pihl, 1995. Responses to hypoxia of plaice, *Pleuronectes platessa*, and the dab, *Limanda limanda*, in the south-east Kattegat: distribution and growth. Env. Biol. Fish. 43: 311–321.

Pickett G. D. & M. G. Pawson, 1994. Sea Bass, Biology, exploitation and conservation Chapman & Hall, London, 337 pp.

Pihl L., 1994. Changes in the diet of demersal fish due to eutrophication-induced hypoxia in the Kattegat, Sweden. Can. J. Fish. aquat. Sci. 51: 321–336.

Pihl L., S. P. Baden & R. J. Diaz, 1991. Effects of periodic hypoxia on distribution of demersal fish and crustaceans. Mar. Biol. 108: 349–360.

Rydberg, L., L. Edler, S. Floderus & W Granéli, 1990. Interaction between supply of nutrients, primary production, sediment and oxygen consumption in SE Kattegat. Ambio 19: 134–141.

Spoor, W. A., 1990. Distribution of fingerling brook trout, *Salvelinus frontinalis* (Mitchill), in dissolved oxygen concentration gradients. J. Fish. Biol. 36: 363–373.

Stott, B.& D. G. Cross, 1973. The reactions of roach (*Rutilus rutilus* L.) to changes in the concentration of dissolved oxygen and free carbon dioxide in a laboratory channel. Water Res. 7: 793–805.

Suthers I. M. & J. H. Gee, 1986. Role of hypoxia in limiting diel spring and summer distribution of juvenile yellow perch (*Perca flavescens*) in a prairie marsh. Can. J. Fish. aquat. Sci. 43: 1562–1570.

Whitmore C. M., C. E. Warren & P. Doudoroff, 1960. Avoidance reactions of salmonid and centrarchid fishes to low oxygen concentrations. Trans. am. Fish. Soc. 89: 17–26.

*Hydrobiologia* **371/372:** 215–224, 1998.
*J.-P. Lagardère, M.-L. Bégout Anras & G. Claireaux (eds), Advances in Invertebrates and Fish Telemetry.*



# Heart rate telemetry to study environmental influences on fish metabolic expenditure

Christel Lefrançois, Guy Claireaux & Jean-Paul Lagardère
*Centre de Recherche en Écologie Marine et Aquaculture, Place du Séminaire, B.P. 5 F-17137 L'Houmeau, France*
*e-mail: clefranc@ifremer.fr*

*Key words:* fish, heart rate, metabolism, telemetry

## Abstract

Routine metabolic rate (RMR) and heart rate (HR) were measured simultaneously in 15 °C-acclimated and free swimming European sea bass (*Dicentrarchus labrax* L.) in normoxic and hypoxic conditions. The RMR fell within the boundaries previously reported for that species, i.e. the standard metabolic rate and the $LOC_{MO_2}$-curve which describes the relationship between maximum sustainable oxygen consumption ($MO_2$) and water oxygen concentration ($C_wO_2$). Heart rates values distributed below a $LOC_{HR}$-curve which figures the relationship between the maximum HR and $C_wO_2$. A mathematical model describing this relationship is proposed. Using the $LOC_{MO_2}$- and $LOC_{HR}$-curves, the range of the correlated changes of HR and $MO_2$ is delimited. The influence of $C_wO_2$ on these limits is also considered and discussed. Two examples illustrating the limiting effect of low ambient oxygenation level on fish metabolic demand are described.

## Introduction

The measurement of instantaneous metabolic dissipation is an important prerequisite to the establishment of the energy budget of a free swimming fish (Kerr, 1982; Boisclair & Sirois, 1993). To reach that goal, the utilisation of telemetered heart rate (HR) as a correlate to predict metabolic rate ($MO_2$) has been attempted several times (e.g. Priede & Young, 1977; Priede & Tytler, 1977; Priede, 1983; Armstrong, 1986; Lucas et al., 1991, 1993; Claireaux et al., 1995a, 1995b; Sureau, 1995) with mixed success. The prediction of metabolic rate from heart rate in unrestrained fish relies upon preliminary calibrations realised under laboratory conditions. Unfortunately, physiological and environmental conditions affecting the relationship between HR and $MO_2$, calibrations curves commonly obtained have had limited range of application and displayed high interstudy variability. In a recent review, Thorarensen et al. (1996) discussed this variability and suggested that the HR vs $MO_2$ relationship was best described by a family of curves than by a single curve. In this context, the objective of the present paper is, using heart rate telemetry, to establish the range of the reciprocal changes in HR and $MO_2$ in 15 °C-acclimated unrestrained sea bass (*Dicentrarchus labrax* L.). In order to consider all of that range, fish energetic demand was manipulated through stress (chasing), feeding (apparent specific dynamic action) and ambient oxygenation.

## Materials and methods

### *Fish holding*

European sea bass weighing 980 to 1430 g were raised in 400 l indoor rearing tanks supplied with recirculated and filtered natural sea water (salinity: 27–31‰; water renewal rate: 30–50% per week). Fish were subjected to natural photoperiod and temperature conditions, and fed twice a week on commercial dry pellets.

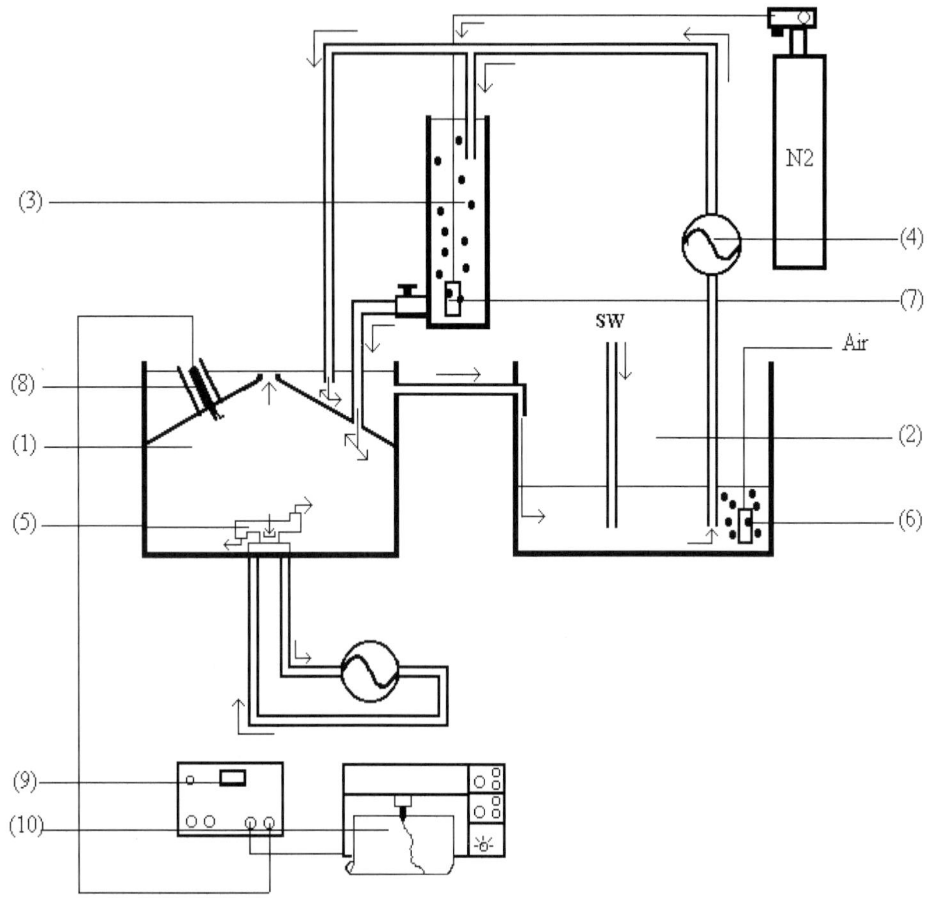


*Figure 1.* Experimental set-up. The respirometer chamber (1) was built from a tank identical to those utilised during the various acclimation procedures. A perspex lid hermetically separated the measuring chamber (240 l) from an upper thermoregulation compartment. Before reaching the chamber, the water pumped (4) from a mixing tank (2) flowed through a counter-current gas equilibration column (3) bubbled with either air, in control condition, or nitrogen to induce hypoxia (6 and 7). A second pump prevented water stratification in the respirometer (5). Oxygen concentration was recorded using a polarographic electrode (8) plugged in an oxygen meter (9) connected to a paper recorder (10).

*Experimental protocol*

Ten days before experiments started, four groups of 3 bass (total weigh 3 kg) were anesthetized (2-phenoxyethanol), weighed and transferred to an acclimation tank situated in a 15 °C-thermoregulated room. In this room, the artificial lighting matched natural hours of daylight and darkness. At the time of the transfer, the temperature difference between the rearing tank and the thermo-acclimation tank was +3.6 °C. Following acclimation, each experimental group followed the same protocol. While two individuals were directly introduced in the respirometer, the third one was fitted with an electrocardiogram (ECG) acoustic tag before joining them in the measuring chamber. Fish were then left undisturbed for another 24 h. The experimental set up is described in Figure 1.

The routine metabolic rate (RMR) of the group and the heart rate (HR) of the tagged individual were monitored in normoxic condition, following a period of disturbance (chasing), while digesting, or during a stepwise decreases in ambient oxygen concentration (decrement 1 $mgO_2$ $l^{-1}$, down to 1.5 $mgO_2$ $l^{-1}$). To measure oxygen consumption, the pump controlling the water supply to the respirometer chamber was shut off and the water oxygen concentration was continuously monitored for the next 10–15 min. The measuring chamber was then washed with air-saturated or hypoxic water (approximately 15 min) before the next measurement was started.

RMR ($mgO_2$ $kg^{-1}$ $h^{-1}$) was calculated using the following formula:

$$\text{RMR} = \Delta C_w O_2 \times \Delta T^{-1} \times \text{VOL}_{\text{resp}} \times M^{-1},$$

where $\Delta C_w O_2$ is the change in water oxygen concentration ($mgO_2\ l^{-1}$), $\Delta T$ is the measuring time (h), $VOL_{resp}$ is the volume of the respirometer minus the volume of fish (l), and $M$ is the mass of fish (kg).

Routine checks of the respirometer showed negligible background oxygen consumption by microorganisms. Feeding was discontinued 24 h before any manipulation of the animals and 48 h ahead of their transfer into the respirometer.

*Tagging procedure*

Anaesthetized bass were placed belly up on a surgery table and their gills perfused with oxygenated water containing a small dose of 2-phenoxyethanol. Animals were then fitted with an acoustic heart rate tag according to the method described by Claireaux & Lefrançois (This volume). ECG electrodes were inserted near the pericardial cavity through the skin and muscles and sutured in place. The tagging procedure was generally completed in less than 15 min. After being ventilated with anesthetic-free water, fish resumed ventilatory activity within 5 min.

*Ultrasonic transmitters*

The ultrasonic transmitters (Vemco VR16-4LR; 10 cm long, 1.6 cm in diameter and 37.5 g in air) were 2.6 to 3.8% of the fish's body weight. The ECG was detected via two differential input electrodes place 1–1.5 cm apart near the pericardial cavity. Heart rate was monitored by recording the pulse rate of the transmitters which was equal to the heart signal rate for signals above a threshold of approximately 120–150 $\mu$V. The ultrasonic receiver (Vemco VR-20) was interfaced with a computer using an RS-232 serial cable.

## Results

Oxygen consumption ($MO_2$) values were distributed within the limits defined by Claireaux & Lagardère (submitted) for 15 °C-acclimated sea bass (Figure 2). These limits are firstly the standard metabolic rate which represents the maintenance metabolic cost of a resting, non maturing and unfed animal, and secondly the limiting oxygen concentration curve ($LOC_{MO_2}$-curve) which describes the relationship between fish's maximum sustainable $MO_2$ and the water oxygenation concentration ($C_w O_2$).

In normoxic conditions, sea bass routine oxygen consumption ranged from 68 $mgO_2\ h^{-1}\ kg^{-1}$ in resting conditions, to 145–150 $mgO_2\ h^{-1}\ kg^{-1}$ during period of disturbance (stress) or post-feeding (peak specific dynamic action). During the development of hypoxia, sea bass $MO_2$ was initially maintained at its original level until the limiting oxygen concentration (LOC) corresponding to that initial metabolic expenditure was reached (i.e. horizontal leftward shift on Figure 2). At this point, any further decrease in $C_w O_2$ resulted in a gradual decrease of fish's routine metabolic expenditure along the $LOC_{MO_2}$-curve. Of course, groups of animals with high metabolic rate were confronted sooner with limiting $C_w O_2$ than those having relatively lower $MO_2$.

In a similar way as for the $LOC_{MO_2}$-curve, we established a $LOC_{HR}$-curve which describes the relationship between fish's maximum heart rate ($HR_{max}$) and ambient oxygen concentration (Figure 3). The general mathematical equation used is the following:

$$HR_{max} = Y_{max} * (1 - e^{(\alpha * C_w O_2 + \beta)}). \quad (1)$$

The model's asymptote ($Y_{max}$) corresponds to the highest heart beat frequency in normoxia and was estimated at 68 beats $min^{-1}$. This estimation was then verified through the analysis of heart beat variability via the construction of HR phase delay maps (Goldberger & Rigney, 1991). During resting periods, the beat to beat variability is important (Figure 4a), suggesting that in such conditions the heart is beating freely at a frequency which is loosely related to metabolic requirements. On the other hand, when oxygen demand rises, heart beat frequency increases and becomes more regular although a certain degree of variability still persists. However, there is an upper limit to heart rate which is illustrated by the gradual change in the shape of the map from a disc to a triangle as HR approached $Y_{max}$ (Figure 4b).

The asymptote of the model being set, the constants $\alpha$ and $\beta$ were then adjusted so that the $LOC_{HR}$-curve fitted the maximum HR observed at each $C_w O_2$. Equation 1 then became:

$$HR_{max} = 68 * (1 - e^{(-0.375 * C_w O_2 + 0.401)}). \quad (2)$$

In normoxic conditions, heart beat frequency ranged from 28.5 to 65 beats $min^{-1}$. During hypoxia, sea bass HR remained approximately constant until the $LOC_{HR}$-curve was approached, and when oxygenation

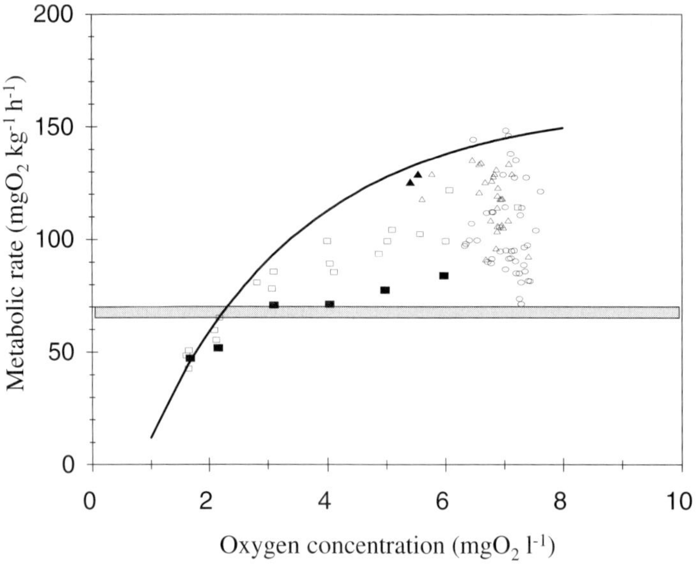


*Figure 2.* Influence of ambient oxygen concentration on the routine metabolic rate of sea bass in normoxia (○), hypoxia (□) and postfeeding (Δ). Closed symbols correspond to the examples presented in Figures 5 and 6. The shaded area indicates the standard metabolic rate (from Claireaux & Lagardère, submitted) and the solid line figures the $LOC_{MO_2}$-curve.

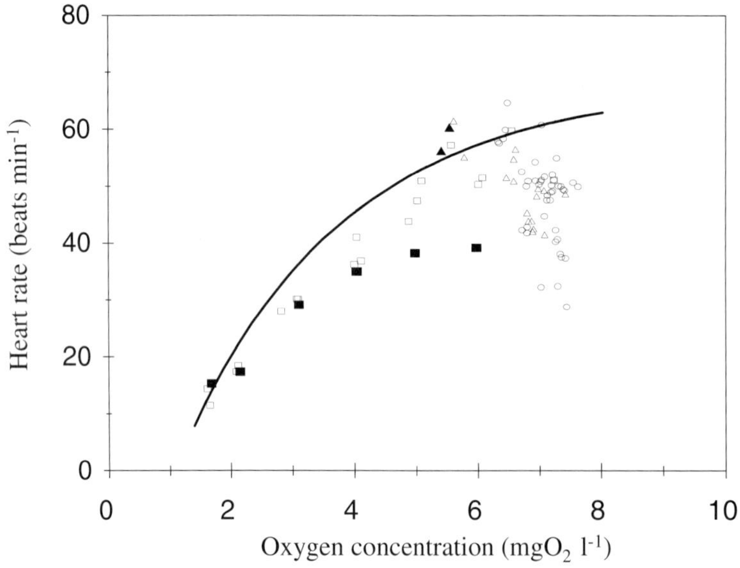


*Figure 3.* Influence of ambient oxygen concentration on the heart rate of sea bass in normoxia (○), hypoxia (□) and postfeeding (Δ). Closed symbols correspond to the examples presented in Figures 5 and 6. The solid line figures the $LOC_{HR}$-curve (equation 2).

conditions worsen further, a gradual bradycardia was then observed. The standard heart rate (SHR), i.e. HR recorded at $C_wO_2 = C_{crit}$, was estimated at 20 beats min$^{-1}$. Here, $C_{crit}$ is considered as the minimal oxygen concentration required for the fish to sustain SMR (2.2 mg l$^{-1}$).

The possible range of heart rate and metabolic rate is constrained by the fundamental characteristics of the organisms and/or by the environmental conditions tested. These limits are illustrated in Figure 5a. The coordinates of point *A* are SHR, SMR and $C_{crit}$. The curves *AB* and *AE* are the $LOC_{MO_2}$- and $LOC_{HR}$-curves respectively. The projection of *AB* and *AE* perpendicular to the planes *YZ* and *XZ* respectively, allows the definition of two new planes (*ABC* and *AEC*) which intersect along the line *AC*. This line defines the

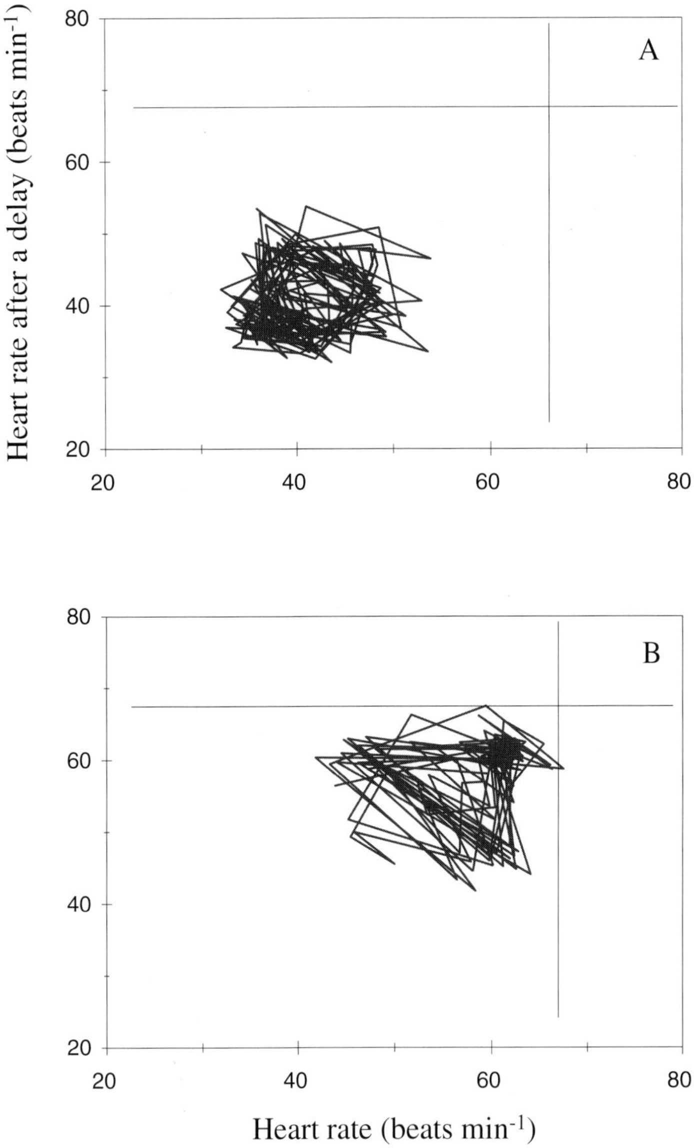


*Figure 4.* Phase space delay map of a the heart rate of a sea bass in resting conditions (A) and postfeeding (B). The delay map shows the trajectories of the heart rate vectors where the first variable is the current heart rate and the second variable is heart rate after a fixed delay of one beat. The solid line indicates the estimated maximum heart rate ($HR_{max}$) used in equation 2. Each map is constructed based on a time series of approximately 5 min.

impact of ambient oxygenation level on the $HR_{max}$ vs $MO_{2max}$ relationship. Thus, the range of the simultaneous changes in HR and $MO_2$ over the whole $C_wO_2$ is figured by the volume *ADCE*. At any given $C_wO_2$, a plane parallel to the plane *XY* can be defined (e.g. *BDE* at air saturation). The intersection of this plane with *ADC* is a line (e.g. *DC* at air saturation) which allows, at the $C_wO_2$ considered, the estimation of $MO_{2max}$ for a given heart rate.

In fully aerated conditions, the HR vs $MO_2$ relationship occurs within the triangle *DCE*. The $MO_{2max}$ of a fish having a heart rate of say 50 beats $min^{-1}$ is given by the projection of *G* onto the plan *YZ* (*I*). Thus, the distance *DI* corresponds to the actual or instantaneous metabolic scope of that animal. If HR is then increased to 60 beats $min^{-1}$, the new metabolic scope is now represented by *DH*. Hence, the lines *IB* and *HB* correspond to the amount of scope remaining and

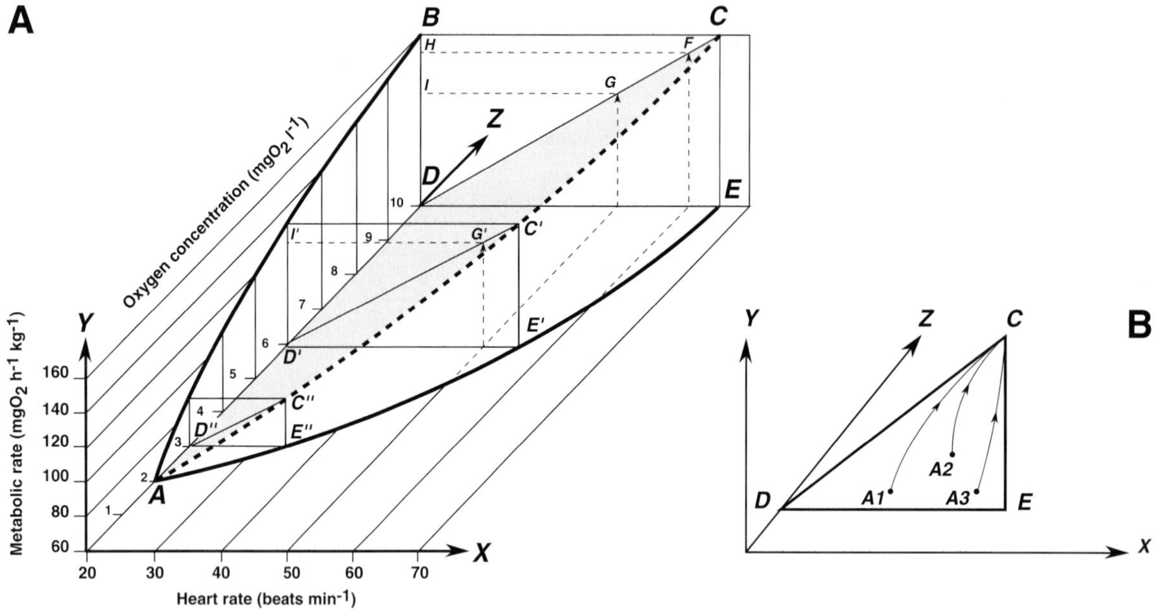


*Figure 5.* A – Relationship between metabolic rate, heart rate and water oxygen concentration. Line *AB*, $LOC_2$-curve; *AE*, $LOC_{HR}$-curve; *CD*, relationship between $Hr_{max}$ and $MO_{2max}$ in normoxic conditions; *CDE* 'triangle of scope'; *D′C′E′* and *D″C″E″*, 'triangles of scope' at decreasing ambient oxygen concentrations. B – Triangle of scope in normoxia where 3 hypothetical situations are presented ($A_1$, $A_2$ and $A_3$). Arrows indicate possible HR vs $MO_2$ relationships corresponding to each situation.

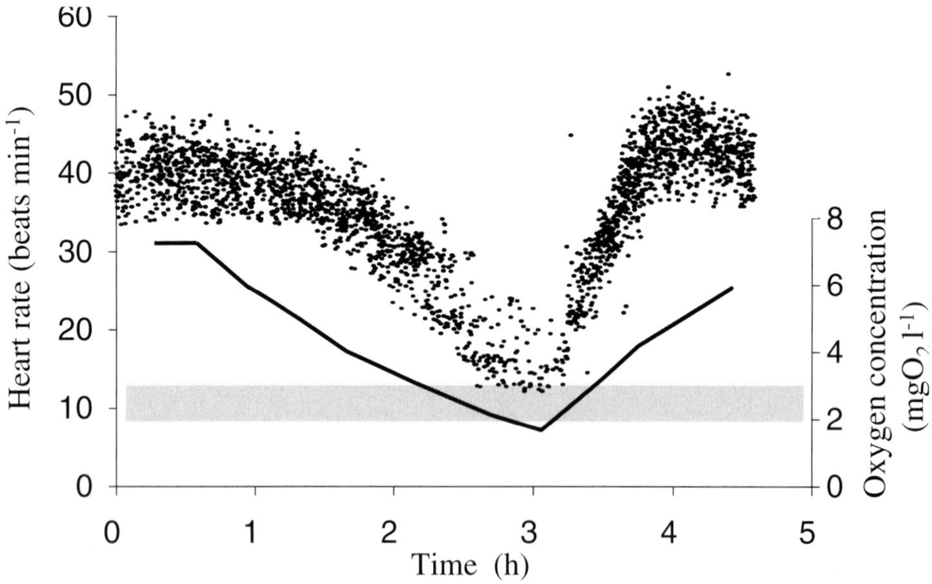


*Figure 6.* Heart rate (dots) and water oxygen concentration (solid line) during hypoxia. This example corresponds to the filled squares in Figures 2 and 3. The bradycardia becomes significant when the water oxygen concentration reaches the limiting oxygen concentration (shaded bar).

thus denote the 'safety margin' still available owing to further tuning of HR.

As $C_wO_2$ decreases, the polygon *DCE* is shifted down the *Z* axis (e.g. $D'C'E'$ and $D''C''E''$). The accompanying decrease in the triangle area and the reduction in the safety margin (e.g. from *IB* to $I'B'$), underlines the constraints exerted by low oxygenation level on the capacity of the animal to regulate its metabolic demand.

A tentative illustration of the correlated changes in HR and $MO_2$ in the neighbourhood of the LOC is given in Figures 6 and 7. However, it must be remembered that due to of our experimental design (mean $MO_2$ of 3 fish / HR of 1 fish), this attempt to correlate directly HR to $MO_2$ obviously needs confirmation through a more appropriate experimental protocol. Nevertheless, these two examples are individualised as filled symbols in Figures 2 and 3.

In normoxia, the mean RMR of the group presented in Figure 6 ranged from 65 to 85 $mgO_2\ h^{-1}\ kg^{-1}$ (filled squares on Figure 2), and the corresponding limiting oxygen concentration thus fell between 2 to 3 $mgO_2\ l^{-1}$. As this critical zone was approached during hypoxia, the member of that group which carried the ECG tag displayed a marked bradycardia.

Oxygen concentration can be limiting not only during hypoxia but also when oxygen demand is high. In the example presented in Figure 7, oxygen concentration was close to air saturation and the metabolic rate of the group was experimentally raised by feeding the 3 fishes to satiation ($\sim 7\%$ body weight). This example has been differentiated as filled triangles on Figure 2 and 3. Due to post-feeding specific dynamic action, metabolic rate peaked at nearly 125 $mgO_2\ h^{-1}\ kg^{-1}$ 3 h after feeding, the corresponding limiting oxygen concentration then reaching approximately 6 $mg\ l^{-1}$ (Figure 2). In these conditions where high oxygen availability is counterbalanced by elevated oxygen demand, any reduction of $C_wO_2$ by as little as 0.5 $mgO_2\ l^{-1}$ imposed limiting oxygenation conditions to the animals and triggered hypoxia-like cardiac responses by the tagged individual.

## Discussion

Claireaux & Lagardère (submitted) modelled the influence of ambient oxygenation and temperature conditions on the aerobic metabolic expenditure of the sea bass. They showed that routine metabolic rates (RMR) distribute within limits which are, the standard metabolic rate (SMR) and the $LOC_{MO_2}$-curve. Data obtained in the present investigation are consistent with the boundaries determined for 15 °C-acclimated animals (Figure 2).

Following stress or during digestion, fish metabolic expenditure temporarily reached 2.2 × SMR, a value which is close to the active metabolic rate (AMR) reported for that species at 15 °C, i.e. 160 mg $O_2\ h^{-1}\ kg^{-1}$ (Claireaux & Lagardère, submitted). Apparent specific dynamic action (SDA) of 2.4 × SMR have been measured in digesting pike (*Esox lucius*) by Armstrong (1986) and this value also falls within the range 1.5–2.5 found for a number of other species (Jobling, 1981).

As water oxygen concentration decreased, sea bass oxygen consumption was not altered until the limiting oxygen concentration (LOC) was reached. This initial preservation of the level of metabolic expenditure can be ascribed to cardiovascular and respiratory adjustments (reviewed by Randall, 1982). However, as oxygen availability dropped below the LOC, the original metabolic rate could no longer be maintained and $MO_2$ decreased together with $C_wO_2$. This relationship between the maximum sustainable $MO_2$ and $C_wO_2$ is described by the $LOC_{MO_2}$-curve (Figure 2). Moreover, the higher the animal's metabolic rate, the higher the LOC and so the more substantial the metabolic adjustments imposed during hypoxia. The point at which the $LOC_{MO_2}$-curve reaches the SMR is referred to as the critical oxygen concentration ($C_{crit}$). In the present investigation, $C_{crit}$ was found to be approximately 2.2 mg $O_2\ l^{-1}$. For comparison, Schurmann & Steffensen (1997) reported a $C_{crit}$ of 2.4 mg $O_2\ l^{-1}$ in 15 °C acclimated Atlantic cod (*Gadus morhua*).

Heart rates were found to distribute below a limit which defined the relationship between the maximum heart rate ($HR_{max}$) and $C_wO_2$ (Figure 3). In normoxia, and using phase delay maps (Goldberger & Rigney, 1991), $HR_{max}$ was confirmed at 68 beats $min^{-1}$ (Figure 4b) while the minimum HR measured at $C_{crit}$ (SHR) was found to be approximately 20 beats $min^{-1}$.

Cardiac activity in sea bass has been very little studied and the high degree of interspecific differences in cardiac activity makes all comparisons hazardous (Farrell, 1993). In the only investigation we found in the literature, the highest heart rate recorded in free swimming bass at temperatures ranging from 8 to 11 °C was 36 beats $min^{-1}$ (Sureau & Lagardère, 1991). Physiological factors such as membrane permeability in ion-pumping activities, conduction of electrical signals and myocardial oxygen supply con-

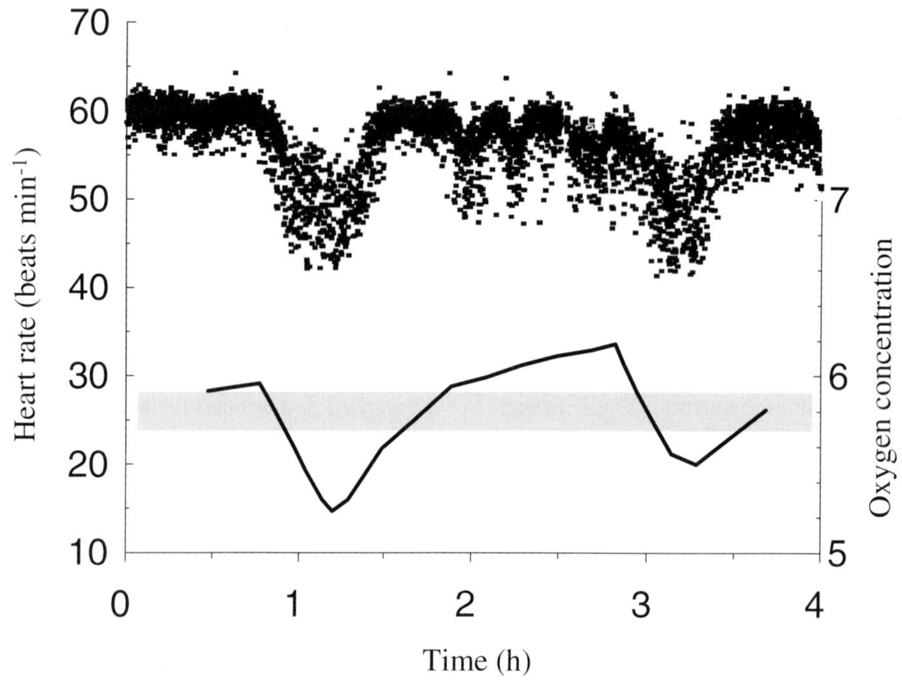


*Figure 7.* Heart rate (dots) and water oxygen concentration (solid line) during digestion in normoxic conditions. This example is figured as filled triangles in Figures 2 and 3. Due to SDA, the limiting oxygen concentration of this group of fish was approximately 6 mg $O_2$ $l^{-1}$ (shaded bar). Each measurements of $MO_2$ and the associated decrease in oxygenation level in the respirometer chamber induced an hypoxia-like bradycardia.

tributes to setting the upper limit for HR. However, the movement of extracellular calcium for the excitation-contraction coupling has been proposed as the major limiting factor of $HR_{max}$ (Farrell & Jones, 1992; Farrell, 1992).

During hypoxia, heart rate was initially maintained, suggesting that reduced oxygen availability was temporarily offset by increased stroke volume (*i.e.* maintenance of cardiac output) and/or by increased difference in arterial-venous blood oxygen content (Holeton & Randall, 1967; Fritsche & Nilsson, 1989). As the LOC was eventually reached, a bradycardia progressively developed and the reduction of HR down the $LOC_{HR}$-curve (Figure 3) was paralleled by a similar decline in fish oxygen demand along the $LOC_{MO_2}$-curve (Figure 2). Similar bradycardia has been reported in numerous species and was observed in tuna swimming through an hypoxic body of water by Bushnell & Brill (1991).

Many authors used HR as a physiological correlate to $MO_2$ (e.g. Priede & Young, 1977; Priede & Tytler, 1977; Priede, 1983; Armstrong, 1986; Lucas et al., 1991, 1993; Claireaux et al., 1995a, 1995b; Sureau, 1995) and their studies have been recently reviewed and discussed by Thorarensen et al. (1996). These authors emphasised the fact that in some cases, fish heart rate can vary over its physiological range with no significant changes in $MO_2$, while in others, $MO_2$ may change although HR remains invariable (see also Randall, 1968). Thorarensen et al., also pointed out that physiological state and environmental factors affecting the relationship between HR and $MO_2$, several curves may exist for a single species.

The inherent difficulties in the use of conventional HR vs $MO_2$ calibration curves have been acknowledge longtime ago by users of heart rate telemetry, and attempts were made to overcome these limitations. Priede & Tytler (1977) proposed a sketching procedure for interpreting heart rate data in terms of metabolic scope. Although our own experimental design was different, Figure 5a is nothing more than a revised version of Priede & Tytler 'quadrangle of scope'. Yet, our representation includes an improvement and a difference. The improvement concerns the third dimension of the graph which allows the interpretation of the influence of $C_wO_2$ on the metabolic scope of tagged animals. The difference relies on the $HR_{max}$ vs $MO_{2max}$ relationship. In the representation given by Priede & Tytler this relationship (line *OEB*; Figure 4) crosses the SMR at HR < SHR, while in our representation the line *CD* crosses SMR at SHR, resulting in a 'triangle of scope'.

The impact of $C_wO_2$ on the framework within which the correlated changes in HR and $MO_2$ take place is noteworthy. As oxygen availability decreases, the line *AC* dips toward point *A* (SHR, SMR, $C_{crit}$) resulting in a decrease of the area of the 'triangle of scope', e.g. from *CDE* to $C'D'E'$ and to $C''D''E''$ at $C_wO_2 = 10$, 6 and 3 mg $l^{-1}$ respectively. Thus, the triangle of scope represents the potentialy achievable metabolic scope as it is set by the environmental conditions (in our case ambiant oxygenation level). On the other hand, the actual or instantaneous metabolic scope is fixed by the amount of oxygen potentially delivered to the tissus, and it is thus linked to the heart beat frequency. In normoxic conditions an increase of HR from 60 beats $min^{-1}$ to $HR_{max}$ ($\approx 65$ beats $min^{-1}$) induces the instant aerobic potential to increase from *DH* to *DB*, the full scope of the fish then being available. Conversely, if HR decreases from say 60 to 50 beats $min^{-1}$, the true scope is then reduced from *DH* to *DI*. Thus, depending on the environmental conditions tested (e.g. Claireaux et al., 1995a and 1995b) but also on the animal's physiological state (i.e. HR vs $MO_2$ starting point) a whole set of HR vs $MO_2$ relationships can be generated. However, this family of curves should theoretically lie within the limits of the triangle of scope (Figure 5b). Furthermore, in situations where fish need to make use of all their scope, these curves should then all be converging on *C* ($HR_{max}$, $MO_{2max}$).

Keeping in mind the limitations resulting from our experimental design, Figures 6 and 7 intend to illustrate the kinetics of the mutual alterations in HR and $MO_2$ in the vicinity of the plane *ADC*. The first example shows the impact of low oxygenation level on a group of fish which had relatively small metabolic requirements (Figure 6). As $C_wO_2$ was reduced, no alterations in either HR or $MO_2$ were observed until the LOC was reached (in this case 2–3 $mgO_2$ $l^{-1}$; Figure 2). On the other hand, when fish had to make use of all their metabolic scope (e.g. during digestion), they became more dependent upon oxygen availability (Figure 7). In this example, although water oxygen saturation was between 75 and 80%, each small drop in $C_wO_2$ associated with the measurements of $MO_2$, triggered metabolic adjustments which were revealed by episodic bradycardia. In both examples, ventilatory and cardiovascular regulation mechanisms were at full stroke and the relation between HR and $MO_2$ became then predictable as it corresponded to the $MO_{2max}$ vs $HR_{max}$ relationship (line *AC*).

In conclusion, by establishing the scatter range of $MO_2$ and HR in free moving bass, the present investigation confirmed the views expressed by various authors (e.g. Randall, 1968; Priede & Tytler, 1977; Thorarensen et al., 1996) concerning the difficulty of using heart beat frequency for predicting metabolic expenditure. However, our work also confirmed that heart rate telemetry was still a powerful tool, allowing the study of the dynamic of the links between heart rate and metabolic rate and the analysis of the environmental constraints on the bioenergetic framework (metabolic scope) within which fish activities take place. In the short run, this approach should allow to test the energetic value of the behavioural adaptative responses observed in fish following alteration of the environmental conditions (e.g. Schurmann et al., this volume).

## Acknowledgements

The authors are indebted to Dr H. Schurmann for his helpful comments on earlier drafts on this manuscript. We also wish to thank M. Guillaut and M. Prineau for their skilful technical help as well as Y. Descatoire for designing some of the graphs (merci les gars!).

## References

Armstrong, J. D., 1986. Heart rate as an indicator of activity, metabolic rate, food intake and digestion in pike, *Esox lucius*. J. Fish Biol. 29: 207–221.

Boisclair, D. & P. Sirois, 1993. Testing assumptions of fish bioenergetics models by direct estimation of growth, consumption, and activity rates. Trans. am. Fish. Soc. 122: 784–796.

Bushnell, P. G. & R. W. Brill, 1991. Responses of swimming skipjack (*Katsuwonus pelamis*) and yellowfin (*Thunnus albacares*) tunas to acute hypoxia, and a model of their cardiorespiratory function. Physiol. Zool. 64: 787–811.

Claireaux, G. & J. P. Lagardère. Influence of temperature, oxygen and salinity on the metabolism of the European sea bass (*Dicentrarchus labrax*). J. Fish. Biol., submitted.

Claireaux, G. & C. Lefrançois, 1998. A method for the external attachment of acoustic tags on roundfish. Hydrobiologia 371-372: 113–116.

Claireaux, G., D. M. Webber, S. R. Kerr & R. G. Boutilier, 1995a. Physiology and behaviour of free swimming Atlantic cod (*Gadus morhua*) facing fluctuating temperature conditions. J. exp. Biol. 198: 49–60.

Claireaux, G., D. M. Webber, S. R. Kerr & R. G. Boutilier, 1995b. Physiology and behaviour of free swimming Atlantic cod (*Gadus morhua*) facing fluctuating salinity and oxygenation conditions. J. exp. Biol. 198: 61–69.

Farrell, A. P., 1992. Cardiac output in fish: regulation and limitations. In Bicudo, J. E. P. W. (ed.), The Vertebrate Gas Transport Cascade, CRC Press, Bocca Raton FL, USA: 208–214.

Farrell, A. P. & D. R Jones, 1992. The heart. In Hoar, W. S., D. J. Randall & A. P. Farrell (eds), Fish Physiology vol XII A, Academic Press, San Diego CA, USA: 1–88.

Fritsche, R. & S. Nilsson, 1989. Cardiovascular responses to hypoxia in the Atlantic cod *Gadus morhua*. exp. Biol. 48: 153–160.

Goldberger, A. L. & D. R. Rigney, 1991. Nonlinear dynamic at the bedside. In Glass, L., P. Hunter & A. McCulloch (eds), Theory of Heart. Springer-Verlag, New York, Berlin, Heidelberg, London, Paris, Tokyo, Hong Kong, Barcelona: 583–605.

Holeton, G. F. & D. J. Randall, 1967. The effect of hypoxia upon partial pressure of gases in the blood and water afferent and efferent to the gills of rainbow trout. J. exp. Biol. 46: 317–327.

Jobling, M., 1981. The influence of feeding on the metabolic rate of fishes: a short review. J. Fish Biol. 18: 385–400.

Kerr, S. R., 1982. Estimating the energy budgets of actively predatory fishes. Can. J. Fish. aquat. Sci. 39: 371–379.

Lucas, M. C., A. D. F. Johnston & I. G. Priede, 1993. Use of physiological telemetry as a method of estimating metabolism of fish in the natural environment. Trans. am. Fish. Soc. 122: 822–833.

Lucas, M. C., I. G. Priede, J. D. Armstrong, A. N. Z. Gindy & L. De Vera, 1991. Direct measurements of metabolic activity and feeding behaviour of pike, *Esox lucius* L., in the wild, by the use of heart rate telemetry. J. Fish Biol. 39: 325–345.

Priede, I. G., 1983. Heart rate telemetry from fish in the natural environment. Comp. Biochem. Physiol. 76: 515–524.

Priede I. G. & P. Tytler, 1977. Heart rate as a measure of metabolic rate in teleost fishes; *Salmo gairdneri*, *Salmo trutta* and *Gadus morhua*. J. Fish Biol. 10: 231–242.

Priede, I. G. & A. H. Young, 1977. The ultrasonic telemetry of cardiac rhythms of wild brown trout (*Salmo trutta* L.) as an indicator of bioenergetics and behaviour. J. Fish Biol. 10: 299–318.

Randall, D. J., 1968. Functional morphology of the heart in fishes. Am. Zool. 8: 179–189.

Randall, D. J., 1982. The control of respiration and circulation in fish during exercise and hypoxia. J. exp. Biol. 100: 275–288.

Schurmann, H. & J. F. Steffensen, 1997. Effects of temperature, hypoxia and activity on the metabolism of juvenile Atlantic cod. J. Fish Biol. 50: 1166–1180.

Schurmann, H., G. Claireaux & H. Chartois, 1998. Change in vertical distribution of sea bass (*Dicentrarchus labrax* L.) during a hypoxic exposure. Hydrobiologia 371/372: 207–213.

Sureau, D., 1995. Télémétrie en milieu maritime, analyse du rythme cardiaque comme indicateur de la relation entre le poisson et son environnement naturel. Etude chez la sole (*Solea solea*) et le bar (*Dicentrarchus labrax*). PhD thesis, University of Poitiers (France): 201 pp.

Sureau, D. & J. P. Lagardère, 1991. Coupling of heart rate and locomotor activity in sole *Solea solea* (L.), and bass, *Dicentrarchus labrax* (L.), in their natural environment by using ultrasonic telemetry. J. Fish Biol. 38: 399–405.

Thorarensen, H., P. E. Gallaugher & A. P. Farell, 1996. The limitations of heart rate as a predictor of metabolic rate in fish. J. Fish Biol. 49: 226–236.

*Hydrobiologia* **371/372:** 225–232, 1998.
*J.-P. Lagardère, M.-L. Bégout Anras & G. Claireaux (eds), Advances in Invertebrates and Fish Telemetry.*



# Effects of thermal environment on Electromyographical signals obtained from Atlantic salmon (*Salmo salar* L.) during forced swimming

T. A. Beddow & R. S. McKinley
*Waterloo Biotelemetry Institute, Dept. of Biology, University of Waterloo, Waterloo, Ontario, N2L 3G1, Canada*

*Key words: Salmo salar*, electromyogram, EMG, temperature, biotelemetry, swimming

## Abstract

Hatchery-reared Atlantic salmon (*Salmo salar*) were acclimated to 8 °C and 18 °C for 1 month prior to surgical implantation of telemetry devices that measured electromyographical (EMG) impulses. Transmitters were 5 cm long, 1.6 cm diameter and weighed 18 g in air. The teflon coated, stainless steel bipolar electrodes were secured in place using gold tips (9 carat, 5 × 1 mm), that acted as anchors in the locomotory musculature. The electrodes were placed in the red muscle band, half way down the body length of the fish, approximately 5–10 mm apart. Transmitters emitted a signal when the electrical difference between the 2 electrodes exceeded a factory determined threshold, thus, EMG pulse interval (ms) was related to the rate of muscle contraction. Fish were maintained at their respective acclimation temperatures and EMG signals were monitored over a 4-month period during critical swimming speed tests ($U_{crit}$'s). Mortality rates were higher in the 18 °C acclimated group (50% after 3 months), compared to the 8 °C group (0%), probably due to the lack of feeding in the former group. No tags were expelled during this period. Critical swim speed did not change over time or with temperature. EMG pulse intervals were correlated to swimming speed, and became significantly shorter (indicating greater activity) at higher speeds. The relationship of 6 out of 7, 8 °C-acclimated fish, compared to only 1 out of 3, 18 °C-acclimated salmon, remained constant over the 4-month period. These results suggest that care needs to be taken when extrapolating calibration data from the laboratory to the field over long periods of time, especially if changes in the muscle properties of the fish are likely to occur.

## Introduction

Recent advances in the field of biotelemetry have come about due to the successful miniaturization of electronic devices capable of emitting signals, pertinent to physiological processes. To date, physiological parameters such as heart rate (Priede, 1983; Sureau & Lagardere, 1991), ventilation rate (Rogers & Weatherley, 1983), tailbeat frequency (Ross et al., 1981) and locomotory muscle activity (Kaseloo et al., 1992 & 1996; Weatherley et al., 1982) have been successfully used as activity indicators in free swimming fish. Most of the focus so far, has centered on estimating the latter parameter as locomotory activity in fish can be related to quantitative estimates of metabolic rate (Weatherley et al., 1982).

Fish swim by the sequential activation of the axial musculature, which is brought about by waves of electrical excitation traveling caudally down the length of the fish (Jayne & Lauder, 1995). These waves of electrical activity are known as electromyograms (EMG's), and it is these electrical impulses that can be measured and transmitted via telemetry devices (Weatherley et al., 1982). Increasingly, the EMG telemetry devices have been utilized in the field to determine activity levels of free swimming fish during daily and seasonal movements (Demers et al., 1996), spawning behaviour (Økland et al., 1997; Weatherley et al., 1996) and during migration through areas of difficult passage (Hinch et al., 1996). Commonly, fish are swum at known speeds in the laboratory and EMG values are determined at each speed. Locomotory activity can also be correlated to oxygen consumption in a respirometer and hence, can be used to obtain quantitative estimates of metabolic rate of free ranging animals in the wild (Briggs & Post, 1997; McKinley & Power, 1992; Weatherley et al., 1982). Few studies have, however, examined the long term effects of the transmitters on the fishes swimming capabilities or how the relationship between EMG signal and fish swim speed may change over time. Kaseloo et al. (1992) monitored telemetered EMG's of rainbow trout, *Oncorhynchus mykiss* Walbaum, over a 2 month period (at swimming speeds of 30–60 cm $s^{-1}$), and found EMG signals during this period did vary, however, transmitters had no effect on growth, swimming behaviour or mortality.

Properties of axial muscle are known to be highly temperature dependent with 2–3-fold decreases in power output and maximum shortening speeds being observed for a drop in temperature from 20 °C to 10 °C (Rome, 1983). Some fish species are known to compensate for this reduction in muscle power output at low temperatures by recruiting more fibres and faster fibre types at lower swimming speeds, termed 'Compression of Recruitment Order' (Rome et al., 1984). Thus, for a fish swimming at a fixed speed, more fibres would be expected to be active at cold- relative to warm-temperatures. The objectives of this study were firstly, to examine long term effects of transmitter implantation on fishes swimming ability, secondly, to assess the changes in signal quality over the long term and, thirdly, to examine the effects of thermal acclimation on the signals obtained from hatchery reared Atlantic salmon.

## Materials and methods

Hatchery reared Atlantic salmon (*Salmo salar* L.) were obtained in April 1996, from the Ontario Ministry of Agriculture and Fisheries (OMAF), research station, Alma, Ontario, where they had been raised at a constant temperature of 8 °C under ambient light conditions. These salmon were large, immature postsmolts (S3), maintained throughout their life in freshwater. Fish were transported to the University of Waterloo's freshwater aquarium, where they were held at 8 °C in 3 m diameter circular tanks prior to experimentation. Fresh, aerated well water was circulated through the tanks in a flow through system and photoperiod was held constant (12 h light: 12 h dark). On 1st August 1996, half the fish ($n = 20$) were moved to a 2 m wide square tank, 0.75 m deep, where the temperature was increased by 0.5 °C per day until a constant temperature of 18 ± 0.5 °C was reached. Fish were acclimated to this temperature for at least 4 weeks. Both groups of fish were fed daily (commercial pellets) but the 18 °C acclimated group declined food throughout the test period.

### *Transmitters and surgical procedure*

Radio telemetry devices used for measuring EMG's were similar to those described by Booth et al. (1995), except that the transmitters could not measure temperature. Transmitters were cylindrical 5 cm long, 1.6 cm diameter and weighed 18 g in air and 8 g in water (Lotek Engineering Inc., Newmarket, Ontario). Muscle activity was detected via 9 carat gold sensing tips (0.5 × 0.1 cm), connected to teflon coated, surgical steel electrodes. A radio pulse was transmitted when the integrated electrical activity exceeded a factory determined threshold. This radio pulse interval (ms), thus, became shorter with increased electrical activity in the muscle. Radio pulses were detected and recorded automatically using a radio receiver (SRX400, Lotek Engineering Inc., Newmarket, Ontario) and downloaded to a computer via a RS232 serial communication port. The receiver was programmed to record 30 signals every 3 hours from each fish on a daily basis and to record continually for an individual fish undergoing a swim trial.

Transmitters were surgically implanted into eight fish from each group on the 23rd–25th September 1996, details of weights and lengths are given in Table 1. Fish were anesthetized in a fresh oxygenated solution of MS-222 (100 mg $l^{-1}$, pH 7), placed on a surgical table, ventral side up, supported on a bed of foam. Freshly oxygenated anesthetic was constantly circulated over the gills. A 2–3 cm incision was made posterior to the pelvic girdle and the transmitter carefully inserted forwards into the peritoneal cavity. Specially designed, stainless steel rods (21 gauge) were used to insert the gold tips of the electrodes into the red muscle band of the fish via the incision. Special care was taken not to damage any internal organs during this process. Electrode tips were inserted approximately 5–10 mm apart, at 0.5 body lengths. The antenna (45 cm long) was exited posterior to the incision using a 16 gauge hypodermic needle and the incision closed using 3–4 separate surgical sutures (3–0 Polydioxanone, Ethicon). The procedure took less than 10 min. Fish were allowed to recover and

*Table 1.* Initial weight and fork length of the Atlantic salmon used in this study

| Fish number | 8 °C acclimation group | | Fish number | 18 °C acclimation group | |
|---|---|---|---|---|---|
| | Weight (g) | Length (cm) | | Weight (g) | Length (cm) |
| 1 | 1156 | 47.5 | 9 | 1629 | 55.0 |
| 2 | 1471 | 49.5 | 10 | 1568 | 54.0 |
| 3 | 757 | 44.0 | 11 | 921 | 47.0 |
| 4 | 737 | 44.5 | 12 | 908 | 43.0 |
| 5 | 803 | 40.5 | 13 | 657 | 39.0 |
| 6 | 1144 | 47.0 | 14 | 507 | 36.0 |
| 7 | 1178 | 51.0 | 15 | 622 | 38.0 |
| 8 | 942 | 42.0 | 16 | 973 | 42.5 |
| Mean ± STD | 1024 ± 258 | 46.0 ± 3.6 | | 973 ± 419 | 44.0 ± 7.1 |

placed back in the tanks at their respective acclimation temperatures.

*Swim trials*

Forced swim trials were conducted on radio-tagged and control (untagged) fish at monthly intervals following transmitter implantation. Trials were conducted in a Blazka-type swim speed chamber (Booth et al., 1995). The chamber was supplied with well water from the same source as the aquarium and maintained at the respective acclimation temperatures (±0.5 °C) using a recirculating heater/cooler unit. Fish were acclimated to the chamber for 30 min at a water velocity of 0.46 m $s^{-1}$ (approx. 1 body length $s^{-1}$) and the velocity was increased by 0.17 m $s^{-1}$ at 10 min intervals. Velocity was increased until fish fatigued and could no longer swim against the current. Time to fatigue was noted and critical swim speed ($U_{crit}$) was calculated using the formula developed by Brett (1964). Fish were anesthetized following each swim trial, weighed and measured.

*Statistical analysis*

Linear regressions were carried out on EMG pulse interval versus swimming velocity. Only signals transmitted during the middle 2 min of the 10 min velocity increment, when fish adjusted to the change in water velocity and were swimming steadily were used for this analysis. Analysis of covariance (ANCOVA) was used to test differences in linear regression relationships over time, for individuals & temperature groups. Paired (for changes in individuals) and unpaired (for differences between groups) *t*-tests were also used to look at changes in the weight, condition factor and $U_{crit}$'s over time.

## Results

Mortality rates were higher in the 18 °C acclimated group (50% after 3 months), compared to the 8 °C group (0%), probably due to the unanticipated lack of feeding in the former group. Weight and condition factor (Figure 1a & b) did not significantly differ between the two temperature groups, however, weight and condition factor decreased between the $2^{nd}$ & $3^{rd}$ months in the 8 °C-group (paired *t*-tests, $P < 0.05$), and progressively over time in 18 °C-acclimated group (paired *t*-tests, $P < 0.05$).

Postmortem dissections of all the electrode positions showed the tips were 7 mm apart in 87.5% of cases with 1 electrode always being found in the red muscle and the second generally being found either in the red or intermediate muscle fibre types. In one fish (no: 7), electrodes were 11.5 mm apart, and a second fish, the signal changed abruptly after the $2^{nd}$ month when the antenna caught in the tank and ripped open the previously healed incision, presumably dislodging the electrodes (> 30 mm apart). Despite the injury, the fish recovered and the wound rehealed within 2 weeks. No correlation was found between distance between electrodes and signal strength. In two fish, one electrode tip pierced the skin and was exposed to the environment creating a potential site of infection, these signals were discarded. One of the tags malfunctioned for unknown reasons in the 8 °C group within

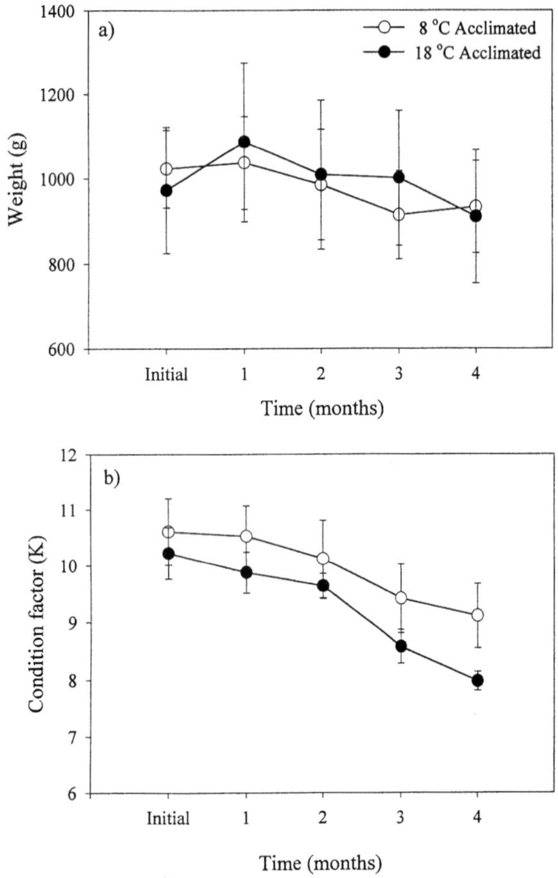


*Figure 1.* a) Weight and b) Condition factor, $K = (\text{weight (kg)}/\text{length}^3\ (\text{m}))$ of the radio-tagged Atlantic salmon over the 4 month period (values are mean ± S.E.).

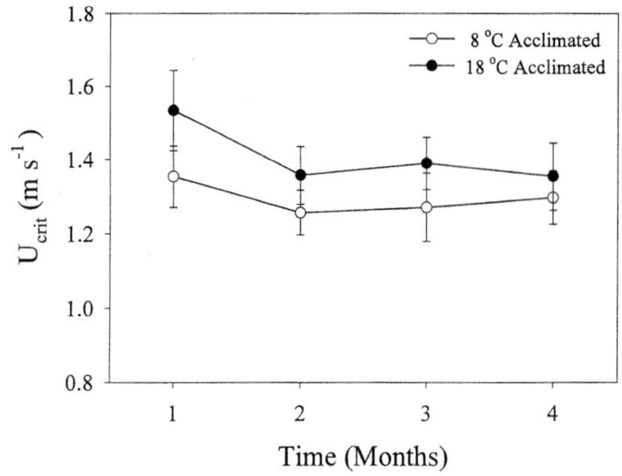


*Figure 2.* Critical swimming speed ($U_{crit}$'s) of Atlantic salmon acclimated to 8 °C & 18 °C over a 4-month period (values are mean ± S.E.).

the first week of implantation probably due to the electrodes touching. At postmortem all tags showed some degree of encapsulation by connective tissue, though no tags were expelled at any point during the test period. Just prior to the battery life of the tags expiring, signals became briefly erratic and then stopped within a 3 hour period, suggesting that the integrity of the data is maintained until near the end of transmitter life.

*Swim performance*

Swimming performance was unaffected by the presence of the radio transmitter over the 4 month period ($P > 0.05$), with no differences found between tagged and untagged individuals. Maximum sustainable swimming speeds were generally higher in the 18 °C- compared to the 8 °C-acclimated salmon, though no significant differences were found between groups (Figure 2). Critical swim speeds were $1.35 \pm 0.23$ m $s^{-1}$ and $1.53 \pm 0.25$ m $s^{-1}$ (mean ± STD) at 1 month decreasing slightly to $1.29 \pm 0.16$ m $s^{-1}$ and $1.36 \pm 0.18$ m $s^{-1}$ at 4 months for 8 °C- and 18 °C-acclimated salmon respectively.

*Individual variation over time*

EMG pulse intervals were plotted against swimming velocity for each swim test and linear regression relationships determined for each individual, during each monthly swim trial (Figures 3 & 4). Data were rejected from swim tests where the fish rested against the backboard or did not swim steadily. In 6 out of 7, 8 °C adapted salmon, no differences were found in the linear regression relationships between EMG pulse interval and swimming velocity over the 4 month period (Figure 3; ANCOVA $P > 0.05$). The relationship did vary over time in one 8 °C acclimated fish (no: 6) being displaced vertically during the 2nd month relative to other months; It was noted at this time that this fish showed visible signs of infection in the antenna region, possibly accounting for the reduced swim performance noted during month 2 ($U_{crit} = 1.04$ m $s^{-1}$). Of the three 18 °C acclimated fish whose transmitters lasted for the entire 4 month period, only 1 set of transmitted signals (no: 12) did not significantly vary with time (Figure 4). Signals from a fourth 18 °C fish that survived the entire test period, did not vary between months 1 and 2, but electrodes were pulled out of position during the 3rd month and therefore the signals from months 3 and 4 were discarded.

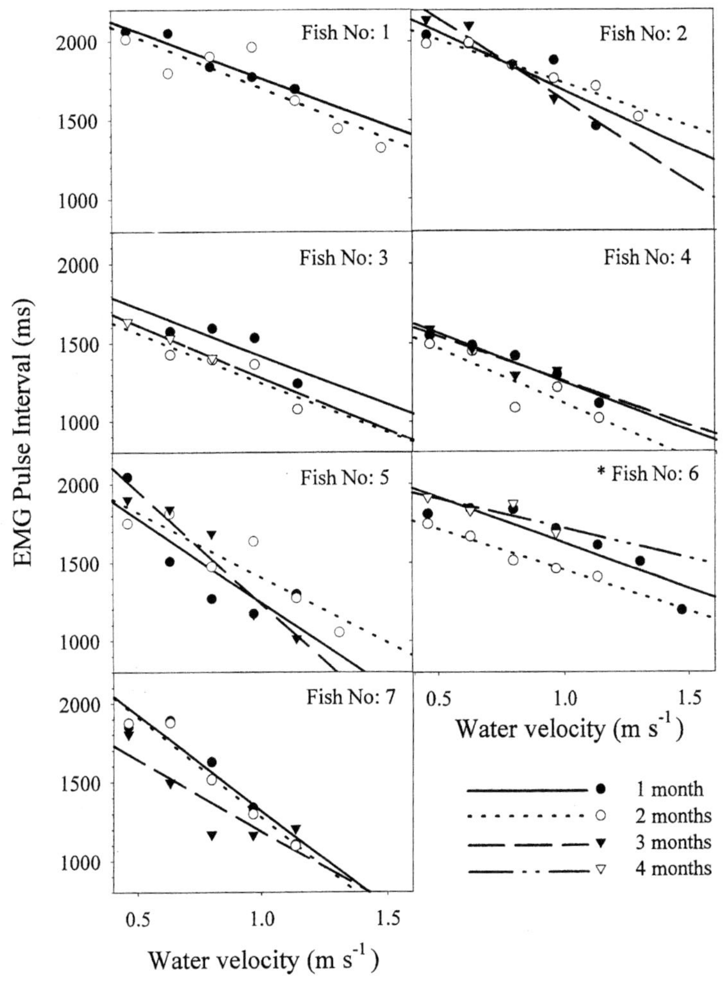


*Figure 3.* EMG pulse interval vs. swimming velocity relation for each individual fish, during each separate swim trial over a 4 month period for 8 °C-acclimated salmon (values are means for 2 min). * Represents individuals whose relationship significantly differed over the 4 month period.

*Variation between individuals*

EMG activity data were pooled for each individual and linear regression relationships were determined for each fish, as a mean for the entire study period, $r^2$ values varied from 0.49–0.83 for the combined data. Relationships of individuals differed in terms of intercept and slope within each temperature group (ANCOVA $P < 0.05$).

*Comparison of temperature groups*

Monthly data were pooled for individuals within each test group, in order to investigate the overall effects of temperature over time (Figure 5). EMG activity regression lines for 8 °C fish did not significantly vary over the test period (ANCOVA $P > 0.05$; $r^2 = 0.33$–0.62), however, regression lines for the 18 °C fish did vary significantly over the 4 months ($P < 0.001$; $r^2 = 0.24$–0.46) becoming less steep and more elevated over time.

Comparing regression lines of the 8 °C- and 18 °C-acclimation groups indicated no difference during the first 2 months, however, by the 3rd month significant changes in the relationships were apparent with less muscle activity being observed in the warm acclimated group during months 3 and 4.

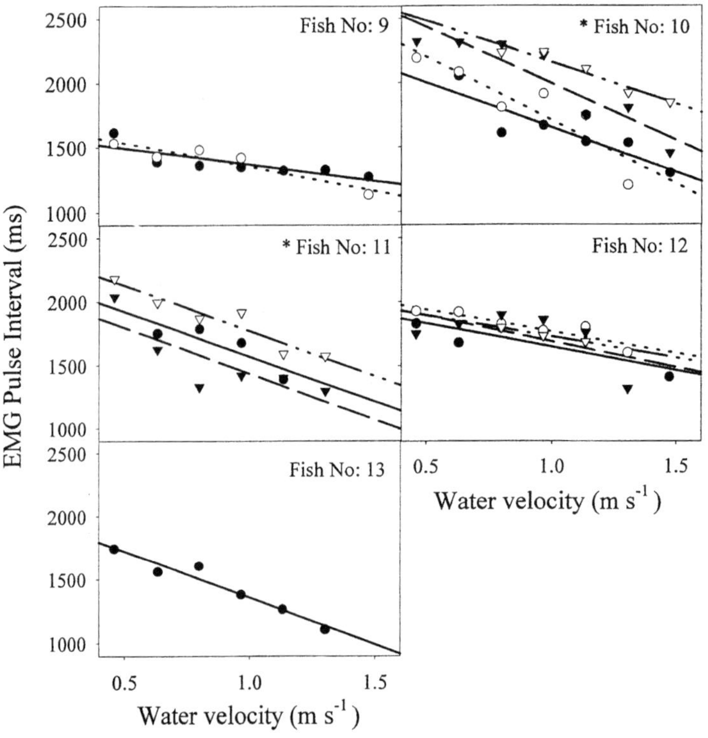


*Figure 4.* EMG pulse interval vs. swimming velocity relation for each individual fish, during each separate swim trial over a 4 month period for 18 °C-acclimated salmon (values are means for 2 min). * represents individuals whose relationship significantly differed over the 4 month period.

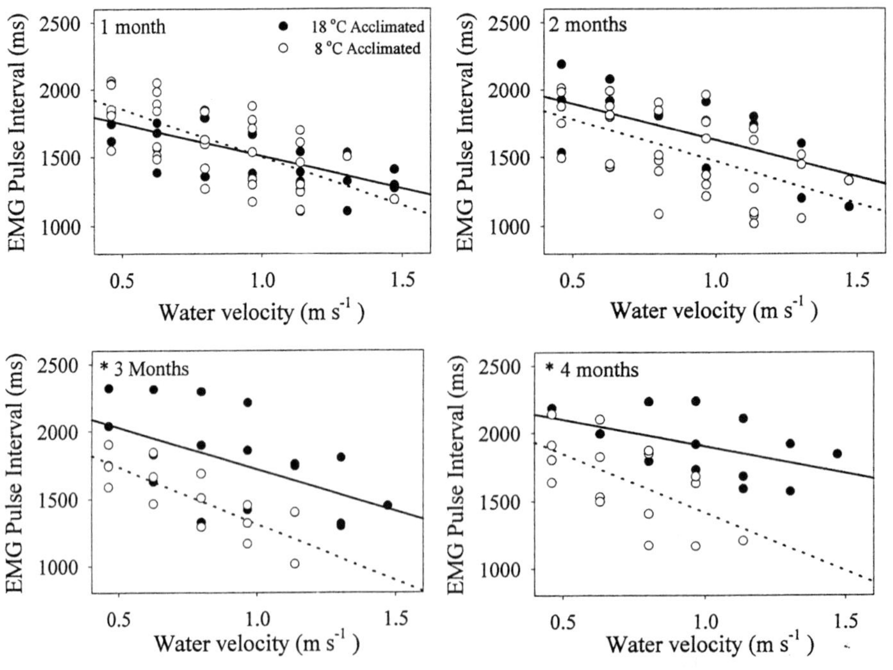


*Figure 5.* Effects of temperature on EMG pulse interval vs. swimming velocity relationship over the 4 month period. * represents significant differences between temperature groups.

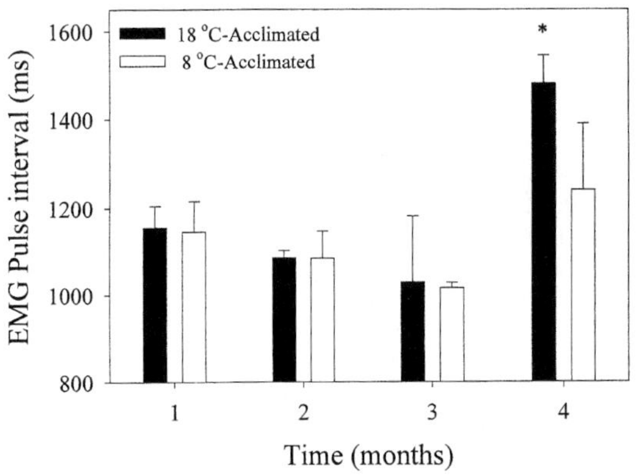


*Figure 6.* EMG pulse interval measured during the last 20 sec of exhaustive exercise in Atlantic salmon (mean ± S.E.). * Indicates significant difference between pulse interval of the 18 °C-acclimated salmon at 4 months compared to 2 months.

*Exhaustive exercise*

EMG pulse intervals were collected for the last 20 sec of forced swimming, just prior to fatigue (Figure 6), when all the muscle fibres are assumed to be active. The intervals did not differ significantly between the temperature groups or over time, except, by month 4, the pulse intervals were significantly longer in 18 °C fish compared to those obtained during month 2, indicating reduced muscle activity at this time ($P < 0.05$). Over the first 3 months burst swimming speeds were indicated by mean EMG pulse intervals of 1094 ± 30 ms or less, and exhaustion was indicated when these intervals were maintained for more than 20 seconds (mean ± S.E.).

## Discussion

Swimming performance was not significantly altered by the presence of the tag over the four month period, indicating that tag implantation would not deleteriously affect the locomotory capabilities of salmon this size. This result is important as it is implicit in many biotelemetry studies that attachment of the transmitter does not affect behaviour or performance of the individual tested. Booth et al. (1995) found similar results for wild Atlantic salmon.

Lack of differences in the performance of salmon, swimming at different temperatures was unexpected considering the thermal dependency of many muscle properties (reviewed by Johnston, 1993), although it is possible that, as response to temperature is a bell shaped curve, the optimum temperature for these fish may fall between the two test temperatures used here. Alternatively, the results suggest that some mechanism of thermal compensation may occur following a period of cold acclimation (Rome et al., 1985) that enables cold acclimated fish to increase their performance relative to acutely cooled fish. Little work, to date, has been done on effects of temperature acclimation on Atlantic salmon. However, recent studies suggest that protein concentrations of red and white muscle increase in cold acclimated salmon as does the oxidative potential of the white muscle fibres (E. Bombardier pers. com.). The steady change in the signal obtained from 18 °C-acclimated fish could also arise from the decreasing protein levels and increased water content of muscle caused by long term starvation (Heap et al., 1986).

Comparisons with a study on swimming performance and EMG activity in wild Atlantic salmon (Booth et al., 1997) indicates differences may exist in performance levels between wild and hatchery groups, with the latter having lower swimming capabilities. The relationship between temperature, swimming performance and EMG activity also differed in wild compared to hatchery salmon, being significantly higher in fish tested at 18 °C compared to 12 °C (Booth et al., 1997). However, these results may in fact be a reflection of muscle content or condition of these fish with the swim tests being conducted in July (18 °C) and September (12 °C) during the spawning migration (Booth et al., 1997). During spawning migrations Atlantic salmon consume no food and the energy required to swim, mature and spawn is derived from stored energy reserves (Jonssen et al., 1991).

The validity of using laboratory calibration procedures to relate EMG pulse interval to swimming speed appears to be dependent on several key factors. Signal viability was maintained in the 8 °C-acclimated salmon, which were feeding normally throughout the test period, in the majority of fish (86%). However, in the 18 °C-acclimated group, which did not feed throughout the experiment, the EMG-swim speed relationship changed significantly with time in all but 1 fish. These data suggest that care should be taken when extrapolating results from the laboratory to the field over long periods of time, and that changes in condition of the fish are of prime consideration when designing biotelemetry studies where the muscle properties are likely to alter.

## References

Booth, R. K., R. S. McKinley, F. Økland & M. M. Sisak, 1995. In situ measurement of swimming performance of wild Atlantic salmon (*Salmo salar*) using radio transmitted electromyogram signals. Aquat. Living Resour. 10: 213–219.

Booth, R. K., D. A. Scruton, R. Goosney & R. S. McKinley, 1995. In C. Amlaner, C. Christalli & J. Neuman (eds), Biotelemetry XIII. Measurement of Red Muscle Activity and Oxygen Consumption in Wild Atlantic Salmon (*Salmo salar*) in Relation to Swimming Speed Using Radio Transmitted EMG Signals. Conf. Proc. Maryland, USA, March 27–31, 1995: 209–215.

Brett, J. R., 1964. The respiratory metabolism and swimming performance of young sockeye salmon. J. Fish. Res. Bd Can. 21: 1183–1226.

Briggs, C. T. & J. R. Post, 1997. Field metabolic rates of rainbow trout estimated using electromyogram telemetry. J. Fish Biol. 51: 807–823.

Demers, E., R. S. McKinley, A. H. Weatherley & D. J. McQueen, 1996. Activity Patterns of Largemouth and Smallmouth Bass Determined with Electromyogram Biotelemetry. Trans. am. Fish. Soc. 125: 434–439.

Heap, S. P., P. W. Watt & G. Goldspink, 1986. Myofibrillar ATPase activity in the carp *Cyprinus carpio*: Interactions between starvation and environmental temperature. J. exp. Biol. 123: 373–382.

Hinch, S. G., R. E. Diewert, T. J. Lissimore, A. M. J. Prince, M. C. Healey & M. A. Henderson, 1996. Use of electromyogram telemetry to assess difficult passage areas for river-migrating adult sockeye salmon. Trans. am. Fish. Soc. 125: 253–260.

Jayne, B. & G. Lauder, 1995. Red muscle motor patterns during steady swimming in the largemouth bass: Effects of speed and correlations with axial kinematics. J. exp. Biol. 198: 1575–1587.

Johnston, I. A., 1993. Phenotypic plasticity of fish muscle to temperature change. In C. Rankin & F. B. Jensen (eds), Fish Ecophysiology. Chapman & Hall, London: 322–340.

Jonssen, N., B. Jonssen & L. P. Hansen, 1991. Energetic costs of spawning in male and female Atlantic salmon (*Salmo salar*). J. Fish Biol. 39: 739–744.

Kaseloo, P. A., A. H. Weatherley, P. E. Ihssen, D. A. Anstey & M. D. Gare, 1996. Electromyograms from radiotelemetry as indicators of reproductive activity in lake trout. J. Fish Biol. 48: 664–674.

Kaseloo, P. A., A. H. Weatherley, J. Lotimer & M. D. Farina, 1992. A biotelemetry system recording fish activity. J. Fish Biol. 40: 165–179.

McKinley, R. S. & G. Power, 1992. Measurement of activity and oxygen consumption for the adult lake sturgeon (*Acipenser fulvescens*) in the wild using radio-transmitted EMG signals. In I. G. Priede & S. M. Swift (eds), Wildlife Telemetry. New York: Ellis-Horwood: 456–465.

Økland, F., E. B. Thorstad, R. S. McKinley, B. Finstad & R. K. Booth, 1997. Radio transmitted electromyogram (EMG) signals as indicators of physical activity in Atlantic salmon (*Salmo salar*). J. Fish Biol. 51: 476–488.

Priede, I. G., 1983. Heart rate telemetry from fish in the natural environment. Comp. Biochem. Physiol. 76A: 515–524.

Rogers, S. C. & A. H. Weatherley, 1983. The use of opercular muscle electromyography as an indicator of the metabolic cost of fish activity in rainbow trout, *Salmo gairdneri* Richardson, as determined by radiotelemetry. J. Fish Biol. 23: 535–547.

Rome, L. C., 1983. The effect of long-term exposure to different temperatures on the mechanical performance of frog muscle. Physiol. Zool. 56: 33–40.

Rome, L. C., P. T. Loughna & G. Goldspink, 1984. Muscle fiber activity in carp as a function of swimming speed and muscle temperature. Am. J. Physiol. 247: 272–279.

Rome, L. C., P. T. Loughna & G. Goldspink, 1985. Temperature acclimation: improved sustained swimming performance in carp at low temperature. Science 228: 194–196.

Ross, L. G., W. Watts & A. H. Young, 1981. An ultrasonic biotelemetry system for the continuous monitoring of tail-beat rate from free swimming fish. J. Fish. Biol. 18: 479–490.

Sureau, D. & J. P. Lagardère, 1991. Coupling of heart rate and locomotor activity in sole, *Solea solea* (L.) and bass, *Dicentrachus labrax* (L.) in their natural environment by using ultrasonic telemetry. J. Fish Biol. 38: 399–405.

Weatherley, A. H., S. C. Rogers, D. G. Pincock & J. R. Patch, 1982. Oxygen consumption of active rainbow trout, *Salmo gairdneri* Richardson, derived from electromyograms obtained by radiotelemetry. J. Fish Biol. 20: 479–489.

Weatherley, A. H., P. A. Kaseloo, M. D. Gare, J. M. Gunn & B. Lipicnik, 1996. Field activity of lake trout during the reproductive period monitored by electromyogram radiotelemetry. J. Fish. Biol. 48: 675–685.

*Hydrobiologia* **371/372:** 233–240, 1998.
*J.-P. Lagardère, M.-L. Bégout Anras & G. Claireaux (eds), Advances in Invertebrates and Fish Telemetry.*



# Remote monitoring of heart rate as a measure of recovery in angled Atlantic salmon, *Salmo salar* (L.)

W. G. Anderson[1], R. Booth[1], T. A. Beddow[1], R. S. McKinley[1*], B. Finstad[2], F. Økland[2] & D. Scruton[3]

[1] *Waterloo Biotelemetry Institute, University of Waterloo, Waterloo, Ontario, Canada, N2L 3G1*
[2] *Norwegian Institute for Nature Research, Trondheim, Norway*
[3] *Department of Fisheries and Oceans, St. John's, Newfoundland, Canada*
* *Author to whom correspondence should be sent: E-mail: rsmckinle@sciborg.uwaterloo.ca*

*Key words:* catch and release, heart rate, telemetry, angle, recovery

## Abstract

The introduction of 'Catch and Release' fishery programs are now widely employed by fisheries managers in most Atlantic Provinces, primarily due to the recent decline of Atlantic salmon stocks on the east coast of Canada. However, there is still considerable debate among special interest groups and regulators as to the effectiveness of the technique. Heart rate telemetry has been utilized as a tool for the assessment of metabolic rate in wild fish by a number of investigators, and was employed in the present study in order to assess recovery following staged angling events in Atlantic salmon. Wild Atlantic salmon were successfully angled at $20 \pm 2$ °C and $16.5 \pm 1$ °C at Noel Paul's Brook, Newfoundland. In addition, hatchery reared Atlantic salmon were angled at the Ontario Ministry of Agriculture and Fisheries Research Station, Alma, Ontario, at a temperature of $8 \pm 1$ °C. Survival rate for the angled salmon was 20% at $20 \pm 2$ °C; 100% at $16.5 \pm 1$ °C; and 100% at $8 \pm 1$ °C. Mean resting heart rate for the fish angled at 16.5 °C and 20 °C was approximately 1.6 and 1.8 times greater than that of fish angled at 8 °C. Heart rate, post angling, was found to increase 1.2 fold in the 8 °C group, 1.3 fold in the 16.5 °C group and approximately 1.15 fold in the 20 °C group. Time to recovery was assessed as a return to observed resting heart rate for each individual fish and was found to be similar for both the 8 °C and 16.5 °C angled groups (approximately 16 h). Although heart rate telemetry in fish is, perhaps, not an ideal measure of metabolic rate, the present study has demonstrated that remote monitoring of heart rate is a good indicator of post exercise physiological activity.

## Introduction

Recent declines in the numbers of Atlantic salmon on the east coast of Canada have prompted federal and provincial governments to impose strict regulations on commercial and recreational fishing activities. One measure employed was the introduction of year round 'Catch and Release' fishery programs that are now widely used throughout most Atlantic Provinces. Although many fishery managers now practice catch and release, there is still considerable debate among special interest groups and regulators as to the effectiveness of the technique. Differences in opinion are largely focused on the limited knowledge available regarding post release survival rates and biological effects of angling on Atlantic salmon. As the importance in using catch and release programs as a management tool grows, particularly in Atlantic salmon sports fisheries, there is a greater need to understand how angling and releasing wild fish influences the recovery of the species.

Previous studies involving measurements of acid-base and metabolic disturbances have greatly assisted in our current understanding of the effects of angling on salmon physiology (Booth et al., 1995; Wilkie et al., 1996). These studies have validated laboratory findings concerned with acid-base and metabolic disturbances in exhaustively exercised fish, such as glycogen store depletion and increases in muscular lactate levels (Wood et al., 1983; Milligan & Wood, 1986). However, they have also raised important management questions with regards to optimal times for the implementation of catch and release programs, with temperature appearing to exert a major influence on the survivability of angled and released salmon.

The development of physiological telemetry techniques has enabled researchers to investigate and corroborate results previously reported in laboratory based studies. Using telemetry devices which measure physiological processes such as EMG activity (Kaseloo et al., 1992), and tailbeat frequency (Johnstone et al., 1992) researchers have attempted to correlate these physiological processes with metabolic rate and thus an estimation of energy consumption and fish activity. However, both these processes require the fish to swim and after exhaustive exercise, such as an angling event, the fish will typically remain motionless until such time when oxygen debt is repaid.

Heart rate telemetry has been utilized as a tool for the assessment of metabolic rate in wild fish by a number of investigators (Priede & Tytler, 1977; Armstrong, 1986; Lucas et al., 1991), with the correlation proving to be good in northern pike, *Esox lucius*, (Armstrong, 1986) but not as close in more aerobically active fish such as rainbow trout (Priede & Tytler, 1977). Although Thorarensen et al. (1996) recently demonstrated the drawbacks in using heart rate in fish as the sole indicator of metabolic rate, at the present moment, remote assessment of heart rate is by far the best indicator of physiological activity post exercise.

The present study utilised heart rate telemetry as a means of obtaining *in situ* measurements of activity during angling of Atlantic salmon, *Salmo salar*, and monitoring recovery following release.

## Materials and methods

Staged angling events were carried out on wild Atlantic salmon grilse, at Noel Paul's Brook on the Exploits river system in Newfoundland in August of 1996. Ambient water temperature was found to vary greatly during the study period with a maximum temperature of 22 °C at the beginning, dropping to 16 °C towards the end of the study period. A total of 20 fish were used in the study in Newfoundland. Due to the nature of the physiological parameter being measured (see below) not all fish were found to exhibit a constant obtainable signal. However, a total of 10 fish were implanted with heart rate tags (Lotek Eng. Newmarket, Ont.) at Noel Paul's Brook and successfully angled. These fish were grouped into two main categories due to the varying temperature at the time of testing; 5 fish (3 males and 2 females) successfully angled at $20 \pm 2$ °C (mean weight $1723 \pm 282$ g, mean fork length $58.8 \pm 4.1$ cm), and 5 fish (2 females 3 males) successfully angled at $16.5 \pm 1$ °C (mean weight $1295.6 \pm 79.6$ g, mean fork length $54.4 \pm 1.2$ cm). Staged angling events for hatchery reared Atlantic salmon were carried out at the Ontario Ministry of Agriculture and Fisheries Research station in Alma, Ontario in September 1996. Temperature at the fish hatchery during the study period was $8 \pm 1$ °C. A total of 6 fish (3 females, 3 males) were angled at the hatchery (mean weight $2888 \pm 185$ g, mean fork length $61.33 \pm 1.1$ cm).

Due to the nature of the pool where the wild fish were angled, the potential was there for these fish to escape into the Exploits river system. Consequently, an anaesthesia that would produce sufficient calmness for a 10–15 min surgical operation and did not demand an extensive withdrawal period post-release, such as MS-222, was required. The natural anaesthetic Clove oil (Anderson et al., 1997), at a concentration of 40 mg $l^{-1}$, was used to induce anaesthesia. Once a sufficient depth of anaesthesia was deemed to have been reached the fish were transferred to a foam bed where a fresh solution of well aerated clove oil (30 mg $l^{-1}$) was irrigated across the gills for the duration of surgery. The heart rate transmitters used in the present study were 18 g in air (8 g in water) and $5 \times 1.6$ cm (LOTEK Eng. Newmarket, Ontario). Each tag was cleaned in ethanol prior to insertion, allowed to airdry, and gold tips were attached to the end of each electrode to aid in electrical contact and anchorage within the pericardial cavity.

A small incision was made on the ventral surface of the fish (approx. 2 cm) and the body of the tag was inserted within the abdominal cavity with the trailing antenna brought out through a small puncture wound on the ventro-lateral side of the fish. The electrode wires were brought through the anterior end of the wound, subcutaneously, towards the pectoral fins. The gold tips were placed on specially designed surgical probes and positioned within the pericardial cavity through small holes in the lepidotrichs bone of the pectoral girdle (Harder, 1975). Due to the nature of the physiological process being measured, constant contact with the heart was necessary to ensure a consistent signal from the transmitter. As a consequence, those fish which gave a constant signal during the operating procedure did not always provide a constant signal once recovered, particularly during periods of high activity such as angling. In an effort to reduce this, the electrode wires leading from the pericardial cavity to the abdominal cavity were tightly secured onto the ventral surface of the fish, with enough slack to allow for free movement of the electrode wires with the movement of the fish. An autopsy was also carried out on each individual post experiment to ensure the electrodes were still positioned appropriately. The abdominal incision was closed with 4–6 sutures and a button tag was sutured anterior to the dorsal fin for future identification. The fish was then removed from the operating table and placed in a recovery tank (2 m × 1 m × 0.5 m) filled with fresh aerated river water at ambient temperature. All fish were allowed a minimum of three days in the recovery tank prior to the angling event. Resting heart rate was recorded between 48 and 72 h post surgery.

The angling events in Noel Paul's Brook were carried out in a controlled channel (25 × 5 × 1 m) at Noel Paul's Brook adjacent to the station. For each individual angling event, fish were removed from the recovery tank, placed in a cooler of fresh aerated water, transferred to the channel, manually hooked (#8) in the upper jaw and released. The fish were then angled to exhaustion on an 1.8 m graphite fly rod using 7 weight line and a 10 weight tippet. Exhaustion was deemed to have been reached when the fish no longer responded to the angler. Upon exhaustion the fish were released from the hook and immediately released into the angling pool. Heart rate of these fish was recorded as soon as the fish was manually hooked and released into the channel using a SRX_400 telemetry receiver fitted with W20 firmware (LOTEK Eng. Newmarket Ontario). Post angling heart rate was continually recorded for a minimum of 3 h and up to 16 h where possible. Hatchery reared fish at the Alma Research Station were operated on, hooked and angled in a similar fashion as those in Newfoundland except that the hatchery reared fish were angled in a large 10 m diameter by 1.5 m deep circular tank.

## Results

Analysis of the heart rate data has concentrated on the angling period and the following recovery period (Figures 1, 2 and 3). Each value presented in figures 1, 2 and 3 is a mean $\pm$ 1 SE and represents a minimum sample size of 4. Although the 16.5 °C fish were angled for an average of 5.1 min, sufficient telemetry data was not obtainable due to the intermittency of signal from the heart rate transmitters (particularly during periods of high activity) and also the low angling times in some of the fish. This also applies to the 20 and 8 °C groups, where angling times were an average of 11.8 and 8.1 min, respectively.

The mean resting heart rate of fish angled at $8 \pm 1$ °C was $40 \pm 0.28$ beats per minute (bpm), and ranged from $34.2 \pm 0.5$ bpm to $48.2 \pm 0.5$ bpm. Mean resting values for the fish angled at $16.5 \pm 1$ °C was $66.9 \pm 0.5$ bpm with a minimum of $54.8 \pm 0.4$ and a maximum of $88.8 \pm 1.3$ bpm. Mean resting values for the fish angled at $20 \pm 2$ °C was $72.3 \pm 0.4$ bpm, and ranged from $53.4 \pm 1.81$ bpm to $87 \pm 1$ bpm (Figure 1).

Mean peak heart rate in the 8 °C fish post angling was $48.77 \pm 0.7$ bpm which occurred 2.5 h after the event (Figure 2). Mean peak heart rate values in the $16.5 \pm 1$ °C group was $90.68 \pm 1.4$ bpm 4 hrs after angling (Figure 2). Heart rate in both the 16.5 and 8 °C groups did not return to basal for as much as 16 h (Figure 2). Due to the high mortality rates experienced in the 20 °C group sufficient numbers for the recording of heart rate beyond 3 h was not obtainable. Mean peak heart rate values in the $20 \pm 2$ °C group was $83.54 \pm 0.17$ bpm 35 min after angling (Figure 3). Furthermore, heart rate of these fish during the three hour observation period appeared to be very unpredictable.

One of the most striking initial findings of the present study was the mortality rate of fish post angling. Of the five fish angled at $20 \pm 2$ °C, only one fish survived the following 72 hour post angling period. All the wild and hatchery reared salmon angled at $16.5 \pm 1$ °C and $8 \pm 1$ °C respectively survived the event and were in good condition when removed for autopsy following the 72 h observation period. Interestingly, exhaustive states at 20, 16.5 °C, and 8 °C were found to be visually very different. The 20 and 16.5 °C fish at the end of the angling period were found to be extremely docile and in some cases unable to maintain equilibrium immediately post release from the hook. However, the 8 °C fish, although exhausted to the extent that handling was not difficult, were still able to maintain equilibrium and in some cases hold station within the circular tank. These visual observations are reflected in the heart rate telemetry data, whereby the heart rate between 1 and 4 min of angling in the 16.5 °C group decreased from basal by a factor of 1.4 (Figure 1). A similar decrease (factor of

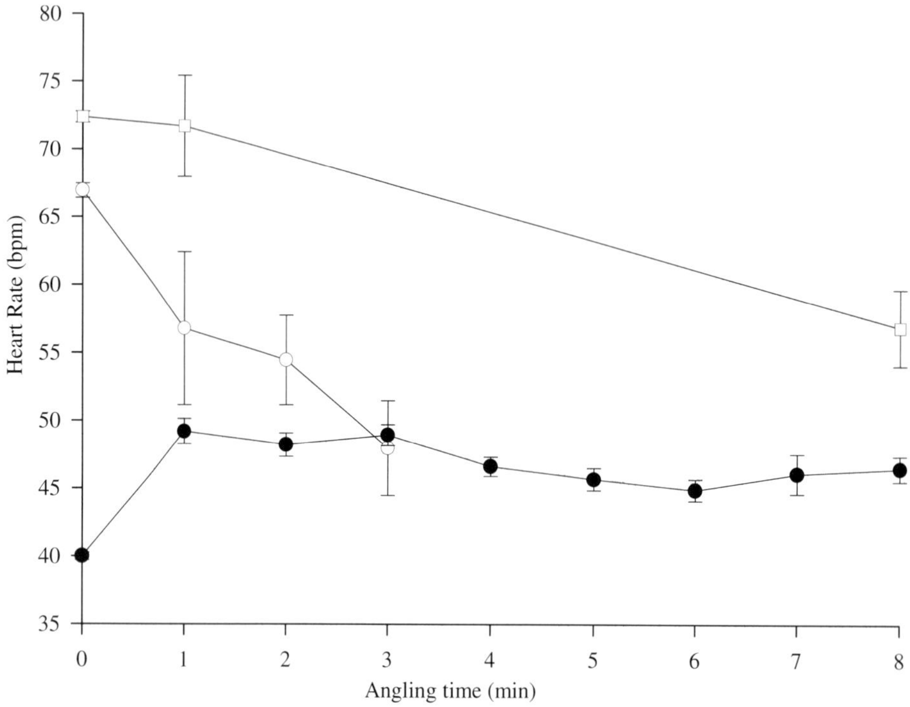


*Figure 1.* Change in heart rate of adult Atlantic salmon during angling at 3 distinct temperatures. Open circles = 16.5 ± 1 °C; Open squares = 20 ± 2 °C; Closed circles = 8 ± 1 °C. Values are expressed as a mean ± 1 SE.

1.3) from 1–8 min of angling in the 20 °C group was observed. However, heart rate increased by a factor of 1.16 over an 8 minute angling period of the 8 °C group (Figure 1).

## Discussion

Results from the present study have demonstrated an increase in resting values with an increase in temperature (Figure 1). Resting values of fish angled at 8 °C and 16.5 °C increased by a factor of approximately 1.6, which increased further to a factor of 1.8 from 8 to 20 °C. This is perhaps not surprising as heart rate in fish has been shown to be closely related to changes in temperature (Farrell & Jones, 1992) with the degree of vagal tone perhaps being a major factor with changes in temperature (Priede, 1974). Although, Gamperl et al., (1994) demonstrated levels of catecholamines and cortisol to remain elevated for up to six days post-operation, the present study used only a 48 h period prior to the collection of basal heart rate data. This was considered to be sufficient as previous studies using heart rate telemetry used a minimum of 24 h post surgery before the onset of experimentation (Claireaux et al., 1995; Lucas, 1994; Nelson et al., 1996). Furthermore, the results reported by Gamperl et al. (1994) were from fish confined in 'black box' experiments. The fish in the present study were allowed to recover in an large (2 m × 1 m × 0.5 m) recovery tank.

Examination of heart rate during angling has demonstrated a decrease in heart rate of the wild fish angled at 20 and 16.5 °C but an increase in heart rate of the hatchery reared 8 °C group. Visual observations during angling demonstrated that the wild fish fought very well with the angler and would often use anaerobic bursts in swimming in an effort to escape from the line. The hatchery reared fish on the other hand did not demonstrate a similar willingness to fight to the extent that one fish had to be dragged through the water to encourage it to struggle. Burst exercise in fish has been reported to induce a bradycardia (decrease in heart rate) (Stevens et al., 1972; Farrell, 1982). This depression is considered to be necessary to avoid hypertension as the violent contractions of skeletal muscle close off the peripheral blood vessels (Farrell & Jones, 1992). It is possible, therefore, that the depression in heart rate in the wild fish is due to the intense activity in these fish in comparison to the hatchery reared fish at 8 °C fish. Whether the reduction

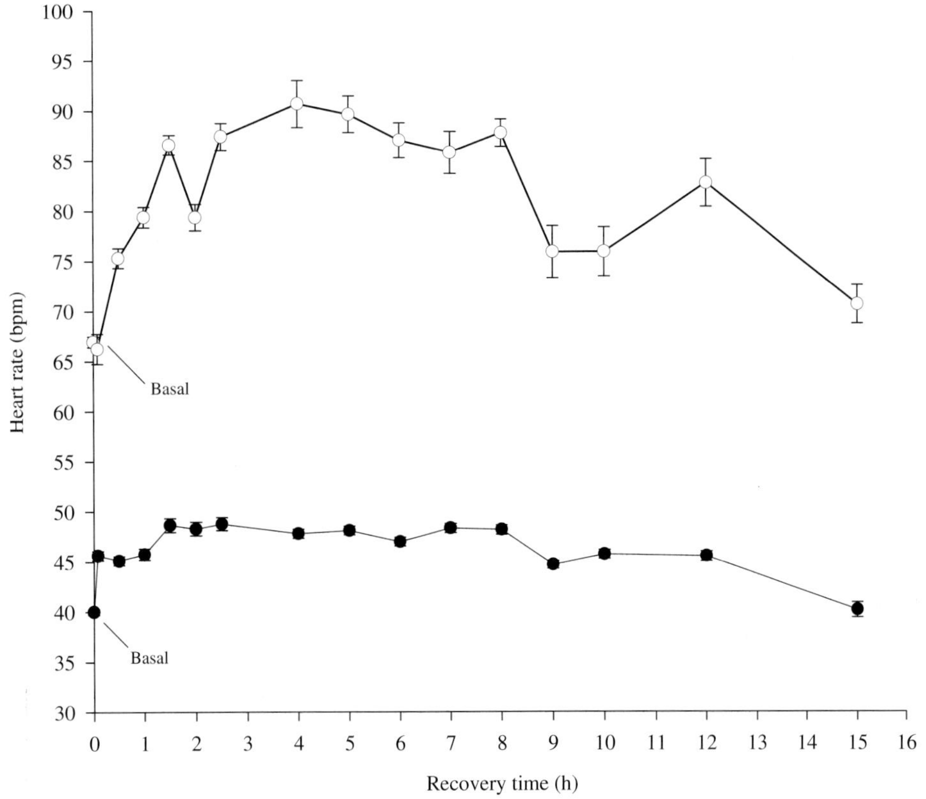


*Figure 2.* Change in heart rate of adult Atlantic salmon during recovery from angling at two distinct temperatures. Open circles = 16.5 ± 1 °C; Closed circles = 8 ± 1 °C. Values are expressed as a mean ± 1 SE.

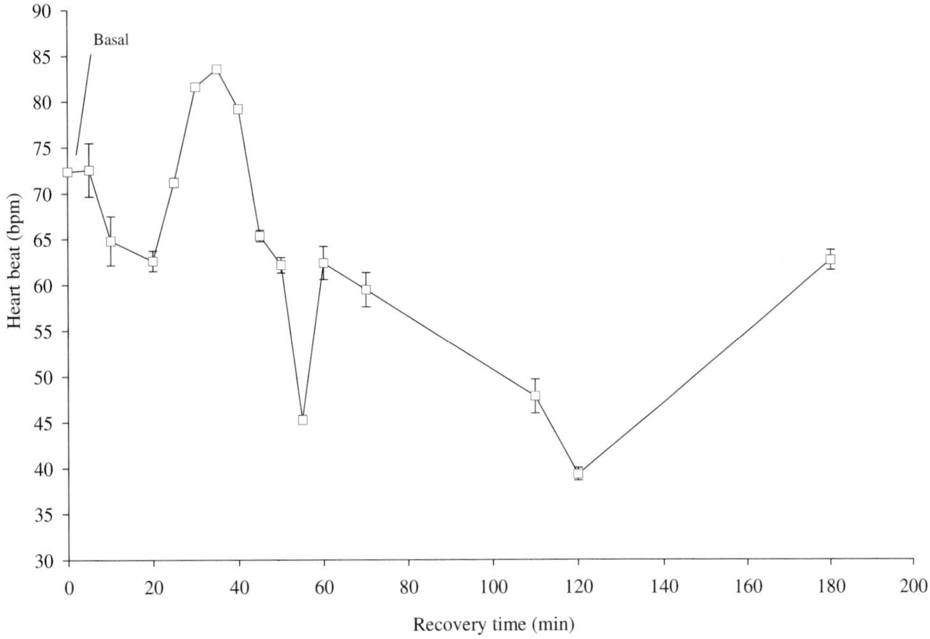


*Figure 3.* Change in heart rate of adult Atlantic salmon during recovery from angling at 20 ± 2 °. Values are expressed as a mean ± 1 SE.

in activity of the 8 °C fish was a function of temperature or their life history is hard to say. However, studies investigating the swimming performance of wild and hatchery reared Atlantic salmon smolts and parr have produced conflicting results with regards to holding ability and swimming performance, with the hatchery reared fish generally having a lower performance than their wild counterparts (Graham et al., 1996; Moore et al., 1995; Peake & McKinley, 1997; Rimmer et al., 1985).

Although trends in heart rate during angling of the 8 and 16.5 °C groups varied considerably, the trend in heart rate during recovery of these two groups did not. Both groups demonstrated an increase in heart rate post angling for up to 16 h. However, the elevation in heart rate above resting levels was greater in the 16.5 °C group than that observed in the 8 °C group. Maximal increase above basal during recovery in the 8 °C group was by a factor of 1.2 after 2.5 h, whereas heart rate increased by a factor of 1.35 in the 16.5 °C group 4 h after angling. Interestingly heart rate peaked in Atlantic salmon recovering from forced exercise within 30 min of the trial by factors of 1.2 and 1.3 for 8 and 18 °C acclimated fish (Anderson, unpublished data). The reasons for the discrepancies in recovery times between the forced exercised fish and the angled fish is unclear. However, the stressful event of being hooked and angled will certainly have exacerbated any cardiovascular changes associated with exhaustive exercise. This demonstrates the usefulness of the heart rate transmitter in varifying and extending our knowledge of the cardiovascular function in fish post exercise.

Unfortunately extended recording of heart rate in fish angled at 20 °C was not possible due to the mortalities in this group. However, what is evident is that the heart rate of these fish immediately after angling and for the following 3 h followed an unusual trend. Typically following exhaustive exercise, fish heart rate increases (Farrell & Jones, 1992) and this is demonstrated in the present study for the 8 and 16.5 °C fish. However, immediately after angling the heart rate of the 20 °C group dropped below basal and remained lower for up to 30 min. The reasons for this are unclear but what is not known is changes in stroke volume and ultimately cardiac output during this time. To compensate the observed dramatic decrease in heart rate following angling in this group (Figure 3), one would expect an even bigger increase in stroke volume to allow for an anticipated increase in cardiac output following exhaustive exercise.

Success of catch and release obviously depends on the survival rate of released salmon and, therefore, the question of recovery rate and delayed mortality is important. Delayed mortalities have been reported in some exhaustively exercised and angled fish (Bouck & Ball, 1966; Beggs et al., 1980; Graham et al., 1982; Wood et al., 1983; Ferguson & Tufts, 1992), but relatively low mortality has been reported for caught and released wild Atlantic salmon (Booth et al., 1995; Tufts et al., 1991). However, Wilkie et al. (1996) reported a mortality rate of 40% in their angled salmon, a figure which is lower than the present study for fish angled at high temperatures (80%) which may be a function of the increased stress and temperature on the fish implanted with a transmitter. However, zero mortalities in both the 16.5 and 8 °C angled groups would indicate that an increase in temperature is the critical factor regarding increases in the likelihood of delayed post angling mortality. Although all the fish angled at $16.5 \pm 1$ °C survived the angling event, immediately post angling the fish were very docile and in some cases were unable to maintain equilibrium. As a consequence, if further excessive demand is placed on the heart in response to a repeated stress in warmer waters, these fish could have a reduced ability to adjust cardiovascular parameters accordingly and, therefore, increase the likelihood of mortality.

Heart rate during recovery in the fish angled at 20 °C was very irregular, the reasons for which are unknown but serve to underline the severe physiological imbalance these fish experience post angling. The cardiovascular parameters in conjunction with the reduction in metabolites (Wilkie et al., 1996) of fish angled at higher temperatures could make them more susceptible to disease, although this would occur over the longer term. Furthermore, fish angled at these temperatures experience an energy loss which could ultimately effect their ability to spawn and/or defend nesting sites. Interestingly however, Booth et al. (1995) demonstrated that fish angled in the late fall (6 °C) very close to the spawning season did not have a reduction in spawning capacity when compared to fish not angled and taken at the same time.

Catch and release programmes were employed by fisheries managers in North America as a early as 1964 (Barnhart, 1989) as year round strategies, in an effort to reduce depletion of fish stocks related to overfishing in rivers and lakes. The suite of catch and release studies carried out in New Brunswick, Canada (Bielak, 1996; Booth et al., 1995; Wilkie et al., 1996), in combination with the present study, demonstrate that catch and release may not be an effective management strategy throughout the entire season. Angling during the mid summer months when water temperatures are close to and in excess of 20 °C may well induce a high percentage of delayed post angling mortality. Unfortunately management decisions based solely on changes in temperature would be difficult to implement as prediction of river temperature is not precise. Furthermore, the dynamic range of temperature throughout any given river system could vary considerably. Consequently, a clearer understanding of the environmental changes in individual watersheds is warranted prior to the implementation of catch and release management decisions based on temperature alone.

## Acknowledgements

Support for this work was provided by a grant from the Canada/Newfoundland Cooperation Agreement on Salmonid Enhancement and Conservation to RSM and is greatly appreciated. The support from Rex Porter, Department of Fisheries and Oceans, St. John's, Newfoundland, and Richard Moccia of the Ontario Ministry of Agriculture and Fisheries Research Station in Alma is gratefully acknowledged.

## References

Anderson, W. G., R. S. McKinley & M. Colavecchia, 1997. The use of clove oil as an anesthetic for rainbow trout and its effects on swimming performance. North am. J. Fish. Mgmt 29: 301–307.

Armstrong, J. D., 1986. Heart rate as an indicator of activity, metabolic rate, food intake and digestion in pike, *Esox lucius*. J. Fish Biol. 29: 207–221.

Barnhart, R. A., 1989. Symposium review: Catch and release, a decade of experience. North am. J. Fish. Mgmt 9: 74–80.

Beggs, G. L., G. F. Holeton & E. J. Crossman, 1980. Some physiological consequences of angling stress in muskellunge, *Esox masquinongy* Mitchell. J. Fish. Biol. 17: 649–659.

Bielak, A. T., 1996. A discussion document on the implications of catch and release angling for Atlantic salmon, with particular reference to water temperature related river closures. DFO, Atlantic Fisheries Research Document, 96/117, 17 pp.

Booth, R. K., J. D. Kieffer, K. Davidson, A. T. Bielak & B. L. Tufts, 1995. Effects of late-season catch and release angling on anaerobic metabolism, acid-base status, survival, and gamete viability in wild Atlantic salmon (*Salmo salar*). Can. J. Fish. aquat. Sci. 52: 283–290.

Bouck, G. R. & R. C. Ball, 1966. Influence of capture methods on blood characteristics and mortality in the rainbow trout (*Salmo gairdneri*). Trans. am. Fish. Soc. 95: 170–176.

Claireaux, G., D. M. Webber, S. R. Kerr & R. G. Boutilier, 1995. Physiology and behaviour of free-swimming Atlantic cod (*Gadus morhua*) facing fluctuating temperature conditions. J. exp. Biol. 198: 49–60.

Farrell, A. P., 1982. Cardiovascular changes in the unanaesthetised ling cod (*Ophiodon elongatus*) during short term progressive hypoxia and spontaneous activity. Can. J. Zool. 60: 933–941.

Farrell, A. P. & D. R. Jones, 1992. In W. S. Hoar, D. J. Randall & A. P. Farrell (eds), Fish Physiology, Academic Press Inc. New York. 12A: 1–73.

Ferguson, R. A. & B. L. Tufts, 1992. Physiological effects of brief air exposure in exhaustively exercised rainbow trout (*Onchorhynchus mykiss*): implications for 'catch and release' fisheries. Can. J. Fish aquat. Sci. 49: 1157–1162.

Gamperl, A. K., M. M. Vijayan & R. G. Boutilier, 1994. Epinephrine, norepinephrine, and cortisol cocentrations in cannulated seawater-acclimated rainbow trout (*Onchorhynchus mykiss*) following black-box confinement and epinephrine injection. J. Fish Biol. 45: 313–324.

Graham, W. D., J. E. Thorpe & N. B. Metcalfe, 1996. Seasonal current holding performance of juvenile Atlantic salmon in relation to temperature and smolting. Can. J. Fish. aquat. Sci. 53: 80–86.

Graham, M. S., C. M. Wood & J. D. Turner, 1982. The physiological responses of the rainbow trout to strenuous exercise: interactions of water hardness and environmental acidity. Can. J. Zool. 60: 3153–3164.

Harder, W., 1975. Anatomy of Fishes. Parts I and II. E. Schweizerbart'sche Verlagsbuchhandlung, Stuttgart.

Johnstone, A. D. F., M. C. Lucas, P. Boylan & T. J. Carter, 1992. Telemetry of tail beat frequency in Atlantic Salmon (*Salmo salar*) during spawning. In I. M. Priede & S. W. Swift (eds), Wildlife Telemetry, Ellis-Horwood, New York: 456–465.

Kaseloo, P. A., A. H. Weatherly, J. Lotimer & M. D. Farina, 1992. A biotelemetry system recording fish activity. J. Fish Biol. 40: 165–179.

Lucas, M. C., 1994. Heart rate as an indicator of metabolic rate and activity in adult Atlantic salmon, *Salmo salar*. J. Fish Biol. 44: 889–903.

Lucas, M. C., I. G. Priede, J. D. Armstrong, A. N. Z. Gindy & L. De Vera, 1991. Direct measurement of metabolism, activity and feeding behaviour of pike (*Esox lucius*) in the wild by use of heart rate telemetry. J. Fish Biol. 39: 325–345.

Milligan, C. L. & C. M. Wood, 1986. Tissue intracellular acid-base status and the fate of lactate after exhaustive exercise in the rainbow trout. J. exp. Biol. 123: 123–144.

Moore, A., E. C. E. Potter, N. J. Milner & S. Bamber, 1995. The migratory behavior of wild Atlantic salmon (*Salmo salar*) smolts in the estuary of the River Conway, North Wales. Can. J. Fish. aquat. Sci. 52: 1923–1935.

Nelson, J. A., Y. Tang & R. G. Boutilier, 1996. The effects of salinity change on the exercise performance of two Atlantic cod (*Gadus morhua*) populations inhabiting different environments. J. exp. Biol. 199: 1295–1309.

Peake, S. & R. S. McKinley, 1997. A re-evaluation of swimming performance in juvenile salmonids relative to downstream migration. Can. J. Fish. aquat. Sci. (in press).

Priede, I. D., 1974. The effects of swimming activity and section of the vagus nerve on heart rate in rainbow trout. J. exp. Biol. 60: 305–319.

Priede, I. D. & P. Tytler, 1977. Heart rate as a measure of metabolic rate in teleost fishes *Salmo gairdneri, Salmo trutta* and *Gadus morhua*. J. Fish Biol. 10: 231–242.

Rimmer, D. M., R. L. Saunders & U. Paim, 1985. Effects of temperature and season on the position holding performance of juvenile Atlantic salmon, *Salmo salar*. Can. J. Zool. 63: 92–96.

Stevens, E. D., G. R. Bennion, D. J. Randall & G. Shelton, 1972. Factors affecting arterial blood pressure and blood flow from the heart in intact, unrestrained ling cod, *Ophiodon longatus*. Comp. Biochem. Physiol. 43: 682–695.

Thorarensen, H., P. E. Gallaugher & A. P. Farrell, 1996. The limitations of heart rate as a predictor of metabolic rate in fish. J. Fish Biol. 49: 226–236.

Tufts, B. L., Y. Tang, K. Tufts & R. G. Boutilier, 1991. Exhaustive exercise in wild Atlantic salmon (*Salmo salar*): Acid-base regulation and blood gas transport. Can. J. Fish. aquat. Sci. 48: 868–874.

Wilkie, M. P., K. Davidson, M. A. Brobbel, J. D. Kieffer, R. K. Booth, A. T. Bielak & B. L. Tufts, 1996. Physiology and survival of wild Atlantic salmon following angling in warm summer waters. Trans. am. Fish Soc. 125: 572–580.

Wood, C. M., J. D. Turner & M. S. Graham, 1983. Why do fish die after severe exercise? J. Fish. Biol. 22: 189–201.

*Hydrobiologia* **371/372:** 241–247, 1998.
*J.-P. Lagardère, M.-L. Bégout Anras & G. Claireaux (eds), Advances in Invertebrates and Fish Telemetry.*




# Using speed-sensing transmitters to construct a bioenergetics model for subadult lemon sharks, *Negaprion brevirostris* (Poey), in the field

L. Fredrik Sundström[1] & Samuel H. Gruber[2]
[1]*University of Linköping, Department of Biology, SE-58183 Linköping, Sweden*
[2]*Bimini Biological Field Station and University of Miami, Miami, FL 33149, USA*
*Correspondence should be addressed to: Fredrik Sundström, öja Böcklinge 8, SE-64040 stora Sundby, Sweden (e-mail: fredrik.sundstrom@jonkoping.mail.telia.com)*

*Key words:* lemon shark, *Negaprion brevirostris*, bioenergetics, telemetry, swimming speed, transmitter

## Abstract

Several methods to indirectly estimate metabolism of aquatic animals have been reported, including heart rate, electromyograms, video recording, and averaging velocity of an animal moving between two or more points. The present study carried out in the lagoon at Bimini Islands, Bahamas, used acoustic, speed-sensing transmitters to indirectly estimate energy consumption of 1.5–2 m subadult lemon sharks (*Negaprion brevirostris*). Speed records from three sharks tracked a total of 170 h, yielded average swimming speeds of 0.44–0.71 m $s^{-1}$. These speeds were converted into energy consumption to obtain metabolism.

By combining the estimates of metabolism with calculated values on assimilation and production, we are able to present a balanced bioenergetics model for the subadult lemon shark: 100C = 7P + 66M + 26E, where $C$ = consumption, $P$ = production, $M$ = metabolism, and $E$ = excretion. When comparing juvenile lemon sharks to subadults, an ontogenetic shift is seen: 0 to 2 yr olds allocate 22% of consumed energy to production and 50% to metabolism while 6 to 9 yr olds show values of 7% and 66%, respectively.

We believe this to be the first field-based, balanced equation presented for any large elasmobranch.

## Introduction

The increased commercial interest in meat and products derived from elasmobranchs (Rose, 1996) has led to over-fishing of some species, e.g. sandbar shark (*Carcharhinus plumbeus*, Nardo) (Branstetter & Burgess, 1996) and caused severe damage to stocks of other species, e.g. soupfin shark (*Galeorhinus galeus*, L.) (Stevens, 1996). With slow growth, late maturation and few offspring, sharks are extremely vulnerable to over-fishing, and in the past depleted stocks have needed several decades to recover from over-exploitation (Sminkey, 1996).

To manage fisheries in a sustainable way, extended and refined bioenergetics models are frequently used (Ney, 1993). However, most of the existing models have been developed for teleost fishes while little attention has been devoted to elasmobranchs. Due to the important difference in life style between most teleosts with high fecundity and fast growth, and the sharks with low fecundity and slow growth, bioenergetics models for teleosts cannot be extrapolated to sharks (Bonfil, 1996). The need for a shark-based bioenergetics model as an aid in predicting such factors as production potential and recovery time after over-exploitation for different shark species is therefore increasing.

The basic bioenergetics model, consumption = production + metabolism + excretion is based on the first law of thermodynamics regarding the conservation of mass and energy, i.e. all energy consumed by a fish must either be incorporated into tissue, used up in metabolism or disposed as feces or urine and other products such as mucus.

Techniques for direct and reliable measurements of the field metabolism of free-ranging fishes in the field have not yet been developed. Instead several indirect methods have been tried. Heart rate provided good estimates for northern pike (*Esox lucius*, L.) and Atlantic salmon (*Salmo salar*, L.) (Lucas et al., 1993) but, when applied to juvenile lemon sharks (*Negaprion brevirostris*, Poey) in captivity, the heart rate showed too much variability between individuals to give a reliable measure of metabolism (Scharold & Gruber, 1991). Another telemetry-based method used to estimate field metabolism is the electromyogram which has been tried on free-swimming rainbow trout (*Oncorhynchus mykiss*, Walbaum) (Kaseloo et al., 1992). Field metabolism can also be estimated by using video recording of fish activity (Krohn & Boisclair, 1994; Koch & Wieser, 1983) and relating the recorded movements to known energetic costs. However, it is impractical to obtain long-term data from larger mobile species in the field.

In the present study, we estimated field metabolism from activity measured with a speed-sensing transmitter. The metabolism together with calculated values for production and assimilation enabled us to present a balanced bioenergetics model for the subadult lemon shark.

## Material and methods

This study was conducted between November 1995 and April 1996, in and around a shallow, subtropical lagoon at Bimini Islands, Bahamas, approximately 85 km east of Miami, Florida. Due to its ease of capture, abundance and size, the lemon shark *Negaprion brevirostris* (family Carcharhinidae), was chosen as the study subject. Water temperature during the study varied between 18–24 °C with an average of 22 °C.

### *Shark capture*

Sharks were captured at the eastern margin of the Bimini lagoon using long fishing lines with multiple hooks baited with pieces of barracuda (*Sphyraena barracuda*, Walbaum). As soon as a shark between 1.5–2.0 m in length was caught, it was tied up along side a small fishing skiff and an ultrasonic identification (ID) transmitter was surgically implanted into the shark's celom. Before the shark was released, the incision was closed with three sutures, the shark's sex was determined and its total and precaudal length was measured.

Sharks that could be relocated by the ID-transmitter, and thus stayed around the Bimini Islands, were recaptured after a recovery period of at least 2 months by chasing and netting. Upon recapture the shark was fitted with a speed-sensing transmitter. The transmitter was attached to a cradle with cable ties and the cradle was either bolted to the shark's first dorsal fin with two steel bolts and a plastic backing plate or fixed to the back with three small steel darts just below the base of the fin.

Immediately after release the shark was tracked from a skiff using a hydrophone connected to an ultrasonic receiver. The number of ultrasonic impulses from the calibrated speed-sensing transmitter was recorded for one minute every fifth minute and later converted to average swimming speed using the calibration curve.

### *Transmitter technical data*

The 12 × 100 mm ID transmitter had an output frequency of 76 kHz and a life span of at least 18 months. It generated a coded signal that could be identified at a distance of 300 m. The 33 × 201 mm cylindrical, speed-sensing transmitter used a paddle wheel with magnets attached to the blades. A reed switch sensor inside the transmitter detected the magnetic field which was electronically converted into an acoustical pulse with an output frequency of 69 kHz and a width of 10 ms. With a higher swimming speed the paddle wheel would turn faster and consequently the signal rate increased. The signal could be detected from about 200 m but the range decreased when the shark was in very shallow water or in dense sea grass. The transmitter's battery was replaceable in the field and had a nominal life of 21 days.

### *Bioenergetics*

#### *Metabolism*

The metabolic component in the bioenergetics equation was calculated through several steps. First, measured swimming speed was converted to oxygen consumption using a power-performance curve arrived at from measurements on juvenile lemon sharks (Bushnell et al., 1989). The equation

$$\log \mathrm{O}_2 = 0.36 \times \frac{S}{\mathrm{BL}} + 1.945, \qquad (1)$$

derived from the curve gave total oxygen consumption $O_2$ (ml $O_2$ kg$^{-1}$ h$^{-1}$), for a 1 kg shark with a body length BL (m) when integrated over swimming speed $S$ (m s$^{-1}$). The average oxygen consumption per h was then extrapolated, to give the consumption for a larger shark, using an allometric equation (Schmidt-Nielsen, 1984):

$$O_2 = A \times W^b, \quad (2)$$

where $W$ = body weight (kg), $A$ = oxygen consumption when $W$ = 1 (ml $O_2$ $kg^{-1}$ $h^{-1}$), and $b$ = body mass exponent. Equation 1 provided $A$, while $W$ was obtained from the weight–length equation for lemon sharks (Gruber, 1984):

$$W = 8.11 \cdot 10^{-6} \times \mathrm{PCL}^{3.08} \; (r^2 = 0.98; N = 103) \quad (3)$$

where $W$ = body weight (kg), and PCL = precaudal length (cm).

The body mass exponent was obtained from the literature as an average value of 0.86 calculated from numerous publications.

The extrapolated oxygen consumption was converted into energy consumption by multiplying with the oxycalorific value of 19.5 J $ml^{-1}$ $O_2$ (Elliot & Davison, 1975). Then multiplying with 24 h $d^{-1}$ thus gave the diel routine metabolism (RM).

*Production*

Production was calculated by combining Equation 3 with the growth curve for lemon sharks presented in Brown & Gruber (1988):

$$\mathrm{PCL}_t = 317.65 \times (1 - e^{-0.057(t+2.302)}) \quad (4)$$

$$(r^2 = 0.99; N = 110),$$

where $\mathrm{PCL}_t$ = precaudal length at time $t$. First the present age of the shark was calculated by inserting the measured PCL in Equation 4. By adding one (1) to the obtained age, and reinserting this into Equation 4, the presumptive length one year later was calculated.

The measured and the calculated lengths were converted into weights with Equation 3 and then the present weight was subtracted from the presumptive weight. In this way the expected annual increase in weight was obtained. This weight increase was converted to the energy equivalent by multiplying with the energy content of a lemon shark, i.e. 5.4 kJ $g^{-1}$ wet weight (Cortes & Gruber, 1994). Then dividing with 365 d $yr^{-1}$ thus gave an estimate of diel production.

*Excretion and consumption*

Excretion was estimated by assuming an absorption efficiency of 80% for energy (Wetherbee & Gruber, 1993). Thus the fecal production was described as $0.2 \times C$, where $C$ is consumption. Adding to this, a 7% loss of assimilated energy due to urine and gill effluents (Brett & Groves, 1979), the total excretion was described by

$$E = 0.27 \times C. \quad (5)$$

Values for production and metabolism together with the relation between consumption and excretion (Equation 5) were inserted into the bioenergetics equation $C = P + M + E$. Only $C$ was unknown and thus could be calculated.

## Results

*Shark capture and transmitter attachment*

One male and two female sharks were captured and equipped with an internal ID transmitter between November 1995 and January 1996 (see Table 1). Within four months after first capture each shark had been recaptured and fitted with a speed-sensing transmitter. When the speed-sensing transmitter was darted to the shark (shark no. 1), the signal could not be heard at day five after attachment indicating that the transmitter had fallen off. Subsequently the speed-sensing transmitter was found on the bottom at day seven after attachment.

When the transmitter was bolted onto a shark, the shark had to be recaptured in order to remove the transmitter. Shark no. 2 was recaptured 24 d after transmitter attachment and shark no. 3 after 12 d.

*Speed measurements and energetics*

Speeds above the shark's fastest monitored swimming speed could be measured but the transmitter's lowest detectable speed was 0.364 m $s^{-1}$. Shark no. 1 spent no time swimming slower than this speed while shark no. 2 and no. 3 spent 3.2% and 3.1% respectively, swimming slower than detectable. During tracking, speeds were measured for one minute giving an accuracy of 0.0146 m $s^{-1}$, i.e. a counting error of one signal per minute.

Signals from the speed-sensing transmitter were consistent and even small changes in signal rate were easily detected by the trackers. A few times the paddle wheel became stuck. This was solved by scaring the shark so it would make a quick burst which created sufficient force for the paddle wheel to disengage. Within seconds after a chase the shark would resume

*Table 1.* Data on capture, morphology and energetics for 3 subadult lemon sharks (Negaprion brevirostris) captured east of Bimini Islands, Bahamas. Each shark was fitted with a speed-sensing transmitter that was tied to a cradle, either by darting the cradle to the back just below the first dorsal fin or attaching the cradle at the base of the first dorsal fin by running two bolts through the fin

| Shark ID | No. 1 | No. 2 | No. 3 |
|---|---|---|---|
| Date of capture | 12 Nov. 95 | 16 Dec. 95 | 29 Jan. 96 |
| Speed-sensing transmitter attachment date | 26 Feb. 96 | 15 Mar. 96 | 30 Mar. 96 |
| Attachment method | darts | bolts | bolts |
| Sex | male | female | female |
| Total length (m) | 1.88 | 1.70 | 1.54 |
| Precaudal length (m) | 1.41 | 1.30 | 1.20 |
| Tracking time (h) | 18.5 | 59.5 | 62 |
| No. of swimming speed measurements | 205 | 648 | 511 |
| Calculated weight (kg) | 33.8 | 26.3 | 20.6 |
| Routine swimming speed (m $s^{-1}$) | 0.71 | 0.60 | 0.57 |
| Routine metabolism (kJ $d^{-1}$) | 1166 | 921 | 760 |
| Production (kJ $d^{-1}$) | 115 | 104 | 94 |
| Consumption (kJ $d^{-1}$) | 1755 | 1404 | 1170 |
| Excretion (kJ $d^{-1}$) | 474 | 379 | 316 |

the speed it swam prior to chase. Still, speeds measured for 10 minutes after a chase were excluded from analysis. This was also the case for measurements directly after speed-sensing transmitter attachment which were excluded until the average swimming speed during one h dropped below 0.4 bl $s^{-1}$. This limit was chosen arbitrarily and resulted in 5 h of speed measurements being excluded from analysis for shark no. 1, 19 h for shark no. 2, and 6 h for shark no. 3.

Valid for analysis were a total tracking time of 140 h and 1364 speed measurements (see Table 1). Average swimming speed for the three sharks was 0.58 m $s^{-1}$ equivalent to an energy consumption of 948.3 kJ $d^{-1}$. Average production was 104 kJ $d^{-1}$ and the calculated average energy consumption of 1442 kJ $d^{-1}$ resulted in 389.3 kJ $d^{-1}$ being excreted each day.

Setting the consumption as 100% a balanced energetic equation for each shark could be constructed (Table 2). Of any consumed energy the subadult lemon shark allocated on average 7.2% to growth and 65.8% for metabolic needs while 27% was disposed as feces and urine.

*Table 2.* Balanced bioenergetics equations for various fish species at different ages. Prefixes are from the equation $C = P + M + E$, where $C$ = consumption, $P$ = production, $M$ = metabolism, and $E$ = excretion. References are [1] Brett & Groves (1979); [2] Gruber (1984); [3] This study; [4] Schmid & Murru (1994); [5] Compiled from data in Boisclair & Leggett (1989)

| Species | Age group | Prefixes |
|---|---|---|
| Herbivores[1] | | $100 = 20 + 37 + 43$ |
| 15 carnivorous teleosts[1] | ? | $100 = 29 + 44 + 27$ |
| *Negaprion brevirostris*[2] | juvenile | $100 = 22 + 50 + 28$ |
| *Negaprion brevirostris*[3] | subadult (no. 1) | $100 = 6.6 + 66.4 + 27$ |
| *Negaprion brevirostris*[3] | subadult (no. 2) | $100 = 7.4 + 65.6 + 27$ |
| *Negaprion brevirostris*[3] | subadult (no. 3) | $100 = 8.0 + 65.0 + 27$ |
| *Carcharhinus leucas*[4] | adult | $100 = 8 + 72 + 20$ |
| *Perca flavescens*[5] | size class I+ | $100 = 21 + 55 + 24$ |
| *Perca flavescens*[5] | size class II+ | $100 = 10 + 66 + 24$ |
| *Perca flavescens*[5] | size class III+ | $100 = 6 + 70 + 24$ |

## Discussion

### *Effects of transmitter attachment*

An externally attached device is likely to increase drag. In response, a fish may either maintain the same speed and experience an increased energy consumption, or maintain the same energy consumption and slow it's swimming speed. The former situation was

*Table 3.* Routine metabolic rates for six shark species of different weights at different temperatures. References are [1] Parsons (1990); [2] Lowe (1996); [3] Scharold et al. (1989); [4] Brett & Blackburn (1978); [5] Graham et al. (1990); [6] Nixon & Gruber (1988); [7] Bushnell et al. (1989); [8] This study

| Species | Routine metabolic rate (kJ $kg^{-1}$ $d^{-1}$ | Body weight (kg) | Temperature (°C) |
|---|---|---|---|
| *Sphyrna tiburo*[1] | 320–63 | 0.0095–4.7 | 25 |
| *Sphyrna lewini*[2] | 105 | 'pups' | not given |
| *Triakis semifasciata*[3] | 43* | 2.2–5.8 | 16 |
| *Squalus acanthias*[4] | 16 | 2 | 10 |
| *Isurus oxyrinchus*[5] | 124 | 3.9 | 18 |
| *Negaprion brevirostris*[5] | 106 | 1.7 | 22 |
| *Negaprion brevirostris*[6] | 68 | 1.0 | 23 |
| *Negaprion brevirostris*[7] | 52 | 1.1 | 22 |
| *Negaprion brevirostris*[8] | 37–34 | 20.3–33.8 | 22 |

*value calculated as average of metabolic rates at resting and maximum swimming speed.

found for juvenile lemon sharks (Scharold & Gruber, 1991) and leopard sharks (*Triakis semifasciata*, Girard) (Scharold et al., 1989) fitted with heart rate transmitters.

As the swimming speeds of the sharks in the present study were in the upper range compared with previous results on lemon shark swimming speeds from laboratory studies (Scharold & Gruber, 1991; Bushnell et al., 1989) and preliminary field studies (Gruber et al., 1988), it was assumed that the increased drag resulted in an increase in energy consumption rather than a decrease in swimming speed. If this assumption is correct the increase in energy consumption due to additional drag does not affect the estimated value for metabolism in this study. This is because the swimming speed was converted to oxygen consumption with a power performance curve obtained from uninstrumented sharks.

### *Metabolic rate*

The lemon shark has a considerably lower metabolic rate than highly active species with internal heat conservation such as the mako shark (*Isurus oxyrhinchos*, Rafinesque) (Graham et al., 1990) or obligate swimmers such as the bonnethead shark (*Sphyrna tiburo*, L.) (Parsons, 1990) and scalloped hammerhead shark (*Sphyrna lewini*, Griffith & Smith) (Lowe, 1996). Less active species such as the leopard shark (Scharold et al., 1989) and the spiny dogfish (*Squalus acanthias*, Smith & Radcliffe) (Brett & Blackburn, 1978) have lower routine metabolic rates (see Table 3). However metabolic rates are greatly influenced by temperature and body size (see e.g. Schmidt-Nielsen, 1990), which makes it difficult to compare results obtained under different conditions for different sized specimens.

### *Energy consumption*

Based on the routine metabolism (Table 1), the yearly consumption for the sharks in this study ranges from 75–113 kg of food equivalent to blue runner, *Caranx crysos* (energy content: 5.71 kJ $g^{-1}$ Cortes & Gruber, 1994). With an energy content of the lemon shark of 5.4 kJ $g^{-1}$ fresh weight (Cortes & Gruber, 1994), the consumption of these sharks would correspond to 3.6–3.8 times their body weight in one year. This is equivalent to a daily ration of 1% body weight which is half the rate of a 1–2 year old lemon shark. These very young, fast growing sharks consume up to 6.5 times their body weight in a year (Bushnell et al., 1989) and have a daily ration of about 2% (Cortes & Gruber, 1990). However, this difference between young and older individuals is expected as a bigger shark uses relatively less energy per unit body weight (Schmidt-Nielsen, 1984).

In a study by Schmid & Murru (1994) bull sharks (*Carcharhinus leucas*, Valenciennes), of twice the weight of the lemon sharks in this study, fed ad libitum in captivity had a daily ration of about 0.5%. Previously Schmid et al. (1990) had found, for other bull sharks, a daily ration close to 0.5% and even lower for sand bar sharks 0.45%, nurse sharks (*Ginglymostoma cirratum*, Bonnaterre) 0.3% and sand tiger sharks (*Eugomphodus taurus*, Rafinesque) 0.26%. The values for these shark species and the lemon shark are very low compared to mako sharks of 79 kg with a calculated daily ration of 3.1% (Stillwell & Kohler, 1982). However, the mako shark is an extremely active, epipelagic species that grows fast and maintains an elevated body temperature 7–10 °C above ambient water (Stillwell, 1991) and thus has a very high metabolic rate (Graham et al., 1990).

### *Validity of the bioenergetics model*

Existing bioenergetics models often include one or more deficiencies. These can be grouped into four major categories (Ney, 1993): (1)-unknown activity costs, (2)-extrapolation of allometric functions, (3)-unjustified borrowing of parameters from other species, and (4)-inadequate estimation of external variables.

By directly measuring the swimming speed with the speed-sensing transmitter the resultant activity rates can be considered as reliable. Unfortunately, techniques are not available to measure oxygen consumption in the field, and laboratory measurement on larger specimens requires elaborate and expensive respirometers. Thus the metabolic rates had to be extrapolated from swimming speed which may introduce an error (Ney, 1993).

In many cases the cost and difficulties associated with field measurements of all the variables included in a bioenergetics equation requires the researcher to use values from other closely related or morphologically similar species. Fortunately there is an extensive database on the lemon shark and as this is a preliminary study, the compiling of more data from other individuals will enhance the reliability of the field data.

In Table 2 balanced bioenergetics models for various fish species are given. Comparing the bioenergetics model presented by Gruber (1984) for a standardized juvenile lemon shark (0.60 m TL and 1.2 kg at 25 °C) the present models, developed for subadult lemon sharks (1.5–1.9 meters TL and 20–34 kg at 22 °C), show that energy allocated to metabolism has increased, while energy allocated to production has decreased. This trend is also obvious, for example when going from size class I+ to class III+ for yellow perch (*Perca flavescens*) (Boisclair & Leggett, 1989). A model similar to those for the subadult lemon sharks has been presented for adult bull sharks in captivity (Schmid & Murru, 1994). Unfortunately there is no model available for juvenile bull sharks to compare with.

The results suggest that a young individual devotes relatively more energy to production than to metabolism. This is not unexpected as young fish normally have relatively higher growth rates than adults.

## Acknowledgements

This paper constitutes a part of a MS thesis by the first author. Constructive comments on that work were given by Sverre Sjölander and Charlotte Borrvall. The authors would also like to thank all the volunteers at the Bimini Biological Field Station who helped in the field. The study was sponsored by the US Office of Naval Research and University of Linköping.

## References

Boisclair, D. & W. C. Leggett, 1989. The importance of activity in bioenergetics models applied to actively foraging fishes. Can. J. Fish. aquat. Sci. 46: 1856–1867.

Bonfil, R., 1996. Can simple fishery models be used for elasmobranch stock assessment?: a Monte Carlo simulation approach. AES 12th Ann. Meet., New Orleans June 13–19: 87.

Branstetter, S. & G. H. Burgess, 1996. Characterization and status of the southeast U.S. commercial shark fishery with recommendations for sustainability. AES 12th Ann. Meet., New Orleans June 13–19: 90.

Brett, J. R. & J. M. Blackburn, 1978. Metabolic rate and energy expenditure of the spiny dogfish, *Squalus acanthias*. J. Fish. Res. Bd Can. 35: 816–821.

Brett, J. R. & T. D. D. Groves, 1979. Physiological energetics. In W. S. Hoar, D. J. Randall & J. R. Brett (eds), Fish Physiology. Academic Press, New York 8: 279–353.

Brown, C. A. & S. H. Gruber, 1988. Age assessment of the lemon shark, *Negaprion brevirostris*, using tetracycline validated vertebral centra. Copeia 3: 747–753.

Bushnell, P. G., P. L. Lutz & S. H. Gruber, 1989. The metabolic rate of an active, tropical elasmobranch, the lemon shark (*Negaprion brevirostris*). Exp. Biol. 48: 279–283.

Cortes, E. & S. H. Gruber, 1990. Diet, feeding habits, and estimates of daily ration of young lemon sharks, *Negaprion brevirostris* (Poey). Copeia 1: 204–218.

Cortes, E. & S. H. Gruber, 1994. Effect of ration size on growth and gross conversion efficiency of young lemon sharks, *Negaprion brevirostris*. J. Fish Biol. 44: 331–341.

Elliot, J. M. & W. Davison, 1975. Energy equivalents of oxygen consumption in animal energetics. Oecologia 19: 195–201.

Graham, J. B., H. Dewar, N. C. Lai, W. R. Lowell & S. M. Arce, 1990. Aspects of shark swimming performance determined using a large water tunnel. J. exp. Biol. 151: 175–192.

Gruber, S. H., 1984. Bioenergetics of the captive and free-ranging lemon shark. AAZPA 340–373.

Gruber, S. H., D. R. Nelson & J. F. Morrissey, 1988. Patterns of activity and space utilization of lemon sharks, *Negaprion brevirostris*, in a shallow Bahamian lagoon. Bull. mar. Sci. 43: 61–76.

Kaseloo, P. A., A. H. Weatherley, J. Lotimer & M. D. Farina, 1992. A biotelemetry system recording fish activity. J. Fish Biol. 40: 165–179.

Koch, F. & W. Wieser. 1983. Partitioning of energy in fish: can reduction of swimming activity compensate for the cost of production? J. exp. Biol. 107: 141–146.

Krohn, M. M. & D. Boisclair, 1994. Use of a stereo-video system to estimate the energy expenditure of free-swimming fish. Can. J. Fish. aquat. Sci. 51: 1119–1127.

Lowe, C., 1996. Swimming metabolism and SDA of juvenile scalloped hammerhead sharks. AES 12th Ann. Meet., New Orleans June 13–19: 204.

Lucas, M. C., A. D. F. Johnstone & I. G. Priede, 1993. Use of physiological telemetry as a method of estimating metabolism of fish in the natural environment. Trans. am. Fish. Soc. 122: 822–833.

Ney, J. J., 1993. Bioenergetics modeling today: growing pains on the cutting edge. Trans. am. Fish. Soc. 122: 736–748.

Nixon, A. J. & S. H. Gruber. 1988. Diel metabolic and activity patterns of the lemon shark (*Negaprion brevirostris*). J. exp. Biol. 248: 1–6.

Parsons, G. R., 1990. Metabolism and swimming efficiency of the bonnethead shark *Sphyrna tiburo*. Mar. Biol. 104: 363–367.

Rose, D. A., 1996. An overview of world trade in sharks and other cartilaginous fishes. TRAFFIC International, Cambridge, UK, 108 pp.

Scharold, J. & S. H. Gruber, 1991. Telemetered heart rate as a measure of metabolic rate in the lemon shark, *Negaprion brevirostris*. Copeia 4: 942–953.

Scharold, J., N. C. Lai, W. R. Lowell & J. B. Graham, 1989. Metabolic rate, heart rate, and tailbeat frequency during sustained swimming in the leopard shark *Triakis semifasciata*. Exp. Biol. 48: 223–230.

Schmid, T. H. & F. L. Murru, 1994. Bioenergetics of the bull shark, *Carcharhinus leucas*, maintained in captivity. Zoo. Biol. 13: 177–185.

Schmid, T. H., F. L. Murru & F. McDonald, 1990. Feeding habits and growth rates of bull (*Carcharhinus leucas* (Valenciennes)), sandbar (*Carcharhinus plumbeus* (Nardo)), sandtiger (*Eugomphodus taurus* (Rafinesque)), and nurse (*Ginglymostoma cirratum* (Bonnaterre)) sharks maintained in captivity. J. Aquac. Aquat. Sci. 5: 100–105.

Schmidt-Nielsen, K., 1984. Scaling, Why is Animal Size so Important? Cambridge University Press, New York, 241 pp.

Schmidt-Nielsen, K., 1990. Animal Physiology: Adaptation and Environment. Cambridge University Press, New York, 602 pp.

Sminkey, T. R., 1996. Demographic analyses of natural and exploited populations of three large coastal sharks. AES 12th Ann. Meet., New Orleans June 13–19: 284.

Stevens, J., 1996. The productivity of *Galeorhinus* and the sustainability of the Australian fishery. AES 12th Ann. Meet., New Orleans June 13–19: 289.

Stillwell, C., 1991. The ravenous mako. In S.H. Gruber (ed.), Discovering Sharks. Underwater Naturalist 19–20: 77–78.

Stillwell, C. E. & N. E. Kohler, 1982. Food, feeding habits, and estimates of daily ration of the shortfin mako (*Isurus oxyrinchus*) in the northwest Atlantic. Can. J. Fish. aquat. Sci. 39: 407–414.

Wetherbee, B. M. & S. H. Gruber, 1993. Absorption efficiency of the lemon shark *Negaprion brevirostris* at varying rates of energy intake. Copeia 2: 416–425.

*Hydrobiologia* **371/372:** 249–262, 1998.
*J.-P. Lagardère, M.-L. Bégout Anras & G. Claireaux (eds), Advances in Invertebrates and Fish Telemetry.*



# Migratory and reproductive activity of radio-tagged Arctic char (*Salvelinus alpinus* L.) in northern Labrador

Toni A. Beddow, Caroline Deary & R. Scott McKinley
*Waterloo Biotelemetry Institute, Dept. of Biology, University of Waterloo, Waterloo, Ontario, N2L 3G1, Canada*

*Key words: Salvelinus alpinus*, Arctic char, anadromous, migration, spawning, radio-telemetry

## Abstract

Migratory and reproductive activities of Arctic Char (*Salvelinus alpinus* L.) were studied using radio-telemetry in two adjacent river systems in northern Labrador, Canada. A total of 154 Arctic char were radio-tagged. Coded radio tags were surgically implanted into the body cavity of char, in the Summer and Fall of 1996. Individuals movements were monitored using 3 solar powered, fixed data logging stations, that recorded the time, date, channel and code of any transmitter within a predetermined distance (150 m) of the antenna (every 5 seconds). Movements of fish throughout the remaining sections of the 2 rivers were monitored on a weekly basis using aerial and manual tracking techniques. Gonadosomatic indices of spawning females increased throughout the Summer to peak between 20–36% in mid September, and dropped to <2% when spent, indicating that spawning occurred in the last week of September. Arctic char radio-tagged within the smaller river, Reid Brook ($n = 125$), exhibited a complex range of migratory behaviors. Initially, the majority of Reid Brook char migrated to spawning areas located throughout the mid to upper reaches of the river, where they remained until spawning was completed. Only 8 radio-tagged fish, (35.9–56.4 cm) successfully negotiated a waterfall called Reid Falls, to enter Reid Pond where they remained for the duration of the study. Immediately following spawning, during river ice up at water temperatures 0–3 °C, most of the radio-tagged Arctic char made abrupt downstream migrations and 61% of Reid Brook radio-tagged char moved through into the adjacent river system, Kogluktokoluk Brook. Final tracking data (29$^{\text{th}}$ October 1996), indicated that 8 fish remained in Reid Pond, 8 in the pool below Reid Falls, 62 char had moved through to adjacent river systems and the remaining char were generally located close to the mouth of Reid Brook. Char radio-tagged in Ikadlivik Brook, dispersed throughout that river system to spawn. Throughout October radio-tagged individuals from both river systems congregated in large deep pools ($\approx$ 30 m deep) in the Ikadlivik and these sites were considered suitable habitats in which to over-winter. This strategy whereby Arctic char spawn in one river system but over-winter in a larger neighboring system (within a single year) has been postulated for other char species but never previously proven, highlighting the value of radio-telemetry for remotely studying complex migratory patterns.

## Introduction

Arctic char (*Salvelinus alpinus* L.) is a salmonid species that occur throughout the circumpolar oceans of the northern hemisphere (40°N–82°N latitude) (Johnson, 1980). Their life history is complex and poorly understood, with Arctic char displaying a high degree of phenotypic plasticity that appears to enable them to adapt to specific habitats and extreme environments (Barbour, 1984). Both 'landlocked' freshwater populations and anadromous forms of Arctic char have evolved, with anadromy becoming more prevalent in populations north of 60°N latitude (Johnson, 1980). Anadromous char populations generally undertake annual migrations from freshwater to the sea in early spring to feed, and then return to freshwater to spawn in the fall, where they remain until spring break up. In some populations however, the fall spawners forgo the feeding migration and remain in freshwater throughout the summer (Johnson, 1989). Since spawning requires high energy output, females often only spawn every 2 or 3 years (Scott & Crossman, 1973) and, therefore, not all the adults are part of the spawning population in a given year.

Detailed work has been carried out on the anadromous Arctic char of northern Labrador (Andrews & Lears, 1956; Dempson, 1984 & 1995; Dempson & Green, 1985; Dempson & Kristofferson, 1987), indicating the existence of several discrete populations along that coast (Dempson, 1984). Extensive studies on the annual migrations of Fraser River Arctic char in Northern Labrador has revealed that seaward migration of char occurs during May and early June coinciding with spring run off and river ice break up (Dempson & Green, 1985). The larger fish, both maturing and non-maturing, usually begin to enter the sea first, followed by smaller adults and juveniles (Dempson & Kristofferson, 1987). The return upstream migrations begin mid July with the peak return occurring in August (Dempson & Green, 1985). Maturing fish, predominantly females tend to leave the ocean first, followed by the smaller and non-maturing adults, and juveniles (Dempson & Kristofferson, 1987; Johnson, 1989). Maximum duration of the seawater phase was 13 weeks for char tagged in both the Nain Bay area and further north in Tikkoatokak Bay, with average ocean duration's of 8.2 weeks and 7.5 weeks respectively (Dempson & Kristofferson, 1987).

Like many cold water species Arctic char are slow growing, with the growth rate decreasing the more northerly the population (Andrews & Lears, 1956) taking many years for char to reach sexual maturity. Average age of sexually mature fish from the Fraser River, northern Labrador, was 5.2 years for males and 6.9 years for females, with age of maturity appearing to increase with increasing latitude (Dempson, 1984). These life history factors, combined with the intense commercial fishery in northern Labrador make char particularly vulnerable to environmental pollution and exploitation.

The main aim of this study was to improve the knowledge of the migratory behaviour and life history strategies of Arctic char in northern Labrador, Canada. Migratory behaviour of char was studied using radio-telemetry, which is an accurate and reliable method, capable of monitoring precise locations and movements of individual fish. Special interest was focused on identifying spawning grounds and the over-wintering preferences of Arctic char.

**Description of sites studied**

Approximately 35 km south of Nain Bay another distinct population of char exists within the Voisey's Bay area of northern Labrador (56°15′ N, 61°12′ W). Three main rivers, Kogluktokoluk Brook, Ikadlivik Brook and Reid Brook flow easterly into the northwest end of Voisey's Bay, the latter two of which were targeted for this study (Figure 1). Ikadlivik Brook flows into Kogluktokoluk Brook about 5 km upstream of the mouth. The drainage area of the Kogluktokoluk and Ikadlivik Brook system is 1095 km$^2$ containing 2 large lakes and 18 tributaries (Anderson, 1985). Adjacent to the mouth of Kogluktokoluk Brook (700 m wide) is the mouth of Reid Brook (100 m wide). Reid Brook is part of a different watershed and is a much smaller river (main stem = 31 km) with a drainage area of 171 km$^2$, with 11 small tributaries (Anderson, 1985). There is a 500 m freshwater flume extending from both rivers into Voisey's Bay with the salt wedge extending up to 1 km from the mouths of these rivers, depending on tidal cycles, weather conditions, precipitation and season.

Ikadlivik Brook is a large, fairly fast flowing river, approximately 75 km long, varying from a few centimeters to 1–2 m in depth for much of its length, with no apparent obstacles blocking fish passage except for waterfalls located approximately 70 km upstream, near the origin of the river. Large pools are situated half way along the Ikadlivik's length (56°21′48 N, 62°23′05 W), 20–30 m deep, 0.5 km at their widest point and about 4.5 km long. Reid Brook on the other hand is a highly meandering, relatively slow flowing river, approximately a third of the length of the Ikadlivik, ranging from a few centimeters to a maximum of 2 m depth throughout most of its length. About 25 km upstream of the mouth, Reid Brook becomes very wide, shallow and braided before reaching a small, deep pool (~ 10 m deep, 2–3 m wide) immediately below Reid Falls (56°22′30 N, 62°07′64 W). The falls are fed by Reid Pond and provide a difficult, if not impossible obstacle for many of the fish attempting to pass them. Reid Pond (56°23′ N, 62°13′ W) itself is about 30 m at its deepest, 0.5–1 km wide and 4.5 km long. This study was conducted throughout the summer of 1996 which was very hot and dry, both rivers dramatically reducing in depth (Reid Brook dropped by about 10 cm between mid August and September).

Although Arctic char are the dominant species in the river systems in this area, other species do coex-

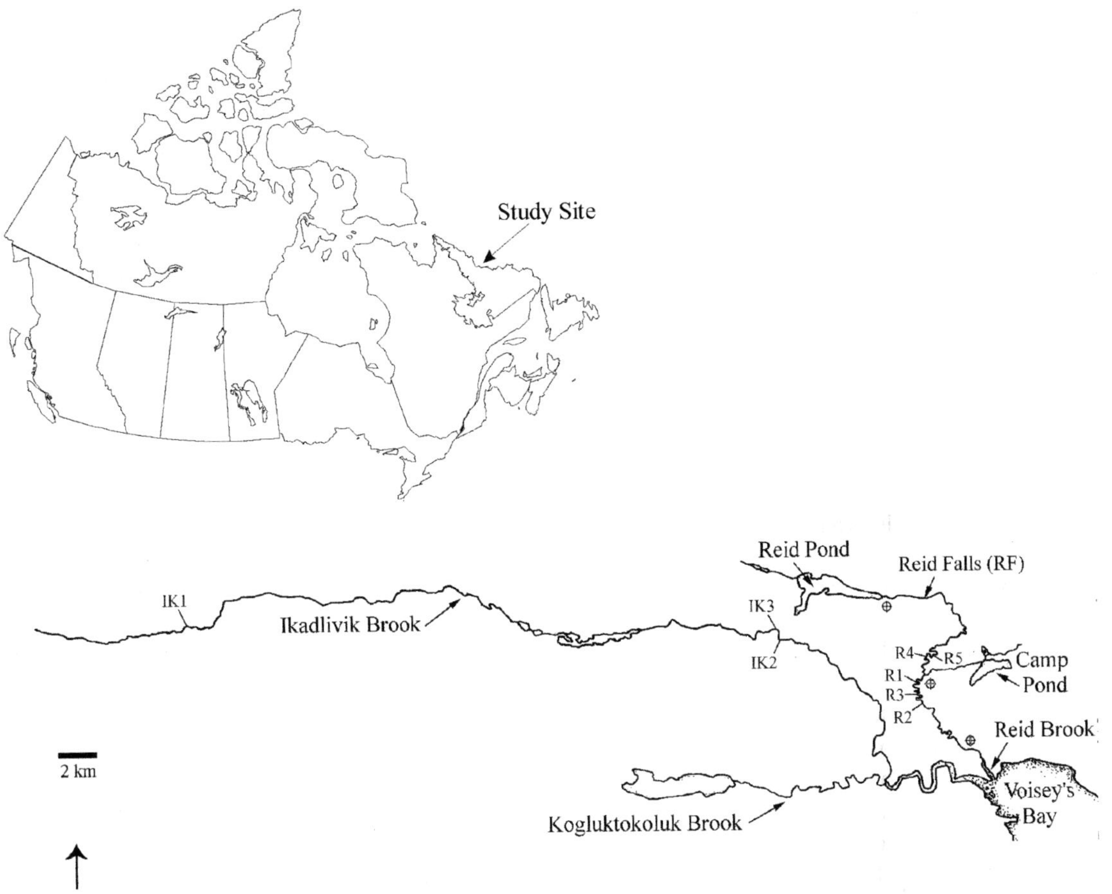


*Figure 1.* Showing the location of the study area in Voisey's Bay, Northern Labrador, Canada. Inset shows detailed map of the river systems under review. RF & R1–R5 indicate tagging sites on Reid Brook, IK1–IK3 indicate tagging sites on Ikadlivik Brook. See Table 1 for more tagging information. ⊕ Indicates the positions of the fixed data logging stations along Reid Brook.

ist. Most common are three-spine stickleback (*Gasterosteus aculeatus*), nine-spine stickleback (*Pungitius pungitius*), Atlantic salmon (*Salmo salar*), brook trout (*S. fontinalis*), lake trout (*S. namaycush*) and occasionally lake whitefish (*Coregonus clupeaformis*).

## Methods

A total of 233 Arctic Char (166 from Reid Brook, 26.0–61.4 cm), (67 from Ikadlivik Brook 33.5–59.0 cm) were caught using hook and line, gill nets (variable mesh size) or seine nets between 1st August and 4th October, 1996. One hundred and fifty-four fish were implanted with radio-transmitters in the abdominal cavity and released (Table 1, Figure 1). Seventy char were sacrificed for biochemical analysis and the remaining char were too small to tag and therefore released.

### *Radio transmitter attachment*

Radio-transmitters were implanted during 3 different sampling periods: 1st–7th August, 22nd–28th August and 1st–4th October 1996. A total of 125 fish were tagged on the Reid Brook system at six different sites between a point 5 km upstream of the mouth and Reid Brook Falls and 29 fish were tagged in the lower and upper reaches of the Ikadlivik (Figure 1; Table 1). Coded tags were 11 mm long, 4.9 mm wide, weighed 8.9 g in air and 4.3 g in water (Lotek Marine, St. Johns, Newfoundland). The tags transmitted on 6 different radio frequencies and each transmitter had a unique code permitting individual identification of fish.

*Table 1.* Location and date for radiotagging Arctic char in Northern Labrador, Canada. Total mortality that occurred at each site is indicated. Location code refers to sites marked on Figure 1

| Date | Location code | Latitude N | Longitude W | Radio-tagged fish (*n*) | Mortalities (*n*) |
|---|---|---|---|---|---|
| 3rd Aug 1996 | I1 | 56°21′54 | 62°42′48 | 9 | 3 |
| 5th Aug 1996 | I1 | 56°21′54 | 62°42′48 | 3 | – |
| 7th Aug 1996 | I1 | 56°21′54 | 62°42′48 | 4 | – |
| 22nd Aug 1996 | I2 | 56°20′20 | 62°13′00 | 9 | 1 |
| 23rd Aug 1996 | I3 | 56°20′93 | 62°12′58 | 4 | – |
| 1st Aug 1996 | RF | 56°22′30 | 62°07′64 | 5 | 2 |
| 2nd Aug 1996 | RF | 56°22′30 | 62°07′64 | 17 | 4 |
| 4th Aug 1996 | RF | 56°22′30 | 62°07′64 | 6 | 2 |
| 24th Aug 1996 | RF | 56°22′30 | 62°07′64 | 19 | 4 |
| 25th Aug 1996 | RF | 56°22′30 | 62°07′64 | 16 | 5 |
| 6th Aug 1996 | R1 | 56°20′20 | 62°07′80 | 11 | 6 |
| 27th Aug 1996 | R2 | 56°19′00 | 62°08′00 | 5 | – |
| 2nd Oct 1996 | R3 | 56°19′90 | 62°08′28 | 14 | – |
| 3rd Oct 1996 | R4 | 56°20′84 | 62°07′41 | 11 | – |
| 4th Oct 1996 | R5 | 56°20′87 | 62°07′19 | 21 | – |

### *Surgery and recovery*

Fish were anaesthetised prior to surgery using a natural anaesthetic (Clove Oil) at concentrations between 20–50 mg $l^{-1}$ depending on the ambient water temperature (Anderson et al., 1997). Fork length and weight were determined at this time, and during surgery, attempts were made to sex fish, based on external morphological characteristics (such as the development of a hooked lower jaw in males) or evidence of eggs or testes in the internal body cavity. Fish were immersed in fresh, aerated water during surgery. Placed ventral side up, supported by a foam cushion, a small incision (2–3 cm) was made just behind the pelvic girdle. The radio-tag was dipped in isopropanol and then allowed to air dry. The tag was pushed through the incision, and gently slid towards the pelvic girdle. A 16 gauge hypodermic needle was used to exit the antenna through the body wall. The incision was sutured, approximately 1 suture per cm (silk, black braid, Ethicon) in a procedure that took less than 5 min. Fish were revived in the river, while being held upright, head on to the direction of the current.

### *Automatic radio tracking*

Fixed stations were established at 3 locations along Reid Brook on 7th August, 1996 (Figure 1). Station 1 was situated approximately 4 km upstream from the mouth (56°18′27 N, 62°05′44 W), Station 2 was situated just downstream from the mouth of Camp Pond tributary (56°20′20 N, 62°07′80 W), and Station 3 was situated on the southeast end of Reid Pond at the head of Reid Brook (56°22′30 N, 62°09′60 W). Each station contains a Lotek SRX_400 receiver configured with W17 firmware, which automatically scan all 6 frequencies and record the date, time, channel and code (every 5 seconds) of any radio-transmitters within a predetermined range. Signal strength of the radio-transmitters was mapped out with regards to the fixed stations position at various distances from the antennas. The receivers were connected to a solar charged battery supply.

### *Manual radio tracking*

Manual tracking was conducted weekly from the air and on foot from 24th August, 1996 to 29th October, 1996. For aerial tracking, 2 external 'H' antennas were attached on each end of a crossbar located at the anterior of the helicopter. The antennas were then connected with a coaxial cable to the Lotek SRX_400 receiver located inside the helicopter. The receiver was set to scan all frequencies while flying along the rivers at heights ranging from 9 m to tree top level at speeds of 16–48 km $h^{-1}$. When a radio-transmitter was detected, the operator would adjust the gain of the receiver until the GPS coordinates of the fish were determined.

Once fish were located through aerial tracking, ground tracking was used to visually confirm the actual site and describe habitat conditions and fish behavior. Ground tracking was conducted with the same receiver that was used for aerial tracking. A hand held 'H' antenna or Yagi antenna was connected to the receiver for detecting transmitters.

*Biological characteristics*

Fish were sampled from both Reid Brook (n = 32; 30.3–55.1 cm) and Ikadlivik Brook ($n = 38$; 37.0–59.0 cm) from the beginning of August until the end of September 1996. Biological information collected included: liver and gonad weight, examination of stomach contents and otolith removal for age determination. Age was determined from one otolith per fish (2 sections), blind replicates were re-aged. Gonadosomatic Indices (GSI's) were calculated using equation (1) and Hepatic Somatic Indices (HSI's) were calculated using equation (2).

$$\text{GSI} = \frac{\text{wet weight gonads} \times 100}{\text{wet weight (fish–gonads)}} \quad (1)$$

$$\text{HSI} = \frac{\text{wet weight liver} \times 100}{\text{wet weight (fish including liver–gonads)}} \quad (2)$$

Differences between the sexes and over time were examined with one-way ANOVA and students $t$-tests (Minitab Inc.).

## Results

There are currently 127 radio-tagged Arctic char remaining within the Reid and Ikadlivik Brook systems. Of 125 fish originally radio-tagged in Reid Brook, 23 (18%) were known to have died between 2nd August, 1996 and 27th October, 1996 (Table 1). Four of the 29 fish (14%) implanted in Ikadlivik Brook also died. These fish all had transmitters implanted during August. There was no observed mortality in char that were implanted with radio-transmitters in October 1996 in Reid Brook ($n = 46$) (Table 1).

*Movements of fish radio-tagged in August – Ikadlivik Brook*

Fish were radio-tagged in the lower and upper reaches of the Ikadlivik Brook system (Figure 1). Within 2 weeks of implantation, the 13 fish tagged in early August on the upper reaches of the Ikadlivik moved downstream to the large pools (56°21′48 N, 62°23′05 W), in the middle reaches of the Ikadlivik, where they remained until late August (Figure 2a). In early September radio-tagged char began to spread along the whole length of Ikadlivik Brook where they remained until after spawning (Figure 2a). Spawning was believed to have occurred during late September and in the first week of October, when water temperatures dropped below 7 °C. The majority of radio-tagged char were observed in groups of at least 10 individuals, in pools ranging in depth from 1–2 m, or in association with large boulders downstream of riffles and on the outside bend of the Brook. Habitat for the majority of sites consisted of 40% boulders, 50% rubble, and 10% sand/gravel. The width of the Brook at these sites was on average about 30 m and water flow approximately 1 m s$^{-1}$. Several fish were located in the deeper pools (56°21′48 N, 62°23′05 W) or at the edge of these pools, prior to and during spawning. Substratum of the pools was indeterminable due to low visibility. Many fish were observed on redds late in September and due to spread of the fish throughout the system it appears that much of the length of Ikadlivik Brook is used for spawning. Regardless of the original tagging locations, in late October during freeze up, many of the Arctic char began congregating in the deep pools (56°21′48 N, 62°23′05 W) in the middle reaches of the Ikadlivik.

*Movements of fish radio-tagged in August – Reid Falls*

A total of 63 fish were tagged at Reid Falls, 46 (73%) of which survived at least until the end of October 1996. Six of these fish moved downstream in late August and early September prior to time of spawning, 1 char remained in lower Reid Brook and the other 5 moved through into Ikadlivik Brook (Figure 2b). These latter fish migrated approximately 70 km to the deep pools half way upstream in the Ikadlivik (56°21′48 N, 62°23′05 W). The sex of these 6 fish was impossible to determine at time of implantation, suggesting they may have been part of the nonmature population in the fall of 1996. Eight radio-tagged fish managed to negotiate the falls and move up into Reid Pond, 4 char in late August, 2 in mid September and 2 in late October (Figure 2b). The quickest ascent of the falls occurred by a 2 kg char who moved into Reid Pond less than 6 days after surgery. Once char entered Reid Pond their movements were fairly limited, being

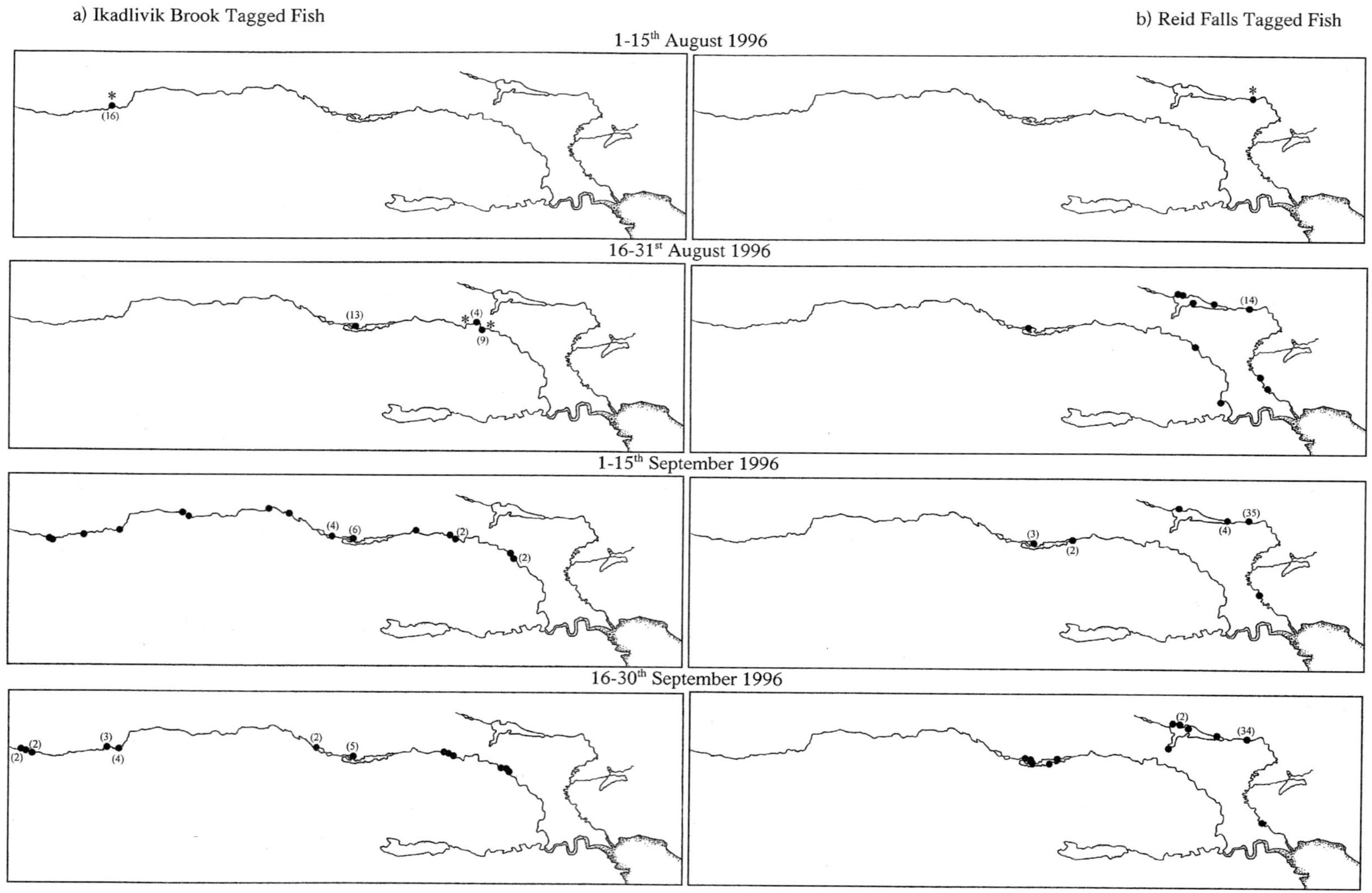


*Figure 2.* Spatial distribution of Arctic char radio-tagged in (a) Ikadlivik Brook and (b) Reid Brook Falls, at two weekly intervals. Black dots indicate fish positions, where more than 1 fish is present the number is indicated in brackets. ∗ Indicates initial tagging location.

generally located near the edge of the Pond, with the majority of fish congregating in the north-west corner in late October (Figure 2b). Visual observations were not possible in this region due to the depth of the water.

The pool immediately below Reid Falls was approximately 10 m deep and 2–3 m in diameter with a predominately rubble and gravel substrate. Divers were sent down into the pool in September 1996 to check on the condition of the fish and no mortalities were observed. Twenty-two of the fish radio-tagged at Reid Falls remained there until after the spawning season and then, in mid to late October, swam downstream into lower Reid Brook ($n = 8$), or out of the mouth of Reid Brook into the adjacent Kogluktokoluk ($n = 2$) and Ikadlivik Brooks ($n = 12$) (Figure 2b). None of the fish that negotiated Reid Falls and entered Reid Pond exhibited a downstream movement. As of October 29th 1996, 8 fish tagged at Reid falls had migrated to the Pond, 8 char had moved downstream to lower Reid Brook, 18 fish were located in Ikadlivik Brook and 2 were found below an impassable falls on Kogluktokoluk Brook, 8 fish remained in the pool for the winter and 2 tagged char were unaccounted for (Figure 2b).

1-15th October 1996

16-31st October 1996

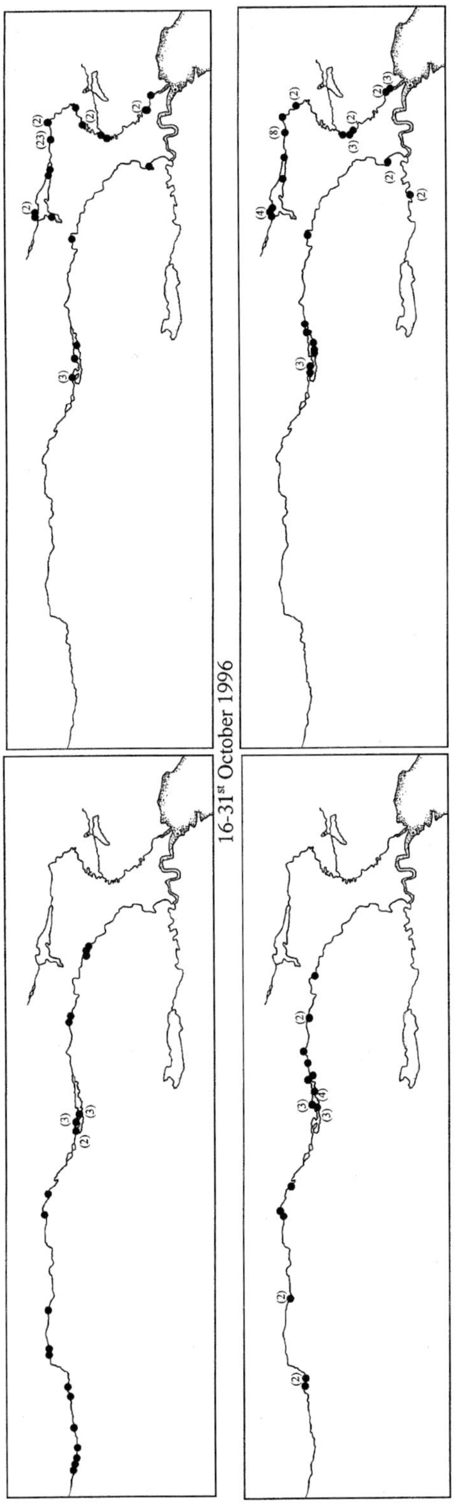

Figure 3a.

*Movements of fish radiotagged in Lower Reid Brook – August*

Sixteen fish were tagged in lower Reid Brook in early August, 6 of which died soon after surgery (Figure 1; Table 1). The 10 remaining char generally remained in areas close to where they were tagged throughout August up until late September. Fish occupied pools 1–1.5 m deep on the outside bends of the river, with substrates comprised of 40% boulders, 40% sand and 20% rubble and gravel, in water velocities ranging from 0.5–1 m $s^{-1}$. Reid Brook appears to provide suitable substratum and conditions for spawning along most of it's length and indeed many groups of spawning individuals were visible on redds throughout late September and early October. Following spawning, during late September and early October, 8 of these 10 char migrated downstream in Reid Brook, moving through into the adjacent river systems (Figure 3a).

*Movements of fish radiotagged in Lower Reid Brook – October*

Forty-six char were implanted with radio-tags in early October, 1996. The majority of these fish appeared to have already spawned as the abdominal flesh appeared flaccid. Twenty (43%) of these char, almost im-

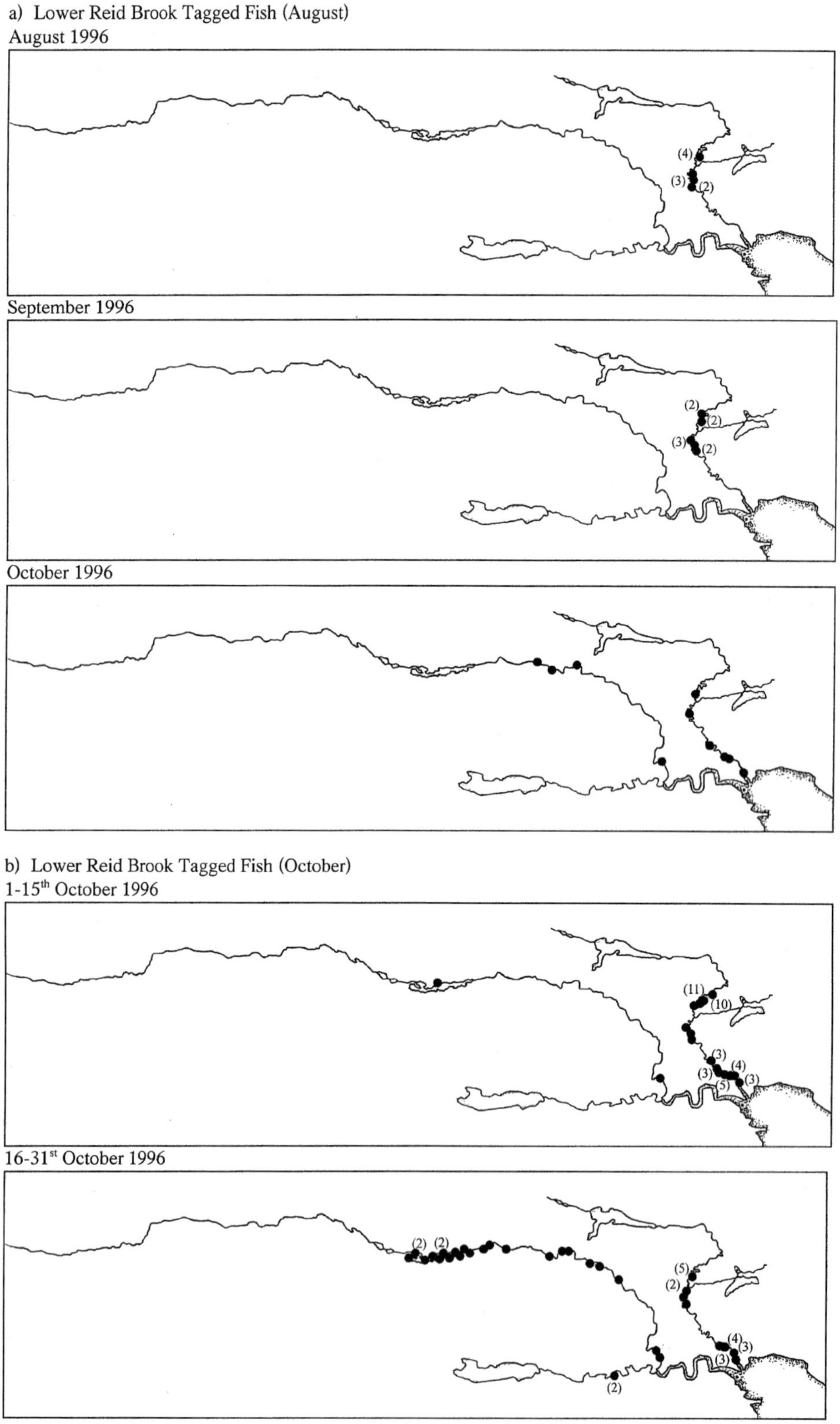


*Figure 3.* Spatial distribution of Arctic char radio-tagged in Lower Reid Brook in (a) August and (b) October. Black dots indicate fish positions, where more than 1 fish is present the number is indicated in brackets.

mediately following surgery, migrated approximately 10 km downstream to Station 1 (around the 9th October, at water temperatures below 3 °C), and subsequently moved through to Ikadlivik ($n=18$) and Kogluktokoluk Brooks ($n=2$) (Figure 3b). This exodus coincided with the main downstream migration observed for Reid Brook fish radio-tagged in August. Untagged char were also observed moving downstream during this time period and radiotagged individuals were seen swimming upstream in the Ikadlivik with groups of 5–10 non-tagged fish. Most of the fish exiting Reid Brook swam towards the large deep pools (56°21′48 N, 62°23′05 W) in the Ikadlivik where many of the char radio-tagged in the Ikadlivik also congregated.

*Summary of overall movements of Reid Brook tagged fish*

As of October 29th, 1996 the final day of tracking, 8 fish had negotiated Reid falls and had moved up into Reid Pond, 8 fish remained in the pool below Reid Falls, 62 fish (61% of all those tagged in Reid Brook) had swum out of the mouth of Reid Brook, and upstream either to just below a waterfalls on Kogluktokoluk Brook ($n=4$) or into Ikadlivik Brook. Two tagged fish were unaccounted for and the remaining radiotagged char were found in lower Reid Brook generally swimming downstream towards the mouth of Reid Brook.

*Biological characteristics*

Weight/length relationship of Arctic char did not significantly differ between the two river systems, being nearly identical for the fish examined. The relationship for Ikadlivik char was $\mathrm{Log}_{10}$ weight (g) = 1.77 + (0.027) length (cm) with an $r^2$ value of 0.84, and for Reid Brook Arctic char it was $\mathrm{Log}_{10}$ weight (g) = 1.76 + (0.028) length (cm), with a $r^2$ value of 0.92. The average length of Arctic char caught in Reid Brook was 44.8 ± 6.3 cm with a weight of 1.1 ± 0.4 kg (mean ± STD). Char caught in the Ikadlivik were generally larger than those caught in Reid Brook in terms of both length ($P<0.005$) and weight ($P<0.05$). When all the data was combined, the most common modal size range for the char within the 2 rivers was 44 ± 2 cm fork length and 1.2 ± 0.1 kg weight. The largest fish found was a 2.6 kg male in full spawning colors, 59.5 cm in length, caught in lower Reid Brook on the 4th October. More females (34%) were caught than males (21%), though by far the greatest proportion of the population were sexually immature or non-maturing adults (46%). The youngest sexually mature char determined was a 4-year-old male, weighing 0.31 kg at 30.3 cm in length. Fish ages were determined with moderate to good confidence levels according to the codes set up by the Ontario Ministry of Natural Resources. Relationship between fork length and age did not significantly differ in fish from the two river systems (Figure 4). Average age of Reid Brook fish ($n=32$) was 7.0 ± 1.9 years and did not differ from mean age of the Ikadlivik char ($n=38$) 7.1 ± 1.5 years. Six year old char were the most common age group in Reid Brook, whereas, 7-year-old char were predominant in Ikadlivik Brook. The oldest fish found was a 12-year-old female, caught in Reid Brook with a fork length of 55.1 cm.

All the stomachs examined were empty, regardless of date and location of capture, nor were any fish observed feeding. Gonadosomatic indices ranged from 0.2% to 35.7% for females, 0.1% to 6.3% for males, and was between 0.0 and 2.1% for fish that were too immature to sex accurately (Figure 5). The wide range of GSI's observed in females occurs due to the progressive maturation of eggs in spawning females and the fact that not all the females sampled were maturing. In spawning females, there was a definite increase in GSI's throughout August and September ($P<0.001$), with GSI's of females immediately prior to spawning ranging from 20–30%, with eggs 3.5–5.0 mm diameter. In late September GSI's declined significantly to pre-maturing levels for spent females, approximately 2% ($P<0.001$). GSI's of sexually mature males on the other hand, followed a much narrower range 1.3–6.3%, with no real trend being apparent over the spawning period.

Hepatic-somatic indices are indicators of stored energy levels, and are affected by gonad maturation. HSI's of char generally decreased throughout the summer, from 2.4 ± 0.8 (females), 1.5 ± 0.3 (males) in early August to 1.4 ± 0.3 (females), 1.2 ± 0.1 (males) in late September ($P<0.01$ females; $P<0.05$ males) (Figure 5). Although the decrease in HSI's was more dramatic in the females (40%) than the males (20%), by the end of September males and females had similar HSI's due to the higher initial values of the females. A significant drop in HSI's occurred in the females sampled in the last week in September ($P<0.005$), suggesting that spawning process, which demands high amounts of energy, occurred during this time.

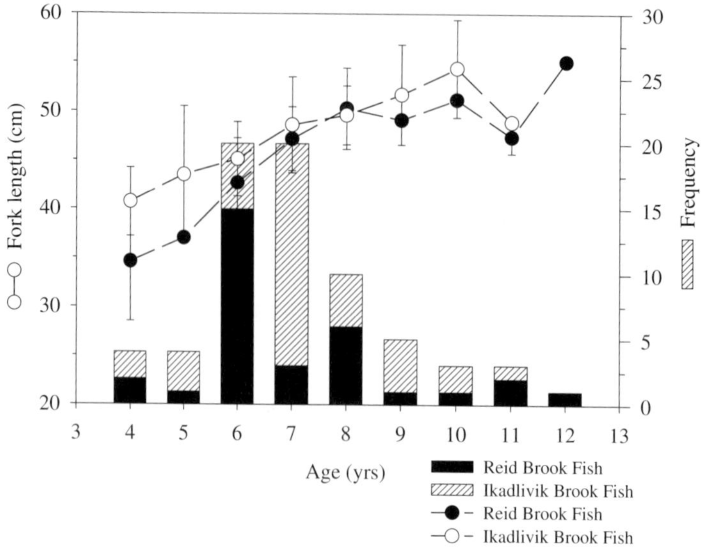


*Figure 4.* Comparison of mean fork length against age for fish from Ikadlivik and Reid Brook river systems. Values are means ± STD. Frequency of age distribution for each river system is shown on the second y axis.

## Discussion

Migratory behavior of anadromous char is known to be highly complex and variable, possibly as an evolutionary adaptation for maximizing habitat utilization and therefore potentially survival in the harsh subarctic environment (Johnson, 1980). Homing has been considered a generalized strategy, evolved to optimize reproductive success in many salmonid species (Northcote, 1978), however, given the complex and opportunistic nature of Arctic char migrations, it is a strategy that has been difficult to identify in this particular species. Arctic char have the capability to home to their natal rivers but the response appears to be very plastic in nature, with fish ranging throughout available river systems (Dempson & Kristofferson, 1987; Gyselman, 1994; Johnson, 1980). It has been postulated that only non-maturing char range into foreign freshwater systems to over-winter but return to their natal streams to spawn (Johnson, 1980), and the movement of non-maturing fish from Reid to Ikadlivik Brook in mid August during this study seems to support this. Extensive studies on char from Nauyuk Lake, N.W.T., Canada, found that general fidelity to a system was low (53%) but that fidelity of fish that had been observed returning previously was high (75%) (Gyselman, 1994). Low returns of post spawning adults was attributed to the higher mortality rate of the latter (Gyselman, 1994). On the basis of the current study it is impossible to comment on homing ability of Voisey's Bay Arctic char, it is possible that Reid and Ikadlivik Brook char, are in fact distinct populations that only intermingle during the sea phase migration (Dempson, 1984) and during over-wintering. Alternatively, it is possible that all the fish in Reid, Ikadlivik and Kogluktokoluk Brooks are from the same population and intermingle during both spawnig and over-wintering. Genetic characterizations and long term radio-tagging studies (3 years and more)/or multiple counting fence installations would be required to determine the true nature of the homing response of Arctic char.

The Arctic char in Newfoundland and Labrador, extending as far south as Pistolet Bay, Newfoundland, at a latitude of 51° N, are the most southerly populations known to undergo anadromous migrations (Scott & Crossman, 1964). The timing of upstream migration in Reid and Ikadlivik Brooks agrees closely with that observed for Arctic char by Dempson & Green (1985) in the Fraser River approximately 50 km North of Voisey's Bay, where the run began in late July and peaked in August. Adult Arctic char were found in the upper reaches of Reid and Ikadlivik Brooks by early August suggesting that the upstream run began prior to this, probably in the latter part of July.

In populations from more northerly regions, it has been reported that the upstream char migrants consist mainly of non-maturing fish, with the maturing fish remaining in the river throughout the summer prior to spawning (Johnson, 1989). In contrast, southerly char populations in Labrador, have migrant popula-

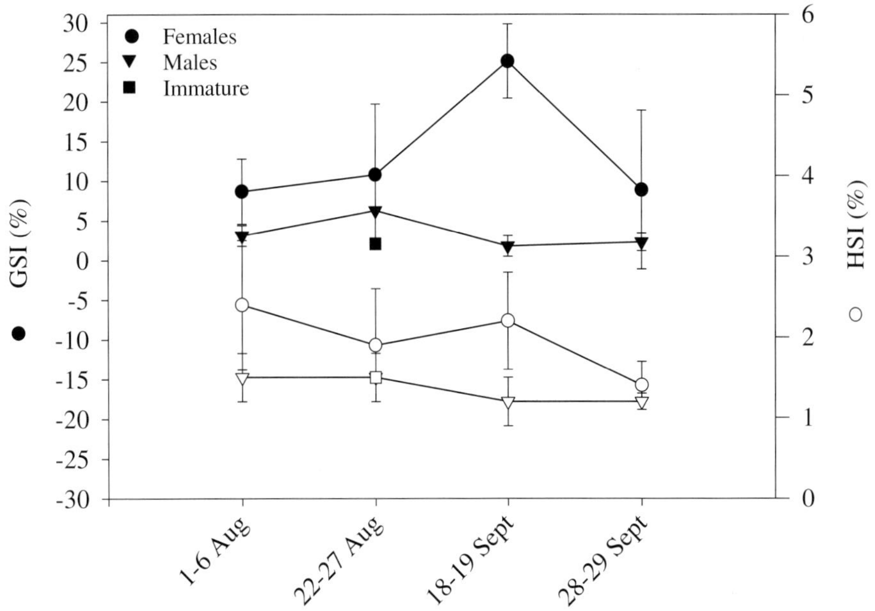


*Figure 5.* Gonadosomatic Index (GSI) (black symbols) & Hepatic Somatic Index (HSI) (white symbols) of male and female Arctic Char from Reid and Ikadlivik Brooks, Northern Labrador, over time (mean ± SD).

tions consisting of a high proportion of maturing fish, with 60% of the females migrating upstream in the Fraser River in 1979 predicted to be spawners (Dempson & Green, 1985). Similar results were found in the present study. Of the maturing fish examined in Ikadlivik and Reid Brooks, 62% were females and of the total number of fish caught during this study it was estimated that 55% of the population would spawn in the fall of 1996. The predominance of females in the spawning migration has been noted many times (Craig, 1978; Dempson & Green, 1985; Nordeng, 1983), though the reason for this phenomena remains unclear. Males tend to dominate the non-migratory portion of char populations in Norway, though equal numbers of males and females occur as first time migrants (Strand & Heggberget, 1994).

Spawning in temperate regions is believed to be largely influenced by photoperiod and water temperature, with spawning being stimulated by the lowering of water temperatures during the final stages of egg maturation (Fabricius, 1950; Scott, 1979). In Reid Brook, spawning was believed to have occurred in the last week of September and the 1st week of October, as water temperatures dropped from 7 °C to 1 °C. Spawning for anadromous char has previously been reported to occur under similar temperature conditions between 2–7 °C in Eastern USSR and at 0.5 °C in Baffin Island (Moore, 1975). The gradual increase of gonadosomatic indices in the female Arctic char, observed during August to mid September, before the dramatic reduction observed in late September combined with the increased utilization of lipid stores by females during this period, confirms spawning was most likely to occur during the last week in September. The results from this study are supported by recent work on female broodstock Arctic char, where following transformation of oocytes from stage V to stage VI in late July, a steady increase in GSI's also occurred, up to a maximum of 18.2% in mid October before falling to around 1% following manual stripping (Frantzen et al., 1997). Gonadosomatic indices in hatchery fish were generally lower than those observed for wild Arctic char, which peaked at 35.7% during this study.

Lack of food contents in the stomach supports previous observations that very little feeding occurs once char re-enter freshwater after the first summer at sea (Johnson, 1980; Boivin & Power, 1989). In other salmonids, such as Atlantic salmon, lipid stores are believed to be the primary energy reserves utilized for the physiological processes of maturation and reproduction (Kadri et al., 1996). Decreasing HSI values observed during August and September in this study are indicative of the utilization of stored lipid deposits, brought about by the increasing energy requirements necessary for reproduction. Maturing, hatchery-reared, female Arctic char were found to have lower lipid reserves than non-maturing females, with females generally having larger lipid reserves than the males (Adams & Huntingford, 1997). This supports our data suggesting that not only do females store more energy rich metabolites during the sea phase migration, but also as expected, the reproductive costs are much higher in spawning females, compared to males.

Fish that migrated from one river system to another early in the study, were probably non-maturing as no sexual characteristics were identified in these char during radio-tagging in early to mid August. This movement may be a behavioral mechanism based on straying (Quinn, 1984) where the fish routinely search for alternative over-wintering or spawning areas. Whereas, homing offers the advantages of a concerted effort in areas of known reproductive success, straying allows for the exploration of new systems, which offers certain evolutionary advantages to a species that lives in an extreme environment (Johnson, 1980). Dempson & Kristofferson (1987) stated that in such areas as northern Labrador, where several rivers of similar characteristics enter the same bay, the interchange of char among rivers may be expected to be higher than in other regions.

The main post-spawning downstream migration of Reid Brook char occurred in late October as opposed to the following spring, and coincided with a decrease in temperature (to below 3 °C), an increase in water level and the onset of ice up. This abrupt downstream migration of Arctic char could be related to the lack of suitable over-wintering habitats within Reid Brook. Indeed the only obvious location in which adult char could over-winter is Reid Pond itself or, to a limited degree, the pool below Reid Falls. Thus, it appears char spawning in Reid Brook may have developed an adaptive behavioral mechanism whereby they spawn in one river, but in the same year to avoid freezing, move into another, deeper river system in which to over-winter. This immediate post-spawning migration from one river to another has never been shown in other Arctic char or salmonid species. Similar movements to those observed for the Ikadlivik radio-tagged char have been observed for anadromous dolly varden char (*S. malma*), where fish spawn in a river and then move to a deeper lake system *within* the same river system (Armstrong & Morrow, 1980). There are obvious advantages to these migratory strategies, firstly, utilizing the maximum number of spawning grounds increases habitat availability for fry and juvenile fish, and therefore, may enhance possible recruitment into the adult population. Secondly, avoiding death from freezing in a system with limited over-wintering habitats for adults. Over-wintering areas favored by char in Reid and Ikadlivik Brooks appear to be large volumes of deep water, with perhaps some fish remaining in shallow areas of the river where some ground water seepage occurs.

Radio-telemetry offers an excellent method to gain long term information on the timing and extent of movements of individual fish within a freshwater system. The most important prerequisite to a telemetry study is that tag implantation doesn't affect the behaviour of the animal under investigation. Tag ratio was well below the accepted rule of 2% of fishes body weight, suggesting that presence of the telemetry device would not adversely affect performance of the fish in question (Mellas & Haynes, 1985). It seems reasonable to assume radio-tag implantation didn't effect migratory behaviour of Arctic char as untagged fish were observed swimming alongside radio-tagged fish during the October exodus. Radio-telemetry has advantages over conventional tag/recapture procedures being less labour intensive, remote (after the initial implantation), providing detailed information on individual fish movements over a long period of time and being able to locate fish positions through ice. Intensive 'Carlin' type tagging studies were carried out on the Labrador coast by Dempson & Kristofferson (1987) with a recovery rate of 23% in the Voisey's Bay area over a 3-year period. The advantages of the intensive studies are that detailed information on the size structure of the population can be gained, in addition to data from the fishery, though unfortunately, only limited information on the migratory routes of the fish can be determined (tagging & recapture locations). Remote data logging stations can collect information during adverse weather conditions that make it difficult or dangerous to track manually, however, these systems though reliable, are expensive and need some basic maintenance procedures in order to be left for long periods of time.

The adult Arctic char were vulnerable to predators such as river otters, black bears and human poachers. Mortality rates were highest for fish radio-tagged at Reid Falls (27%), a small percentage of which (about 3 fish) were attributed to a combination of unusually high water temperatures (> 20 °C) and the stress associated with surgery. In these cases the fish were found downstream of the tagging site 1–2 days after surgery, with the transmitter still present. Surgery was therefore suspended on days where the water temperature rose above 19°C. The other mortalities in that area were probably caused by predators, as radio-transmitters were found isolated in the water close to the banks of the rivers, or next to carcasses of half eaten fish on the banks of the river or in the bush. Overall mortality rates for radio-tagged fish were 17.5%, with no mortality observed for char implanted with radio-transmitters in October. This may be partially due to the much reduced ambient temperature (< 5 °C) and more importantly to the reduced predation in the area at that time.

Over the last 20 years Arctic char from the Voisey's Bay stock unit have contributed an estimated 10 000 fish per annum to the commercial fishery, with a total of 370 tons being removed from Voisey's Bay between 1974–1992 (Dempson, 1993). This represents a significant proportion of the commercial catch in that region, but despite the intense exploitation of the stocks in the Nain region of northern Labrador, the size/age structure of the population has remained relatively stable (Dempson, 1995). Total population of the Voisey's Bay stock is unknown, however it is probably at least 2 to 3 times larger than the 10 000 annually removed, suggesting that a minimum of 20 000 adult fish inhabit the Kogluktokouk, Ikadlivik and Reid Brook systems. The results from this study suggest that while all three rivers contain suitable spawning areas and juvenile rearing grounds, only Ikadlivik Brook provides reliable over-wintering sites for the majority of the char in this area. The complex nature of migratory journeys of Arctic char, combined with its slow growth rate and advanced age at sexual maturity, renders it highly susceptible to exploitation, and longer term radio-tagging studies are required to determine the true extent of homing versus straying in this and other populations of char.

## References


Adams, C. E. & F. A. Huntingford, 1997. Growth, maturation and reproductive investment in Arctic charr. J. Fish. Biol. 51: 750–759.

Anderson, T. C., 1985. The rivers of Labrador. Can. Spec. Publ. Fish. aquat. Sci. 81: 389 pp.

Anderson, W. G., R. S. McKinley & M. Colavecchia, 1997. The use of clove oil as an anesthetic for rainbow trout and its effects on swimming performance. North am. J. Fish. Mgmt 17: 301–307.

Andrews, C. W. & E. Lear, 1956. The biology of Arctic Char (*Salvelinus alpinus* L.) in Northern Labrador. J. Fish. Res. Bd Can. 13: 843–860.

Armstrong, R. H. & J. E. Morrow, 1980. The dolly varden char, *Salvelinus malma*. In E. K. Balon (ed.), Chars. Dr W. Junk Publishers, The Hague, Netherlands: 99–140.

Barbour, S. E., 1984. Variation in life history, ecology and resource utilization by Arctic charr *Salvelinus alpinus* (L.) in Scotland. Doctoral Thesis. University of Edinburgh, Scotland, 181 pp.

Boivin, T. G. & G. Power, 1989. Winter conditions and proximate condition of anadromous arctic charr (*Salvelinus alpinus*) in eastern Ungava Bay, Quebec. Can. J. Zool. 68: 2284–2289.

Craig, P. C., 1978. Movements of stream resident and anadromous Arctic char (*Salvelinus alpinus*) in a perennial spring on the Canning River, Alaska. J. Fish. Res. Bd Can. 35: 48–52.

Dempson, J. B., 1984. Identification of anadromous Arctic char stocks in coastal areas of Northern Labrador. In L. Johnson & B. L. Burns (eds), Biology of the Arctic Char, Winnipeg, Manitoba, May 1981. Univ. Manitoba Press, Winnipeg: 143–162.

Dempson, J. B., 1993. Summary of catch statistics for the northern Labrador Arctic charr and Atlantic salmon fisheries in 1992. Dept. Fish. Oceans Atlantic Fisheries Res. Doc. 93/5, 31 pp.

Dempson, J. B., 1995. Trends in population characteristics of an exploited anadromous Arctic charr, Salvelinus alpinus, stock in northern Labrador. Nordic J. Freshwat. Res. 71: 197–216.

Dempson, J. B. and J. M. Green, 1985. Life history of anadromous arctic charr, *Salvelinus alpinus*, in the Fraser River, northern Labrador. Can. J. Zool. 63: 315324.

Dempson, J. B. & A. H. Kristofferson, 1987. Spatial and temporal aspects of the ocean migration of anadromous Arctic char. In M. J. Dadswell, R. J. Klauda, C. M. Moffitt, R. L. Saunders, R. A. Rulifson & J. E. Cooper (eds), Common Strategies of Anadromous and Catadromous Fishes. Proceedings of an international symposium held in Boston Massachusetts USA March 913 1986. Bethesda MD USA. Am. Fish. Soc. Symp. 1: 340–357.

Fabricius, E., 1950. Heterogeneous stimulus summation in the release of spawning activities in fish. Inst. Freshwat. Res. Drottningholm Rep. 31: 57–99.

Frantzen, M., H. K. Johnsen & I. Mayer, 1997. Gonadal development and sex steroids in a female Arctic charr broodstock. J. Fish. Biol. 51: 697–709.

Gyselman, E. C., 1994. Fidelity of anadromous Arctic char (*Salvelinus alpinus*) to Nauyuk Lake, N.W.T., Canada. Can. J. Fish. aquat. Sci. 51: 1927–1934.

Johnson, L., 1980. The Arctic Char, *Salvelinus alpinus*. In E. K. Balon (ed.), Chars. Dr W. Junk Publishers, The Hague, Netherlands: 15–98.

Johnson, L., 1989. The anadromous arctic charr, *Salvelinus alpinus*, of Nauyuk lake, N.W.T., Canada. Physiol. Ecol. Japan, Spec. Vol. 1: 201–227.

Kadri, S., D. F. Mitchell, N. B. Metcalfe, F. A. Huntingford & J. E. Thorpe, 1996. Different patterns of feeding and resource accumulation in maturing and immature Atlantic salmon. Aquaculture 142: 245–257.

Mellas, E. J. & J. M. Haynes, 1985. Swimming performance and behavior of rainbow trout (*Salmo gairdneri*) and white perch (*Marone americana*): effects of attaching telemetry transmitters. Can. J. Fish. aquat. Sci. 42: 488–493.

Moore, J. W., 1975. Reproductive biology of anadromous arctic char, *Salvelinus alpinus* (L.), in the Cumberland Sound area of Baffin Island. J. Fish Biol. 7: 143–152.

Nordeng, H., 1983. Solution to the 'char problem' based on Arctic char (*Salvelinus alpinus*) in Norway. Can. J. Fish. aquat. Sci. 40: 1372–1387.

Northcote, T. G., 1978. Migratory strategies and production in freshwater fishes. In S. D. Gerking (ed.), Ecology of Freshwater Fish Production. Blackwell Scientific Publications, Oxford, England: 326–359.

Scott, D. B. C., 1979. Environmental timing and the control of reproduction in teleost fish. Symp. Zool. Soc. Lond. 44: 105–132.

Scott, W. B. & E. J. Crossman, 1964. Fishes occurring in the fresh waters of insular Newfoundland. Dept. Fish. Ottawa.

Scott, W. B. & E. J. Crossman, 1973. Freshwater Fishes of Canada. Ottawa: Fisheries Research Board of Canada.

Strand, R. & T. G. Heggberget, 1994. Growth and sex distribution in an anadromous population of Arctic char in Northern Norway. Trans. am. Fish. Soc. 123: 377–384.

Quinn, T. P., 1984. Homing and straying in Pacific salmon. In J. D. McCleave, G. P. Arnold, J. J. Dodson & W. H. Neill (eds), Mechanisms of Migration in Fishes. Plenum, New York, USA: 357–362.

*Hydrobiologia* **371/372:** 263–274, 1998.
*J.-P. Lagardère, M.-L. Bégout Anras & G. Claireaux (eds), Advances in Invertebrates and Fish Telemetry.*



# Environmental unpredictability rules the autumn migration of brown trout (*Salmo trutta* L.) in the Belgian Ardennes

M. Ovidio, E. Baras, D. Goffaux, C. Birtles & J. C. Philippart
*University of Liège, Laboratory of Fish Demography and Aquaculture, Chemin de la Justice, 10, B-4500 Tihange, Belgium. E-Mail: M.Ovidio@ulg.ac.be; E.Baras@ulg.ac.be*

*Key words:* spawning migration, temperature, flood, *Salmo trutta*, radio tracking, River Meuse Basin

## Abstract

This telemetry study aimed to document the mobility of *Salmo trutta* in the River Ourthe sub-basin (tributary of the River Meuse) during summer and autumn, and to analyse the environmental factors which trigger spawning migration or limit their extension. Nine trout (233–2217 g and 26.6–55.2 cm FL) were radio-tagged with intraperitoneal radio transmitters and positioned daily, from 14 August 1996 to 15 January 1997. Until 1 October, fish showed restricted movements: daily journeys never exceeded 300 m and corresponded to displacements by high floods or to routine home range movements. From 7 October to 15 November, seven of the nine trout travelled upstream over distances from 5.6 to 22.95 km, into tributaries and sub-tributaries. Migration speed was fast during the early days, when trout could travel over more than 5 km per night, then progressively decreased as they were approaching putative spawning redds under lower temperature. Both in the River Ourthe and in the Aisne stream, all migrations started within less than three weeks (early October) and were found to be triggered by the combination of three environmental factors: high variations of water temperature and water level between consecutive days, within a thermal range of 10–12 °C. From the trout point's of view, these may be signs that the environment becomes unpredictable, as its variability increases within a thermal range which no longer enables them to achieve high growth rates. These results are discussed within the context of foraging strategies, life history strategies and management of trout population.

## Introduction

The migration of trout *Salmo trutta* L. has been abundantly documented (Shetter, 1968; Solomon and Templeton, 1976; Harcup et al., 1984; Baglinière et al., 1987; Hesthagen, 1988; Jonsson, 1991). However, recent studies questioned the nature of restricted movements of brown trout and urged the need for a detailed and precise analysis of their space utilisation strategies at the individual level (Gowan et al., 1994; Fausch & Young, 1995; Ovidio et al., 1998). This analysis at the individual level becomes more crucial during the spawning migrations, when mobility also corresponds to a flow of genes through the ecosystem (Philippart & Baras, 1996) and when interferences with genuine mobility patterns may affect the structure of the population.

In Belgian waters, most trout populations have been deeply affected by restocking practices and by hydraulic works, which caused the extinction of most anadromous species or ecotypes, including the Atlantic salmon *Salmo salar* L. and the sea trout *Salmo trutta trutta* L. (Philippart, 1987; Philippart et al., 1988, 1994). In these ecosystems, there is a dearth of knowledge on the actual mobility patterns of trout, the location of their spawning redds in tributaries or sub-tributaries, and on their actual chances to access these, considering the potential impact of small weirs and dams which are often regarded as obstacles with no or minor effect. This study aimed to study the spawning migrations of trout in the Belgian Ardennes, focusing on the timing of spawning migrations, and on the role of environmental variables in triggering, enhancing or inhibiting the spawning migration of trout.

These may represent key factors in the efficient management of rivers within the scope of ecohydraulic development (Leclerc et al., 1996). Considering the need for detailed information on trout movements, the study relied on the day-by-day location of individuals equipped with radio telemetry transmitters.

## Study area

The River Ourthe is the main sub-basin of the River Meuse in Southern Belgium. It runs through the Belgium Ardennes where it meets typical salmonid tributaries such as the Amblève or Aisne streams (Figure 1). The study area was located in between the villages of Chanxhe (Ourthe) and Erezée (Aisne). This section is devoid of any major dam but contains small weirs, of less than 1 m in height, which are presumed to have little impact on the free circulation of trout (Figure 1). The R. Ourthe in this area has a mean slope of 1.5‰, a mean width and flow of 25 m and 32 $m^3\ s^{-1}$, respectively. It is typical of the barbel and grayling zones and hosts a mixed cyprinid-salmonid fish assemblage of which the main constituents are: the barbel *Barbus barbus* (L.), chub *Leuciscus cephalus* (L.), nase *Chondrostoma nasus* (L.), grayling *Thymallus thymallus* (L.) and trout *Salmo trutta* (L.). The Aisne is a typical trout stream (slope 2.78‰, mean width and flow: 14 m and 2.41 $m^3\ s^{-1}$) which meets the R. Ourthe in Bomal (Figure 1). The fish assemblage mainly consists of trout, grayling, sculpin *Cottus gobio* and stone loach *Barbatula barbatula*. The water temperature in the R. Ourthe and Aisne stream varies over the annual cycle, from 0 °C to 26 °C (mean: 10.5 °C), and from to 0 °C to 19 °C (mean: 9.4 °C), respectively.

## Materials and methods

In spring and summer 1996, trout were captured by electric fishing (EPMC, 2.5 kVA) or caught in a fish trap (controlled daily) at the weir of Bomal on the Aisne stream, 400 m upstream of the confluence with the R. Ourthe. All trout were tagged according to the methodology evaluated by Birtles et al. (1995). They were anaesthetised with a 0.25 ml $l^{-1}$ solution of 2-phenoxy ethanol, then placed ventral side up into a v-shaped support adjusted to their morphology. A mid ventral incision was made between the pelvic girdle and the anus and an alcohol sterilised transmitter (40 MHz, internal coiled antenna; Table 1) was inserted into their body cavity. The weight of the transmitter ranged from 3.7 to 22.0 g depending on trout body weight, making sure that the transmitter to fish body weight ratio in the air would not exceed 2.0%. The incision was closed by two or three separate stitches, 9–10 mm apart, using sterile plain catgut (2.0, 3.0 or 4.0 Dec., depending on fish size) on cutting needles. Fish were released precisely at their capture site as soon as they had recovered posture and spontaneous swimming (about 3 min after surgery), in order to minimise possible biases originating from long term post operative care.

Fish were located at least once a day until the end of their transmitter battery life or latest until 15 January 1997, in order to document all movements during the spawning period. Locations were made by triangulation from markers on the rivers' banks, using a mobile FieldMaster radio receiver and a loop antenna (ATS). All locations were made during daytime, with an accuracy from 1 to 4 $m^2$, depending on the distance between the fish and the observer. Net daily journeys were defined as the distance between diurnal locations at two consecutive days and were used to quantify migratory activity. The relationships between the mobility of trout and environmental variables were investigated by stepwise multiple regression analyses. In order to identify the key variables, we compared the relative frequency (%) of net daily journeys longer than the extension of the daily activity area (30 m) at the weekly level over 13 weeks, to the variations of day length, water temperature and level (mean, standard deviation, weekly and daily variation), using linear variables, semi-logarithmic and logarithmic transformates. The relevant variables were later categorised and tested with multiple analyses of variance to identify more precisely the sets of environmental conditions which favoured, enhanced or inhibited the migration of trout.

Temperature in the River Ourthe and Aisne stream was logged every 30 min on data loggers installed in Hamoir and Juzaine, respectively, with an accuracy of 0.1 °C. Water levels were measured daily with an accuracy of 1 cm, on limnimetric scales installed near the temperature loggers (Figure 1).

## Results

### *Aisne stream*

Trout in the Aisne stream showed contrasted mobility patterns depending on season. From mid-August to

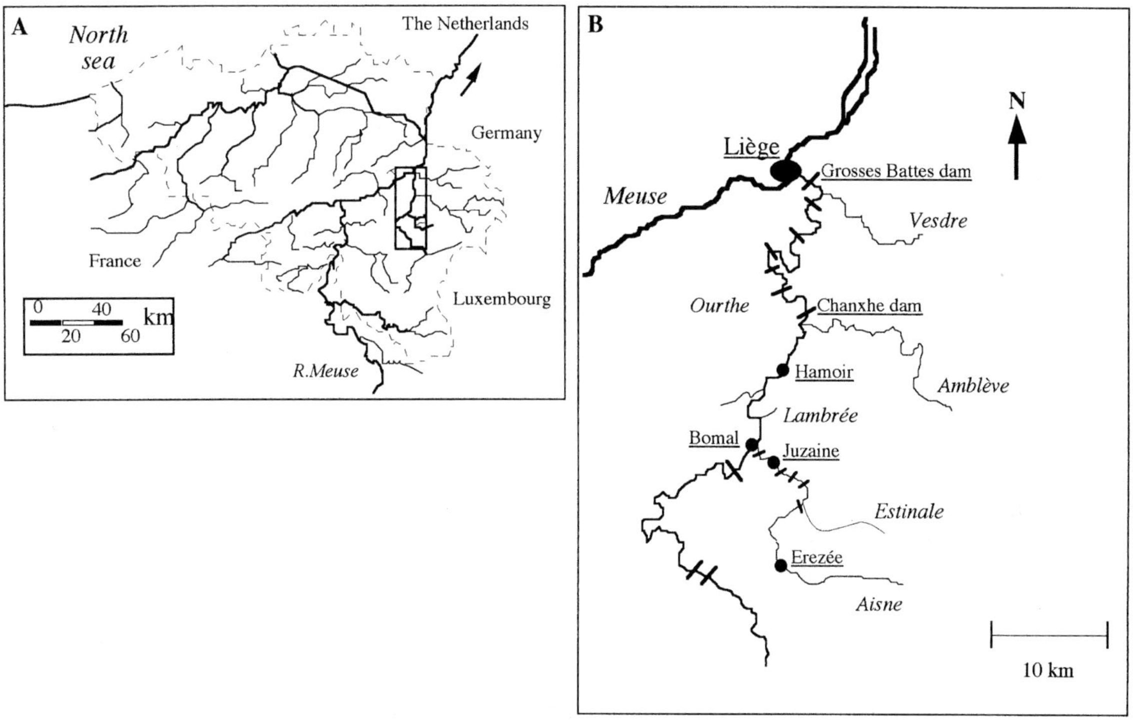


*Figure 1.* Location of the study area on the River Ourthe and the Aisne stream in Southern Belgium (Wallonie), in between Liège and Erezée. Transverse bars on the River Ourthe and on the Aisne stream on map B are weirs and dams which may interfere with the free circulation of fish.

*Table 1.* Characteristics of the radio-tagged trout in 1996. Fish code TA and TO stand for fish released in the Aisne stream and River Ourthe, respectively. TW and TBWR are the transmitter weight and transmitter to body weight ratio in the air. Trout captured in fish pass (FP) at weirs or by electric fishing (EF). All fish released on the day of capture, except for TO2 (∗)

| Fish code | Length (FL, cm) | Weight (g) | Sex | T W (g) | TBWR (%) | Date of capture | Place, method of capture | |
|---|---|---|---|---|---|---|---|---|
| TA1 | 32.5 | 398 | M | 3.7 | 0.9 | 22 Apr | Bomal | FP |
| TA2 | 28.0 | 305 | F? | 3.7 | 1.2 | 14 Aug | Juzaine | EF |
| TA3 | 28.8 | 271 | | 3.7 | 1.4 | 14 Aug | Juzaine | EF |
| TA4 | 26.6 | 233 | | 3.7 | 1.6 | 14 Aug | Juzaine | EF |
| TA5 | 27.2 | 233 | M | 3.7 | 1.6 | 14 Aug | Juzaine | EF |
| TA6 | 29.5 | 287 | | 3.7 | 1.3 | 19 Sep | Bomal | EF |
| TO1 | 43.8 | 1083 | | 12.5 | 1.1 | 08 May | Grosses Battes | EF |
| TO2 | 48.0 | 1550 | | 22.4 | 1.4 | 04 Jun | Chanxhe | EF∗ |
| TO3 | 55.2 | 2217 | M | 20.0 | 0.9 | 19 Sep | Chanxhe | EF |

early October, they showed most restricted movements which never exceeded 300 m. The longest movements were observed during a spate in late August, which coincided with a sudden summer rainstorm which raised the water level by 1 m and presumably displaced the trout from their residences (Figure 2). Trout A1, which was located downstream of the weir at Bomal consistently occupied this place during autumn and winter and made no migration, even when the weir could be cleared at any time under high flood. The five other trout in the Aisne stream made much more extensive upstream movements, which all started from early to mid-October 1996.

Trout A6, which occupied the most downstream position, upstream of Bomal's weir was the first fish to initiate its migration on 6 October. The four trout in Juzaine all started migrating within the same week: 14 October for TA2, 17 October for TA3 and 20 October for TA4 and TA5. TA3 was lost two days after it left its summer home range and had travelled more than 2.40 km. All other trout were located daily and migrated over considerable distances compared to the length of the Aisne stream (app. 50 km). Trout TA5 entered a sub-tributary, the Estinale stream, whereas all other trout remained within the Aisne stream. Small weirs in the Aisne stream (see Figure 1) sometimes caused short migration delays (maximum two days) but turned out to be cleared easily by the fish, at least under rising water levels: trout TA6 and TA5 cleared five weirs each, TA4 cleared four weirs and TA2 cleared two. The signals of TA4 and TA5 were lost in early November, when their battery failed. TA2 which had covered 5.2 km within 3 days, abruptly stopped moving whilst other fish were still migrating. Its transmitter was found lost on the gravel substratum in early December. TA6 also abruptly stopped migrating and was consistently located at the same place until the end of its transmitter battery life, in early December. Frequent walks in the stream in late November never caused any movement by the fish, which was presumed to have lost its transmitter as well. Transmitter losses in fish which normally have healed their incision may suggest a papilla exit at the time trout lay their eggs. This however could not be confirmed as high turbidity in the Aisne stream did prevent any direct observation of trout spawning during autumn.

TA6, 5, 4 and 2 travelled over 12.94 km within 13 days, 19.68 km within 22 days, 22.95 km within 13 days and 5.3 km within 3 days, respectively, corresponding to mean upstream progression speeds from 0.99 to 1.56 km $d^{-1}$ (3373 to 5856 body length – BL – $d^{-1}$). Migrating trout were not travelling continuously over 24 hours. Except for some small scale movements during daytime, the migration took place at night. Three net daily journeys were longer than 5.00 km (5.97 km for TA5; 5.98 and 7.20 km for TA4). Considering a mean nocturnal duration of 14 hours in mid-October, the mean upstream progression rate of trout during these long journeys ranged from 0.43 to 0.54 BL $s^{-1}$. As the mean water velocity in the Aisne stream ranged from 30 to 50 cm $s^{-1}$, depending on environment, trout TA5 and TA4 presumably had swimming speeds from 1.53 to 2.42 BL $s^{-1}$ over the night.

Stepwise multiple-regression analyses indicated that mean water temperature, the amplitude of daily variations of water temperature and level (regardless of whether such variable were increasing or decreasing), were significantly correlated to the frequency of fish movements ($P = 0.002$, 0.021 and 0.053, respectively). These three variables were categorised for a 3-way ANOVA over the length of net daily journeys (Table 2A). This analysis clearly indicates that trout show a significantly higher propension to migrate under varying water level and temperature between 10 and 12 °C than under any other set of environmental conditions. The comparison between the occurrence of this special set of environmental conditions throughout summer and autumn and the mobility of trout, indicates that this condition was fulfilled only three times before all trout started their migration (October 2, 3 and 20). TA6 left its summer home range on the first day when it happened and TA5 and TA4 together on the third occasion. TA2 started moving under similar thermal conditions, but under a stable water level. This finding substantiates the influence of environment variability on the propension of trout to migrate.

*River Ourthe*

As in the Aisne stream, all three trout showed restricted movements during summer and early autumn but were displaced downstream by summer high floods (Figure 3). TO1 consistently remained downstream of the Grosses Battes dam until mid-December, when the water temperature decreased down to 0 °C. It left this place down to the River Meuse, which was 2 to 3 °C warmer but obviously made no spawning migration as no spawning areas are available in the R. Meuse. TO2 made an early 2 km upstream movement on 11 October then resumed its upstream migration on 22 October, after TO3 had arrived in this place.

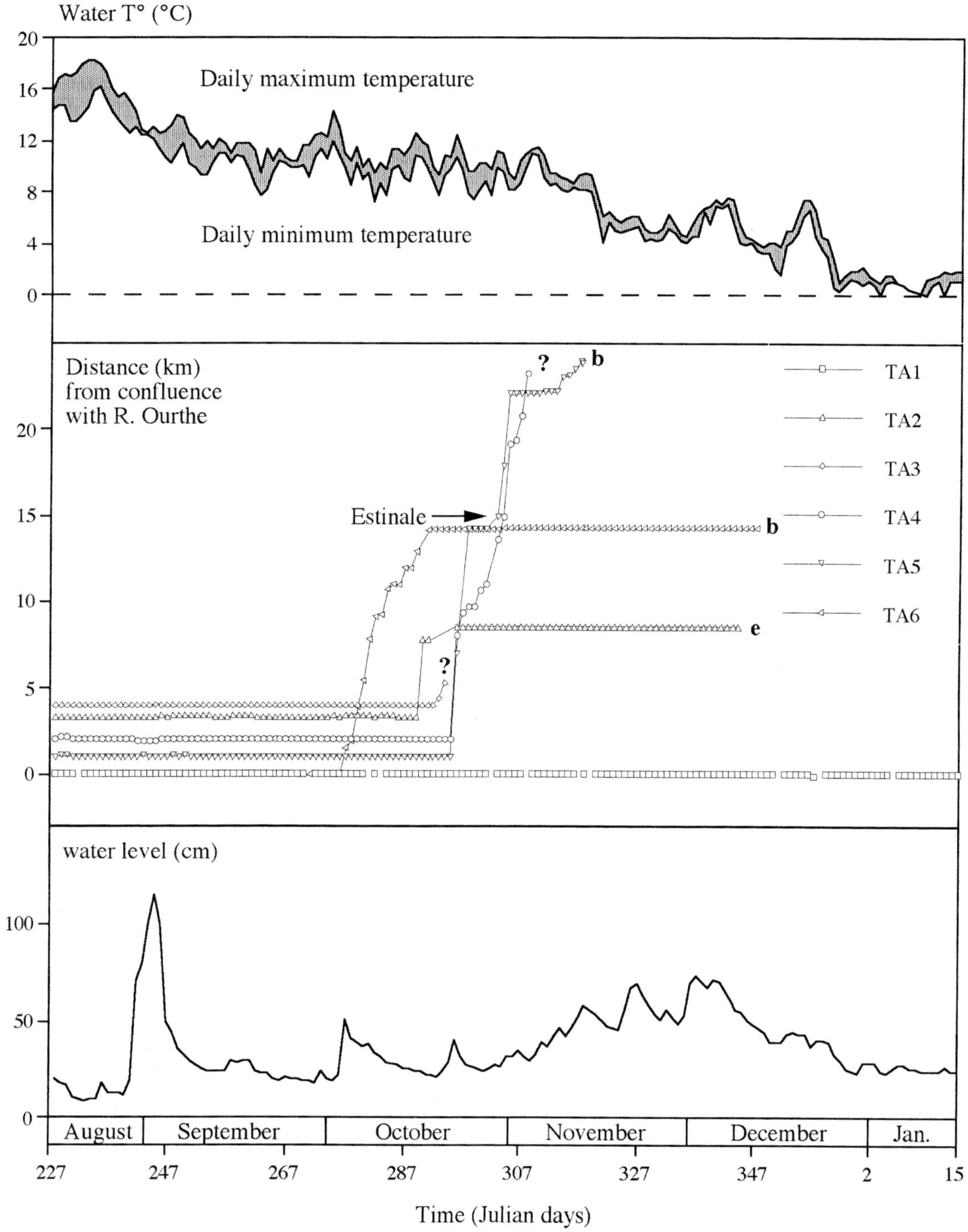


*Figure 2.* Daily variations of water temperature (°C) and water level in the Aisne stream, from 14 August 1996 (JD 227) to 15 January 1997 (JD 15). Symbols on the central graph are positions of radio-tagged trout in the stream, with the confluence with the River Ourthe sited at 0 km. b: end of the transmitter battery life, ?: fish lost; e: expulsion of the transmitter.

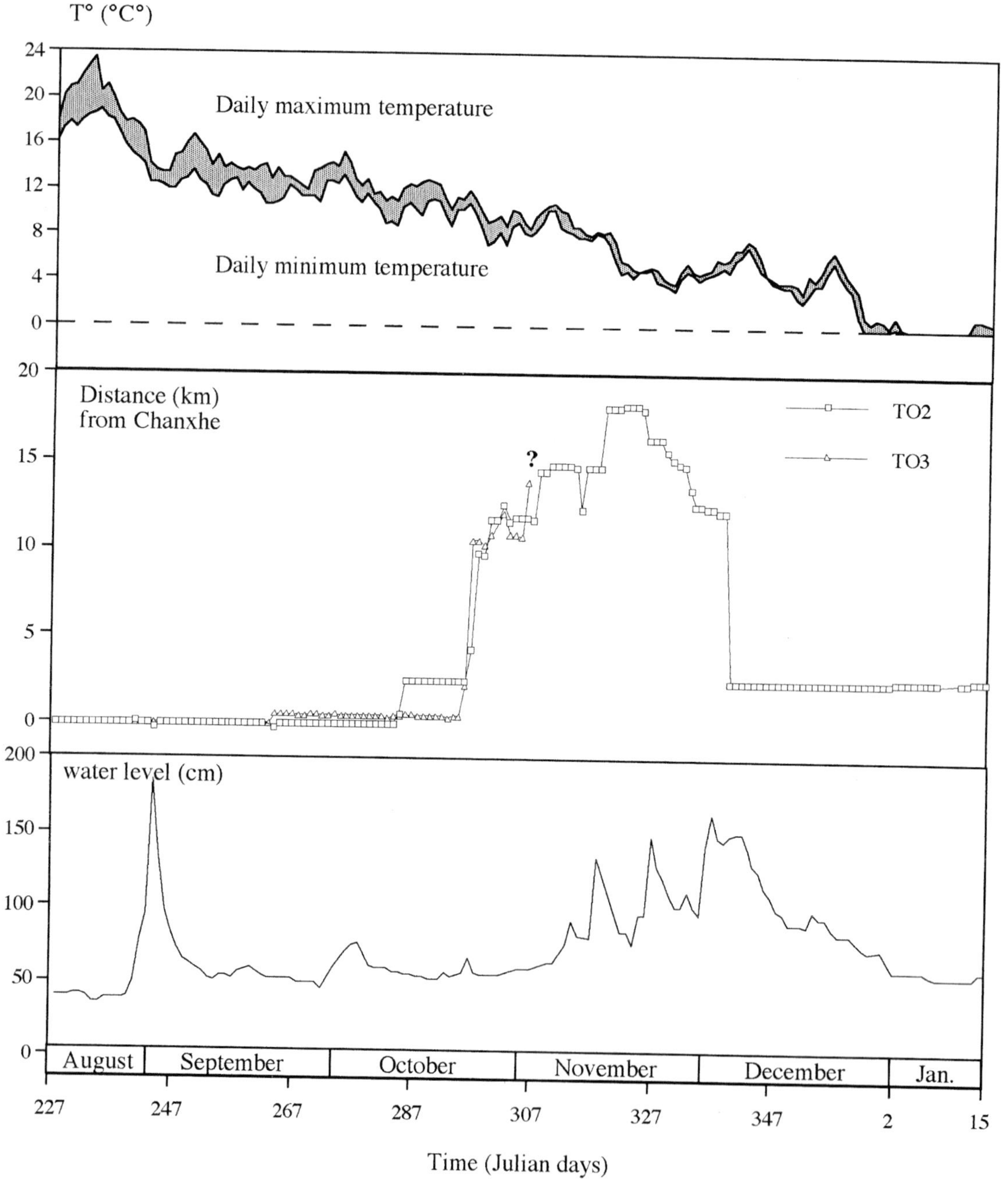


*Figure 3.* Daily variations of water temperature (°C) and water level in River Ourthe, from 14 August 1996 (JD 227) to 15 January 1997 (JD 15). Symbols on the central graph are positions of radio-tagged trout in the River Ourthe, with the dam of Chanxhe sited at 0 km. ?: fish lost.

*Table 2.* Mobility of trout in the Aisne stream (A) and River Ourthe (B) depending on water temperature (WT°), variations of water temperature (ΔWT°) and of water level (ΔWL) between consecutive days, regardless of the sense of the variation (decrease or increase). Trout TA1 and TO1, which made no migration, were excluded from the analyses. Values are means (± standard deviation) of the length of net daily journeys (m) over N observations. The values sharing at least one common upperscript (a or b) do not differ at the 0.05 level of significance (Scheffe *F*-tests)

| A | | ΔWL > 5 cm | *N* | ΔWL < 5 cm | *N* |
|---|---|---|---|---|---|
| WT° > 12 °C | ΔWT° > 1 °C | 6.7 (±6.7)$^{a}$ | 3 | 7.0 (±11.4)$^{a}$ | 10 |
| | ΔWT° ≤ 1 °C | 21.5 (±6.9)$^{a}$ | 24 | 8.1 (±5.0)$^{a}$ | 27 |
| 10 °C ≤ WT° ≤ 12 °C | ΔWT° > 1 °C | 1587.1 (±591.0)$^{b}$ | 15 | 442.5 (±282.0)$^{a}$ | 26 |
| | ΔWT° ≤ 1 °C | 11.6 (±3.2)$^{a}$ | 21 | 39.1 (±21.3)$^{a}$ | 78 |
| WT° < 10 °C | ΔWT° > 1 °C | – | 0 | 284.1 (±129.2)$^{a}$ | 24 |
| | ΔWT° ≤ 1 °C | 215.0 (±215.0)$^{a}$ | 2 | 331.4 (±122.8)$^{a}$ | 31 |

3-Way ANOVA: probabilities not calculated due to incomplete matrix

| B | | ΔWL > 5 cm | *N* | ΔWL < 5 cm | *N* |
|---|---|---|---|---|---|
| WT° > 12 °C | ΔWT° > 1 °C | 0.7 (±0.7)$^{a}$ | 3 | 8.3 (±3.9)$^{a}$ | 9 |
| | ΔWT° ≤ 1 °C | 39.9 (±16.7)$^{a}$ | 17 | 15.4 (±6.4)$^{a}$ | 31 |
| 10 °C ≤ WT° ≤ 12 °C | ΔWT° > 1 °C | 3887.5 (±1850.7)$^{b}$ | 4 | 138.3 (±95.8)$^{a}$ | 6 |
| | ΔWT° ≤ 1 °C | 20.0 (±28.3)$^{a}$ | 2 | 146.2 (±79.4)$^{a}$ | 32 |
| WT° < 10 °C | ΔWT° > 1 °C | 1403.6 (±1379.0)$^{a,b}$ | 7 | 243.8 (±155.1)$^{a}$ | 8 |
| | ΔWT° ≤ 1 °C | 350.2 (±141.3)$^{a}$ | 30 | 492.4 (±164.4)$^{a}$ | 31 |

3-Way ANOVA: WT° (A): $P = 0.0024$; ΔWT° (B): $P = 0.009$; ΔWL (C): $P = 0.008$
Interactions: A–B: $P = 0.006$; A–C: $P = 0.0142$; B–C: $P = 0.0002$; A–B–C: $P = 0.0074$

TO3 was lost after it had migrated over 15 km in 11 days and was never recovered, despite intensive tracking surveys over more than 60 km. TO2 went further upstream and entered the Lambrée stream on 13 November, when the water level in the River Ourthe increased by more than 50 cm and temperature dropped to less than 6.0 °C. On 16 November it moved to a gravel redd 250 m upstream where it remained for three days but could not be observed spawning, due to high turbidity and water level. On 20 November, it left the tributary, then progressively migrated down to the dam of Chanxhe. The corresponding rates of upstream migration were 741 and 1364 m $d^{-1}$ (1537 and 2470 BL $d^{-1}$), for TO2 and TO3, respectively. As in the Aisne stream, little activity was observed during daytime and most movement presumably took place at night. The longest upstream net daily journeys of TO2 and TO3 were 5500 and 8260 m, respectively. Considering a mean nocturnal duration of 14 hours in mid-October and a mean water velocity of 0.5 to 0.7 m $s^{-1}$ in the River Ourthe, these journeys would have been made at swimming speeds from 1.21 to 1.67 BL $s^{-1}$, thus at a similar speed as the trout in the Aisne stream. A 3-way ANOVA over the lengths of net daily journeys, using the same categories of environmental variables (water temperature, daily variations of water temperature and level) as in the River Aisne, revealed that trout of the River Ourthe travelled over significantly longer distances under highly variable water levels and fluctuating temperatures within the 10–12 °C range, than in any other set of environmental conditions (Table 2B). This compares with the situation observed in the Aisne stream, except for the distances, which were about twice longer in the River Ourthe. Similarly, the day when TO3 and TO2 made their first long journey (22 October), was the first day when this set of environmental conditions was observed under decreasing day length in the River Ourthe.

## Discussion

This study provided evidence that the migratory behaviour of trout in Belgian waters was ruled by precise environmental conditions. The chance that the tagging procedure or presence of the transmitter would have interfered with fish behaviour can not systematically be excluded based on healing dynamics (see Baras et al., 1998), but is presumably at a minimum as all fish had been tagged several weeks or months before they started their migration. Furthermore, we used extremely low transmitter to body weight ratios comparatively to other studies and these are most unlikely to have impaired the swimming capacity of the fish. Two fish eventually lost their intraperitoneally implanted transmitter in the latest stage of their migration, possibly because these transmitters, which were relatively small units, were expelled at the same time as the eggs.

During summer and early spring, when water temperature was above 12 °C, no or few journeys were observed and mainly corresponded to routine home range movements or changes of residence within the same daily activity area. Similar short range summer movements of brown trout, in comparison with autumn mobility, have already been described in other telemetry studies (Clapp et al., 1990; Meyers et al., 1992; Young, 1994). Some trout showed higher mobility at the time when they were displaced by high floods but eventually homed under lower water level. This observation suggests that their low mobility during summer reflects their attachment to a precise residence. This may be interpreted within the context of optimal foraging strategies in a territorial species (Bachman, 1984; Hart, 1986), especially within a thermal range (12–16 °C) which is considered as optimal for trout growth (Elliott, 1975). At a time of the year which permits fast growth and energetic investment into reproductive products, trout would presumably gain advantage by remaining in an environment which they have experienced and recognised as suitable than by exploring new patches of which the quality is uncertain.

Although no observation of spawning activity could be made due to high and turbid waters, it can be reasonably assumed that autumn migrations were related to reproduction. In this study, all trout migrating during autumn moved upstream and two of them entered sub-tributaries. Except for some studies which documented downstream migrations of spawners (e.g. *Oncorhynchus clarki* Richardson, Brown & Mackay, 1995), this finding is consistent with the abundant literature on trout spawning runs (Solomon & Templeton, 1976; Harcup et al., 1984; Baglinière et al., 1987, Jonsson, 1991). Beyond the availability of suitable physical habitat features, the reason why trout move upstream is still debated. Upstream movements by spawners may compensate the natural drift of alevins with rising waters in spring (Margalef, 1963). A complementary hypothesis refers to the effect of high floods on spawning redds. High water levels and velocities interfere with the digging of spawning pits by salmonids (Baglinière et al., 1979), and high sediment loads affect the permeability of gravel beds, making them less suitable for the embryonic development (Tappel & Bjornn, 1983). It is generally admitted that, due to their geomorphology, upper streams undergo much more intense but shorter variations of water level than lower streams and rivers. During periods of naturally high waters, such as late autumn and early winter, upstream spawners may thus find suitable conditions to spawn and for their eggs to develop, much more easily than downstream migrants.

The seasonality of trout spawning runs has been abundantly documented in various ecosystems (Jensen, 1968; Euzenat & Fournel, 1976; Solomon & Templeton, 1976; Baglinière et al., 1979, 1989; Jonsson, 1985; Euzenat et al., 1991; Maisse & Baglinière, 1991; Meyers et al., 1992; Evans, 1994; Young, 1994; Arnekleiv & Krabol, 1996; Marmulla & Ingendahl, 1996). The final ascent of spawning tributaries and sub-tributaries is known to take place in autumn, with spawning from late autumn to mid winter, consistently with the findings that their gonads only undergo the final maturation stage under decreasing photoperiod (Billard et al., 1982; Breton et al., 1983). Within this migration period, environmental variables such as temperature, flow and their variations have been shown to influence the movements of winter spawning trout (*Salmo trutta*) and of spring spawning trout (cutthroat trout *Oncorhynchus clarki*). It has been known for some time that the initial spawning movements of cutthroat trout are stimulated by increased flows (Rayner, 1942; Johnson, 1963; Allan, 1978; Brown & Mackay, 1995). Similarly, Baglinière et al. (1987) provided evidence that the spawning runs of brown trout in French rivers are conditioned by the timing of high floods. More recently, Ovidio et al. (1996) documented that flow was a major factor allowing spawners, blocked in the downstream reaches of dams and weirs which could not be cleared at low flow, to move. However, the influence of flow alone is still debated. Evans (1994) observed considerable movements of brown trout under low flow. Baglinière et al. (1987), Whelan et al. (1988) and Jonsson (1991) identified water temperature as an important factor in triggering, enhancing or inhibiting trout movement. Clapp et al. (1990) suggested that brown trout were migrating in response to seasonal changes of temperature. Meyers et al. (1992) indicated that rising temperature more than water levels was influencing the spring migration of brown trout.

This study clearly showed that it is a combination of factors, rather than a single variable, which triggered and enhanced the spawning runs of brown trout. Both in the River Ourthe and in the Aisne stream, trout started moving and travelled the longest distances when both flow and temperature varied considerably between consecutive days, within a precise thermal range (10–12 °C). From an ecological point of view, these two factors may be regarded as cues of that the environment becomes unpredictable and/or hazardous, as its variability increases within a thermal range which is below the optimum, and no longer enables trout to achieve high growth rates (Elliott, 1975). This interpretation completes the aforementioned hypothesis on the restricted mobility of brown trout during summer. A suitable and stable environment would favour the resident behaviour of trout, whilst high or increasing environmental unpredictability would presumably cause them to trade off reproduction versus feeding and growth. This is consistent with the theory and findings on life-history strategies in teleost fishes, which increase their reproductive effort per unit of time when the environment becomes hazardous (e.g. Baras, 1995; synthesis in Wootton, 1990). Our analyses further pinpoint that a single cue of unpredictability is not enough to trigger trout migration, although this is not really surprising from a functional point of view. As rivers and streams are naturally variable environments, each of the three subsets, the combination of which triggers migration, happens on several occasions at different times of the year, including in summer (e.g. summer spate in August 1996 documented in this paper). Trout responding to a single environmental cue would thus start migrating earlier in the season and would lose a part of the energetic advantage provided by the summer sit and wait strategy, that could prove detrimental to their fitness at the time they would participate to spawning. An additional cue of environmental unpredictability could be the income of early migrants, that locally increase the population density in a reach. However, the hypothesis of a density-dependent migration pressure still is speculative without additional information on the variation of population structure along time, which could not be collected at the time of the study as catch efforts probably would have strongly interfered with the genuine mobility pattern of radio-tagged fish.

This study provided new insights on the way trout adapt their behaviour and mobility pattern to environmental variations, which could have hardly been achieved without positioning the fish each and every day. Most tracking studies which provided contrasted results on the environmental cues triggering trout migration relied on the use of automatic listening stations (e.g. Marmulla & Ingendahl, 1996) or on a much looser positioning frequency by mobile tracking: once a month (Young, 1994), once every two weeks (Clapp et al., 1990; Meyers et al., 1992), once a week (Økland et al., 1996), twice a week (Meyers et al., 1992; Young, 1994), once every two days (Brown & Mackay, 1995). As the determinant cues documented in this study referred to variations between consecutive days, their influence could have hardly be revealed by these studies. The only study which provided detailed, day-by-day, information on the spawning runs of brown trout came from Evans (1994). However, it only documented the final days of the spawning migration and gave no information on their triggering.

The day-by-day positioning in this study also permitted to provide much more detailed information on the home range, migration pattern, extension of net daily journeys and on the swimming speeds which were presumably achieved by the trout. Trout in the River Ourthe and Aisne stream made seasonal upstream migrations ranging from 5.60 to 22.95 km, with a mean upstream progression rate ranging from 0.74 to 1.56 km $d^{-1}$. Long autumn migrations of brown trout are not surprising since Jensen (1968) and Jonsson (1985) provided evidence that brown trout could travel over more than 100 km. Our results compare with those of Clapp et al. (1990) who reported that five of six trout in a Michigan stream travelled over more than 10 km during autumn, with a maximum of 22.73 km in two weeks (1.62 km $d^{-1}$). Marmulla and Ingendahl (1996) documented migration rates of sea trout up to 3.30–4.00 km $d^{-1}$, over three to four days. This exceeds the mean rate of our trout but compares with some parts of their migration as we observed that four fish made net daily journeys longer than 5 km $d^{-1}$. Not surprisingly, we detected very little migratory activity during daytime (two occasions) and all long range movements of trout in autumn were made at night. This finding goes along with the observations by Clapp et al. (1990) and by Heggenes et al. (1993), who reported that brown trout tended to become highly nocturnal during winter. The migration pattern of trout in the Aisne stream and River Ourthe resembled a log curve: the daily journeys were longer or more frequent in the first days of the migration than when the fish came closer to its presumed spawning place. Meyers et al. (1992) also found that the migration of trout in Northern Wisconsin slowed down in its second part. Evans (1994), who focused on the migratory behaviour of trout in their spawning tributary, on their last days of ascent, reported no upstream progression rate higher than 2 km $d^{-1}$. The reason for reduced migration speeds at the end of the ascent may refer to lower temperatures, higher floods or search for spawning redds, but remains to be investigated.

Contrary to the findings by Young (1994), we found no major difference between the distances travelled by fish of different size in the River Ourthe and Aisne stream, neither for the mean upstream progression rates nor for the maximum length of net daily journeys. This indicates that small trout in the Aisne stream travelled over longer distances comparatively to their body length than larger fish in the River Ourthe. However, when the mean water velocity, relative to fish length, was added to the progression rate, we found that the presumed mean swimming speeds during long journeys in the two rivers were quite close (2.42 and 1.67 BL $s^{-1}$, respectively). Furthermore, the maximum swimming speed in the Aisne stream compares with the maximum record in the River Ourthe, which was documented in a previous study (Ovidio et al., 1998): a 489 mm trout made a nocturnal upstream migration over 15 km in late December 1995 (8 L–16 N), at a time of the year when the water velocity averaged 1 m $s^{-1}$ in the river, corresponding to an upstream progression rate of 0.53 BL $s^{-1}$ and an estimated swimming speed of 2.57 BL $s^{-1}$. These swimming speed estimates compare with the maximum sustained speeds over long periods, documented in the literature on salmonid swimming (e.g. *O. nerka*: 2.8 BL $s^{-1}$ over 10 h at 10 °C; Brett, 1964). This comparison suggests that trout move as fast and as far as they possibly can (limitations by day-night alternation and swimming speed) during their early migration days. As swimming speed basically depends on water temperature, early fast movements may prove an energetically advantageous solution for long range migrations.

## Conclusion

This study provided evidence that the spawning migration of trout in Belgian waters is triggered by environmental unpredictability, which causes the fish to abandon a sit and wait strategy and to move as fast as possible to their spawning places, under water temperatures which still permit fast swimming speeds. The consistency of the results in the two rivers supports the idea that spawning migration of trout in the River Ourthe sub-basin is well focused, despite the probable heterogeneity of the fish stock. It should be noted however that some fish living in the downstream reaches of dams and weirs did not migrate at all in 1996. The meaning of this hyper-resident behaviour still remains to be investigated.

All data were collected during a single year and demand further confirmation. The strong synchrony in the spawning migrations does not permit to determine to what extent the combination of environmental variables which triggered the migration of trout in 1996 was accidental. In any case, it is obvious that results and interpretations would have definitely been different with a positioning frequency less than once a day. Looser sampling would have made us unable to understand how trout adapt to their environment, and as a corollary, how they would presumably react to a natural or man-made modification of their environment. These often represent major shortcomings of models and should be considered in future studies. From a management point of view, this study suggests that any hydraulic work which buffers the variability of the environment may suppress key cues for trout migration, delay their migration and possibly impact on their reproductive success and populations.

## Acknowledgements

M. Ovidio received a Ph.D. research grant from FRIA ('Fonds pour la Formation à la Recherche dans l'Industrie et dans l'Agriculture') for a project on the behavioural ecology of trout. This research is part of the 'Meuse Saumon 2000' project, which is supported by the Ministry of Environment (D.G.R.N.E) of the Walloon Government. The authors wish to thank G. Rimbaud for field support and M. Blaude, who kindly permitted us to access his private properties when tracking trout. J. C. Philippart is a research associate of the Belgian FNRS.

## References


Allan J. D., 1978. Trout predation and the size composition of stream drift. Limno. Oceanog. 23: 1231–1237.

Arnekleiv, J. V. & M. Krabol, 1996. The effects of induced floods on the upsteam migration of adult brown trout, and the effects of water release on the postspawning downstream migration in a regulated Norwegian River. In E. Baras & J. C. Philippart (eds), Underwater Biotelemetry, Proceedings of the first conference and workshop on fish telemetry in Europe, University of Liège, Belgium: 172.

Bachman R. A., 1984. Foraging of free ranging wild and hatchery brown trout in a stream. Trans. am. Fish. Soc. 113: 1–32.

Baglinière, J. L., A. Champigneulle & A. Nihouarn, 1979. Le frai du saumon atlantique (*Salmo salar*) et de la truite commune (*Salmo trutta*) sur le bassin du Scorff. Cybium. 3: 75–96.

Baglinière, J. L., G. Maisse, P. Y. Le Bail & E. Prevost, 1987. Dynamique de la population de truite commune (*Salmo trutta* L.) d'un ruisseau breton (France) – les géniteurs migrants. Acta oecol., oecol. Appl. 8: 201–215.

Baglinière, J. L., G. Maisse, P. Y. Lebail & A. Nihouarn, 1989. Population dynamics of brown trout (*Salmo trutta* L.) in a tributary in Brittany (France): spawning and juveniles. J. Fish Biol. 34: 97–110.

Baras, E., 1995. Thermal related variations of seasonal and daily spawning periodicity in *Barbus barbus*. J. Fish Biol. 46: 915–917.

Baras, E., C. Birtles, L. Westerloppe, X. Thoreau, M. Ovidio, D. Jeandrain & J. C. Philippart, 1998. A critical review of surgery techniques for implanting telemetry devices into the body cavity of fish. In Y. Le Maho (ed.), Proceedings of the 5th European Conference on Wildlife Telemetry. Strasbourg, France, 26–30 August 1996: 10 pp.

Billard, R., A. Fostier, C. Weil & B. Breton, 1982. Endocrine control of spermatogenesis in teleost fish. Can. J. Fish. aquat. Sci. 39: 65–79.

Birtles, C., E. Baras, P. Poncin, G. Goessens & J. C. Philippart, 1995. A behavioural and histological assessment of post-tagging stress in brown trout *Salmo trutta* L. equipped with surgically implanted biotelemetry transmitters. In: Abstracts of the second Benelux congress of zoology. University of Leiden, The Netherlands: 95.

Breton, B., A. Fostier, Y. Zohar, P. Y. Le Bail & R. Billard, 1983. Gonadotropine glycoprotéique maturante et oestradiol-17 B pendant le cycle reproducteur chez la truite fario (*Salmo trutta*) femelle. Gen. Comp. Endocrinol. 49: 220–231.

Brett, J. R., 1964. The respiratory metabolism and swimming performance of young sockeye salmon. J. Fish. Res. Bd Can. 21: 1183–1226.

Brown, R. S. & C. Mackay, 1995. Spawning ecology of cutthroat trout (*Oncorhynchus clarki*) in the Ram River, Alberta. Can. J. Fish. aquat. Sci. 52: 983–992.

Clapp, D. F., R. D. Clark & J. S. Diana, 1990. Range, activity and habitat of large, free-ranging brown trout in a Michigan Stream. Trans am. Fish. Soc. 119: 1022–1034.

Elliot, J. M., 1975. The growth rate of brown trout (*Salmo trutta* L.) fed on maximum ration. J anim. Ecol. 44: 805–821.

Euzenat, G. & F. Fournel, 1976. Recherche sur la truite commune (*Salmo trutta* L.) dans une rivière de Bretagne, le Scorff. Thèse de troisième cycle Biol. Anim. Univ Rennes 1, 213 pp.

Euzenat, G., F. Fournel & A. Richard, 1991. La truite de mer en Normandie-Picardie. In J. L. Baglinière & G. Maisse (eds), La truite: biologie et écologie. INRA Editions, Paris: 183–213.

Evans, D. M., 1994. Observations on the spawning behaviour of male and femele adult sea trout, *Salmo trutta* L., using radiotelemetry. Fish. Mgmt Ecol. 1: 91–105.

Faush, K. D. & M. K. Young, 1995. Evolutionary significant units and movement of resident stream fishes: A cautionary tale. Am. Fish. Soc. Symp. 17: 360–370.

Gowan, C., M. K. Young, K. D. Fausch & S. C. Riley,. 1994. Restricted movements in resident stream salmonids: a paradigm lost. Can. J. Fish. aquat. Sci. 51: 2626–2637.

Harcup, M. F., R. Williams & D. M. Ellis, 1984. Movements of brown trout, *Salmo trutta* L., in the River Gwyddon, South Wales. J. Fish Biol. 24: 415–426.

Hart, P. J. B., 1986. Foraging in teleost fishes. In T. J. Pitcher (ed.), The Behaviour of Teleost Fishes. Croom Helm, London: 211–235.

Heggenes, J., O. M. W. Krog, O. R. Lindas, J. G. Dokk & T. Bremnes, 1993. Homeostatic behavioural responses in a changing environment: brown trout (*Salmo trutta*) become nocturnal during winter. J. anim. Ecol. 62: 295–308.

Hesthagen, T., 1988. Movements of brown trout, *Salmo trutta*, and juvenile Atlantic salmon, *Salmo salar*, in a costal stream in northern Norway. J. Fish Biol. 32: 639–653.

Jensen, K. W., 1968. Sea trout (*Salmo trutta* L.) of the river Istra, western Norway. Rep. Inst. Freshwat. Drottningholm 48: 187–213.

Johnson, H. E., 1963. Observations on the life history and movement of cutthroat trout, *Salmo clarki*, in Flathead River drainage, Montana. Proc. Mont. Acad. Sci. 23: 96–110.

Jonsson, B., 1985. Life history patterns of freshwater resident and sea-run migrant brown trout in Norway. Trans. am. Fish. Soc. 114: 182–194.

Jonsson, N., 1991. Influence of water flow, water temperature and light on fish migration in rivers. Nordic J. Freshwat. Res. 66: 20–35.

Leclerc, M., H. Capra, S. Valentin, A. Boudreault & Y. Coté (eds), 1996. Proceedings of the Second IAHR Symposium on Habitat Hydraulics, Ecohydraulics 2000, published by INRS-Eau, Québec (Canada), Vol. A: 893 pp, Vol. B: 995 pp.

Maisse, G. & J. L. Baglinière, 1991. Biologie de la truite commune (*Salmo trutta*). In J. L. Baglinière & G. Maisse (eds), La truite: biologie et écologie. INRA, Versailles, France: 25–45.

Marmulla, G. & D. Ingendahl, 1996. Preliminary results of a radio telemetry study of returning Atlantic salmon (*Salmo salar* L.) and sea trout (*Salmo trutta trutta* L.) in River Sieg, tributary of River Rhine in Germany. In E. Baras & J.C. Philippart (eds.), Underwater Biotelemetry, Proceedings of the First Conference and Workshop on Fish Telemetry in Europe, University of Liège, Belgium: 109–117.

Margalef, R., 1963. On certain unifying principles in ecology. Am. Naturalist 97: 357–374.

Meyers, L. S., T. F. Thuemler & G. W. Kornely, 1992. Seasonal movements of brown trout in Northeast Wisconsin. N. am. J. Fish. Mgmt 12: 433–441.

Økland, F., J. A. Jensen & B. O. Johnsen, 1996. Winter habitat and seaward migration of a Norwegian brown trout population. In E. Baras & J. C. Philippart (eds), Underwater Biotelemetry, Proceedings of the First Conference and Workshop on Fish Telemetry in Europe, University of Liège, Belgium: 161–171.

Ovidio, M., C. Birtles, E. Baras & J. C. Philippart, 1996. A preliminary telemetry investigation on the obstacles to anadromous Salmonids migration in spawning streams of the Belgian Ardennes (river Meuse basin). In M. Leclerc et al. (eds), Ecohydraulics 2000, Proceedings of the Second IAHR Symposium on Habitat Hydraulics. INRS-Eau, Vol A: 83–88.

Ovidio, M., C. Birtles, E. Baras, J. C. Philippart & F. Giroux, ????. Daily and seasonal movements of radio-tagged brown trout (*Salmo trutta*) in the Belgian Ardennes: mobility versus residency. In Y. Le Maho (ed.), Proceedings of the Fifth European Conference on Wildlife Telemetry in Europe. Strasbourg (France), 25–30 August 1996: in press, 6 pp (in press).

Philippart, J. C., 1987. Histoire de l'extinction et de la problématique de la restauration des salmonidés migrateurs dans la Meuse. In M. Thibault et R. Billard (eds), La restauration des rivières à saumons, INRA, Paris, France: 125–137.

Philippart, J. C., A. Gillet & J. C. Micha, 1988. Fish and their environment in large European river ecosystems. The River Meuse. Sciences de l'Eau 7: 115–154.

Philippart, J. C., J. C. Micha, E. Baras, C. Prignon, A. Gillet & S. Joris, 1994. The Belgian project 'Meuse Salmon 2000'. First results, problems and future prospects. Water Sci. Technol. 29: 315–317.

Philippart, J. C. & E. Baras, 1996. Comparaison of tagging and tracking studies to estimate mobility patterns and home range in *Barbus barbus*. In E. Baras & J. C. Philippart (eds), Underwater Biotelemetry, Proceedings of the First Conference and Workshop on Fish Telemetry in Europe, University of Liège, Belgium: 3–12.

Rayner, H. J., 1942. The spawning migration of rainbow trout at Skaneatles Lake, New York. Trans. am. Fish. Soc. 71: 180–183.

Shetter, D. S., 1968. Observations on movements of wild trout in two Michigan streams drainage. Trans. am. Fish. Soc. 97: 472–480.

Solomon, D. J. & R. G. Templeton, 1976. Movements of brown trout in a chalk stream. J. Fish Biol. 9: 411–423.

Tappel, P. D. & T. C. Bjornn, 1983. A new method of relating size of spawning gravel to salmonid embryo survival. N. am. J. Fish. Mgmt 3: 123–135.

Whelan, G. E., S. Marod & W. W. Taylor, 1988. Fisheries Report, E.L.F. communications system ecological monitoring program. U.S. Navy Electronics Systems Command, Technical Report E06548-8, washington, D.C.

Wootton, R. J., 1990. Ecology of Teleost Fishes. Chapman & Hall, London: 392 pp.

Young, M. K., 1994. Mobility of brown trout in south-central Wyoming streams. Can. J. Zool. 72. 2078–2083.

*Hydrobiologia* **371/372:** 275–281, 1998.
*J. P. Lagardère, M.-L. Bégout Anras &G. Claireaux (eds), Advances in Invertebrates and Fish Telemetry.*



# Spawning migration of sea trout (*Salmo trutta* (L)) in a Danish river

Kim Aarestrup & Niels Jepsen
*Danish Institute for Fisheries Research Dept. of Inland Fisheries, Vejlsøvej 39 DK-8600 Silkeborg, Denmark*
*Correspondence to: Kim Aarestrup, Danish Institute for Fisheries Research Dept. of Inland Fisheries, Vejlsøvej 39, DK-8600 Silkeborg, Denmark; e-mail: kaa@DFU.min.dk*

*Key words:* sea trout, *Salmo trutta*, radio-telemetry, migratory behaviour, spawning

## Abstract

From September to November in 1995 a total of 49 mature sea trout were caught and radio tagged in the estuary (Randers fjord) or at the river mouth of the River Gudenå in Eastern Jutland. The tagged trout were between 2 and 6 yr old with total body length of 56–85 cm. Twenty-five of the tagged trout ascended the river. They were tracked every third day, for up to six months, until death or descent. Great variation was found in migration pattern and duration of river residence. Some fish spawned and left the river, some died after spawning, while others died unspent. The sea trout preferred to stay on the southern side of the main river, and Males spent significantly more time of the freshwater stay in spawning tributaries than females. Most of the trout ascended the main spawning tributary, the River Lilleå, where none passed a weir, 2 km upstream the confluence, despite the presence of a fish ladder.

## Introduction

The River Gudenå is a major lowland stream in Denmark. In 1918 the construction of a hydroelectric plant and dam inhibited upstream migration of the sea trout, resulting in a 45% decrease in the sea trout population (Poulsen, 1935). Drainage, channelisation, obstructions, boat traffic and pollution today heavily influence the river system. Therefore no spawning seems to take place in the main stream of the River Gudenå.

A major problem for anadromous fish is the dam at Tange hydro-power station situated 36 km upstream the outlet, creating a 12 km long reservoir. A fish ladder (Denil-type) was constructed in 1980, but still very few trout and salmon (*Salmo salar* (L)) make their way past the dam. As a result, the entire sea trout population today, originates from small tributaries below the dam, of which the River Lilleå is believed to be the most important spawning stream.

Annual stockings of juveniles have been carried out for more than 30 yr in the system, and the run of large mature sea trout is the basis of an intensive recreational fishing in the region. Several investigations have provided information on population structure, return rates and marine migration (Dieperink, 1992), but still little is known about survival, migratory behaviour and spawning success of the sea trout in the watershed. Few investigations on individual behaviour of sea trout during spawning migration have been published. This study was initiated as an attempt to describe some of these traits in detail.

## Material and methods

### *Study area*

With 146 km the River Gudenå is the longest river in Denmark. It empties into a long (30 km) narrow estuary (Randers Fjord) and into the Kattegat (Figure 1). The fjord is shallow (< 8 m) and brackish, with salinities varying from 0–30 ppt. The mean annual discharge is 29 $m^3$ $s^{-1}$, and the elevation from spring to outlet is 69 m. Gudenå was once a productive river supporting not only large runs of salmon and anadromous brown trout, but also sturgeon (*Acipenser sturio* (L))

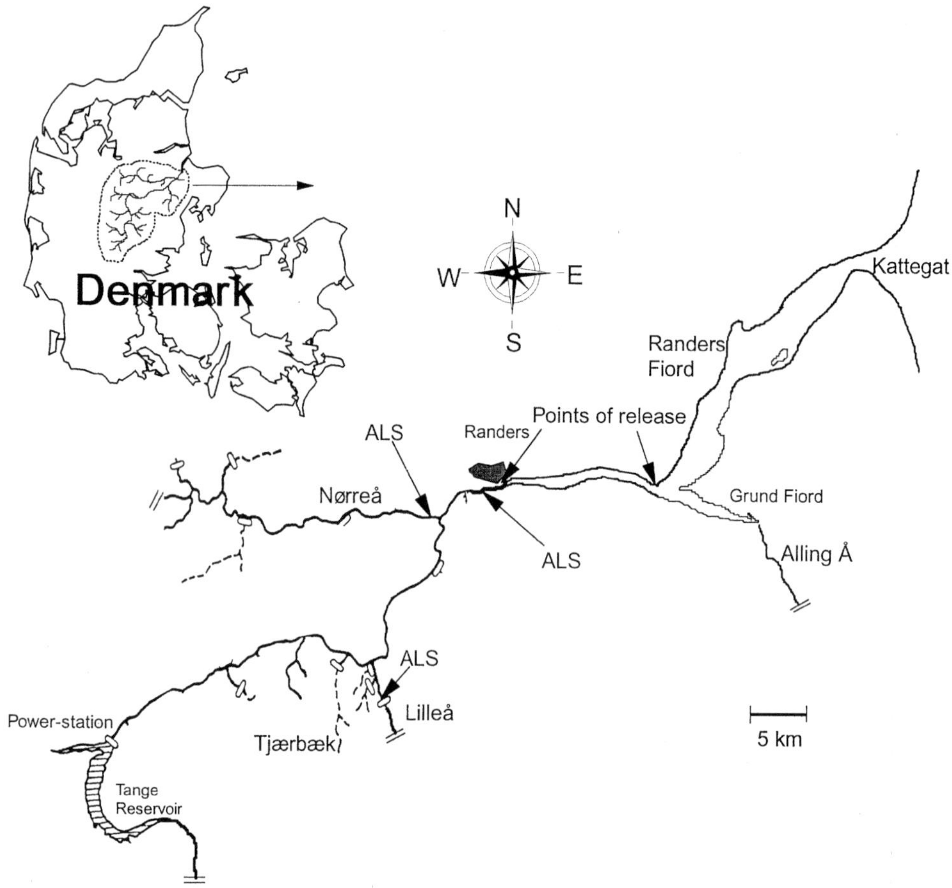


*Figure 1.* Map of the Study area, showing positions of ALS and sites of release.

before human activities made its impact on the river. Today there are recreational fishing for brown trout (both residential and sea run), salmon (sea ranched), pike (*Esox lucius* (L)), pikeperch (*Stizostedion lucioperca* (L)), eel (*Anguilla anguilla* (L)), perch (*Perca fluviatilis* (L)), whitefish (*Coregonus lavaretus* (L)), and various cyprinids. The only larger tributaries below the hydro station are the River Lilleå (mean annual discharge 2.6 $m^3$ $s^{-1}$) and the River Nørreå (mean annual discharge 5 $m^3$ $s^{-1}$). Just 2 km upstream the confluence, the River Lilleå has a weir in connection with a fish farm. A denil type fish passage is present. All main spawning areas are situated above the weir. The River Nørreå is heavily influenced by channelisation and drainage.

*Radio tracking*

The sea trout were caught in pound nets in the estuary and in a trap situated just above the tidal limit. Upon capture the sea trout were tagged and immediately released at the site of capture (Figure 1). From the net or trap the fish were carried in a plastic bag and gently placed in a water-filled tube. All fish were sized (cm), sexed, and a small scale sample was taken. During tagging, the head end of the tube was covered with a black plastic sheet, keeping the head of the fish in darkness. This kept the trout calm during tagging and sampling, and use of anaesthesia was avoided.

The radio transmitters (Model 16M and Model 7PN, standard flat salmon transmitters, Advanced Telemetry System (ATS)) were attached beside the dorsal fin (Mellas & Haynes, 1985). Model 16M is 6.2 cm in length, weighing app. 17 gr in air and 5 gr in water. Model 7PN is 5.8 cm long and weighs app. 14 gr in air and 4 gr in water. Transmitters of this size are reported to have no major effect on spawning activity of salmon larger than about 60 cm (Mellas & Haynes, 1985; Heggberget, 1988). The sea trout were selected for tagging based on 2 criteria, weight and sexual characteristics. Sea trout weigh more than salmon at the same length, therefore only sea trout larger than 55 cm were selected for tagging. Additionally, sea trout were only tagged if the secondary sexual characteristics indicated that the fish would spawn within the same season.

For individual recognition each transmitter operated on a unique combination of pulse rate and frequency. The frequencies were within the 142.000–142.500 MHz range. The tagged trout could not be tracked in the estuary, since radio signals can not be detected in salt water. To measure the time from release to river-entry, an automatic listening station (ALS) (Model D5041, ATS) was established 500 m above the trap at Randers. Two additional identical ALS's were established in the River Lilleå at a weir app. two km from confluence and in the River Nørreå (Figure 1). In the main river the fish were tracked manually by boat every third day from arrival until death or river-decent. Their position was determined with high accuracy ($\pm$ 5 m). When a fish had left a tributary or was believed dead, an attempt to retrieve it by electric fishing or netting, was made. The retrieved fish were examined for wounds and gonadal status. Live fish were returned to the river.

## Results

### *Fate of the tagged fish*

A total of 49 sea trout were radio tagged between September 13th and November 8th; 35 in the fjord and 14 in the trap. Of the 35 individuals tagged in the fjord, 11 ascended the river Gudenå, 14 were reported caught in the fjord and two were caught in a neighbour river. All fish tagged in the trap ascended the river, giving a total of 25 sea trout recorded in the river (Table 1). Of these 25, three left the river shortly after ascent and were not later recorded and two were caught on rod and line. Eight fish left the river in or after the spawning period (November 1st), and 10 died in the river. Of the two latter groups 10 were retrieved, six with empty gonads, and four with intact gonads. Additionally, one fish suddenly disappeared after four months, presumably from transmitter malfunction. Lastly one left the river before the beginning of the spawning period and re-entered later as a spent fish.

### *Time in the estuary*

The sea trout tagged in the fjord, entered the river from one to 147 d (mean: 52 d) after tagging (Table 2). One entered the river, as a spent fish, after 153 d. Maximum speed of migration through the Randers Fjord, based on time between release and entry of the river, recorded by the ALS, was 12.8 km/d. Time in estuary showed no correlation with date of tagging and seaage (Simple regression $P > 0.05$). Additionally no difference of time in estuary was found between males and females or new and previous spawners (Mann Whitney. $P > 0.05$).

Most sea trout, tagged in the trap at Randers, entered the river less than four hours after tagging, except for two that entered one and 18 d after tagging respectively. There was large variation in behaviour of individual fish. Some fish ascended and left the river again shortly after (Table 2). Others entered the river to spawn in the tributaries and as mentioned above, one fish entered after the spawning period was over, presumably as a spent fish.

### *Timing of river entry*

The first sea trout entered on September 27th, two h after it was tagged in the trap at Randers. The first sea trout tagged in the fjord entered the river on October 7th, 22 d after it had been tagged. The last sea trout entered on February 28th. The last sea trout to enter a tributary ascended the river on February 17th, 147 d after the tagging in the fjord (Table 2). No correlation were found between date of river entry and date of tagging or seaage (Simple regression $P > 0.05$). Additionally no difference of date of river entry was found between males and females or new and previous spawners (Mann Whitney, $P > 0.05$). Timing of river entry showed no pattern regarding time of day (data not shown). Because of the hydropower station the river has a rapidly changing flow regime, and it was therefore not possible to test the influence of flow on river entry. Influence of tide could not be tested, because of insufficient data on water levels.

### *Final distribution in tributaries*

A total of 17 sea trout could be designated to a tributary; 13 in the River Lilleå, two in the River Nørreå,

*Table 1*. Size and fate of the tagged sea trout.

| | Tagged (n) | Mean length (cm) | River ascend (n) | Caught in fjord (n) | Caught in river (n) | Died or left river (n) | Spawned (n) |
|---|---|---|---|---|---|---|---|
| Male | 20 | 69.6 | 10 | 9 | 1 | 4 | 3 |
| Female | 29 | 70.3 | 15 | 5 | 1 | 4 | 8 |
| Total | 49 | 70.0 | 25 | 14 | 2 | 8 | 12 |

*Table 2*. Selected data for radiotagged sea trout in River Gudenå 1995. Fish age is determined from scale samples; g = spawning. Asterisk (*) denotes trout believed to have spawned. Fish code + e: tagged in estuary, fish code + t: tagged in the trap at the river mouth.

| Fish code | Sex | Length (cm) | River- and sea age | Date of tagging | Date of river- entry | River stay (days) | Speed of migration in river (km/day) | Migration length (km) | Tributary |
|---|---|---|---|---|---|---|---|---|---|
| 3t | f | 74 | 2.3+ | 27–9 | 27–9 | 163 | 0.53 | 85.5 | Lilleå* |
| 6t | m | 77 | ? | 1–10 | 1–10 | 6 | 3.57 | 21.4 | Left river |
| 7e | f | 60 | ?.2+ | 15–9 | 7–10 | 33 | 1.28 | 42.3 | Rod caught 9–11 |
| 11t | f | 76 | 2.2+ | 18–10 | 18–10 | 93 | 0.92 | 86 | Lilleå* |
| 12t | f | 80 | 2.1.g.g.g+ | 18–10 | 18–10 | 41 | 1.22 | 50.1 | Hagenstrup brook* |
| 13t | f | 79 | 2.2+ | 19–10 | 19–10 | 42 | 0.99 | 41.5 | Lilleå |
| 14e | m | 76 | 2.2.g+ | 16–10 | 18–10 | 8 + 66 | – | – | Left 26–10, reentered 17–1 |
| 15e | f | 71 | 2.2.g+ | 18–10 | 19–10 | 22 | 1.33 | 29.2 | Lilleå* |
| 16t | m | 85 | ? | 20–10 | 21–10 | 11 | 2.76 | 30.4 | Left river |
| 17t | f | 70 | ? | 20–10 | 20–10 | 12 | 2.57 | 30.8 | Left river |
| 18t | f | 70 | ?.1.g.g+ | 20–10 | 20–10 | 26 | 1.68 | 43.8 | Lilleå |
| 22t | f | 62 | 2.2+ | 28–10 | 28–10 | 72 | 0.42 | 30.2 | Disappeared |
| 23t | f | 80 | 1.3+ | 26–10 | 26–10 | 67 | 0.83 | 55.4 | Lilleå* |
| 24t | f | 65 | 2.2+ | 30–10 | 30–10 | 115 | 0.35 | 40.2 | Lilleå* |
| 25t | f | 67 | 2.2+ | 30–10 | 30–10 | 40 | 0.81 | 32.2 | Lilleå |
| 26e | m | 60 | ?.1.g+ | 13–9 | 28–10 | 143 | 0.20 | 29.2 | Lilleå* |
| 27t | f | 76 | 1.3+ | 2–11 | 2–11 | 28 | 2.75 | 77 | Lilleå* |
| 29e | m | 72 | ?.g.g+ | 15–9 | 25–11 | 67 | 0.44 | 29.2 | Lilleå* |
| 30e | m | 72 | 1.2.g.g+ | 9–10 | ? | > 62 | – | (> 31.5) | Lilleå |
| 36e | m | 56 | 1.1+ | 23–9 | 14–1 | 23 | 1.44 | 33.2 | Tjærbæk brook* |
| 37e | f | 61 | 2.2+ | 23–9 | 17–2 | 38 | 1.01 | 38.2 | Lilleå* |
| 38e | m | 76 | 2.3+ | 9–10 | 1–3 | 4 | – | 13.8 | – |
| 39e | m | 61 | 2.2+ | 29–9 | 28–10 | 122 | – | – | Nørreå* |
| 40t | f | 76 | 2.3.g+ | 8–11 | 26–11 | 83 | – | – | Nørreå* |
| 41e | m | 66 | 1.2+ | 23–9 | 26–10 | 2 | – | – | Rod caught 28–10 |

one in Tjærbæk brook and one in Hagenstrup brook. All tributaries except the River Nørreå are known to be spawning tributaries. The fish that ascended the River Nørreå were not further tracked due to lack of manpower, and therefore not included in the following analysis.

Of the 13 fish, which ascended River Lilleå, the ALS at the weir recorded 12. No tagged fish managed to pass and continue their upstream migration.

### *Migration in river*

Duration of river stay was not correlated with fish size, date of tagging, date of river entry or seaage (Simple regression $P > 0.05$). Additionally, no difference in duration of river stay was found between males and females or new and previous spawners (Mann Whitney, $P > 0.05$). The fish which ascended a tributary ($n = 17$) had a mean time in river of 70 d (range 22–163 d) (Table 2). The large variation makes it difficult to generalise about the migration in the river. How-

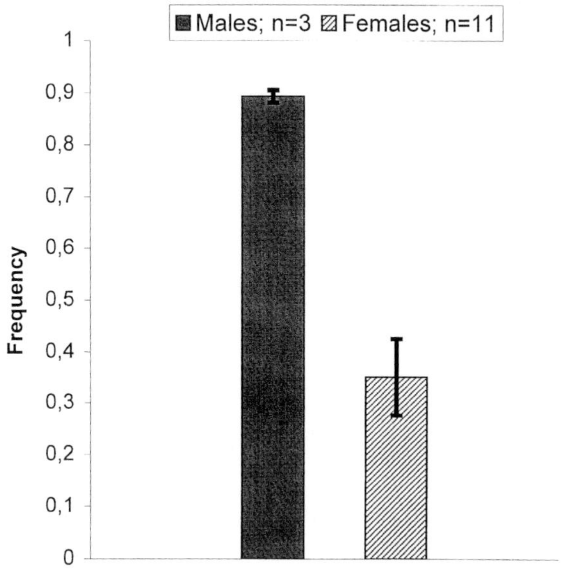


*Figure 2.* Proportion of sea trout freshwater stay, spend in spawning tributary from ascend in main river till descent from the tributary +SE.

ever a difference between males ($n = 3$) and females ($n = 11$) that entered tributaries was observed. Males spent shorter time in the main river (Mann-Whitney test, $W = 29.5$, $P = 0.025$) and longer time in tributaries (not significant) than did females. Males spend a larger part of their freshwater stay in the tributaries, measured as days from ascent to last registration in spawning tributaries (Figure 2). (Mann-Whitney test, $W = 3.0$, $P = 0.021$). No behavioural difference between new and previous spawners was observed.

*Position in river*

Sea trout preferred position on the southern side of the main river (chi square test = 33.5, df = 1, $P < 0.01$) (Figure 3). Additionally, the positioning revealed that sea trout predominantly was situated closer than 5 m to the riverbank. This was also significant (chi square test = 67.1, df = 2, $P < 0.01$).

Another aspect of the observed behaviour in the river is the phenomenon called overshooting. This was observed in eight out of 14 sea trout, that entered a spawning stream. Overshooting fish migrated further upstream the main River than the confluent of their spawning tributary. Females overshooted significantly more than did males (chi-square = 13.3; $P < 0.001$). The overshooting usually only lasted a few days with a mean of 15.8 d (range 3–69 d). The length of the overshooting had a mean of 4.6 km (range 0.1–20.6).

## Discussion

The tagging method used in this experiment, external without anaesthesia, worked very well, and has been used before with good results (Økland et al., 1994). A further indication of this is that 12 of the 14 sea trout tagged in the trap ascended further upstream within 4 h after tagging. This gives reason to assume that the handling and tagging mortality was very low.

As opposed to salmon, indigenous sea trout is still present in the River Gudenå. This may partly be due to the exceptional ability of this species to adapt to varying conditions (Nall, 1930). Such adaptability will naturally give large variation in the life history of individuals. This study has shown that the large variation is expressed also during the spawning migration. Large variation was found in timing of ascent and spawning. This could be interpreted as a way to insure spawning success of at least part of the population in a given year.

The present study has also shown that the 'standing crop' of mature sea trout in River Gudenå is not necessarily inherent to the system. The Gudenå is the largest river in the area, and its large freshwater discharge probably attracts mature sea trout from other populations, which later leaves the river to return to their home rivers. This phenomenon of attraction seems to occur even after spawning. However, the 'foreign' sea trout only ascended the lower reaches of the river in this study.

Of the 10 fish that were retrieved from the river in or after the spawning period, nine had ascended the River Lilleå. Four of the nine were retrieved dead and unspent, while the remaining fish had spawned. This indicates that some of the sea trout unable to pass an obstacle die, probably as a result of stress and injuries associated with the unsuccessful attempts to pass.

Stress from handling and tagging probably do not increase this mortality, since the unspent fish all died more than 26 d after tagging. Additionally, we found at least 5 dead, untagged and unspent trout during tracking in the river Lilleå below the weir. Finally, none of the tagged fish in the other tributaries and the main river were found unspent.

The mean river residence time in the present study is considerably shorter than found by Berg & Berg (1989). However, the present study has also shown large variation in river residence time. We found that

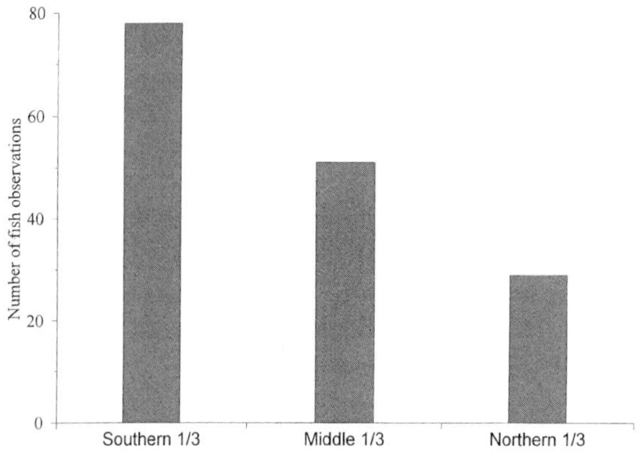


*Figure 3.* River position of sea trout. River width is divided in 3 sections.

males spent relatively longer time in spawning tributaries, than females. This is also seen in other studies (Craig, 1982). Other experiments in Denmark have shown the difference in residence time between males and females (Rasmussen, G. pers. com.). The explanation for this is probably that spawning success of males depend on availability of females on the spawning grounds, while this is not the case for females. Therefore males can optimise their spawning success by spending as much time as possible on the spawning grounds, waiting for ripe females. Fleming et al. (1996) also found males of salmon to be active for longer periods than females on the spawning grounds.

The phenomenon of overshooting observed in this study is interesting. Homing to nursery areas in salmonids is well-known (Heggberget 1989, Heggberget et al., 1986; Hovey et al., 1989).

Overshooting is also known from studies of salmon (Heggberget, pers. com.). It therefore opens the old question if salmonids navigate upstream to the 'home tributary' by following some (olfactory) clues, until it is no longer present, then reverses the migration and moves downstream until the clue is again present and the right tributary is found. Because the fish were only tracked every third day, it is possible that all fish have displayed this behaviour before entering a tributary.

The preferred position close to the riverbanks was expected, as sea trout is known for its preference for hiding (Baras & Philippart 1996). However, the fact that observations of trout in the present study were more frequent close to the southern riverbank than close to the northern riverbank and in the middle is surprising. Approximately 80–90% of the sea trout population in the River Gudenå originates from tributaries running to the Gudenå on the southern side, and the trout might keep to that side to ensure detection of the often very limited flow from the spawning tributary. Another explanation could be that the southern side of the river offers more shaded areas than the northern side, due to the position of the sun during autumn and winter.

The present study confirms the status of the River Lilleå, as the most important spawning tributary for the sea trout population of the River Gudenå. At the weir in the Lilleå there is a fish passage of the denil type, but none of the 12 sea trout registered just below the weir managed to pass the obstacle. Other studies have also indicated that denile type fish passages can be insufficient to insure passage of sea trout (Koed et al., 1996). For the sea trout population in the River Gudenå, it is crucial that the 40 km long tributary is readily accessible, not only on the lowermost 2 km.

## References


Baras, E. & J. P. Philippart (eds.), 1996. Underwater Biotelemetry, Proceedings of the first Conference and workshop on Fish telemetry in Europe, University of Liège, Belgium: 101–105.

Berg, O. K. & M. Berg, 1989. The duration of sea and freshwater residence of the sea trout, *Salmo trutta*, from the Vardnes River in northern Norway. Env. Biol. Fish. 15: 23–32.

Craig, J. F. 1982. A note on growth and mortality of trout, *Salmo trutta* (L)., in afferent streams of Windermere. J. Fish. Biol. 20: 423–429.

Dieperink, C. 1992. Opvandring af ørred og laks i Gudenåen. IFF-Report Nr. 7. Danish Institute freshwater fisheries research (in Danish).

Fleming, I. A, B. Jonsson, M. R. Gross & A. Lamberg, 1996. An experimental study of the reproductive behaviour and success of farmed and wild salmon (*Salmo salar*). J. Appl. Ecol. 33: 893–905.

Heggberget, T. G., 1989. The population structure and migration system of Atlantic salmon (*Salmo salar*). In the River Alta, North Norway. A summary of the studies 1981–86. Salmon migration and – distribution symposium (2: 1987: Proceedings of the salmonid migration and distribution). University of Washington, Seattle: 124–129.

Heggberget, T.G., 1988. Timing of spawning in Norwegian Atlantic salmon (*Salmo salar*). Can. J. Fish. aquat. Sci. 45: 845–849.

Heggberget, T. G., R. A. Lund, N. Ryman & G. Ståhl, 1986. Growth and genetic variation of Atlantic salmon (*Salmo salar*) from different sections of the River Alta, North Norway. Can. J. Fish. aquat. Sci. 43: 1828–1835.

Hovey, S. J., D. P. F. King, D. Thompson & A. Scott, 1989. Mithochondrial DNA and allozyme analysis of Atlantic salmon, *Salmo salar* L., in England and Wales. J. Fish Biol. 35 (Suppl. A): 253–260.

Koed, A. K., G. Rasmussen, G. Holdensgaard & C. Pedersen, 1996. Tangetrappen. DFU-Report Nr. 8. Danish Fisheries Research (in Danish).

Mellas, E. J. & J. M. Haynes, 1985. Swimming performance and behaviour of Rainbow trout (*Salmo gairdneri*) and White Perch (*Morone americana*): Effects of attaching telemetry transmitters. Can. J. Fish. aquat. Sci. 42: 488–493.

Nall, G. H., 1930. The Life of the Sea Trout. Seeley, Service & Co. Ltd. London.

Poulsen, E. M., 1935. Nye undersøgelser over Gudenåens lakse- og havørredbestand. Beretning til Ministeriet for Landbrug og Fiskeri fra Den Danske Biologiske Station, nr.15. C.A. Reizel. København. 1935 (in Danish).

Økland, F., T. G. Heggberget & B. Jonsson, 1994. Migratory behaviour of wild and farmed Atlantic salmon (*Salmo salar*) during spawning. J. Fish Biol. 46: 1–7.

*Hydrobiologia* **371/372:** 283–293, 1998.
*J. P. Lagardère, M.-L. Bégout Anras &G. Claireaux (eds), Advances in Invertebrates and Fish Telemetry.*



# A radio telemetry study of the migration of Atlantic salmon (*Salmo salar* L.) and sea trout (*Salmo trutta trutta* L.) in the upper Rhine

Matthieu Gerlier[1] & Pascal Roche[2]
[1] *Association Saumon-Rhin, 33a, rue de la Tour, F-67200 Strasbourg, France*
[2] *Conseil Supérieur de la Pêche, 18 rue de Nomeny, F-57950 Montigny-lès-Metz, France*
*e-mail: pascalroche@compuserve.com*

*Key words: Salmo salar, Salmo trutta,* spawning migration, radio telemetry, restoration

## Abstract

Atlantic salmon disappeared from the Rhine River in the mid-fifties when pollution and canalization of the main course of the river blocked access to their spawning areas. After water quality improved in the 1980's, a reintroduction program was begun in the French upper Rhine basin in 1992. To better define the difficulties encountered by the returning salmon on their way between the Iffezheim dam, the first main obstacle in the Rhine, and the upriver spawning areas, we undertook a radio-tracking study. Among the fish captured by electro-fishing below the dam in 1996, we tagged 20 salmon and 10 sea trout with radio transmitters, released them upriver and followed them with portable receivers and automatic recording stations. While the salmon captured during the summer interrupted their migration or even migrated downstream after being released, those captured in the autumn migrated immediately upstream. We were able to study the behavior of salmon confronting 15 obstacles. Only two salmon were recorded in the Rhine above the confluence of the Ill River, the main tributary of the Rhine on the French side. Most of the salmon made their way in the Ill toward the Bruche, a tributary of the Ill where spawning was observed. Four sea trout regurgitated their transmitters within three months. Most of the trout migrated upstream to the second obstacle in the river, the Gambsheim dam, while two of them chose to migrate up the Ill. We observed that the fish blocked by the dams in the Rhine often explored and possibly spawned in tributaries which confluence was located a short way downstream. The initial phase of our study provided useful information on the behavior of salmon and sea trout returning to the upper Rhine. It confirmed the need to improve migration passages at most of the dams and enabled us to better define migration equipment priorities. It also confirmed the good prospects of the Bruche as a spawning stream for salmon.

## Introduction

Little is known about the developing sea trout population in the Rhine and the recently reintroduced Atlantic salmon. Questions about the possibilities for these fish to reach suitable spawning grounds and the possible hatchery or stray origin of the Rhine sea trout often arise (Olbricht, 1984; Roche, 1992; Cazemier, 1994). Salmon, once very abundant in the Rhine, rapidly declined in the first half of this century when pollution increased and canalization of the main course of the river blocked access to their spawning areas (Roule, 1921; Vibert, 1950). Beginning in 1892, 21 hydroelectric and navigation dams, 10 to 15 m high, were built in the upper Rhine (German–French and German–Swiss border). They are considered to be important obstacles to fish migration (Staub, 1988), especially to salmon (Roche, 1989 and 1996) which eventually disappeared from the river in the mid 1960's. In the late 1980's, following successful efforts to restore water quality, an international project to reintroduce migratory species was coordinated by the International Commission for the Protection of the Rhine against Pollution (CIPR, 1994). The first successful attempt to bring salmon back to the river by stocking juveniles was accomplished in 1987; adult salmon returned to the Sieg basin in the middle Rhine in 1990 (Steinberg et al., 1991). Following a survey of salmon habitat and obstacles to migration in the French part of the river (Roche, 1991), a program was also introduced in 1992 in the upper Rhine. The release of 83,000 fry in 1992 and 111,000 fry in 1993 in the Ill-Bruche basin resulted in adult salmon returning to the area in 1995. That year a few salmon spawned in the Bruche, the main tributary of the lower Ill (Roche et al., 1996; Largiarder et al., 1997), and nine salmon were caught by electro-fishing at the Iffezheim dam, a site where most fish are blocked or delayed. The construction of a new fish ladder at the dam will be achieved by 1999 or 2000. Radio telemetry studies were begun in the estuary (bij de Vaate, 1996) and in the Middle Rine (Marmulla & Ingendhal, 1996) to assess the migration difficulties met by the fish to reach their spawning grounds. This paper presents the results of the initial phase of a project completed in 1997 and which was designed to study the spawning migration of sea trout and Atlantic salmon in the upper Rhine.

### *Description of the study area*

The study was carried out from June 1996 to January 1997 in the upper Rhine and in its main French tributary in Alsace, the Ill River (Figure 1). The characteristics of the environment are described in Roche (1991). The mean water flow is approximately 1100 $m^3$ $s^{-1}$ in the Rhine at Gambsheim, 40 $m^3$ $s^{-1}$ in the lower Ill and 8 $m^3$ $s^{-1}$ in the Bruche. The water temperature and flow of the Rhine, Ill and Bruche during the 1996 study are given in Figure 2. The water conductivity was in the range of 520–900 $\mu$S $cm^{-1}$ in the Rhine, 540–640 $\mu$S $cm^{-1}$ in the Ill and 170–200 $\mu$S $cm^{-1}$ in the Bruche. The Iffezheim dam, situated 689 km from the sea and 40 km north of Strasbourg, is the first obstacle blocking ascending fish in the main stem of the Rhine (Figure 1). The next dam (Gambsheim) is situated 25 km upriver and 2 km above the Ill-Rhine confluence. Access to the Ill is thus possible once the fish are above the Iffezheim dam. Spawning grounds are available in the tributaries of the Ill, mainly in the Bruche (Roche, 1994) where most of the salmon fry were introduced. The Rhine near Iffezheim and the Ill have few suitable spawning habitats.

## Material and methods

Twenty-two salmon and 63 sea trout were caught in 1996 at Iffezheim by electro-fishing from a boat; one salmon was caught by electro-fishing in the Bruche. Twenty salmon and 10 sea trout with no evidence of external injury were tagged with radio-transmitters (Table 1). Trout with total length (TL) under 500 mm were not tagged. Nineteen salmon were grisle with an average total length of 695 mm (591–756 mm, SD = 50) and one salmon caught in July was a two-sea-winter fish (TL = 815 mm). The sea trout had spent one to two winters at sea and had an average total length of 631 mm (505–750 mm, SD = 75). The fish were anesthetized with 2-phenoxyethanol (2 to 4 ml per 10 l of water depending on temperature) before they were tagged. Scale samples were taken for age determination. A radio transmitter was placed in the stomach with the help of two plastic tubes introduced into the esophagus and a 23 cm long antenna was brought out through the lower part of the gill (McLeave et al., 1978). The fish were transported in soft plastic containers filled with 50 l of water aerated with oxygen and were released 1 to 3 h after their capture. The first fish tagged were released 4 km upstream of the Iffezheim dam. The fish caught after September were released 22 km upstream of the dam after we suspected that the lack of current in the dam reservoir might be responsible for the downstream migration of three of the four first tagged salmon. Later in the season, two salmon were released directly into the Bruche in order to observe their migration in the upper stretch of the stream.

The radio transmitters (Advanced Telemetry Systems, ATS model 5) had a mortality option set up after 12 h without movement. Their battery life was 180 to 225 d and they operated in the 48–49 MHz range. The size and weight of the transmitters (55 mm and 21 g) compared to that of the fish are such that the stress due to the transmitter is considered to be negligible (Solomon & Storeton-West, 1983; Baril & Gueneau, 1986). The fish were followed with mobile receivers. Yagi car antennas were used to find the fish (field range 0–1 km) and directional loop antennas were used to locate them precisely (field range 0–0.8 km). Because of the high water conductivity we could not detect fish deeper than 4 m in the Rhine. The fish were located once a day when possible. Some fish actively migrating were followed more precisely, which allowed swimming speed estimates and observation of their behavior when facing obstacles. Two

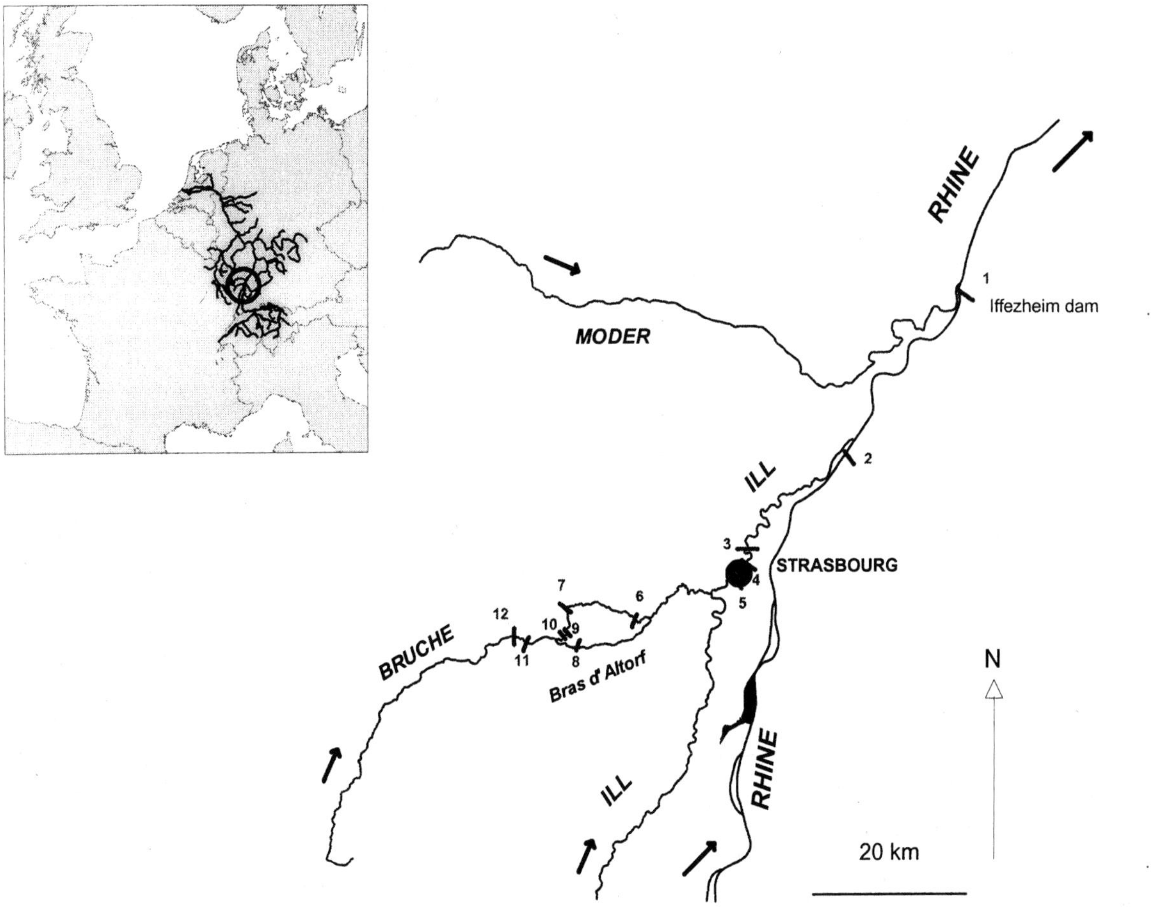


*Figure 1.* Map of the study area. Numbers refer to dams listed and documented in Figure 2.

ATS Automatic Listening Stations using Data Collection Computers were installed in the lower Ill and at dam 5 (Abattoirs; Table 2).

## Results

*Migration patterns and swimming speed*

Out of 20 salmon equipped with transmitters, two returned downstream of the Iffezheim dam after their release. One was caught in a net and died and the other disappeared before reproduction. We tracked 18 salmon upstream of Iffezheim until the spawning season.

- Two salmon entered the Ill but seemed to hesitate between the Ill and the Rhine (Figure 3). One (S1) spent the whole summer in the lower Ill, 2 km upstream of the mouth of the river, without being blocked by any obstacle. On October 3, it moved downstream back to the Rhine and tried to pass the Gambsheim dam during 12 d (October 17–28). Its signal was then recorded from November 29 to December 3 in the lower Ill where we lost it. The second (S5) slowly migrated 12 km upstream in the Ill (October 4–17) before turning back and trying to pass the Gambsheim dam in the Rhine (October 28–29). It went again up the Ill and was blocked by dam 3 in Strasbourg where it stayed from November 8 to November 22.
- Two salmon (S18, S19) migrated straight in the Ill but were also blocked by dam 3.
- Twelve salmon reached spawning habitats in the Bruche. Two others (S16, S20) were transported by car in this stream where they continued their upstream migration, one fish (S16) after descending 5 km. On November 5, a couple of tagged

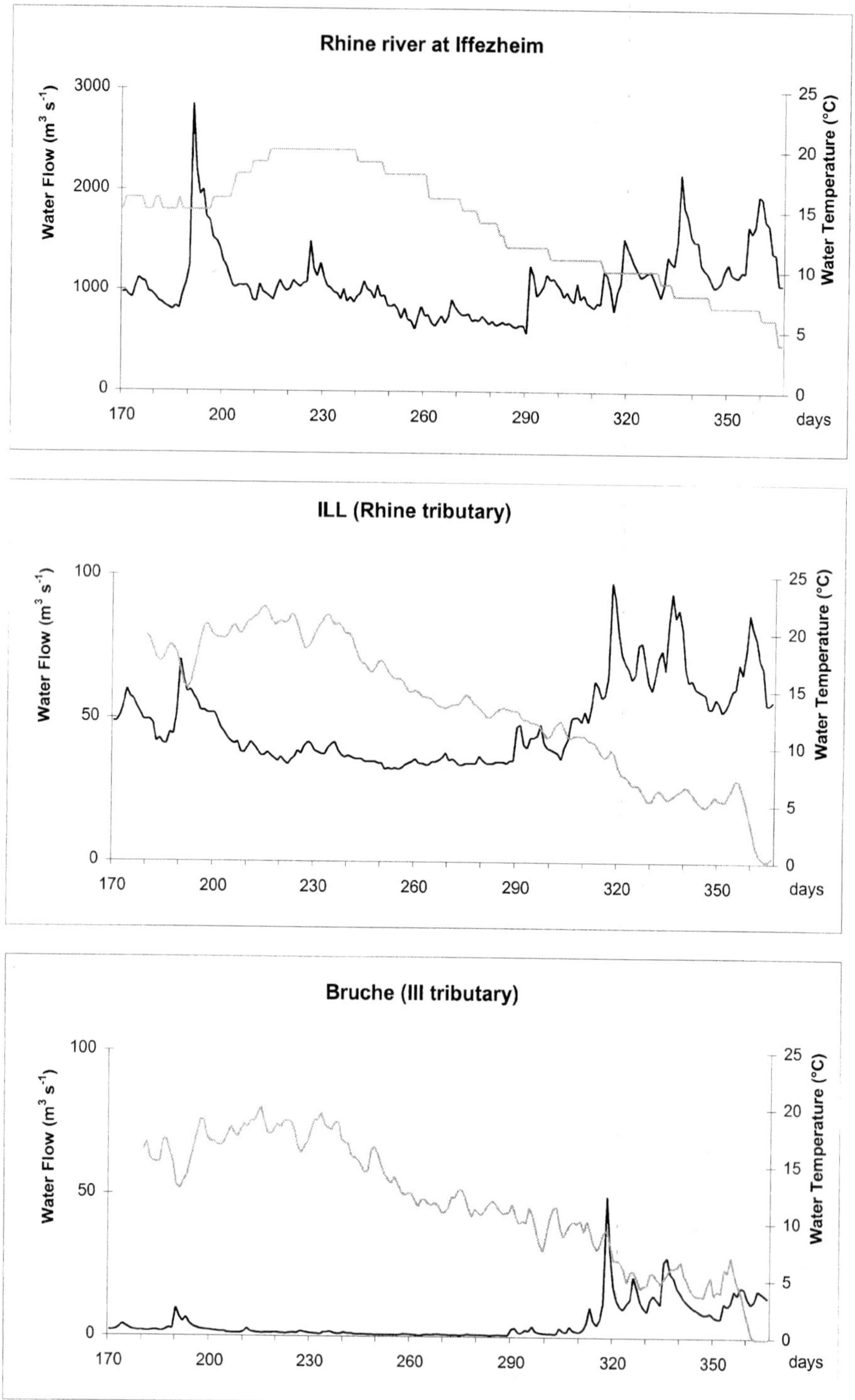


*Figure 2*. Water flows (black line) and water temperatures (grey line) during the study.

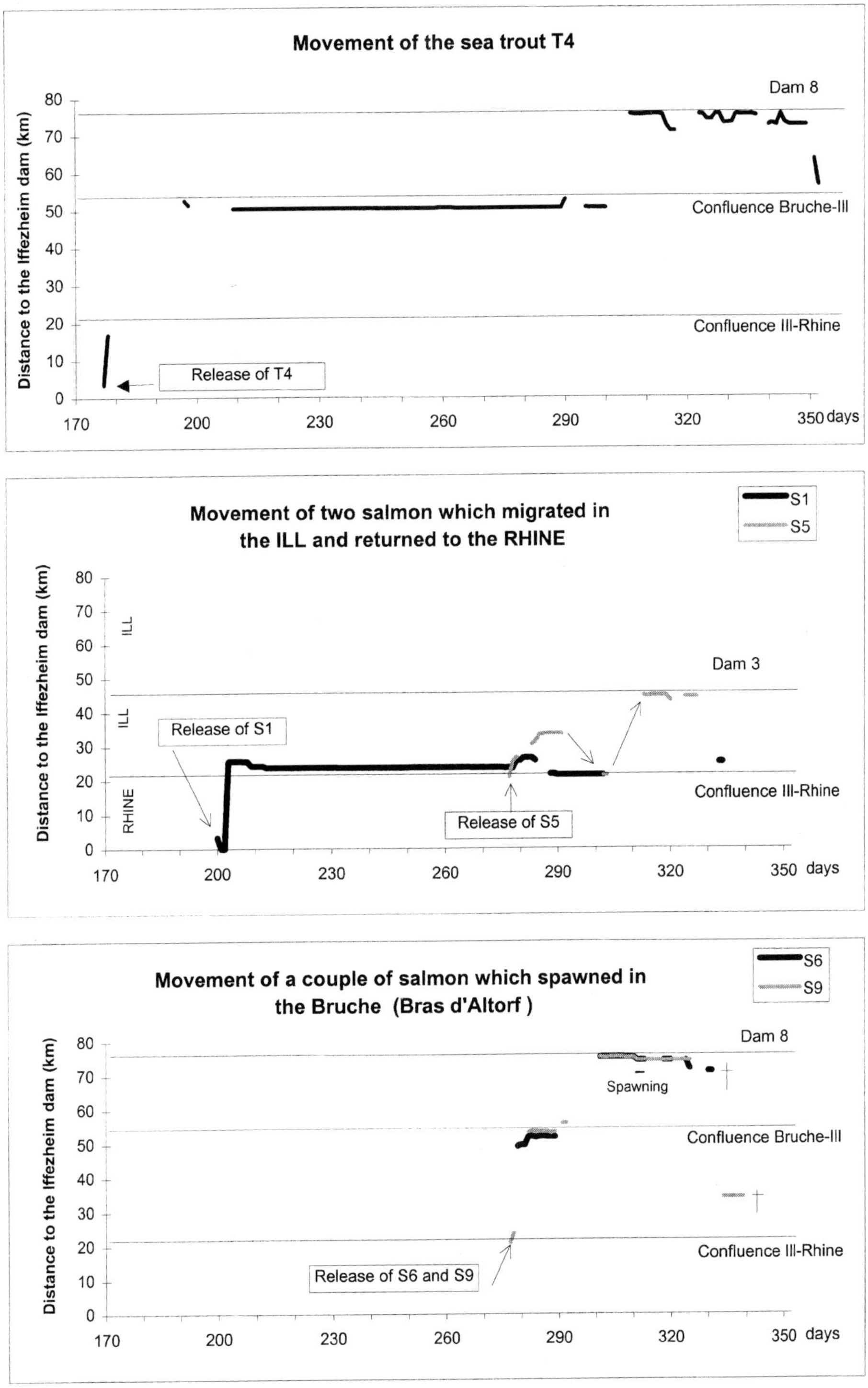


*Figure 3.* Exemples of the migratory behavior of one sea trout and four Atlantic salmon caught in the Rhine below the Iffezheim dam and released above the dam after being radio-tagged.

*Table 1*. Characteristics of the radio-tagged fish in 1996.

| Species | Date of capture | Total length (mm) | Weight (kg ) | Sexe | Age | ID | Release site: distance from Iffezheim dam (km) | Time tracked (days) | Reason of end of tracking |
|---|---|---|---|---|---|---|---|---|---|
| Salmon | July 18 | 815 | | | 2,2+ | S1 | Rhine: +4 | 139 | Ls |
| | July 23 | 591 | 1.8 | F | 1,1+ | S2 | Rhine: +4 | 84 | C |
| | August 20 | 700 | | F | 1 or 2,1+ | S3 | Rhine: +4 | 63 | L |
| | September 4 | 735 | 3.0 | F | 2,1+ | S4 | Rhine: +4 | 80 | Ds |
| | October 3 | 691 | 2.4 | F | 1,1+ | S5 | Rhine: +22 | 50 | Ls |
| | October 3 | 700 | 2.5 | F | 2,1+ | S6 | Rhine: +22 | 54 | Ds |
| | October 3 | 701 | 2.7 | F | 2,1+ | S7 | Rhine: +22 | 70 | Ls |
| | October 3 | 672 | 2.0 | M | 1 or 2,1+ | S8 | Rhine: +22 | 83 | Ds |
| | October 3 | 710 | 2.4 | M | 2,1+ | S9 | Rhine: +22 | 63 | Ds |
| | October 9 | 657 | 2.3 | M | 2,1+ | S10 | Rhine: +22 | 67 | Ds |
| | October 9 | 704 | 2.6 | F | 2,1+ | S11 | Rhine: +22 | 107 | Ds |
| | October 9 | 676 | 2.3 | M | 1,1+ | S12 | Rhine: +22 | 73 | Ds |
| | October 9 | 769 | 3.4 | M | 1,1 or 2+ | S13 | Rhine: +22 | 99 | Ds |
| | October 14 | 661 | 1.6 | M | 2,1+ | S14 | Rhine: +22 | 23 | D |
| | October 21 | 756 | 3.2 | M | 2,1+ | S15 | Rhine: +22 | 24 | Ds |
| | October 23 | 661 | 2.0 | M | 2,1+ | S16 | Bruche: +77 | 56 | Ds |
| | October 23 | 721 | 2.5 | F | 2,1+ | S17 | Bruche: +76 | 24 | Ds |
| | October 28 | 698 | 2.2 | M | 1,1+ | S18 | Rhine: +22 | 88 | Ls |
| | October 30 | 657 | 1.8 | F | 2,1+ | S19 | Rhine: +22 | 70 | Ls |
| | November 7 | 628 | 1.5 | F | 2,1+ | S20 | Rhine: +80 | 30 | Ds |
| Mean | | 695 | | | | | | | |
| SD | | 50 | | | | | | | |
| Sea trout | June 25 | 665 | | | 2,2+ | T1 | Rhine: +4 | 28 | R/D |
| | June 25 | 595 | | | 2,2+ | T2 | Rhine: +4 | 168 | R |
| | June 25 | 680 | | | 1,2+ | T3 | Rhine: +4 | 2 | L |
| | June 25 | 505 | | M | 1,1+ | T4 | Rhine: +4 | 176 | Ls |
| | July 5 | 555 | | | 1,1+ | T5 | Rhine: +4 | 46 | R |
| | July 10 | 677 | | | ?,2+ | T6 | Rhine: +4 | 1 | L |
| | July 10 | 660 | | | 1,2+ | T7 | Rhine: +4 | 9 | R |
| | July 18 | 667 | 3.7 | | 1,2+ | T8 | Rhine: +4 | 34 | R |
| | November 27 | 552 | 1.8 | F | 2,1+ | T9 | Rhine: +22 | 42 | R |
| | November 27 | 750 | 4.7 | M | 1,2+ | T10 | Rhine: +22 | 31 | Ls |
| Mean | | 631 | | | | | | | |
| SD | | 76 | | | | | | | |

L: Lost before the spawning season (November–December); Ls: Lost after the spawning season; D: Dead before spawning; Ds: Dead during or after spawning; C: Captured by a fisherman; R/D: Regurgitation of the transmitter or death; R: Regurgitation of the transmitter.

salmon started spawning. A total of five tagged salmon with three other untagged salmon were observed on five redds before flow and turbidity rose on November 12, which prevented further observation. Five salmon migrated in the main stem of the Bruche while seven migrated in a secondary channel (bras d'Altorf) fed by less than a third of the total flow.

Two tagged sea trout were lost almost immediately after being released and one regurgitated its transmitter before reaching the Gambsheim dam and Ill confluence. Seven trout provided tracking results.

*Table 2.* Impact of 15 ocstacles met by radio-tagged salmon and sea trout in their 1996 spawning migration.

| River/ stream | Dam ID | Dam Name | | Height (m) | Fishway | Number of tagged fish at the dam or group of dams | Number of fish that passed | Minimal time before passage | Maximal time before passage |
|---|---|---|---|---|---|---|---|---|---|
| Rhine | 1 | Iffezheim | | 11 | Fish sluice at the spillway | 2 salmon | 1 | – | – |
| | 2 | Gambsheim | | 10 | Fish sluice at the spillway | 2 salmon | 0 | not passed | – |
| | | | | | | 5 sea trout | 0 | not passed | – |
| Ill | 3 | strasbourg North group | Mill | 1.8 | None | 14 salmon | 12 | < 8 h 30 min | 13 d |
| | | | Aar dam | 2.6 | None | 1 sea trout | 2 | unknown | unknown |
| | | | Main dam | 1.5 | Old pool fish pass | | | | |
| | 4 | Strasbourg aiguilles | | 0.5 | None | 12 salmon | 12 | 0 | 29 h 20 min |
| | | | | | | 1 sea trout | 1 | unknown | unknown |
| | 5 | Strasbourg Center group | Old mills | 1.8 | None | 12 salmon | 12 | 20 min | 5 d 2 h |
| | | | Abattoirs dam | 1.8 | Old pool fish pass | 1 sea trout | 1 | unknown | unknown |
| Bruche | 6 | Kolbsheim | | 1.5 | None | 8 salmon | 5 | 2 d | 6 d |
| | 7 | Avolsheim | | 2.5 | None | 5 salmon | 0 | not passed | – |
| | 8 | Dorlisheim | | 1.2 | None | 7 salmon | 1 | unknown | unknown |
| | | | | | | 1 sea trout | 0 | not passed | – |
| | 9 | Molsheim a | | 0.5 | None | 1 salmon | 1 | 8 d | – |
| | 10 | Molsheim b | | 2.5 | None | 1 salmon | 0 | not passed | – |
| | 11 | Mutzig | | 2 | None | 3 salmon | 0 | not passed | – |
| | 12 | Dinsheim | | 1.2 | None | 1 salmon | 0 | not passed | – |

- Five trout migrated up the Rhine and were blocked by the Gambsheim dam. Among them, T3 first migrated downstream back to the Iffezheim power station screens before migrating upstream. In mid-December, T9 and T10 left the Gambsheim dam and possibly spawned in the lower Ill. The three other trout regurgitated their transmitters, after spending 27, 31 and 145 d respectively below the dam.
- Two trout migrated up the Ill. T5 lost its transmitter after staying from July 16 to August 19 at the same place, 16 km upstream of the river's confluence. T4 reached the confluence of the Bruche where it spent three months (July 15 to October 15) before moving up to dam 8, 23 km upstream in the Bruche (Figure 3).

Based on daily recordings, the salmon migrated at a minimum average speed of 8.0 to 22.0 km $d^{-1}$ in the Rhine and the Ill, and 2.7 to 4.0 km $d^{-1}$ in the Bruche. When tracked while actively migrating, salmon were swimming at speeds of 1.3 to 2.6 km $h^{-1}$ in the Ill (Table 3). S6 was followed twice upstream on the same stretch, at 1.8 and 2.1 km $h^{-1}$. Sea trout migrated at swimming speeds of 0.8 to 1.3 km $h^{-1}$ in the Rhine and 0.8 to 1.0 km $h^{-1}$ in the Ill (Table 3). We observed higher speeds in the headwater above the Iffezheim dam where the water velocity was less than 10 cm $s^{-1}$: 8.4 km $h^{-1}$ for salmon S3 and 2.7 km $h^{-1}$ for sea trout T3.

*Behavior at obstacles and assessment of migration difficulties*

We gathered behavior data at two hydroelectric and navigation dams in the Rhine, four navigation dams and two abandoned power houses in the Ill, and seven dams or sloping weirs in the Bruche (Table 2).

In the Rhine, the two salmon (S2, S3) that returned downstream of the Iffezheim dam and tried to pass it

*Table 3.* Swimming speed over ground of salmon and sea trout obtained from continuous tracking data.

| Fish ID | Fish TL (mm) | River | Distance of continues tracing (km) | Water temperature (°C) | Swimming speed ($km\ h^{-1}$) |
|---|---|---|---|---|---|
| S3 | 700 | Rhine Iffezheim | 3.5 | 16.7 | 8.4 |
| S7 | 701 | Lower Ill | 1.2 | 13.6 | 1.6 |
| S4 | 735 | Ill Strasbourg | 0.8 | 16.1 | 1.9 |
| S4 | 735 | Ill Strasbourg | 3.5 | 16.0 | 1.3 |
| S4 | 735 | Ill Strasbourg | 1.3 | 16.0 | 2.6 |
| S9 | 710 | Ill Strasbourg | 2.0 | 14.4 | 2.0 |
| S6 | 700 | Ill Strasbourg | 2.0 | 13.4 | 1.8 |
| S6 | 700 | Ill Strasbourg | 3.7 | 13.1 | 2.1 |
| S7 | 701 | Ill Strasbourg | 1.7 | 12.6 | 2.0 |
| S13 | 769 | Ill Strasbourg | 2.3 | 12.1 | 1.4 |
| T4 | 505 | Rhine Iffezheim | 3.0 | 15,0 | 1.2 |
| T2 | 595 | Rhine Iffezheim | 8.6 | 15,0 | 1.5 |
| T3 | 680 | Rhine Iffezheim | 2.0 | 15,0 | 2.7 |
| T10 | 750 | Rhine Gambsheim | 2.0 | 9,0 | 0.9 |
| T5 | 555 | Lower Ill | 2.9 | 15,5 | 0.8 |
| T5 | 555 | Lower Ill | 1.3 | 15,5 | 1.0 |

had the following behavior: Both fish were most often found in the power station channel which had the main discharge at this time, but they also explored a wide area downstream of the dam. They often stayed in a gravel pit on the right bank, 1.2 km downstream of the power station. They explored the lower Moder, a tributary of the left bank, 1.6 km downstream of the dam. They also explored a side channel fed by Rhine water and ground water on the right bank, where S2 migrated as far as 19 km upstream of Iffezheim before it died in a fisherman's net. S3 made numerous attempts to pass the power station and eventually found the navigation channel 0.8 km downstream and made its way upstream through the locks. It turned back downstream after 6 km and was found the next day below the power station. The same behavior was observed a second time 14 d later with an upstream migration of 10.5 km and the same final result. The signal of this fish was finally lost after spending 59 days below the obstacle.

Except for S3 followed while it was passing the Iffezheim obstacle through the locks, we obtained evidence that at least six untagged salmon passed the dam; one was caught (S17) and five were seen at dams or on redds in the Bruche. No evidence that fish can pass the Gambsheim dam was obtained during this study. In the lower Ill and the Bruche, no dam had an effective fish way. Two dams in Strasbourg have an old fish pass, but both are too small and lack maintenance (Table 2). Six obstacles, 1.2 to 2.5 m high, were not passed by any of the tagged fish. The salmon that passed the other obstacles did it most often by swimming under an opened sluice or over the sloping weir. Two weirs (6 and 8) were passed by jumping 1.2 and 1.5 m in height, but these dams were not passed when the water temperature dropped under 10.5 °C.

One salmon (S4) passed dam 4 in the Ill through navigation locks under special circumstances. We followed this fish for 10 h as it confronted three obstacles. At 3:00 p.m. S4 was resting 0.8 km downstream of dam 4 after two attempts at passing this dam. We opened the sluices in the locks gates which are rarely used. The discharge at the dam and at the locks were 16 and 2 $m^3\ s^{-1}$ respectively. The current lifted mud inside the locks and S4 moved upstream as soon as the turbid water reached it (1.9 $km\ h^{-1}$). It explored the four openings in the 50 m wide needle dam for 35 min before swimming through the locks sluices. After 95 min hesitating in the headwater pond, S4 migrated upstream (1.3 $km\ h^{-1}$) up to the old mills at obstacle 5. S4 tried to swim through each of four discharge outlets without success, making several attempts of 2–5 min at each outlet. After 35 min at the mills S4 backed downstream 1.3 km to the confluence of the channel leading to dam 5, where it moved rapidly upstream (2.6 $km\ h^{-1}$). Both

channels had a rather similar flow. S4 reached the dam at 1 a.m., explored it for 45 min before finding its way upstream by swimming under an opened sluice.

## Discussion

*Catching and tagging method*

Electro-fishing and insertion of transmitters in the stomach had no apparent effect on the survival of the 20 salmon tagged. However, a questionable effect of catching the fish with electricity and radio-tagging them is raised by the fact that three salmon (S1, S2, S3) and one trout (T3) caught in the summer migrated downstream at least 4 km (back to the Iffezheim dam) after being released. S1 and T3 were blocked by the power station screen where they stayed less than a day before migrating upstream. S2 and S3 found a way through the dam (S3 through the locks) and may have migrated further downstream. They were again found below the power station 43 and 15 d respectively after their release, trying to migrate upriver. Because these fish were released in the headwater reservoir 4 km upstream of the dam, the lack of current may have disoriented them since none of the fish released 22 km upstream were detected downstream. The approaching spawning season may also have stimulated the upstream migration of the latter (Mills, 1989; Baglinière & Porcher, 1994). Water temperature is also known to inhibit upstream migration when it exceeds 20 °C (Thioulouse, 1972; Mills, 1989). However, it could have acted only as an added factor (adding to the stress) since the fish were all captured in the power channel while actively migrating. The three salmon that moved downstream were captured and released at water temperatures between 19.3 °C and 20.5 °C. The fourth salmon (S4) tagged and released at a slightly lower temperature (18.4 °C) migrated straight upstream. Our results suggest that electro-fishing, insertion of transmitters and transportation had no apparent effect on the migratory behavior of the fish. The lack of current in the release area and/or high water temperatures have added some stress and triggered the downstream movement.

Regurgitation of the transmitter did not occur with salmon but was frequent among sea trout. Apart from two trout lost after tagging, four of the remaining six trout tagged in June and July regurgitated their transmitter before reproduction. Bach & Travade (unpublished data) had the same problem in a 1987 study in the Dordogne River. Insertion into the stomach was successfully used on Rhine sea trout in autumn by Marmulla & Ingendhal (1996). We chose this method over surgical insertion to avoid the risk of mortality in the summer (Bach et al., 1996). However, surgical insertion seems necessary with sea trout tagged in spring or summer, if the objective is to track them until reproduction.

*Migratory behavior*

When ascending rivers, Atlantic salmon and sea trout usually have a stationary period which can last several month before they complete their final upstream migration (Thioulouse, 1972; Le Cren, 1985; Hawkins & Smith, 1986). Among the tagged fish in our study, only S1, S4, T4 and T5, all caught during summer, stopped migrating without being blocked by an obstacle. The other fish tagged in summer were blocked by obstacles before they entered their stationary phase. S1 and T5 stopped in the lower Ill in July when the water temperature reached 21.5 °C and 20.7 °C respectively. S1 stayed there for 91 days and T5 for a minimum of 34 d before it regurgitated its transmitter. S4 and T4 both stopped migrating upstream in the Ill when they reached the confluence of the Bruche, which can be interpreted as a precise homing behavior. Both fish awaited higher discharge levels at the outlet of another cooler tributary (1 °C less than the Ill on average) situated 1 km downstream, before they entered the Bruche. We noticed a different behavior between the salmon and the trout: S4 entered at the first rise in water discharge after the end of June, while T4 entered the Bruche only at the third rise seven days later. The first rise made T4 leave its resting place and move to the confluence but it returned. The final upstream movement of T4 in the Bruche seems to have been triggered by a sharp increase in the water temperature of the Ill and of the Bruche (+0.5 °C and +1.5 °C respectively), just before an increase in water discharge (from 1.0 $m^3$ $s^{-1}$ to 2.4 $m^3$ $s^{-1}$). The triggering role of water flow for salmon movement is sometimes cited, especially in estuaries and in small rivers (Mills, 1989; Baglinière & Porcher, 1994), but the role of water temperature is less clear except for extremes (< 3 °C and > 20 °C) which inhibit migration (Baril & Gueneau, 1986; Thioulouse, 1972). A temperature increase is also known to stimulate upstream migration of salmon in the spring (Mills, 1989).

*Swimming speed*

The swimming speeds of salmon observed on a daily basis include possible stops or downstream migration. Our results (3 to 22 km $d^{-1}$) are in the range of average speeds of 4 to over 20 km $d^{-1}$ given in a review by Mills (1989). Hawkins & Smith (1986) recorded speeds of 14–37 km $d^{-1}$ in Scotland and Baril & Gueneau (1986) found an average speed of 19 km $d^{-1}$ in the River Loire, with one salmon migrating at 37 km $d^{-1}$ for two days. Our continuous recordings over short distances indicate that salmon can swim actively at speeds of 1.3 to 2.6 km $h^{-1}$ except for one value of 8.4 km $h^{-1}$ in the Iffezheim reservoir. Sea trout had lower swimming speeds (0.8–1.5 km $h^{-1}$) also with the exception of one fish recorded at 2.7 km $h^{-1}$ in the Iffezheim reservoir. Interestingly, both fish in the reservoir had migrated downstream on this stretch prior to their upstream migration. Le Cren (1989) mentions speeds of 0.7 to 1.1 km $h^{-1}$ for sea trout moving in estuaries in England and Wales and speeds up to 2 km $h^{-1}$ in the River Fowey in England. We found no correlation between the length of the fish and their swimming speed.

*Impact of dams on upstream migration and behavior of fish confronting obstacles*

In the main stem of the Rhine, our goal was to study the behavior of the fish arriving at the Gambsheim dam, constructed on a similar scheme similar to Iffezheim. After their arrival at the dam area, two trout were first found below the power station and three below the operating dam. The latter were attracted in this direction by a flow of 100 to 250 $m^3$ $s^{-1}$ versus 1100 $m^3$ $s^{-1}$ in the power channel. No fish migrated toward the operating dam without any discharge in this channel. Neither did we find any tagged fish in the navigation channel where there is no current since the locks are emptied in the power channel. The situation looks slightly different at Iffezheim despite a similar type of dam, since some salmon succeeded in passing the obstacle through the navigation locks. Apart from the few untagged salmon observed in the Bruche in this study, evidence that salmon and sea trout can pass the dam were obtained in previous years: salmon redds were found in the Bruche in 1995 (Roche et al., 1996) and sea trout are often caught upstream of Iffezheim. The only other possible migration way apart from the locks is the fish sluice (Borland type) on the left bank of the operating dam, but it only operates from March to June. Fish sluices are not often efficient for salmon (Larinier & Gosset, 1994) and this one was shown to be of little use for other fish in a former assessment with camera recordings (Roche, 1990). The probability of the fish finding the navigation locks is small since there is little (at Iffezheim) or no (at Gambsheim) current in the navigation channel (S2 spent 44 d below the obstacle without passing). For this reason the locks cannot be considered as a satisfactory migration possibility (Travade & Larinier, 1992).

The migration in the Ill basin provided important information on delays caused by obstacles and the search for alternative migration routes. Except for one, the dams in the Ill were less passable than we had thought and resulted in delaying most fish, some finally forced to reproduce below the dams. The difficult passage and the resulting delays were often due to the attractiveness of old mills channels where salmon encountered impassable obstacles before looking for other migration routes. A similar observation was made by Bach et al. (1996) who also used the method with success for a global assessment of migration problems in a 108 km long river stretch with 28 obstacles.

Our results also showed that salmon arriving in the upper Rhine in autumn can be blocked by obstacles that were passed by other salmon earlier in the year. Water temperature was the main factor affecting the result of attempts to pass obstacles, by swimming or jumping. Water temperature below 10.5 °C was shown to render three obstacles impassable (dams 3, 5, 6). A similar temperature (11 °C) is given in Mills (1989) as a threshold for salmon to overcome high natural waterfalls. The lack of success is related to the swimming speed of salmon, which usually varies from 5 to 8 m $s^{-1}$ depending on temperature but can be less than 3 m $s^{-1}$ in cold water (Larinier & Gosset, 1994). In the Loire River, a threshold of 5–6 °C was observed under which salmon did not attempt to pass an obstacle (Baril & Gueneau, 1986). The second main factor affecting success was the lack of adequate pools at the foot of obstacles. Dams 3 (Aar), 7, 8, 9 and 10 would be passable by salmon at a favorable temperature if they had a bottom pool closer to the waterfall since their height is not higher than 2.5 m (Larinier & Gosset, 1994). Also critical is the fish condition after a long freshwater migration without feeding and numerous efforts at passing obstacles.

## Conclusion

The first year of this study provided important information on possible migration routes to the spawning

grounds and on the behavior of fish when confronted with obstacles. This information should prove helpful in future phases of the salmon and sea trout restoration project. Findings included:

1) Both the Gambsheim operating dam and its power station attracted fish during periods of overflow at the dam, even with relatively low discharges. Similarly in the Ill River, old power channels with low flows also attracted salmon. These results suggest a need for the installation of two fish passes at these types of obstacles, especially for long rivers with many dams, since cumulative delays and injuries leading to mortality may significantly reduce the spawning population.
2) Some obstacles easily passed before November could not be overcome when water temperature dropped under 10.5 °C. Since a significant number of salmon and trout migrate to the upper Rhine in November and December, these obstacles will require fish passage devices. In the next phase of our study we will use additional automatic recording stations to document arrival times at the Gambsheim dam and at the north group of dams in the Ill. Sea trout will be surgically tagged, since this is a more reliable method for obtaining information from early summer until spawning.

## Acknowledgements

This study was made possible with the financial help of the European Union. We would also like to thank G. Edel, F. Hufschmitt and J. Adam of the Association Saumon Rhin and C. Brugel and T. Clauss of Conseil Supérieur de la Pêche for their field and technical assistance. We are also grateful to DIREN Alsace and Service de la Navigation in Strasbourg for the hydraulic data.

## References

Bach J. M., M. Chanceau & M. Larinier, 1996. Suivi par radiopistage de la migration anadrome des salmonidés amphihalins sur le Gave de Pau. Rapport GHAAPPE Toulouse, 46 pp.

Baglinière, J. L. & J. P. Porcher, 1994. Caractéristiques des stocks reproducteurs et comportement lors de la migration génésique. In 'Le saumon atlantique, biologie et gestion de la ressource.' IFREMER: 101–122.

Baril, D.& P. Gueneau, 1986. Radio-pistage de saumons adultes (*Salmo salar*) en Loire. Bull. fr. Piscic. 302: 86–105.

bij de Vaate, A., 1996. Project migratie zeeforel. Newsletter 1, 4 pp.

Cazemier, W. G., 1994. Present status of the salmonids Atlantic salmon and sea-trout in the Dutch part of the River Rhine. Wat. Sci. Tech. Vol. 29: 37–41.

CIPR, 1994. Saumon 2000. Commission Internationale pour la Protection du Rhin, 29 pp.

Hawkins, A. D. & G. W. Smith, 1986. Radio-trasking observations on Atlantic salmon ascending the Aberdeenshire Dee. Scottish Fisheries Research Report N° 36, 24 pp.

Larinier, M. & C. Gosset, 1994. La libre circulation des migrateurs. In 'Le saumon atlantique, biologie et gestion de la ressource.' IFREMER, Brest: 197–217.

Largiader, C. R., R. Gyomard & P. Roche, 1997. Mise en évidence de la reproduction naturelle du saumon atlantique (*Salmo salar* L.) dans un affluent français du Rhin par analyse génétique d'oeufs prélevés dans des frayères. Bull. fr. Pêche Piscic. 343: 183–188.

Le Cren, E. D., 1985. The biology of sea trout. Summary of the symposium held at Plas Menais, 24–26 October 1984. Atlantic Salmon Trust.

McCleave, J. D., J. H. Power & S. A. Rommel Jr, 1978. Use of radio telemetry for studying upriver migration of adult salmon (*Salmo salar*). J. Fish Biol. 12: 549–558.

Marmulla, G.& D. Ingendhal, 1996. Preliminary results of a radio telemetry study of returning Atlantic salmon (*Salmo salar* L.) in River Sieg, tributary of River Rhine in Germany. In Barras, E. & J. C. Philippart (eds), Underwater Telemetry: Proceedings of the first Conference on Fish Telemetry in Europe, Liège, 4–6 April 1995.

Mills, D., 1989. Ecology and Management of Atlantic Salmon. Chapman and Hall Ed. London, 351 pp.

Olbricht, P., 1984. Untersuchungen zum Wiedererscheinen der Meerforelle (*Salmo trutta* L.) im oberen Niederrheingebiet. Der Fischwirt, 33: 22–24.

Roche, P., 1989. Le saumon du Rhin: Données historiques. CSP Metz, Ministère Environnement, 65 pp.

Roche, P., 1990. Contrôle des passes à poissons des barrages d'Iffezheim et de Gambsheim sur le Rhin par caméra vidéo en 1989. CSP Metz, 5 pp.

Roche, P., 1991. Le Saumon du Rhin: Habitats et stocks potentiels en partie française. CSP Metz, Min. Environnement, Sandoz, 36 pp.

Roche, P., 1992. Mise en évidence de l'écotype truite de mer dans les captures de grandes truites (*Salmo trutta* L.) du Rhin. Bull. fr. Pêche Piscic. 324: 36–44.

Roche, P. 1994. Habitat availability and carrying capacity in the French part of the Rhine for Atlantic Salmon (*Salmo salar* L.). Wat. Sci. Tech. 29: 257–265.

Roche, P., G. Edel & M. Gerlier, 1996. Premières captures de saumons (*Salmo salar* L.) dans le Rhin franco-allemand et mise en évidence de frayères dans la Bruche. Rapport CSP Metz, 7 pp.

Roule, L., 1921. La biologie migratrice du saumon dans le Rhin. Compte rendu Association française pour l'avancement des sciences, Paris, 44: 273–276.

Solomon, D. J. & T. J. Storeton-West, 1983. Radiotracking of migratory salmonids in rivers: Development of an effective system. Fish. Res. Tech. Rep. 75: 11 pp.

Staub, E., 1988. Passes à poissons des centrales électriques du Haut-Rhin. La migration, facteur de compensation ? Bulletin de l'Office Fédéral de Protection de l'Environnement, 4 (88), Berne: 25–30.

Steinberg, L., G. Marmulla, G. W. Schmidt & L. Lehmann, 1991. Erster gesicherter Nachweis des Laches *Salmo salar* L. in den Rhein Nebenfluss Sieg. Fischökologie 5 (Teil 1): 3–18.

Thioulouse, G., 1972. Le comportement du saumon. Essai d'éthologie du saumon de l'Allier. Plein air service, Edit. Scient., Clermont-Ferrand, 279 pp.

Travade, F. & M. Larinier, 1992. Ecluses et ascenceurs à poissons. Bull. fr. Piscic. 326–327: 95–110.

Vibert, R., 1950. La régression du saumon du Rhin. Bull. fr. Piscic. 156: 131–134.

*Hydrobiologia* **371/372:** 295–304, 1998.
*J.-P. Lagardère, M.-L. Bégout Anras & G. Claireaux (eds), Advances in Invertebrates and Fish Telemetry.*



# The migratory behaviour of wild Atlantic salmon (*Salmo salar* L.) smolts in the River Test and Southampton Water, southern England

A. Moore[1]*, S. Ives[1]*, T. A. Mead[1] & L. Talks[2]
[1] *Centre for Environment, Fisheries and Aquaculture Science, Lowestoft Laboratory, Pakefield Road, Lowestoft, Suffolk NR33 0HT, UK*
[2] *Environment Agency, Sarum Court, Sarum Road, Winchester, Hampshire SO22 5DP, UK*
**Author for correspondence (e-mail: a.moore@cefas.co.uk)*

*Key words:* Atlantic salmon, migration, smolt, telemetry, behaviour, estuary

## Abstract

Thirty wild Atlantic salmon *Salmo salar* (L.) smolts, tagged with miniature acoustic transmitters, were tracked in the River Test and Southampton Water, to describe the freshwater and estuarine patterns of migration. Migration in freshwater was predominantly nocturnal (mean time 23:42) and smolts tagged earlier in the study spent significantly longer in the river before migrating into the estuary than fish tagged later in the study. Movement in the upper estuary section of the River Test was indicative of a nocturnal ebb tide transport pattern of migration. The majority of smolts migrated seawards at night and during an ebb tide (mean time after High Water +1 h 54 min), close to the surface and within the fastest moving section of the water column. Smolts continued to emigrate seawards through the lower estuary and Southampton Water during the ebb tide (mean distance 4908 ± 609 m; mean ground speeds of 35 cm $s^{-1}$), although the diurnal pattern of movement was not evident. There was also a significant seaward migration of smolts during the latter part of the flood tide (mean distance 3469 ± 522 m; mean ground speeds of 14 cm $s^{-1}$), suggesting active directed swimming. Smolts emigrated rapidly through Southampton Water, and there was no apparent period of acclimation required when moving from fresh to saltwater. The results are discussed in relation to the environmental cues controlling smolt migration.

## Introduction

Atlantic salmon *Salmo salar* (L.) smolts move seaward through river estuaries during their spring migration. The movement from the fresh water to the marine environment may be considered a particularly critical period in their life history, during which time they encounter increasing salinities, variable tidal currents and novel prey and predators. The success and timing of this movement has been suggested to be important to the survival of the smolts within the marine environment and their subsequent return as adult fish (Hansen & Jonsson, 1989). An understanding of the mechanisms and cues controlling emigration and orientation of salmon smolts through the estuarine environment is therefore a basic requirement for the successful management of salmonid stocks.

There have been a number of telemetry studies which have attempted to describe the estuarine behaviour of both Atlantic (McCleave, 1978; Tytler et al., 1978; Greenstreet, 1992; Moore et al., 1992; Moore et al., 1995) and Pacific salmon smolts *Oncorhynchus kisutch* (Walbaum) (Moser et al., 1991), as they negotiate the transition from the fresh to saltwater. The investigations on wild Atlantic salmon smolts (Moore et al., 1995, 1996) have suggested a seasonal change in the nocturnal pattern of movement and a selective ebb tide transport component to the migration within the estuary (Moore et al., 1995). The behaviour patterns of wild salmon smolts appear to have evolved to permit rapid, energetically favourable migration through the estuary at the optimum time whilst minimising the risk of predation (Moore et al., 1995).

Atlantic salmon smolts from the River Test, are known to emigrate seaward through Southampton Water, a busy commercial and military port in southern England. A proposed scheme to extend the port and dredge the maintained channel to permit access for larger commercial shipping was considered to have the potential to adversely impact on salmon smolt behaviour and survival. Consequently a tracking study was undertaken to describe the movements of wild Atlantic salmon smolts in this estuary. Smolts were trapped in fresh water as they migrated seawards and tagged with miniature acoustic transmitters. The subsequent movements of the smolts within the study area were monitored using acoustic sonar buoys and by tracking individual fish from a small research vessel. This paper describes the temporal and spatial patterns of migration of the salmon smolts in the River Test and Southampton Water, and discusses the results in relation to the possible environmental and physiological cues controlling estuarine migration.

## Material and methods

### *Study site*

River Test, 139 km in length, has a relatively stable flow and during the study period the minimum and maximum daily flows, recorded at a gauging station, were 6.81 and 9.60 $m^3$ $s^{-1}$ respectively. Southampton Water is the drowned lower portions of the Rivers Test and Itchen, and is an approximately 10 km long and 2 km wide extension of the Central Solent (Figure 1). The area is bordered by broad intertidal mudflats with shingle and sand on the eastern side, and a saltmarsh to the west. The estuary has been classified as partially mixed (Dyer, 1970), but its salinity structure depends very much on tidal state and, to a lesser degree, the seasonal cycle in freshwater flow. At high water there is little stratification and the surface salinity in the lower estuary normally exceeds 30‰. However, in the upper estuary the salinity frequently covers a range from almost zero to about 25‰ and stratification occurs. A more detailed physical description of the Test estuary and Southampton Water has been provided by Webber (1980).

### *Smolt trapping and transmitter attachment*

The study was carried out between 1–30 April 1996. Migrating salmon smolts were trapped in fresh water using a modified eel rack built into a weir (Figure 1). All migrating smolts studied were trapped during the hours of darkness between 21:30 and 22:45 and at river water temperatures of 8.7 °C and above.

Salmon smolts were tagged with miniature 300 kHz acoustic transmitters, 17 mm long × 8 mm in diameter, and which weighed 0.35 g in water (Moore et al., 1990a). The transmitter circuitry and battery were enclosed in a cylindrical polycarbonate case with hemispherical ends. They had a maximum range in the estuarine environment of around 100 m and, transmitting at 60 pulses $min^{-1}$, a life of approximately 30 days.

The transmitters were surgically implanted into the peritoneal cavity of the smolts using the techniques previously described by Moore et al. (1990b). This technique of transmitter attachment has been shown to have negligible physiological and behavioural effects on salmon smolts (Moore et al., 1990b; Moore et al., 1995; Moore et al., 1996). After transmitter attachment, the smolts were transferred to a tank of oxygenated water and allowed to recover fully from the effects of anaesthesia as described by Moore et al. (1990b). Recovery was usually within 5–10 minutes, and the fish were released after a period of 30 min together with a number of untagged smolts trapped at the same time.

### *Sonar buoy tracking system*

The movement of the smolts within the freshwater and upper estuary sections of the River Test were monitored using an array of 300 kHz acoustic sonar (Moore & Potter, 1994; Moore et al., 1995). The sonar buoys were positioned at 2 locations in freshwater, and at 7 locations within the downstream portion of the upper estuary (Figure 1). Four automatic listening stations (ALS) were deployed adjacent to these sites to record signals from the sonar buoys. At various intervals throughout the study, a hand-held 300 kHz acoustic hydrophone and a Lowe HF-150 communications receiver (Lowe Electronics Ltd, England) were used to record the positions of smolts holding position in the river during daylight hours.

### *Migration of smolts within the river estuary and Southampton Water*

The movements of salmon smolts within the river estuary and Southampton Water were monitored by manually tracking individual fish as they were detected moving past the sonar buoys deployed within the upper estuary. A small tracking vessel was fitted

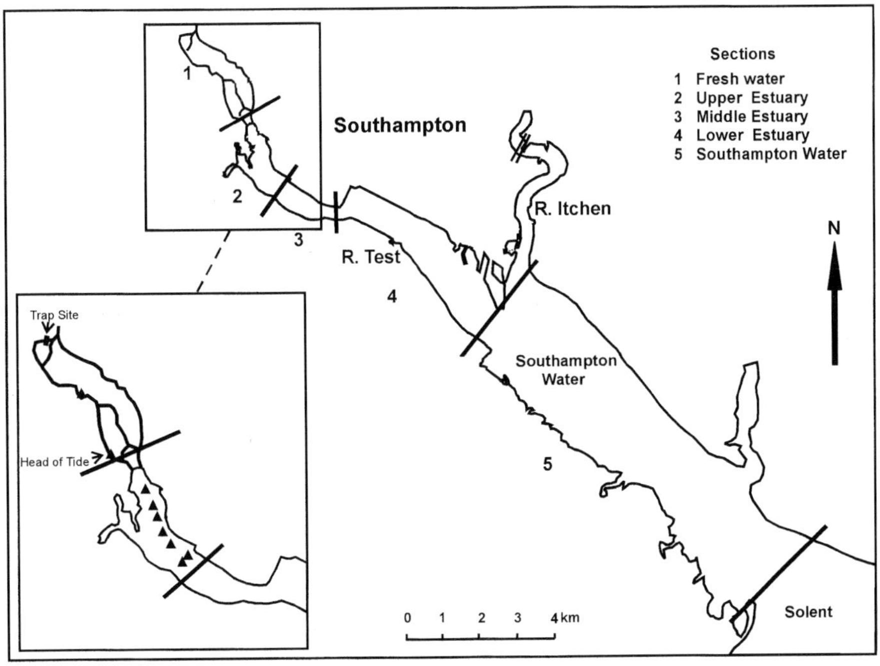


*Figure 1.* Map of the River Test and Southampton Water indicating the position of the trap site, the positions of the acoustic sonar buoys (▲) in the freshwater and upper estuary and the five sections of the river referred to in the text.

with 300 kHz acoustic hydrophone and Lowe HF-150 receiver. The hydrophone could be operated in both the vertical and horizontal modes, and the transducers extended 50 cm below the keel of the vessel to provide a 360° horizontal field of view within the water. Operation of the hydrophone in the vertical mode provided an indication of the depth of the smolts.

Seaward movements of previously tagged smolts, as they moved through this section, were detected by monitoring the radio signals from the appropriate acoustic sonar buoys. When a smolt was detected moving within range of the sonar buoy adjacent to the vessel, the acoustic hydrophone was operated and used to locate the signal from the transmitter inside the fish. The vessel was then manoeuvred to remain constantly within range of the fish and monitor its subsequent movements as it passed through Southampton Water. The position of the boat was fixed at two minute intervals using a Magellan GPS NAV 5000 PRO Satellite Receiver. Water depth was also continually monitored and recorded using the vessel's echo sounder.

### *Data analyses*

The River Test and Southampton Water have been divided into 5 sections to assist in the analyses of the movements of the smolts within the study area (see Figure 1). These sections, which are based on the physical and saline characteristics of the estuary are:

1. Fresh water (1.95 km).
2. Upper Estuary (2.05 km).
3. Middle Estuary (1.20 km).
4. Lower Estuary (4.50 km).
5. Southampton Water (9.45 km).

The mean time of day at which the smolts were detected by the sonar buoys positioned at the two freshwater sites and at the upstream upper estuary section were calculated using vector analysis (see Batschelet, 1981). The mean time in the tidal cycle that the smolts were detected by the upstream sonar buoy within the upper estuary section was also calculated. In addition, the mean time of day and mean time within the tidal cycle that smolts were detected moving through the middle and lower estuary sections, derived from the manual tracking of individuals, were also calculated. There was insufficient data to calculate the movements of smolts within Southampton Water. The length of the mean vectors were further used to test whether movement of the smolts within the relevant section was random with respect to time and state of tide using the Rayleigh test ('r' value) (see Batschelet, 1981). The second high water at Southampton has been used as a reference point for tidal cycles.

*Table 1.* Summary of information relating to Atlantic salmon smolts tagged in the River Test in 1996

| Length (mm) | Date | Capture time | Release time | River temperature (°C) | Residency in freshwater (hours min) | Duration of estuary track (hours min) | Distance of estuary track (metres) | Last recorded position |
|---|---|---|---|---|---|---|---|---|
| 148 | 05/04/96 | 22:10 | 22:50 | 8.7 | 191 52 | 8 30 | 85 | Middle Estuary |
| 159 | 05/04/96 | 22:30 | 22:50 | 8.7 | ** | 9 00 | 5750 | Southampton Water |
| 192 | 07/04/96 | 22:30 | 23:00 | 9.7 | ** | – | – | Upper Estuary |
| 182 | 08/04/96 | 22:35 | 23:05 | 9.7 | 195 04 | 10 30 | 3800 | Lower Estuary |
| 173 | 09/04/96 | 21:20 | 22:00 | 9.8 | 336 50 | – | – | Upper Estuary |
| 173 | 09/04/96 | 21:50 | 22:50 | 9.8 | 118 40 | 24 40 | 3550 | Lower Estuary |
| 184 | 09/04/96 | 22:00 | 22:20 | 9.8 | 173 13 | – | – | Upper Estuary |
| 178 | 14/04/96 | 21:45 | 22:05 | 9.2 | * | – | – | Freshwater |
| 146 | 14/04/96 | 22:40 | 23:00 | 9.2 | 1 59 | 16 15 | 1825 | Middle Estuary |
| 136 | 14/04/96 | 22:50 | 23:00 | 9.2 | 49 57 | – | – | Upper Estuary |
| 157 | 16/04/96 | 21:30 | 22:10 | 11.3 | 209 06 | – | – | Upper Estuary |
| 180 | 16/04/96 | 21:30 | 22:10 | 11.3 | 24 06 | – | – | Upper Estuary |
| 168 | 16/04/96 | 21:30 | 22:10 | 11.3 | 113 37 | 10 16 | 14250 | Solent |
| 146 | 18/04/96 | 21:30 | 22:45 | 10.6 | 74 40 | – | – | Upper Estuary |
| 144 | 18/04/96 | 22:00 | 22:45 | 10.5 | 50 39 | – | – | Upper Estuary |
| 157 | 18/04/96 | 22:10 | 22:45 | 10.5 | 48 56 | 0 45 | 150 | Middle Estuary |
| 153 | 18/04/96 | 22:40 | 23:10 | 10.4 | * | – | – | Freshwater |
| 173 | 20/04/96 | 21:50 | 22:35 | 12.3 | 30 24 | – | – | Upper Estuary |
| 180 | 20/04/96 | 21:50 | 22:35 | 12.3 | 13 16 | – | – | Upper Estuary |
| 162 | 20/04/96 | 22:05 | 22:35 | 12.3 | 101 32 | – | – | Upper Estuary |
| 154 | 20/04/96 | 22:05 | 22:35 | 12.3 | 38 28 | 15 10 | 12250 | Southampton Water |
| 152 | 20/04/96 | 22:05 | 22:35 | 12.3 | 24 59 | – | – | Upper Estuary |
| 173 | 25/04/96 | 21:15 | 22:30 | 11.2 | 35 26 | – | – | Upper Estuary |
| 170 | 25/04/96 | 21:30 | 22:30 | 11.6 | * | – | – | Freshwater |
| 170 | 25/04/96 | 21:45 | 22:30 | 11.2 | 24 18 | 13 12 | 6960 | Upper Estuary |
| 173 | 25/04/96 | 22:00 | 22:30 | 11.6 | 1 46 | – | – | Upper Estuary |
| 154 | 27/04/96 | 20:10 | 20:30 | 13.8 | 5 14 | 14 42 | 2950 | Lower Estuary |
| 163 | 27/04/96 | 22:15 | 22:40 | 13.6 | 24 43 | – | – | Upper Estuary |
| 173 | 27/04/96 | 22:45 | 23:05 | 13.6 | 24 18 | 13 45 | 5450 | Southampton Water |
| 161 | 27/04/96 | 22:45 | 23:05 | 13.6 | 5 13 | – | – | Upper Estuary |

* Located downstream of Nursling trap during study.
** Data unavailable.

## Results

### *Smolt trapping and release*

Between 5 and 27 April 1996, a total of 30 salmon smolts were tagged and released immediately downstream from the trap site (Table 1). Subsequently, a total of 27 smolts were detected by the sonar buoys located at two sites in freshwater and 11 of these were tracked continuously by the vessel for periods of up to 25 h and over distances of up to 14.25 km as they migrated seawards through Southampton Water. The other 16 fish which were detected by the sonar buoys in the estuary migrated during periods when other fish were being tracked by the vessel. Three tagged fish did not migrate into the estuary during the study and remained within the freshwater section of the River Test, 400 m downstream of the trap site.

### *Residency within the River Test and Southampton Water*

After capture and tagging, smolts were resident for 76 ± 16.9 h (mean ± S.E.M.) in the freshwater section of the River Test before migrating past the head of tide and into the upper estuary (Table 2). There was a relationship between the day of release of the tagged smolts and the total residency within the fresh-

*Table 2.* The mean residency times (hours) of Atlantic salmon smolts in each of the sections of the River Test and Southampton Water. The details of each section are given in the text. The times that smolts were migrating within the maintained channel of the estuary are also indicated

| | Freshwater | Upper Estuary | Middle Estuary | | Lower Estuary | | Southampton Water | |
|---|---|---|---|---|---|---|---|---|
| | | | Total | Channel | Total | Channel | Total | Channel |
| Mean time (h) | 76.7 | 28.3 | 4.4 | 2.6 | 7.5 | 4.6 | 4.6 | 1.2 |
| S.E.M. (h) | 16.9 | 9.5 | 1.3 | 0.7 | 3.1 | 1.2 | 1.5 | 0.7 |
| $n$ | 25 | 27 | 11 | 11 | 6 | 6 | 3 | 3 |

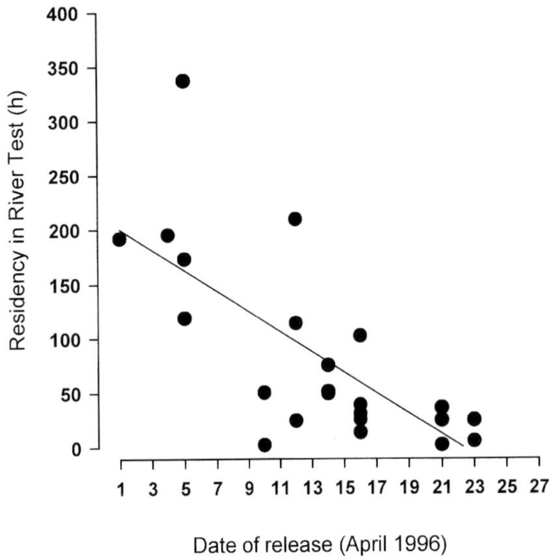


*Figure 2.* The relationship between transit time within the River Test after tagging and day of release of salmon smolts. Details of the fitted regression line are given in the text.

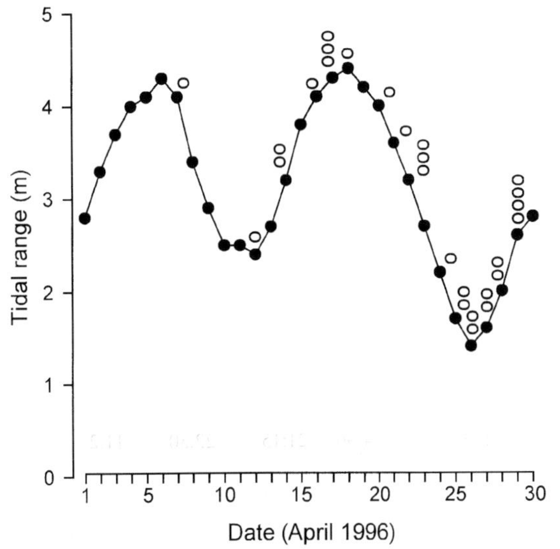


*Figure 3.* The migration of salmon smolts (○) into Southampton Water in relation to the Spring – Neap tidal cycle. The tidal range (●) is the difference between the high and low waters each day.

water section of the River Test downstream of the trap site (Figure 2). As the smolt season progressed, the released smolts spent less time within the freshwater section before migrating into the estuary (least squares regression: $r^2 = 0.539$: $p < 0.01$). A third (33%) of the smolts left the freshwater section of the River Test and entered the upper estuary during a six day period (24–29 April) coinciding with a neap tidal cycle (Figure 3).

There did not seem to be an apparent relationship between river flow measured at the river gauging station and the movement and residency times of smolts within the freshwater section of the River Test (Figure 4).

Smolts were resident in the upper estuary for $28.3 \pm 9.5$ h (mean ± S.E.M.) before moving downstream (Table 2). Smolts moved rapidly through the section of the estuary monitored by the sonar buoys taking $3.1 \pm 0.71$ h (mean ± S.E.M.) to migrate into the middle estuary section. Smolts continued to migrate relatively rapidly seawards through Southampton Water.

### *Diurnal patterns of migration*

The movement of salmon smolts within the freshwater section of the River Test was predominantly nocturnal (Table 3). The mean times that salmon smolts passed the two sonar buoys within the freshwater section were 23:42 and 00:32 respectively, and the movement was non-random with respect to time. Fish continued to migrate predominantly under the cover of darkness as they passed through the upper estuary section.

*Table 3.* The downstream movements of Atlantic salmon smolts through each section of the River Test and Southampton Water in relation to the time of day. The mean times that fish were recorded passing each position have been calculated from the mean vectors (Batschelet, 1981). The '*r*' values provide a measure of randomness of movement with respect to time calculated using the Rayleigh test. The value *n* is the total number of fish movements through the respective sections

| | Freshwater | | Upper Estuary | Middle Estuary | Lower Estuary | Southampton Water |
|---|---|---|---|---|---|---|
| | Testwood | Testwood Pool | | | | |
| Mean time | 23:42 | 00:32 | 00:18 | Random movement | Random movement | Insufficient data |
| *r* value | 0.57 | 0.50 | 0.49 | 0.40 | 0.09 | |
| *P* | < 0.001 | < 0.001 | < 0.05 | > 0.05 | > 0.05 | |
| *n* | 27 | 27 | 22 | 9 | 8 | |

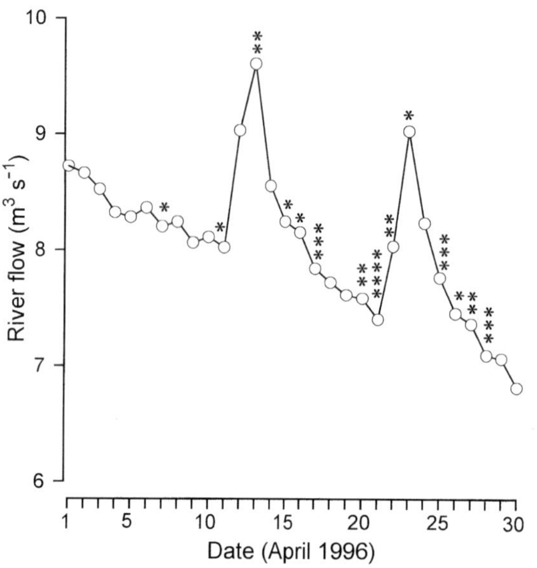


*Figure 4.* The timing of movement of salmon smolts (★) within the freshwater section of the River Test in relation to the river flow measured at a gauging weir.

The mean time that fish were detected passing the sonar buoys located at the upstream and downstream extents of this section were 23:26 and 00:18 respectively. However, the subsequent seaward movement of smolts through the downstream sections of Southampton Water was random with respect to the time of day. There was insufficient data to calculate the mean times that smolts migrated through the subsequent section of Southampton Water, but fish were again recorded emigrating throughout the 24-hour cycle.

## *Tidal patterns of migration*

There was a strong tidal component to the direction of movement of salmon smolts within the river estuary and Southampton Water (Figure 5). Smolts migrated downstream continuously throughout the ebb tidal cycle. The mean times that smolts were recorded moving through the upper and lower estuary sections were 1 h 54 min and 3 h 50 min after High Water respectively (Table 4). During the ebb tide smolts migrated distances of 4908 ± 609 m (mean ± S.E.M.) at mean ground speeds of 35 cm $s^{-1}$. Release of inert objects (oranges) adjacent to the vessel during the period that the smolts were tracked indicated that the fish were moving at speeds in excess of the seaward current. It was also evident from operating the acoustic hydrophone in the vertical mode and the relative signal strength of the tag that the smolts were migrating downstream in the upper water column close to the surface.

During the early part of the flood tide there was an inital upstream movement by a proportion of the smolts (5) in the direction of the prevailing current (mean distance 1836 ± 506 m) (Table 5). During this period, the fish were recorded holding position for periods > 10 min for 28% of the time and it was evident from the operation of the hydrophone that the fish had moved closer to the bed of the estuary. The remaining fish moved away from the main channel to areas of lower flow close to the shore of the estuary. There was a significant net downstream migration of all smolts during the later part of the flood tide and the period of slack water between the two high water marks. During the flood tide smolts migrated downstream distances of 3469 ± 522 m (mean ± S.E.M.) at ground speeds of 14 cm $s^{-1}$ (Table 5), which were significantly different to the ground speeds of smolts migrating during the ebb tide cycle (df=9; $t = 6.01$; $p < 0.001$ – two tailed $t$-test, unequal variances). As a result of the net downstream migration throughout the tidal cycle, the smolts moved relatively rapidly seaward through Southampton Water (Table 2).

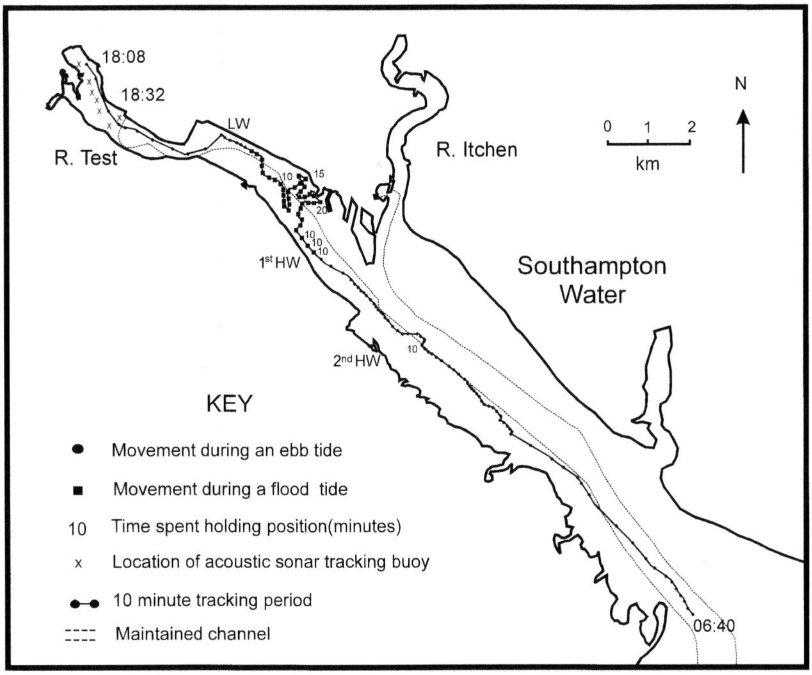


*Figure 5.* The track of a salmon smolt through the upper estuary of the River Test and Southampton Water. Each data point is the position of the tracking vessel in contact with smolt obtained by GPS.

*Table 4.* The downstream movements of Atlantic salmon smolts through the sections of the River Test and Southampton Water in relation to the tidal cycle. The mean times after HW that fish were recorded passing each position have been calculated from the mean vectors (Batschelet, 1981). The '*r*' values provide a measure of randomness of movement with respect to the tidal cycle were calculated using the Rayleigh test. The value *n* is the total number of fish movements through the respective sections. The second high water at Southampton has been used as a reference point

| | Upper Estuary | Middle Estuary | Lower Estuary | Southampton Water |
|---|---|---|---|---|
| mean time | 01:54 | Random movement | 03:50 | Insufficient data |
| $r$ value | 0.381 | 0.356 | 0.78 | |
| $p$ | $< 0.05$ | $> 0.05$ | $< 0.01$ | |
| $n$ | 23 | 9 | 8 | |

*Table 5.* The relationship between tidal cycle, distance migrated, periods holding position (> 10 min), and ground speeds of Atlantic salmon smolts in the River Test estuary and Southampton Water

| | EBB | | FLOOD | |
|---|---|---|---|---|
| | Upstream | Downstream | Upstream | Downstream |
| Mean distance (m $\pm$ S.E.M.) | $655 \pm 224$ | $4908 \pm 609$ | $1836 \pm 506$ | $3469 \pm 522$ |
| Periods holding position (% $\pm$ S.E.M.) | $14.93 \pm 2.13$ | $5.17 \pm 2.15$ | $28.44 \pm 5.39$ | $8.04 \pm 2.10$ |
| Ground speed (cm s$^{-1}$ $\pm$ S.E.M.) | $17.17 \pm 3.08$ | $34.74 \pm 2.97$ | $29.63 \pm 5.21$ | $14.30 \pm 1.63$ |

## Discussion

### *Freshwater migration*

The migration of salmon smolts in the fresh water section of the River Test was predominantly nocturnal as previously reported in the studies on the Rivers Avon, Conwy and Tawe (Moore et al., 1992; Moore et al., 1995; Moore et al., 1996), and other river systems (Thorpe & Morgan, 1978; Thorpe et al., 1981; Lundqvist & Eriksson, 1985; Hansen & Jonsson, 1985; Greenstreet, 1992). There was no indication in the present study of a seasonal change in the diurnal pattern of migration of salmon smolts demonstrated in the River Conwy (Moore et al., 1995) or movement during the day as demonstrated by Bagliniére, (1976). However, there was a significant seasonal change in the residency time of all smolts, with fish tagged later in the season spending less time in the River Test before migrating downstream into the estuary. This behaviour may be related to either a requirement for the smolts to migrate from the River Test into the marine environment during an optimal time or window of opportunity or alternatively, may reflect a seasonal change in the level of smoltification and physiological adaptation to salt water.

The nocturnal pattern of migration of the salmon in the River Test may have been the result of an endogenous rhythm of swimming activity which resulted in fish moving into the upper water column after dusk and migrating downstream (Moore et al., 1995). A number of fish were detected holding station in deep pools during the daytime, but subsequently moved downstream during the hours of darkness. Nocturnal migration of salmon smolts during the smolt run may reduce the chances of avian predation particularly by cormorants (*Phalacrocorax carbo*) and herons (*Ardea cinerea*) which are considered to prey on smolts in the River Test. As the majority of these birds feed visually during the day (Kennedy & Greer, 1988), refuging of salmonid smolts close to the bed of the river during the day and migrating at night would reduce predation pressures on the fish.

A previous study on a chalkstream by Solomon (1978), has suggested that a sudden increase in river flow rate is an important environmental factor initiating smolt migration. In the River Test, there was no observable correlation between the downstream movement of the smolts and river flow, although Russell & Ives (1994) have suggested that peaks of smolt movement earlier in the run tend to be linked with increased flow. Thorpe & Morgan (1978) have argued that during smolt migration in freshwater, the fish become less willing to hold station against the current, and as a result any increase in river flow will result in the downstream displacement of the smolts. Although it is apparent that other factors such as temperature (Osterdahl, 1969; McCleave, 1978) and diurnal patterns of swimming behaviour (Bagliniére, 1976; Moore et al., 1995) are implicated in the control of downstream migration in freshwater, increasing river flow may also be important in permitting the rapid seawards migration of smolts in the River Test.

### *Estuarine migration*

The emigration of salmon smolts in the upper estuary section of the River Test was indicative of a nocturnal ebb tide assisted type of migration similar to that described in the River Conwy (Moore et al., 1995). The smolts migrated seawards close to the surface and within the section of the water column with the highest velocities (Webber, 1980). This type of movement is energetically the most advantageous strategy for migration through an estuary (Moore et al., 1995). The subsequent seaward movement of smolts within the lower sections of the River Test and Southampton Water Smolt was indicative of active directed swimming with smolts migrating seawards during both ebb and flood tides. Movement was at speeds significantly in excess of the current and during a flood against the direction of the prevailing flow. LaBar et al. (1978) have previously reported active swimming against the current by salmon smolts in the lower sections of the Penobscot River estuary. Moore et al. (1995) suggested that during the transition from fresh to saltwater there is significant change in the smolts behaviour where they switch from moving passively with the current and actively swim seawards. The exact cue controlling such behaviour is unknown but is probably related to a particular salinity threshold. Although there was also a significant tidal component to the upstream movement of smolts, this may be explained by fish migrating seawards within the main channel being unable to hold position against the prevailing current (Fried et al., 1978). However, it was evident that these fish subsequently moved away from higher currents located within the main channel towards the lower flows close to the estuary shore and again migrated seaward.

The nocturnal pattern of migration within freshwater and the upper estuary section was not evident during the subsequent migration through Southampton Water, and smolts emigrated during both day and night. The significance of this behaviour together with the active swimming throughout the tidal cycle is that smolts migrated rapidly through the estuary and into coastal waters. A consequence of the seasonal change in transit time of the smolts through the estuary and into coastal waters was that the majority of the smolts migrated out of the estuary over a 13-day period. This period may represent the optimal time or window of opportunity for smolts to migrate from the River Test into the marine environment. Similar behaviour has been demonstrated in Atlantic salmon smolts in the River Conwy (Moore et al., 1995), and it has been previously observed that the timing of migration from fresh to salt water is considered important to their subsequent survival and return as adults (Hanssen & Jonsson, 1989).

The movement of salmon smolts through the estuary and into coastal waters was unaffected by changes in water quality. Increasing salinity did not appear to affect the behaviour of the smolts and there was no apparent period of salt water acclimation required as previously suggested in other salmonid smolts (Moser et al., 1991). This implies that the smolts were physiologically adapted prior to migration into the saline environment and that this had occurred in fresh water. Evidence for this could be provided by a series of seawater challenge tests on the smolts. However, it is not clear in salmon whether the cues initiating smolt migration are related to a physiological requirement to move to a saline environment, or a physiological requirement to leave the freshwater environment (Primett et al., 1988; Thorpe & Moore, 1997).

## Acknowledgements

The project was principally funded by the Ministry of Agriculture, Fisheries and Food and partly funded by Associated British Ports, Southampton. We thank the following for their commitment and expertise in assisting with the study; Ian Russell (CEFAS), Mark Sidebottom (EA), Peter Rees (EA), David Scranney (EA), Dave Hunter (EA), Heb Leman (EA), Adrian Fewings (EA), Trevor Draper (EA), Ian Spillar (EA) and Richard Redsull (EA). Thanks also to the consummate boat skills of the skippers from Southern Professional Training during the tracking of individual smolts through the heavy traffic of Southampton Water.

## References

Baglinićre, J. L., 1976. Etude des populations de saumon Atlantique (*Salmo salar* L., 1766) en Bretagne-Basse-Normandie. II Activité de dévalaison sur la riviere Elle. Ann. Hydrobiol. 7: 159–177.

Batschelet, E., 1981. Circular Statistics in Biology. Academic Press, London, 371 pp.

Dyer, K. R., 1970. Some aspects of coastal and estuarine sedimentation. PhD Thesis, University of Southampton.

Fried, S. M., J. D. McCleave & G. W. LaBar, 1978. Sea ward migration of hatchery-reared Atlantic salmon, Salmo salar, smolts in the Penobscot River estuary, Maine: Riverine movements. J. Fish. Res. Bd Can. 35: 76–87.

Greenstreet, S. P. R., 1992. Migration of hatchery reared juvenile Atlantic salmon, Salmo salar L., smolts down a release ladder. 3. Reactions on exiting the ladder. J. Fish. Biol. 40: 683–694.

Hansen, L. P. & B. Jonsson, 1985. Downstream migration of hatchery-reared smolts of Atlantic salmon (*Salmo salar* L.) in the River Isma. Aquaculture 45: 237–248.

Hansen, L. P. & B. Jonsson, 1989. Salmon ranching experiments in the River Imsa; Effect of timing of Atlantic (*Salmo salar*) smolt migration on the survival to adults. Aquaculture 82: 367–373.

Kennedy, G. J. A. & J. E. Greer, 1988. Predation by cormorants, *Phalacrocorax carbo* (L.), on the salmonid populations of an Irish river. Aquat. Fish. Mgmt 19: 159–170.

LaBar, G. W., J. D. McCleave & S. M. Fried, 1980. Seaward migration of hatchery-reared Atlantic salmon, *Salmo salar*, smolts in the Penobscot River estuary, Maine: open water movements. J. Cons. perm. int. Exp. Mer 38: 257–269.

Lundqvist, H. & L.-0. Eriksson, 1985. Annual rhythms of swimming behaviour and sea water adaptation in young Baltic salmon, *Salmo salar*, associated with smolting. Envir. Biol. Fish. 14: 259–267.

McCleave, J. D., 1978. Rhythmic aspects of estuarine migration of hatchery reared Atlantic salmon (*Salmo salar*) smolts. J. Fish. Biol. 12: 559–570.

Moore, A., T. Storeton-West, I. C. Russell, E. C. E. Potter & M. J. Challiss, 1990a. A technique for tracking Atlantic salmon (*Salmo salar* L.) smolts through estuaries. ICES C.M. 1990/M:18: 1–5.

Moore, A., I. C. Russell & E. C. E. Potter, 1990b. The effects of intraperitoneally implanted dummy acoustic transmitters on the behaviour and physiology of juvenile Atlantic salmon, *Salmo salar* L. J. Fish Biol. 37: 713–721.

Moore, A., E. C. E. Potter & A. A. Buckley, 1992. Estuarine behaviour of migrating Atlantic salmon (*Salmo salar* L.) smolts. In I. G. Priede & S. M. Swift (eds), Wildlife Telemetry Remote Monitoring and Tracking of Animals: 389–399.

Moore, A. & E. C. E. Potter, 1994. The movements of sea trout (*Salmo trutta* L.) smolts through the estuary of the River Avon, Southern England. Fish. Mgmt Ecol. 1: 1–14.

Moore, A., E. C. E. Potter, N. J. Milner & S. Bamber, 1995. The migratory behaviour of wild Atlantic salmon (*Salmo salar* L.) smolts in the estuary of the River Conwy, North Wales. Can. J. Fish. aquat. Sci. 52: 1923–1935.

Moore, A., R. Stonehewer, L. T. Kell, M. J. Challiss, M. Ives, I. C. Russell, W. D. Riley & D. M. Mee, 1996. The movements of emigrating salmonid smolts in relation to the Tawe barrage, Swansea. In N. Burt & J. Watts (eds), Barrages: Engineering Design & Environmental Impacts. HR Wallingford Ltd. John Wiley & Sons Ltd.: 409–417.

Moser, M. L., A. F. Olson & T. P. Quinn, 1991. Riverine and estuarine migratory behaviour of coho salmon (*Oncorhynchus kisutch*) smolts. Can. J. Fish. aquat. Sci. 48: 1670–1678.

Osterdahl, L., 1969. The smolt run of a small Swedish river. In T. G. Northcote (ed.), Salmon and Trout in Streams. H.R. Macmillan Lectures in Fisheries, University of British Columbia, Vancouver, B.C.: 205–215.

Primmett, D. R. N., F. B. Eddy, M. S. Miles, C. Talbot & J. E. Thorpe, 1988. Transepithelial ion exchange in salmon. 1. Seaward migration may be caused by ion regulatory stress. Fish Physiol. Biochem. 5: 181–186.

Russell, 1. C. & M. J. Ives, 1994. Report of the River Test smolt trapping programme 1994. CEFAS Internal Report, 19 pp.

Solomon, D. J., 1978. Some observations on salmon smolt migration in a chalk stream. J. Fish Biol. 12: 571–574.

Thorpe, J. E. & R. I. G. Morgan, 1978. Periodicity in Atlantic salmon *Salmo salar* L. smolt migration. J. Fish Biol. 12: 541–548.

Thorpe, J. E., L. G. Ross, G. Struthers & W. Watts, 1981. Tracking Atlantic salmon smolts *Salmo salar* L. through Loch Voil, Scotland. J. Fish Biol. 19: 519–537.

Thorpe, J. E. & A. Moore, 1997. The migratory behaviour of juvenile Atlantic salmon. In H. Ueda & H. A. Bern (eds), Reviews of Physio-ecological Research on Fish Migration Hoikkaido Research Press, Hoikkaido, Japan.

Tytler, P., J. E. Thorpe & W. M. Shearer, 1978. Ultrasonic tracking of the movements of Atlantic salmon smolts (*Salmo salar*) in the estuaries of two Scottish rivers. J. Fish Biol. 12: 575–586.

Webber, N. B., 1980. Hydrography and water circulation in the Solent. In: The Solent Estuarine System. NERC Publication Series C, Number 22.

*Hydrobiologia* **371/372:** 305–308, 1998.
*J.-P. Lagardère, M.-L. Bégout Anras & G. Claireaux (eds), Advances in Invertebrates and Fish Telemetry.*



# Movements of a bream (*Abramis brama* (L.)), rudd × bream hybrid, tench (*Tinca tinca* (L.)) and pike (*Esox lucius* (L.)) in an Irish canal habitat

R. E. Donnelly, J. M. Caffrey & D. M. Tierney
*Central Fisheries Board, Mobhi Boreen, Glasnevin, Dublin 9, Republic of Ireland*
*(tel: +353 1 8379206; fax: +353 1 8360060; e-mail: richard.donnelly@cfb.ie)*

*Key words:* telemetry, coarse fish, home range, canals

## Abstract

Radio telemetry was used to monitor the movement of four species of coarse fish in a 11.2 km long section of canal from January to June 1996. One specimen of bream, rudd × bream hybrid, tench and pike were captured using electrofishing apparatus and Lotek CFRT-3EMXT pulse coded aquatic transmitter tags, operating at 173.3 Mhz (with individual ID codes), were sutured externally. The movements of the four radio tagged fish were monitored approximately every four days during the course of the study. The results of the study indicated that all four species did migrate to and from specific locations along the canal. These movements are discussed in terms of home range behaviour for each of the species.

## Introduction

Within Irish water courses there is only limited data on migration patterns of bream (*Abramis brama* (L.)) (Whelan, 1983) and apart from reports in relation to this study (Caffrey et al., 1995, 1996) no literature is available on radio or acoustic telemetric studies of bream, cyprinid hybrids, tench (*Tinca tinca* (L.)) or pike (*Esox lucius* (L.)). For this reason a bream, rudd (*Scardinius erythrophtalmus* (L.)) × bream hybrid, tench and pike were chosen for this study.

The 19th Level of the Grand Canal (N °69′31 to N 78°25′) is 11.2 km in length, 15 m–20 m in width with a depth of *c.* 1.5 m and is bound at both ends by lock gates. A branch line, the Barrow Line, of similar dimensions to the Grand Canal flows southwards from Lowtown. A lock gate is present on this branch line 3.4 km south of Lowtown (Figure 1). Details of the physical and chemical characteristics of the Grand Canal are given in Caffrey & Allison (in press). The study site is a renowned angling location and has an average biomass of fish *c.* 250 kg ha$^{-1}$ (Caffrey et al., 1995).

## Methods

The four fish were obtained at the same time and date (January 18th, 1996) from a 0.2 km stretch of canal west of Ticknevin Bridge (Figure 1) using boat mounted electrofishing apparatus (Connolly et al., 1991). They were then anaesthetised in 100 litre drum containing and 1–2 mls of quinaldine. Once the fish was fully anaesthetised the radio tag was sutured to the fish directly beneath the dorsal fin.

Radio transmitters were used as opposed to acoustic transmitters because they are more suitable for canal systems (Armstrong et al., 1988). The tags used were Lotek CFRT-3EMXT pulse coded aquatic RF transmitter tags operating at 173.300 Mhz (with individual ID codes). The tags were 19 mm by 55 mm with a weight in water of 8.6 g. The receiver used was a Lotek SRX 400A automatic scanning telemetry receiver with W5 firmware.

The tags were monitored at least every four days during daylight hours between January 18th and June 18th 1996. Good vehicle access along the canal towpath allowed recordings of the tags to be taken with a car mounted aerial linked to the telemetry receiver. Upon locating a signal the vehicle was stopped and

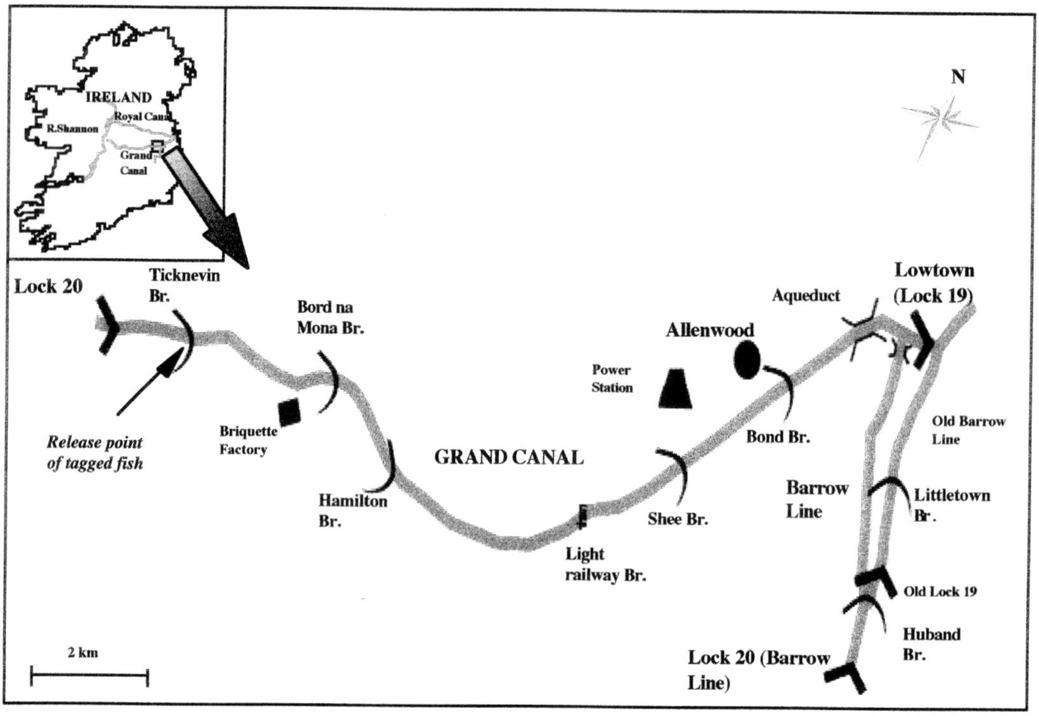


*Figure 1.* Grand Canal Level 19.

*Table 1.*

| Species | Fork-Length (cm) | Weight (g) |
|---|---|---|
| Bream | 38 | 1005 |
| Rudd × Bream Hybrid | 41 | 1400 |
| Tench | 39 | 1170 |
| Pike | 81 | 3400 |

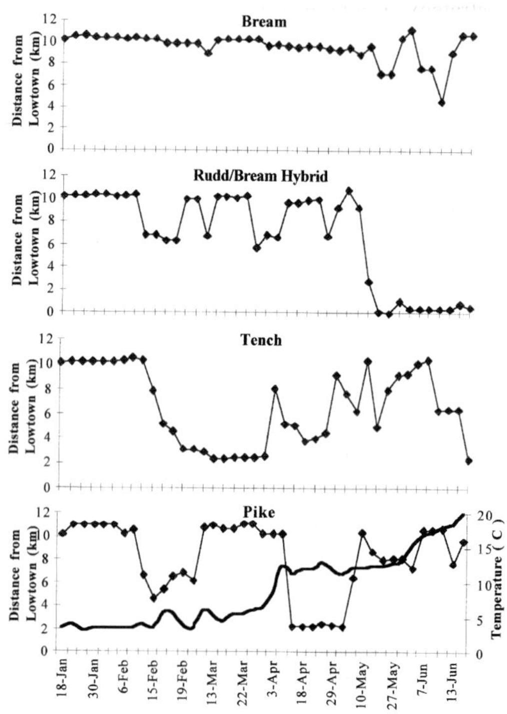


*Figure 2.* Movements of radio tracked fish and water temperature data from January to June 1996.

location of fish was located too within *c.* 5 m by positioning the aerial and receiver at the canal bank. On all monitoring excursions all the tags were recorded within a two-hour period. Water temperature readings were also taken at regular intervals throughout the study period.

## Results

The fork-length and weight of each tagged fish is presented in Table 1. The movements of the four fish are shown in Figure 2 together with water temperature data for the study period.

## Discussion

The bream in this study remained predominately in the Ticknevin area but also migrated to and from this location on a number of occasions. This would suggest that this fish had a home range. The establishment of a home range by bream has previously been documented (Astrauskas, 1971; Whelan, 1983 and Goldspink, 1978a). Although no literature is available on the establishment of a home range by rudd × bream hybrids the present results indicated that this hybrid did establish a home range but it was not as distinct as that of the bream. The bream also migrated significantly shorter distances overall than the hybrid this may be explained by the migratory patterns observed for bream by Goldspink (1978b) and Whelan (1983). These authors documented that the majority of a bream population show localised migrations and a small proportion 12% (Goldspink, 1978b) to 32% (Whelan, 1983) migrate over much longer distances. It is possible that the hybrid was part of the shoal which migrate over these longer distances.

There was no evidence of a distinct home range for the tench. This fish covered a wide range of the available canal habitat, this may indicate foraging activity. Tench are believed to be relatively lethargic in their feeding strategy (Kennedy & Fitzmaurice, 1970). This study shows that this tench was quite active particularly when the temperature increased. It is probable that the tench utilised a greater area of canal due to a shortage of suitable food. Tench feed predominantly on invertebrates in Irish waters (Kennedy & Fitzmaurice, *op. cit.*). The canal in this area has a low abundance and diversity of invertebrate life (Tierney pers. com.) and this may account for the necessity of the tench to avail of a relatively large area of canal habitat in order to sustain itself. This hypothesis is further supported by the fact that as temperature increases the activity rate also increases. This relationship between foraging activity and temperature is also reported by Kennedy & Fitzmaurice, *op. cit.*

The pike also covered an extensive area of the available habitat. This is characteristic of a predatory species which generally has less food available than the herbivore (Clutton-Brock & Harvey, 1984). The pike did show a preference for the Ticknevin area returning to it on four separate occasions. Pike are believed to be non-territorial outside the spawning season (Bregazzi & Kennedy, 1980; Diana et al., 1977). Therefore, this preference of the pike for the Ticknevin area may not necessarily be related to territorial home range behaviour, but could probably be related to the fact that the bream and its associated shoal is located in this area thus providing a ready food supply for the pike. After 152 days the tags on the bream and hybrid had become detached from the fish. Two days prior to the loss of the tag the hybrid had been observed swimming with a shoal of bream, rudd (*Scardinius erythrophtalmus* (L.)) and cyprinid hybrids, and apparently behaving normally. This tag was recovered from the canal bed. The tag on the bream was believed to have become detached at the same time as no observation of the fish was made despite good water clarity and no further movements were noted after this date. This tag was not recovered. At this stage the study was terminated and readings were only taken from the last known date when the tags were attached (June 18th).

Although only one individual of each species was studied this work does provide baseline information on habitat usage. A knowledge of the factors affecting these movements and actual habitat usage is important for establishing the criteria for habitat conservation of these sections of canal. More intensive studies need to be conducted and the number of samples needs to be increased to identify more precisely the habitat requirements in the canal for each of the species. Work is currently underway to address these problems.

## Acknowledgements

This project is financed by the Department of Arts Culture and Gaeltach, Waterways Division and the Tourist Operational Programme of the European Regional Development Fund.

## References

Armstrong, J. D., M. Lucas, J. French, L. Vera & I. G. Priede, 1988. A combined radio and acoustic transmitter for fixing direction and range of freshwater fish (RAFIX). J. Fish Biol. 33: 879–884.

Astrauskas, A. S., 1971. Results of the marking of bream (*Abramis brama* (L.)) and roach (*Rutilus rutilus* (L.)) in the cooling reservoir of the Lithunian hydroelectric power station. J. Ichthyol. 2: 143–146.

Bregazzi, P. R. & C. R. Kennedy, 1980. The biology of pike *Esox lucius* L., in a southern eutrophic lake. J. Fish Biol. 17: 91–112.

Caffrey, J. M., R. E. Donnelly & D. Tierney, 1995. Canal Fisheries Development Programme 1990–1995. Office of Public Works commissioned report, Central Fisheries Board, Dublin, 73 pp.

Caffrey J. M. & J. Allison, 1997. Eutrophication in canals. In: Eutrophication in Irish Waters. Royal Irish Academy, Dublin, in press.

Caffrey J. M., J. J. Conneely & B. Connolly, 1996. Radio telemetric determination of bream (*Abramis brama* L.) movements in Irish canals. In E. Baras & J. C. Philippart (eds), Underwater Biotelemetry, proceedings of the First Conference on Fish Telemetry in Europe: 61–67.

Clutton-Brock, T. H. & P. H. Harvey, 1984. Comparative approaches to investigating adaptation. In J. R. Krebs & N. B. Davis (eds), Behavioural Ecology. Blackwell Scientific Publications, Oxford: 7–29.

Connolly, B., J. Conneely & J. M. Caffrey, 1991. Canal Fisheries Development Programme, Annual Report, 1990–1991. Office of Public Works commissioned report, Central Fisheries Board, Dublin, 41 pp.

Diana, J. S., W. C. Mackay & M. Ehrman, 1977. Movements and habitat preference of northern pike (*Esox lucius*) in Lac Ste. Anne, Alberta. Trans. am. Fish. Soc. 106: 560–565.

Goldspink, C. R., 1978a. The population density, growth and production of bream *Abramis brama* (L.) in Tjeukemeer, The Netherlands. J. Fish Biol. 3: 499–517.

Goldspink, C. R., 1978b. A note on the dispersion pattern of marked bream *Abramis brama* (L.) released into Tjeukemeer, The Netherlands. J. Fish Biol. 3: 493–497.

Kennedy, M. & P. Fitzmaurice, 1970. The biology of the tench *Tinca tinca* (L.) in Irish waters. Proc. Roy. Ir. Acad. 69B: 31–82.

Whelan, K. F., 1983. Migratory patterns of bream Abramis brama L. shoals in the River Suck system. Ir. Fish. Investigat. 23: 11–15.

*Hydrobiologia* **371/372:** 309–319, 1998.
*J. P. Lagardère, M.-L. Bégout Anras & G. Claireaux (eds), Advances in Invertebrates and Fish Telemetry.*



# Radio telemetry as a tool to study habitat use of nase (*Chondrostoma nasus* L.) in medium-sized rivers

Martin Huber & Arthur Kirchhofer
*Institute of Zoology, University of Berne, Baltzerstr. 3, CH-3012 Berne, Switzerland*
*E-mail: martin.huber@zos.unibe.ch*

*Key words: Chondrostoma nasus*, radiotelemetry, habitat use, regulated River, seasonal, home-range

## Abstract

Due to a massive population decline of nase (*Chondrostoma nasus* L.) in the river Aare (Switzerland), a radiotelemetry study was designed to detect the habitat use of this endangered species. Although big individual differences in behaviour occurred for adult nase, habitat preferences for runs with currents between 0.5 and 1 m $s^{-1}$, coarse sediments and depths between 1 and 2 m could be shown. The fish used well defined home-ranges for certain time periods and migrations were mainly found in correlation with spawning or seasonal changes in discharge and temperature. Seasonal and daily differences in home-range size and habitat preferences were discussed in relation to habitat structure.

## Introduction

Knowledge of habitat requirements is of high importance for stock management and conservation of endangered fish species in medium sized rivers. This is especially true for rheophilic and lithophilic Cyprinids, a taxon which includes a high proportion of endangered species in central Europe, for which such information is still scarce (Nelva, 1988; Garner & Clough, 1996; Keckeis et al., 1996; Baras, 1997; Clough & Ladle, 1997; Lucas & Frear, 1997). Although habitat protection is the most effective method for species conservation, small scale and simple habitat improvements can be useful too. However, the needs of the target species must be known (Burgess, 1985; Hermansen & Krog, 1985).

Habitat use can substantially vary both at a daily and seasonal time scale, depending on species and environment (Baras, 1992; Arnekleiv & Krabol, 1996; Bourke et al., 1996). In addition, the habitat requirements vary substantially over the live-cycle. Therefore studies on the larvae, juveniles and adults are needed to cover the whole life-cycle of endangered species (Copp, 1992; Bischoff, 1995; Scholten, 1995; Keckeis et al., 1996). Radio telemetry can be a useful tool to gain such information using only a small number of individuals (Priede, 1991) and therefore can be used on small, threatened populations.

In this contribution the habitat use of adult nase is presented and discussed in relation to habitat structure, daytime and season. The habitat characteristics of short segments of a 7$^{th}$ order stream were recorded and compared with the habitat use. It is assumed that using this technique, the small scale distribution of the nase can be identified and the information used for habitat restoration.

## Study area

The study was performed in the river Aare near Berne (Switzerland), a medium sized 7$^{th}$ order stream. The flow regime is characterised by the alpine snow melt during spring and summer and low flow during the cold season, but occasional winter floods occur during heavy rainfall in the lowlands. The yearly median of daily mean discharge ($Q_m$) is 121 $m^3$ $s^{-1}$. Upstream lakes and reservoirs reduce the differences in discharge between summer and winter (Aschwanden & Wein-

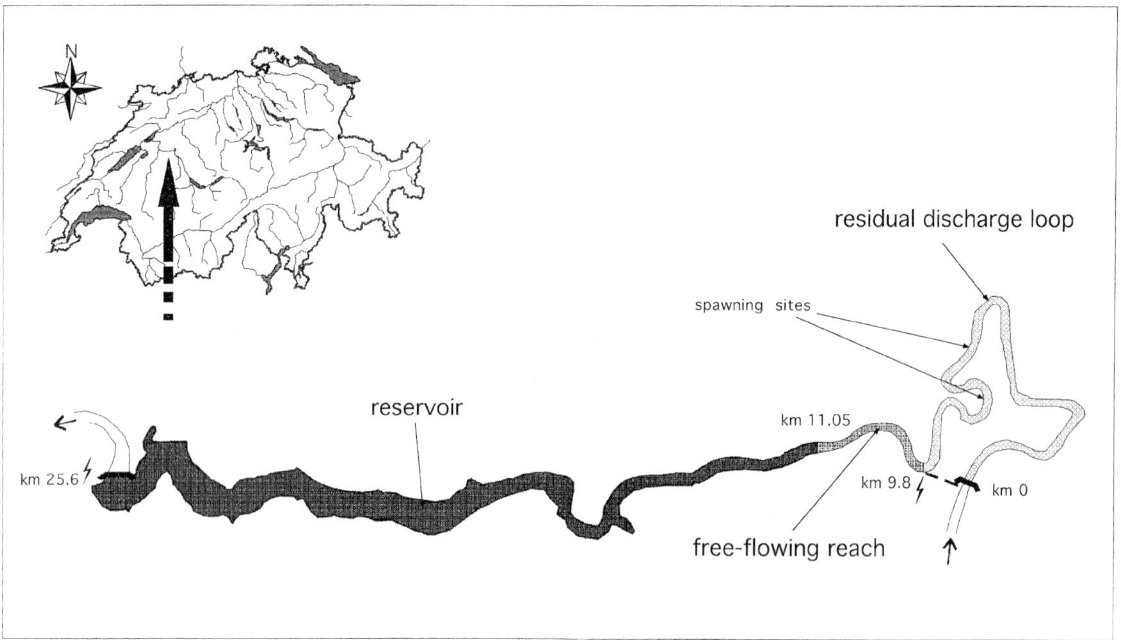


*Figure 1.* Map of the study site, showing the three major sections. Residual discharge section from km 0 to km 9.8, free-flowing reach from km 9.8 to 11.05 and reservoir from km 11.05 to km 25.6.

gartner, 1985; Spreafico et al., 1992; Hydrologisches Jahrbuch der Schweiz, 1994).

The study site is bordered by two dams and associated hydroelectric power-stations (Figure 1). The upper dam is equipped with a fishpass which is not used by nase (Guthruf, 1996), but the lower dam has no fishpass. This system with a residual discharge section (mean annual flow reduction 41–60%), a free-flowing reach and an impounded section is typical for medium sized rivers in Switzerland.

The residual discharge section has a natural morphology with a variety of riffles, runs and pools. The width varies between 35 and 70 metres (average 50 m). The free-flowing reach consists only of run sections, its width is about 75 m (from 55 m to 100 m) and the shore is mostly fixed by rip-rap. The impoundment (width from 75 m to > 500 m) has mainly soft sediment and low current. Due to this modifications the river characteristics have changed in the lower parts from the grayling to the barbel and bream region (Huet, 1949). In spite of the effluents of two water purification plants into this section of the river, water quality is not regarded as problematic for nase (Dedual, 1990; Hydrologisches Jahrbuch der Schweiz, 1994). However, the impact of medium to high nutrient loads on food organisms, especially diatoms, is unknown.

## *Material and methods*

In total 21 adult nase were tagged and tracked between October 1995 and September 1997 (Table 1). In autumn 1995 the fish were caught by rod and line and haul-seine, in spring 1996 by electrofishing on the spawning site and in winter 1996 by electrofishing from a boat at night.

Transmitters (Lotek FRT-3/4, FSM-6) were implanted in the body cavity of the anaesthetised (MS 222) fish (Hart & Summerfelt, 1975). The incision was made ventrally just behind the ventral fins, the external antennas (40 cm) were trailing outside the body.

The tags weighed between 8.9 to 23.7 g in air and did not exceed 2.8% of body weight; battery life was between 295 and 390 days. Frequencies between 148.100 and 148.775 MHz and pulse rates between 40 and 60 bpm were used. Individual fish could be identified by frequency and/or pulse rate.

In order to detect differences between habitat availability and habitat use, the study area was divided in segments of 25 m length. Within each segment the habitat types (riffle, run, pool, reservoir or whirl) and some important habitat parameters (current, substrate and depth) were estimated. This was done by visual observation from the banks during periods of

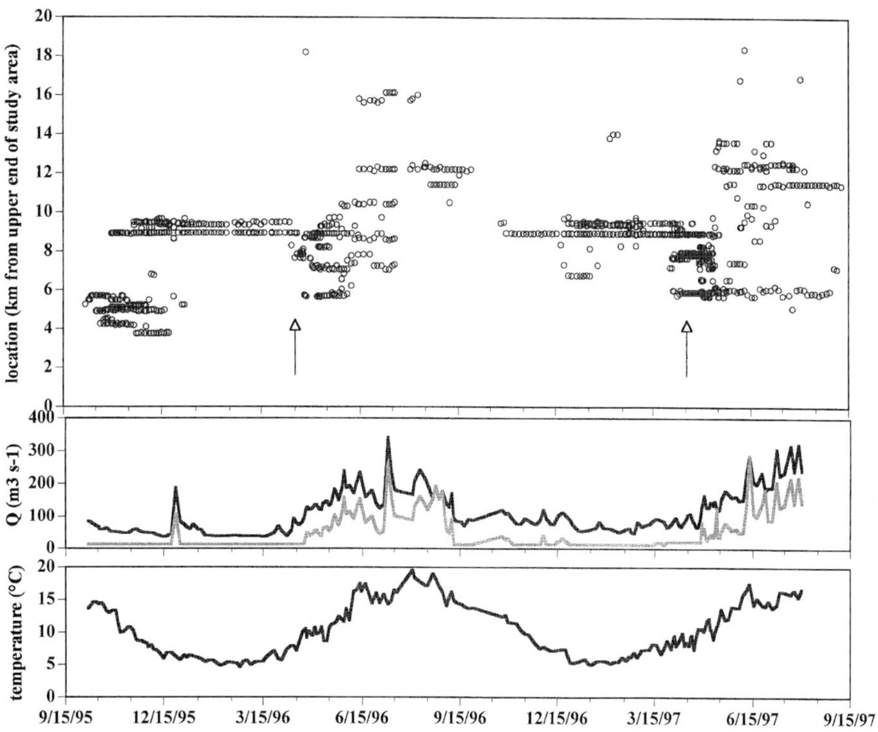


*Figure 2.* Locations of all tracked nase during the whole experimental time in distance (km) from the upper end of the study area. Mean daily discharge Q ($m^3$ $s^{-1}$) at the measuring station Schönau just upstream the study area (upper) and in the residual discharge section (lower). Temperature (°C) at measuring station Schönau. Arrows indicate spawning dates (23.4.96 and 5.5.97).

favourable conditions by selecting the best fitting category of a given set (Table 2). The positions of fish were assigned to each of these segments and if possible confirmed by visual observation. For segments consisting of more than one morphodynamic unit, the major habitat type was used for analysis and the variety recorded. This was rare for the impoundment ($< 1\%$) and the free-flowing reach ($< 7\%$) compared to the residual flow section (27%). Habitat mapping was performed in winter during low discharge and clear water, in order to be able to estimate coarseness of sediment and depth. This might lead to an underestimation of current strength during summer, i.e. segments with low current will be rarer if discharge is high. Similarly for some segments the major habitat type is likely to change between periods with high and low discharge. Therefore the momentary habitat structure was recorded during the tracking of the fish. Values of habitat preference (D) were calculated according to Jacobs (1974)

$$D = r - p/(r + p) - 2rp,$$

where $p$ = proportion of resource available, $r$ = proportion of resource used.

Tracking was done manually (Lotek SRX_400 receiver, three element Yagi antenna), twice a week for most of the time and more frequently during spawning season and at the beginning of the study. At six occasions the positions of 18 adult nase were recorded every two hours in order to study the habitat use on a time scale of 24 hours.

Definitions of the seasons were set after the periodicity of the tracked nase's activity pattern, temperature boundaries and spawning dates.

Autumn: 5.10.95 to 17.11.95 and 16.9.96 to 11.11.96 (Start of the study, respectively 16th of September until the water temperature dropped below 10 °C).

Winter: 20.11.95 to 31.3.96 and 14.11.96 to 31.3.97 (Water temperature below 10 °C until end of march).

Prespawning: 1.4.96 to 25.4.96 and 1.1.97 to 5.5.97 (Start of April until peak spawning).

Postspawning: 26.4.96 to 31.5.96 and 6.5.97 to 31.5.97 (From peak spawning until end of May).

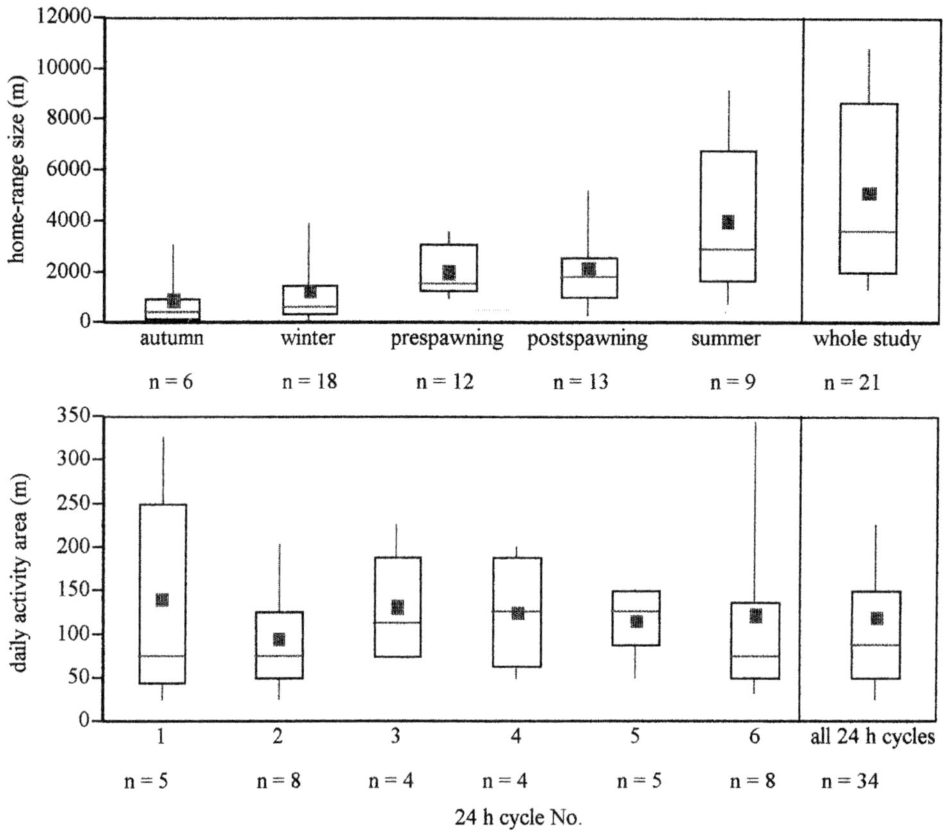


*Figure 3.* Home-ranges for seasons and the whole study time (upper graph); and daily activity areas during the different 24 h cycles (lower graph). 10%, 25%, median, 75% and 90% intervals; means as symbols; $n$ = number of individual nase included in each season or each 24 h cycle.

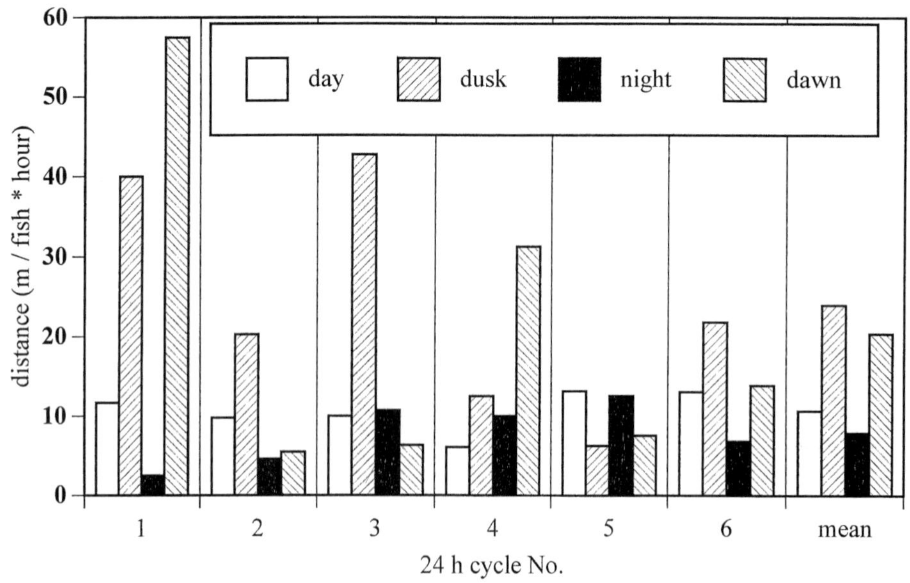


*Figure 4.* Minimal distances swum between two consecutive locations during the 24 h cycles for differnt times of the day.

*Table 1.* Length, weight, age and sex of the 21 adult nase tracked during the study. Age was determined by scale reading, sex by the strength of the first fin ray of the pectoral fin and confirmed by running milk or eggs during spawning season. The presence of the individual fish during the six 24 h cycles and tracking time are indicated on the right-hand-side (cycle 1: 15./16.11.1995; 2: 28./29.11.1995; 3: 21./22.3.1996; 4: 18./19.4.1996; 5: 20./21.6.1996 and 6: 20./21.2.1997)

| Fish | Length | Weight | Age | Sex | Presence in 24 h cycle No. | | | | | | Tracking time |
|---|---|---|---|---|---|---|---|---|---|---|---|
| # | (cm) | (g) | (y) | | 1 | 2 | 3 | 4 | 5 | 6 | |
| A | 46.0 | 1145 | 14 | – | × | × | | | | | 05.10.95–30.11.95 |
| B | 45.5 | 980 | 11 | – | × | × | | | | | 13.10.95–18.12.95 |
| C | 46.3 | 1145 | 15 | – | × | × | | | | | 19.10.95–22.12.95 |
| D | 45.0 | 840 | 11 | – | × | × | | | | | 19.10.95–05.01.96 |
| E | 45.8 | 1115 | 15 | male | × | × | × | × | | | 27.10.95–23.04.96 |
| F | 48.4 | 1350 | 16 | – | | × | × | × | × | | 20.11.95–16.09.96 |
| G | 48.0 | 1355 | 12 | – | | × | × | × | | | 20.11.95–23.04.96 |
| H | 47.4 | 1070 | 13 | – | | × | × | × | × | | 27.11.95–17.07.96 |
| I | 46.0 | – | 17 | female | | | | | × | | 30.04.96–17.07.96 |
| J | 52.0 | 1430 | 14 | female | | | | | | | 02.05.96–17.05.96 |
| K | 46.3 | 1230 | 13 | male | | | | | × | × | 02.05.96–24.07.97 |
| L | 49.8 | 1265 | 12 | female | | | | | × | | 09.05.96–17.07.96 |
| M | 52.0 | 1955 | 18 | female | | | | | | × | 20.12.96–04.08.97 |
| N | 48.0 | 1325 | 15 | male | | | | | | × | 20.12.96–30.11.97 |
| O | 45.0 | 845 | 7 | male | | | | | | × | 20.12.96–07.08.97 |
| P | 43.0 | 895 | 11 | male | | | | | | × | 20.12.96–04.09.97 |
| Q | 51.0 | 1520 | 13 | female | | | | | | | 20.12.96–17.02.97 |
| R | 45.4 | 1125 | 13 | male | | | | | | × | 20.12.96–10.05.97 |
| S | 47.1 | 1190 | 12 | female | | | | | | × | 20.12.96–10.05.97 |
| T | 46.2 | 890 | 12 | female | | | | | | | 20.12.96–10.01.97 |
| U | 41.1 | 860 | 7 | male | | | | | | × | 20.12.96–26.05.97 |
| *Mean* | *46.9* | *1177* | *13* | | *5* | *8* | *4* | *4* | *5* | *8* | |

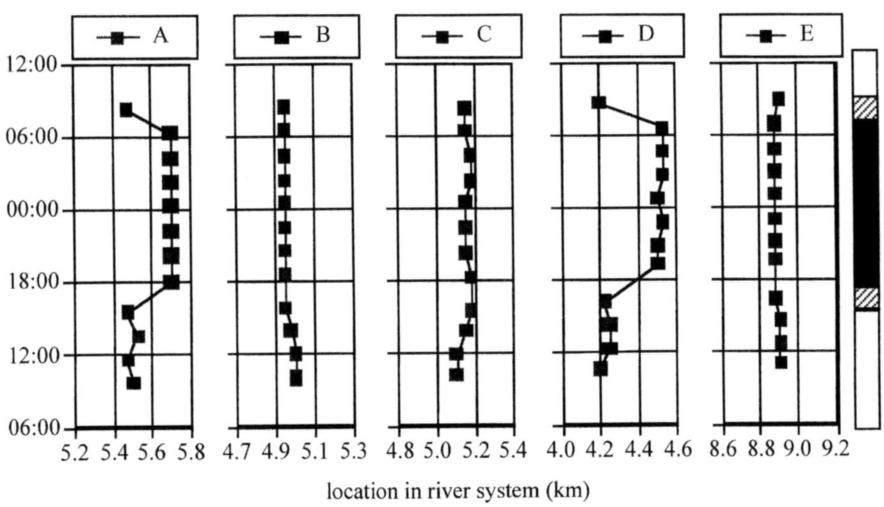


*Figure 5.* Positions of the tracked nase during the first 24 h cycle on the 15./16.11.1995 in distance (km) from the upper end of the study area. The light regime is indicated by the bar at the right-hand side.

*Table 2.* Proportions of habitat availability and habitat use of the recorded habitat variables for the whole study. Preference index (D) (Jacobs, 1974) for the different seasons and for the whole study

| Habitat variable | Habitat availability | Habitat use whole study | (D) Autumn | (D) Winter | (D) Post-spawning | (D) Pre-spawning | (D) Summer | (D) Whole study |
|---|---|---|---|---|---|---|---|---|
| **Habitat type** | | | | | | | | |
| Riffle | 0.07 | 0.04 | 0.00 | –0.76 | 0.37 | –0.76 | –1.00 | **–0.29** |
| Run | 0.67 | 0.88 | 0.60 | 0.84 | 0.35 | 0.38 | 0.47 | **0.57** |
| Pool | 0.03 | 0.01 | –1.00 | –0.51 | –0.51 | 0.15 | –0.51 | **–0.51** |
| Reservoir | 0.22 | 0.02 | –1.00 | –1.00 | –1.00 | –0.93 | –0.48 | **–0.87** |
| Whirl | 0.01 | 0.04 | 0.68 | 0.34 | 0.51 | 0.87 | 0.73 | **0.61** |
| **Current** | | | | | | | | |
| 0.0–0.1 m $s^{-1}$ | 0.05 | 0.01 | –1.00 | –0.68 | –1.00 | –0.68 | –0.68 | **–0.68** |
| 0.1–0.5 m $s^{-1}$ | 0.42 | 0.22 | –0.51 | –0.17 | –0.68 | –0.53 | –0.61 | **–0.44** |
| 0.5–1.0 m $s^{-1}$ | 0.22 | 0.48 | 0.67 | 0.67 | 0.25 | 0.55 | 0.35 | **0.53** |
| $>$ 1 m $s^{-1}$ | 0.31 | 0.29 | –0.20 | –0.75 | 0.46 | 0.02 | 0.33 | **–0.05** |
| **Substrate** | | | | | | | | |
| Silt | 0.39 | 0.03 | –1.00 | –0.97 | –1.00 | –0.85 | –0.68 | **–0.91** |
| Sand | 0.03 | 0.02 | –1.00 | –0.51 | –0.51 | –0.20 | 0.26 | **–0.20** |
| Gravel | 0.01 | 0.00 | –1.00 | –1.00 | –1.00 | –1.00 | 0.00 | **–1.00** |
| Pebbles | 0.23 | 0.19 | 0.50 | –0.55 | –0.37 | 0.13 | 0.29 | **–0.12** |
| Cobbles | 0.31 | 0.73 | 0.43 | 0.89 | 0.86 | 0.35 | 0.29 | **0.72** |
| Boulders | 0.03 | 0.04 | –0.51 | –0.51 | –0.51 | 0.75 | 0.15 | **0.15** |
| **Depth** | | | | | | | | |
| 0.0–0.2 m | 0.00 | 0.00 | – | – | – | – | – | **–** |
| 0.2–0.5 m | 0.01 | 0.01 | –1.00 | –1.00 | 0.61 | 0.00 | 0.00 | **0.00** |
| 0.5–1.0 m | 0.06 | 0.06 | 0.00 | –0.52 | 0.36 | 0.22 | 0.08 | **0.00** |
| 1.0–2.0 m | 0.17 | 0.45 | 0.60 | 0.67 | 0.72 | 0.53 | 0.00 | **0.60** |
| $>$ 2.0 m | 0.77 | 0.47 | –0.57 | –0.58 | –0.79 | –0.53 | –0.05 | **–0.58** |

Summer: 1.6.96 to 12.9.96 and 1.6.97 to 8.9.97 (1st of June until mid-September respectively end of study).

Home-range was estimated as the longitudinal distance between the uppermost and the lowermost locations.

## Results

The tagged nase did not use the whole study area, the inhabited part reached from km 3.75 to km 18.4 and covered all three major sections (Figure 2). The sections were not used equally (Chi-square-test, dF = 2, $P < 0.001$), with 87.88% of all 1229 locations made in the residual discharge section, 1.71% in the free flowing part and 10.41% in the reservoir.

For most of the time the specimens did not exhibit long distance movements. Ninety% of all fish were found in the same 25 m segment of the river or at maximum two segments up- or downstream on the following tracking. This does not mean that individual fish remained stationary all over the year, they rather occupied different stretches of the river as home-ranges for certain periods. Migrations were mainly correlated with spawning or seasonal changes of discharge and temperature (Figure 2). In longitudinal dimension, individual fish used river sections between 625 and 11250 m length over the whole observation time (Figure 3).

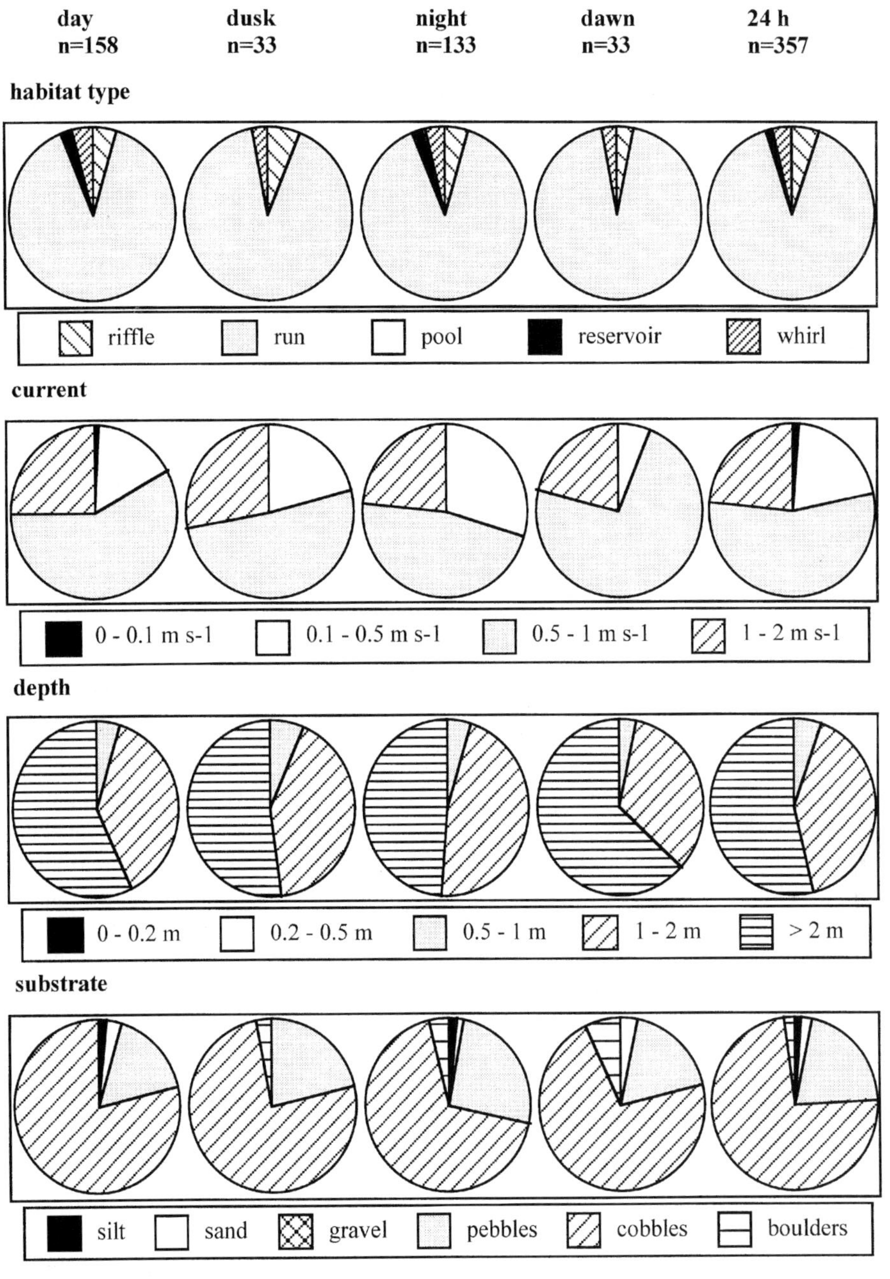


*Figure 6.* Habitat characteristics of all places used during the 24 h cycles for day, dusk, dawn, night and during the whole 24 h surveys ($n$ = number of trackings).

Home-range size was different between the seasons (Kruskal-Wallis-test, $n = 58$, dF = 4, $P = 0.005$). During winter nase used significant smaller home-ranges than in summer (Kolmogorov-Smirnov-test, $n = 27$, $P = 0.011$), during prespawning ($n = 30$, $P = 0.005$) and postspawning ($n = 31$, $P = 0.031$). Home-range size in autumn was smaller than during prespawning too ($n = 18$, $P = 0.008$). During the 24 hour cycles the specimens used stretches between 25 and 425 m with an average daily activity area of 120.5 m (Figure 3). Seasonal differences in the size of daily activity areas were not found.

Minimal distances swum during 24 hours ranged from 50 to 750 m. Movements between segments were more frequent at twilight, the furthest displacements also occurred during dusk and dawn (Figure 4). Significant differences could be found between night and day (paired t-test, dF = 17, $P = 0.049$), night and dusk ($P = 0.010$), night and dawn ($P = 0.036$) and day and dusk ($P = 0.022$) in spite of the apparent inactivity of the majority of the fish. However, with the transmitters used, swimming activity or inactivity could not be distinguished.

Eight of 34 fish used distinct stretches at day and night, with movements between stretches taking place during twilight (fish A and D, Figure 5). Fish A used exclusively run segments for the whole 24 h cycle, but preferred deeper water (> 2 m) and coarser sediment (cobbles of 6.3–20 cm) during the day than during the night (depth 1–2 m, pebbles of 0.63–2 cm). Currents stronger than 1 m $s^{-1}$ were only used once during daylight. Fish B also exclusively used run sections and deeper water during the day. It used segments with currents between 0.1 and 0.5 m $s^{-1}$ during night and 0.5 to 1 m $s^{-1}$ during the day and a higher proportion of finer sediment during darkness (night: 57 % pebbles, 43% cobbles, day: 100% cobbles). However, considering all 24 h cycles only small differences in habitat use according daytime could be found (Figure 6).

There were obvious differences in the behaviour of individual fish even within a single shoal (overlap of seasonal home ranges and at least once in the same 25 m segment). Some fish showed distinct day and night habitats, others did not. Some specimens remained in one stationary shoal, whereas others changed between sites with stationary shoals.

Compared with the available habitat structures in the inhabited part of the river (km 3.75–km 18.4, Figure 2), the tagged nase preferred runs and whirls with currents between 0.5 and 1 m $s^{-1}$, hard substratum consisting mainly of cobbles (6.3 to 20 cm diameter) and depths between 1 and 2 m (Table 2). Soft substrates, shallow depths and slow flowing water were avoided. Pools and the reservoir were rarely used.

As nase did not use their habitat equally all year round, different seasonal habitat preferences could be expected. Riffles are important as spawning grounds for nase (Lelek & Penaz, 1963; Prokes & Penaz, 1978), but were hardly used outside spawning season. During the prespawning period riffles with depths between 0.2 and 0.5 m and currents stronger than 1 m $s^{-1}$ were preferred (Table 2). Pools and whirls were used more often after spawning, probably in search for shelter (high discharge). In summer some fish also used the reservoir, with fine sediment (sand) and deep water (> 2 m). The apparent preference for strong current (> 1 m $s^{-1}$) is presumably due to the high discharge and corresponding higher availability during summer compared with winter, when habitat characteristics were recorded. No big differences occurred between general and autumnal habitat use. During winter most of the tagged fish were aggregated in a relative short reach at the lowest part of the residual flow loop. Although shallow areas were available there, depths less than 1 m were avoided. Currents over 1 m $s^{-1}$ were rarely used and low currents (0.1–0.5 m $s^{-1}$) were less avoided than in the warmer seasons.

## Discussion

Adult nase preferred the residual discharge loop, presumably due to its high morphological variability. This is in accordance to the findings of other authors (Keckeis et al., 1996). The free flowing reach and the reservoir were only used in the warmer seasons during high discharge when the situation in the residual discharge loop is dominated by fast runs and only few sites with reduced current velocity are available.

Although nase are known for their spawning and post-spawning migrations (Penaz, 1996), they remain stationary over longer periods demonstrating pronounced home-ranges (Philippart; 1981; Fielenbach, 1996). The size of the home-ranges must be large enough to provide sufficient food for the animals. Daily activity areas of nase in the river Sieg (Germany) (Fielenbach, 1996) were similar in size to that of the Aare, in spite of the differences between the two rivers.

Home-range size was larger in the warmer season in the river Aare. Fielenbach (1996) also recorded a significant higher activity with activity-sensitive transmitters in the river Sieg. However, the size of daily home-ranges was not reduced in both rivers during winter. Food availability could force the fish to use an area of similar size, producing enough periphyton. Poikilotherm animals have a reduced metabolism during winter and therefore need less food, but the nase is known to feed during all seasons except spawning season (Gyurko, 1959). Feeding with less intensity in winter but in areas of similar size could be the reason for the reduced activity in the same home-range compared to the warmer season.

Lusk (1967) described the nase as active, changing its position several times a day. He also described different feeding and resting habitats in the Rokytna river (Czech Rep.). In the Aare only a part of the fish selected separate day and night habitats. Although clear differences in habitat quality between these places could be observed for single fish and during some seasons, no such differences could be shown regarding all fish during the whole study.

Until now no satisfying explanation of the observed diel activity patterns of nase was described. Gyurko (1959) found full stomachs throughout the day, with a minimum in early morning. For grazers feeding is possible all day and there is no need for diurnal or nocturnal activity. On the other hand, Fielenbach (1996) described a clear diurnal activity pattern, and in the river Aare adult nase covered the furthest distances during dusk and dawn. In case of the barbel (*Barbus barbus*), Baras (1995) showed that these movements were correlated with the shift between their resting and feeding habitats. However, the reasons for the diel activity pattern of adult nase are still unclear, because feeding can not explain these results. If the search for resting places during the night is responsible for the habitat shift, the preference for weaker current, corresponding with finer sediment and shallower water, as observed in some fish, makes sense, but there is still a lack of data.

For juvenile nase daily activity with a maximum during dawn was shown (Bischoff, 1995; Scholten, 1995). Changes in the diel activity pattern related to ontogenetic changes have been shown for brook char (Bourke et al., 1996) and were related to feeding and search for cover. Seasonal variations of the diel activity pattern have been documented in several studies on adult fish of several species, depending on temperature and day length (Heggenes et al., 1993; Baras, 1995).

Individual differences in behaviour were observed, even within the same shoal, which are difficult to explain. One explanation could be that shoals are remaining together predominantly by visual orientation (Pitcher, 1979). During twilight fish will lose visual contact with each other and start to disperse over a wider area.

The nase is known as a shoaling species for juvenile as well as for adult fish; different size and age-classes use separate river sections in size specific shoals (Lusk, 1967; Nelva, 1988; Dedual, 1990). Adult fish are reported to occur in monospecific shoals of a dozen to some hundred individuals depending on season and habitat (Lusk, 1967). Mechanisms of shoaling of nase are unknown so far, except for reproductive behaviour (Prokes & Penaz, 1978; Dedual, 1990). Baras & Cherry (1990) proposed individual relations between members of a shoal for the barbel, but many authors deny such relationships for the nase (Lusk, 1967; Fielenbach, 1996). In the present study shoaling was dynamic and individual fish regularly changed shoals. We suggest that outside spawning season, nase are non-territorial and aggregate at the most favourable habitats within the river. Recent data presented by Baras (1997) demonstrated that, at least for the barbel, the presence of other individuals of the same species is an important factor for habitat selection.

Habitat preferences indicate that optimal sites are related to food and cover (Bourke et al., 1996; Perrow et al., 1996). Only coarse substrate can be grazed by nase (Freyhof, 1995). Cobbles as a good substrate for periphyton and good to scrape for adult nase, only occur if some minimal current prevents siltation. Strong currents are avoided which reduces the energy demand required to maintain position. Shallow sites are also avoided due to anglers or predators like large piscivorous birds (Carss & Marquiss, 1991; Perrow et al., 1996). Deeper water minimises the risk to be seen, but at the same time reduces periphyton growth due to lack of light. Depths between one and two meters, as preferred in this study, can be seen as trade-off between risk avoidance and food occurrence.

During spawning season habitat use is focused on the availability of suitable spawning grounds. Other factors like food occurrence, cover or energy demand are less relevant then. Shallow riffles with strong currents, clearly avoided during the rest of the year, are searched for and the fish are taking high risks for reproduction (Lelek & Penaz, 1963; Prokes & Penaz, 1978). After spawning the fish dispersed over a greater area. Some fish even used the reservoir and the free flowing reach then. It is not known whether they actively selected these habitats or were just flushed downstream with the current. Rising discharge certainly changed the habitat structures, but at least one fish was found dying in the reservoir and others were lost at that time. The nase obviously invested a lot of energy in reproduction and suffered a higher mortality afterwards (Luskova et al., 1995). Tagging at this time of the year proved to be problematic (Huber & Kirchhofer, 1997). In Winter the nase were often seen in larger aggregations (Ulmann, 1997) but concentrated on fewer stretches. During low discharge and clear water the fish avoided shallow sites with less than one meter depth. Strong currents were rarely used and low currents much less avoided than during the warmer seasons, presumably to save energy in times of reduced metabolic activity.

The nase shows high demands on habitat variability, as well regarding different stages in the life cycle, as for seasonal and to some extent even daily preferences. This might be one reason for the observed population decline all over its area of distribution, as a loss in morphological variability due to channelisation and damming is observed all over central Europe (Penaz, 1996).

## Acknowledgements

We thank the Bernese Fisheries Authority, especially M. Flück, H. Walter and M. Tschan for support; M. Escher for implanting the transmitters and delivery of fish, T. von Mandach, M. Mezzera, M. Breitenstein, V. Saladin, H. Malli and M. Günter for help during data collection and H. Minder for the haul-seine. The project is supported by the Swiss National Science Foundation (Project Nr. 31-40885.94) and by the Federal Office of Environment, Forest and Landscape (Project Nr. F-95-5252).

## References

Arnekleiv, J. V. & M. Krabol, 1996. The effects of induced floods on the upstream migration of adult brown trout, and the effects of water release of the postspawning downstream migration in a regulated Norwegian River. In E. Baras & J. C. Philippart (eds), Underwater Biotelemetry, Proceedings of the First Conference and Workshop on Fish Telemetry in Europe, Liège, Belgium: 172 pp.

Aschwanden, H. & R. Weingartner, 1985. Die Abflussregime der Schweiz. Publ. Gewässerkunde 15.

Baras, E., 1992. Etude des stratègies d'occupation du temps et de l'espace chez le barbeau fluviatile, (*Barbus barbus* L.). Cah. Ethol. 12: 125–442.

Baras, E., 1995. Seasonal activities of *Barbus barbus* (L.) Effect of temperature on time-budgeting. J. Fish Biol. 46: 816–828.

Baras, E., 1997. Environmental determinants of residence area selection and long term utilisation in a shoaling teleost, the common barbel (*Barbus barbus* L.). Aquat. Living Resour. 10: 195–206.

Baras, E. & B. Cherry, 1990. Seasonal activities of female barbel *Barbus barbus* (L.) in the River Ourthe (Southern Belgium), as revealed by radio tracking. Aquat. Living Resour. 3: 283–294.

Bischoff, A., 1995. Verteilungsmuster und Nahrungsökologie von Jungfischen in den Rauschen der Sieg. Diplomarbeit Friedrich-Wilhelms-Universität Bonn: 116 pp.

Bourke, P., P. Magnan & M. A. Rodriguez, 1996. Diel locomotor activity of brook charr, as determined by radiotelemetry. J. Fish Biol. 49: 1174–1185.

Burgess, S. A., 1985. Some Effects of Stream Habitat Improvement on the Aquatic and Riparian Community of a Small Mountain Stream. In J. A. Gore (ed.), The Restoration of River and Streams. Butterworth Publishers, Stoneham: 223–246.

Carss, C. N. & M. Marquiss, 1991. Avian predation at farmed and natural fisheries. In M. C. Lucas, I. Diack & L. Laird (eds), Interactions Between Fisheries and the Environment, Proceedings of the Institute of Fisheries Management 22nd Annual Study Course, Institute of Fisheries Management, Nottingham: 179–196.

Clough, S. & M. Ladle, 1997. Diel migration and site fidelity in a stream-dwelling cyprinid, *Leuciscus leuciscus*. J. Fish Biol. 50: 1117–1119.

Copp, G. H., 1992. Comparative microhabitat use of cyprinid larvae and juveniles in a lotic floodplain channel. Envir. Biol. Fishes 33: 181–193.

Dedual, M., 1990. Biologie et problemes de dynamique de population du nase (*Chondrostoma nasus*) dans la Petite Sarine. Ph.D. Thesis, Univ. Fribourg (Switzerland): 159 pp.

Fielenbach, J., 1996. Zeitliche und räumliche Verteilungsmuster der Nase *Chondrostoma nasus* (L.) in der Sieg. Diplomarbeit Friedrich-Wilhelms-Universität Bonn: 112 pp.

Freyhof, J., 1995. Feeding traces as an indicator of habitat choice in nase, (*Chondrostoma nasus*). Folia Zool. 44: 57–61.

Garner, P. & S. Clough, 1996. Habitat use by dace *Leuciscus leuciscus* (L.), in a side channel of the River Frome, England. Fish. Mgmt Ecol. 3: 349–352.

Guthruf, J., 1996. Populationsdynamik und Habitatwahl der Aesche (*Thymallus thymallus* L.) in drei verschiedenen Gewässern des schweizerischen Mittellandes. Diss. ETH Nr. 11720.

Gyurko, S., 1959. Beiträge zur Ernährungsbiologie der Nase (*Chondrostoma nasus* L.). Arch. Hydrobiol. 56: 93–101.

Hart, L. G. & R. C. Summerfelt, 1975. Surgical Procedures for Implanting Ultrasonic Transmitters into Flathead Catfish (*Pylodictis olivaris*). Trans. am. Fish. Soc. 1: 56–59.

Heggenes, J., O. M. W. Krog, O. R. Lindas, J. G. Dokh & T. Bremner, 1993. Homeostatic behavioural responses in a changing environment; brown trout (*Salmo trutta*) become nocturnal in winter. J. Animal Ecol. 62: 295–308.

Hermansen, H. & C. Krog, 1985. A review of brown trout (*Salmo trutta*) spawning beds, indicating methods for their reestablishment in Danish lowland rivers. In J. S. Alabaster (ed.), Habitat Modification and Freshwater Fisheries. FAO., Butterworth.

Huber, M. & A. Kirchhofer, 1997. Habitat use of radiotagged adult nase (*Chondrostoma nasus*) in a regulated river. Folia Zool. 46 (suppl. 1): 67–77.

Huet, M., 1949. Aperçu des relations entre la pente et les populations piscicoles des eaux courants. Schweiz Z. Hydrol. XI: 332–351.

Hydrologisches Jahrbuch der Schweiz, 1994. Landeshydrologie und geologie. EDMZ, Bern.

Jacobs, S., 1974. Quantitative measurement of food selection: a modification of the forage ratio and Ivlevs electivity Index. Oecologia 14: 413–417.

Keckeis, H., P. Frankiewicz & F. Schiemer, 1996. The importance of inshore areas for spawning nase *Chondrostoma nasus* (*Cyprinidae*) in a free-flowing section of a large river (Danube, Austria). In D. W. Sutcliffe (ed.), The ecology of large rivers, Archiv. Hydrobiol. Suppl. 113, Large Rivers 10: 51–64.

Lelek, A. & M. Penaz, 1963. Spawning of *Chondrostoma nasus* (L.) in the Brumovka River. Folia Zool. 12: 121–134.

Lucas, M. C. & P. A. Frear, 1997. Effects of a flow-gauging weir on the migratory behaviour of adult barbel, a riverine cyprinid. J. Fish Biol. 50: 382–396.

Lusk, S., 1967. Population dynamics of *Chondrostoma nasus* (Linnaeus, 1758) in the Rokytna river. Acta sc. nat. Brno 1: 473–522.

Luskova, V., S. Lusk & K. Halacka, 1995. Yearly dynamics of enzyme activities and metabolite concentrations in blood plasma of *Chondrostoma nasus*. Folia Zool. 44: 75–82.

Nelva, A., 1988. Profil écologique du Hotu, *Chondrostoma nasus* (*Ostéichtyen, Cyprinidé*) dans le Haut-Rhone français: structure d'habitat ponctuel et stationnel. Acta Oecologica/OEcol. Applic. 9: 275–296.

Penaz, M., 1996. *Chondrostoma nasus* – its reproduction strategy and possible reasons for a widely observed population decline – a review. In A. Kirchhofer & D. Hefti (eds), Conservation of Endangered Freshwater Fish in Europe, Birkhäuser.

Perrow, M. R., A. J. D. Jowitt & S. R. Johnson, 1996. Factors affecting habitat selection of tench in a shallow eutrophic lake. J. Fish Biol. 48: 859–870.

Philippart, J.-C., 1981. Démographie du Hotu, (*Chondrostoma nasus* L.) (Teleostei: *Cyprinidae*) dans l'Ourthe (Bassin de la Meuse, Belgique). Ann. Soc. r. Zool. Belg. 110: 199–219.

Pitcher, T. J., 1979. Sensory information and the organisation of behaviour in a shoaling cyprinid fish. Anim. Behav. 27: 126–49.

Priede, I. G., 1991. Telemetry in assessment of environmental effects on individual fishes. In M. C. Lucas, I. Diack & L. Laird (eds), Interactions Between Fisheries and the Environment, Proceedings of the Institute of Fisheries Management 22nd Annual Study Course, Institute of Fisheries Management, Nottingham: 179–196.

Prokes, M. & M. Penaz, 1978. The course of spawning, early development and longitudinal growth of the nase carp, *Chondrostoma nasus*, in the Rokytna and Jihlava rivers. Folia Zool. 27: 269–278.

Scholten, M., 1995. Verteilungsdynamik und Nahrungökologie von Jungfischen in den Buchten der Sieg. Diplomarbeit Friedrich-Wilhelms-Universität Bonn.

Spreaficio, M., R. Weingartner & C. Leibundgut, 1992. Hydrologischer Atlas der Schweiz. Landeshydrologie und geologie. EDMZ, Bern.

Ulmann, P., 1995. Die räumliche Verteilung juveniler und adulter Nasen (*Chondrostoma nasus* L.) in einem anthropogen beeinflusstem Aare-Abschnitt bei Bern. Diplomarbeit, Zoologisches Institut der Universität Bern.

*Hydrobiologia* **371/372:** 321–333, 1998.
*J. P. Lagardère, M.-L. Bégout Anras &G. Claireaux (eds), Advances in Invertebrates and Fish Telemetry.*



# The migratory behaviour of juvenile and adult salmonids in relation to an estuarine barrage

I. C. Russell[1], A. Moore, S. Ives[1], L. T. Kell[1], M. J. Ives[1] & R. O. Stonehewer[2]
[1] *Centre for Environment, Fisheries and Aquaculture Science, Lowestoft Laboratory, Pakefield Road, Lowestoft, Suffolk, NR33 0HT, UK*
[2] *Environment Agency (Welsh Region), Llwyn Brain, Ffordd Penlan, Parc Menai, Bangor, Gwynedd LL57 4DE, UK*
*E-mail of corresponding author: i.c.russell@cefas.co.uk*

*Key words:* Atlantic salmon, sea trout, migration, adult, smolt, telemetry, barrage, estuary

## Abstract

There is wide concern regarding the effects that estuarine barrages may have on estuarine environments and the species which utilise them. Of particular concern are the possible effects on diadromous fish which move between fresh and salt water. This paper presents results from a 3-year investigation (1993–95) of the migratory behaviour of both juvenile and adult salmonids undertaken in the vicinity of a barrage on the River Tawe in South Wales, which was completed in 1992. The localised upstream movements of acoustically tagged returning adult salmonids ($n = 28$) were investigated in the immediate vicinity of the barrage and associated fish pass using a High Resolution (HiRes) tracking system. The behaviour of salmonid smolts ($n = 52$), tagged with miniature acoustic transmitters, were also monitored during their spring seaward emigration within the impounded estuary.

The upstream movement of adult salmonids appeared to be delayed by the presence of the barrage with many fish holding station close to the barrage structure. Of these, about half were clearly attracted by the plume of water from the fish pass, but few fish ascended the barrage by this route. Approximately one third of those fish detected below the barrage were subsequently detected upstream in the impoundment, most moving past when the barrage was over-topped. The movement of fish over the barrage was significantly unimodal with respect to the tidal cycle with a mean passage time of 4 minutes after high water. Once in the impoundment fish often remained close to the barrage ($< 100$ m) for several hours or days during which time movements appeared largely random.

The movement of salmonid smolts through the impounded estuary and into the lower estuary, downstream of the barrage, was not continuous. Smolts held position within the impoundment immediately upstream of the barrage. The movement of smolts upstream of the barrage was predominantly random, and fish that successfully negotiated the structure did so by migrating through the ship lock or over the weir during an over-topping flood tide. The migratory behaviour patterns of juvenile and adult salmonids are compared with previous studies in unmodified river estuaries, and discussed in relation to the possible impact of barrages on salmonid migration.

## Introduction

In recent years there has been increasing interest in the UK in the construction of estuarine barrages for urban regeneration, hydroelectric power generation and flood defence purposes. However, there has been wide concern that these constructions may have a significant impact on the estuarine environment and the species that inhabit these ecosystems. Of particular concern are the possible effects an estuarine barrage may have on diadromous fish species which move between the fresh and salt water environments. The construction of a tidal barrage across the lower sections of an estuary may affect the behaviour of migratory species such as the Atlantic salmon and sea trout in a number of ways (Gough, 1996). Firstly, by modifying the tidal regime within the estuary it may significantly reduce the ebb and flood tidal cues which are used by salmonids to orientate (Potter, 1988; Milner, 1990; Potter et al., 1992; Moore et al., 1996). Secondly, by reducing saline intrusion and 'flushing' during the ebb tide cycle both the aquatic environment and water quality will be modified. Thirdly, the barrage may form a physical barrier to the passage of both juvenile and adult salmonids.

The River Tawe estuary in South Wales was significantly modified by the construction of a barrage in 1992. This was the first such estuarine amenity barrage to be completed in the UK. The barrage, which does not exclude all tides, has resulted in the formation of a largely freshwater lake upstream of the barrage which runs much of the length of the original estuary, with only limited saltwater intrusion and tidal influence. There were only limited environmental provisions within the Act authorising the barrage and very little pre-construction monitoring of fish behaviour was possible. However, the promoters of the scheme co-funded a two year post impoundment monitoring programme which was carried out by the then National Rivers Authority (now Environment Agency). The results of this adult salmonid tracking programme, conducted in 1993 and 1994, are reported by Stonehewer et al. (1995) and Mee et al. (1996). Concurrent with this investigation, but for only a part of each year, the Directorate of Fisheries Research (now CEFAS) investigated the detailed movements of some of these acoustically tagged adult fish in the immediate vicinity of the barrage using a high resolution (HiRes) tracking system. The downstream migration of wild salmon and sea trout smolts was investigated during a collaborative CEFAS/EA tracking programme which was commissioned in 1993 and carried out in the springs of 1994 and 1995.

The migratory behaviour patterns of juvenile and adult salmonids are compared with previous studies in unmodified river estuaries, and discussed in relation to the possible impact of barrages on salmonid migration.

## Materials and methods

### *Study site*

The River Tawe is a small river located in South Wales. It rises in the Black Mountains at Moel Feity (590 m above sea level) and has a total length of 48 km, flowing in a south-westerly direction to meet the Bristol Channel at Swansea. The river drains a heavily urbanised catchment of about 272 km$^2$ and has an average daily flow (1982 to 1992) at the tidal limit of 12.7 m$^3$ s$^{-1}$. The salmon and sea trout fishery has improved significantly over recent decades due to the demise of heavy industry, improvements in pollution control, the removal of barriers to migration, habitat improvement and stocking programmes.

The Tawe barrage is located in the lower estuary upstream of the commercial docks (Figure 1). The barrage consists of a primary and secondary weir, a navigation lock, and a 'pool and traverse' fish pass. The crest of the primary weir is 8.05 m above Chart Datum, with the secondary weir 30 cm higher at 8.35 m. The Tawe estuary experiences a large tidal range and approximately 71% of tides overtop the barrage and allow the influx of salt water into the impoundment. However, inundation only accounts for about 16% of the total time (Mee et al., 1996). During overtopping there is limited mixing of fresh water and salt water in the impoundment, and some saline stratification may extend up to 5 km upstream of the barrage. The surface area of water within the impoundment is about 22 ha.

The barrage incorporates a small HEP generator in the east abutment, the outlet for which is situated immediately below the fish pass entrance at the base of the barrage structure. Both the inlet and outlet are screened to prevent the entrainment of fish. An abstraction licence stipulates that the operation of the lock and generator must not reduce the flow through the fish pass to below 1 m$^3$ s$^{-1}$.

### *Fish trapping and tracking – adults*

The adult tracking programme was carried out in 1993 and 1994. Difficulties in obtaining adequate numbers of fish for the study meant that adult salmon and sea trout for tagging were obtained from a number of different sources. Where possible, the capture of fish in salt water was preferred, either by seine netting in the area immediately below the barrage (1994 only), from a trap within the barrage fish pass or from a bag net operated in Swansea Bay (1993 only). However, some salmon and sea trout which had already entered the river were also obtained from a trap located at Panteg, 23 km upstream of the barrage, and were tagged and relocated to the estuary some 0.6 km below the barrage. Further details of the methods of capture and of the size and age of tagged fish are provided by Stonehewer et al. (1995).

Captured fish which were considered to be large enough for tagging ($>$ 60 cm fork length) and in good physical condition, were firstly anaesthetised using 2-phenoxy ethanol (0.4 ml l$^{-1}$) and then tagged with

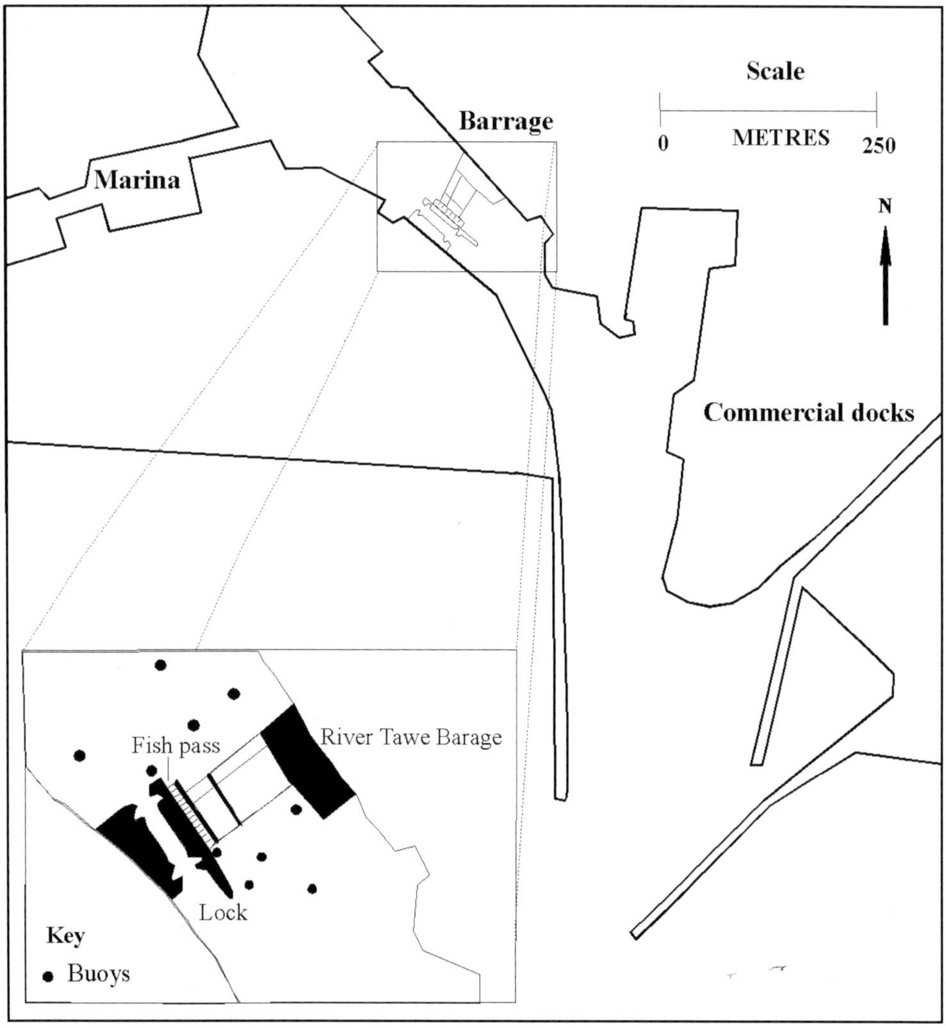


*Figure 1.* Diagram of the River Tawe estuary showing the position of the barrage and the location of the sonar buoys during 1994.

combined acoustic and radio tags (CARTs). In all, 127 adult salmonids were tagged over the two years; 52 in 1993 (35 salmon and 17 sea trout) and 75 in 1994 (46 salmon and 29 sea trout). The CARTs incorporate a 76 kHz acoustic transmitter (Mitson & Storeton-West, 1971), and a 173.7–174.0 MHz radio transmitter (DTI type approval cert. no. RTD3435). The transmitter circuitry and battery are housed in a polycarbonate case with hemispherical ends, 70 mm long and 16.5 mm in diameter and weighing 4.5 g in water. Tags are distinguished by their radio frequency and pulse rate. Both transmitters are powered by a single 800 mA h capacity cell. In order to conserve power (the acoustic transmitter has a heavy power drain) and to minimise the difficulties of signal interpretation from a number of 'active' acoustic tags in the same area, the acoustic stage is programmed to switch off after a pre-set time (3 to 35 days in this study). The radio transmitter operates continuously for a period of about 6 months. Tags were inserted into the stomach of the fish (Home Office Project Licence No. PPL 80/730) using the method described by Solomon & Storeton-West (1983). Fish were then measured and an external T-bar anchor tag attached before being allowed to recover from the anaesthesia. The fish caught in salt water were released at the capture site to avoid unnecessary handling, usually within 30 minutes of capture. In the case of relocated fish, the time from tagging to release was longer (~ 3 hours), allowing for transportation and a period of phased acclimation to salt water prior to release in the lower estuary.

The movements of fish through the estuary and impoundment and into the river were monitored by the Environment Agency using sonar buoys and automatic listening stations (as described by Solomon & Potter, 1988). The results of this study have been described by Stonehewer et al. (1995) and Mee et al. (1996). This paper reports upon the more detailed examination of the movements of some of the same fish in the immediate vicinity of the barrage using a high resolution (HiRes) tracking system (Russell et al., 1990; Kell et al., 1994). The HiRes system depends upon the detection of a signal from an acoustic tag (it can be operated at either 76 kHz or 300 kHz) by at least three sonar buoys placed at known co-ordinates in the study area. The time delays between pairs of buoys receiving the tag signal are generated by a timer gate interface linked to a PC and the position is estimated using a common statistical technique, non-linear least squares. In 1993, the HiRes system was only deployed below the barrage (July 22 to November 15), but in 1994 the downstream system was deployed for longer (9 June to 31 December) and a duplicate system was also deployed above the barrage for most of the study period (7 July to 31 December). Deployment of tracking systems both below and above the barrage enabled more detailed tracks to be recorded and provided a clearer picture of where and at what time fish crossed the barrage. The position of the sonar buoys in relation to the barrage, fish pass, lock and weirs is illustrated in Figure 1 for 1994.

In presenting the tracking results from this study, movements are described for three phases of migration: approach to the barrage, crossing the barrage, and behaviour immediately above the barrage in the impoundment.

*Fish trapping and tracking – smolts*

The smolt tracking programme was conducted during 1994 and 1995. Migrating salmon and sea trout smolts were trapped in fresh water 2.6 km above the head of tide using a modified fyke net. Over the two years, 52 smolts (31 salmon and 21 sea trout) were tagged. Smolts were tagged using miniature 300 kHz acoustic transmitters. These tags are also housed in polycarbonate cases with hemispherical ends, measure 17 mm long by 8 mm in diameter, and weigh 0.35 g in water (Moore et al., 1990a). Smolts were anaesthetised using 2-phenoxy ethanol (0.4 ml $l^{-1}$) and the transmitters surgically implanted into the peritoneal cavity (Home Office Project Licence No. PPL 80/612; Moore et al., 1990b); following full recovery ( 30 minutes from capture) fish were released to resume their downstream migration. This technique of transmitter attachment has been shown to have negligible physiological and behavioural effects on salmon smolts (Moore et al., 1990b), and has been used to successfully track both salmon and sea trout smolts in a number of investigations (Moore et al., 1992; Moore & Potter, 1994; Moore, 1995; Moore et al., 1995; Moore et al., 1996).

The movement of smolts within the lower river, impoundment and estuary were monitored using an array of sonar buoys operating at 300 kHz. The buoys were positioned at ten locations between the limit of tidal influence and immediately below the barrage. The movements of smolts were subsequently analysed within each of six defined sections of the Tawe, ranging from freshwater through the impoundment to the estuary (Moore et al., 1996). The HiRes tracking system was also utilised in the area above the barrage to examine the movements of the smolts as they approached the obstruction.

## Results

*Adult movements*

Analysis of all the tracks of adult salmon and sea trout over the two years of the study carried out by the Environment Agency indicated no statistical difference between the fates (i.e. entered the impoundment, detected in the estuary below the impoundment, not detected) of the tagged fish in relation to species or capture and tagging location (i.e. whether fish were relocated from the in river trap or caught below the barrage), although there was a significant difference in the fate of the fish between the two years studied (Mee et al., 1996). The results presented here are for the fish that were tracked with the HiRes system during the study period. Since these are a sub-set of the full study, the number of tracks is relatively small, and in light of the above findings about the fate of the fish, no attempt has been made to relate these detailed movement patterns to fish species or capture location.

During the period of deployment of the HiRes tracking system in 1993, only six relatively short fish tracks (all salmon) were recorded in the area below the barrage as acoustically tagged fish approached the Tawe barrage. These fish were predominantly detect-

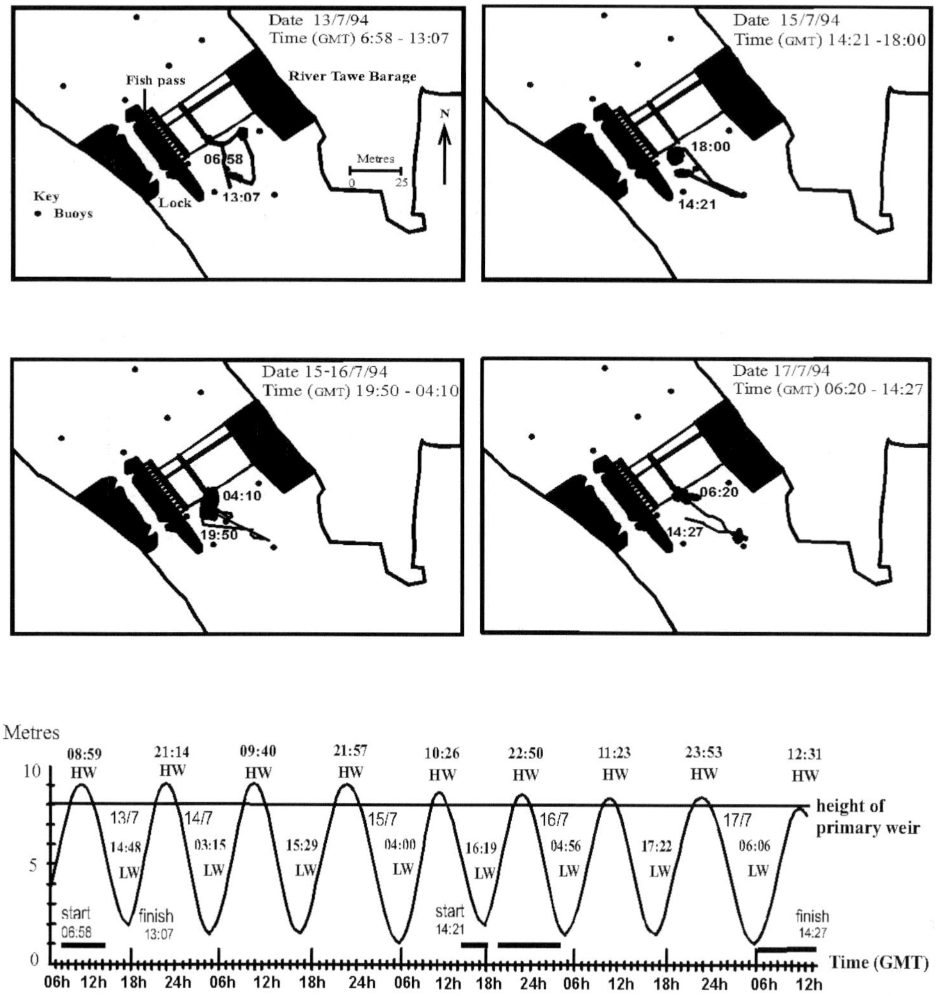


*Figure 2.* High resolution track of an adult salmon in the area immediately below the River Tawe barrage over the period 13 to 17 July 1994 (the fish was not detected on 14 July). The fish is attracted to the plume of water from the fish pass and holds station in this for varying periods of time over three consecutive days, but does not subsequently cross the barrage. The dark spots on the track indicate periods of extended residence in a localised area. The horizontal lines on the graph at the foot of the Figure indicate the duration of tracks relative to the tidal cycle.

ed some distance below the barrage in the area below the lock and fish pass, although one fish was observed to traverse across the face of the barrage. In 1994, significantly more fish (75) were tagged and a higher proportion of these were either recorded below the barrage (50.7%) or entered the impoundment (42.7%) than in 1993 (52 tagged, 28.8% recorded below the barrage and 11.5% entered the impoundment) (Stonehewer et al., 1995). Thus the results presented in this paper concentrate primarily on only one year, 1994.

In all, 22 tracks were recorded with the HiRes system in 1994, of which 15 were for fish below the barrage (14 salmon and 1 sea trout) and 7 describe movements both below and above the barrage (5 salmon and 2 sea trout). However, 5 of the downstream tracks only record movements for a short period immediately after tagging and release at the barrage site; these have been disregarded here.

The majority of the fish (73%) which approached the barrage were found to hold station for relatively long periods of time, often fairly close (< 10 m) to the barrage structure. Many of these fish holding station (~ 50%) were clearly attracted by the plume of water discharging from the fish pass and spent continuous periods of sometimes some hours within a radius of a few metres adjacent to or in the plume. One of these fish demonstrated a similar pattern of movement on a number of consecutive days (Figure 2). Three other

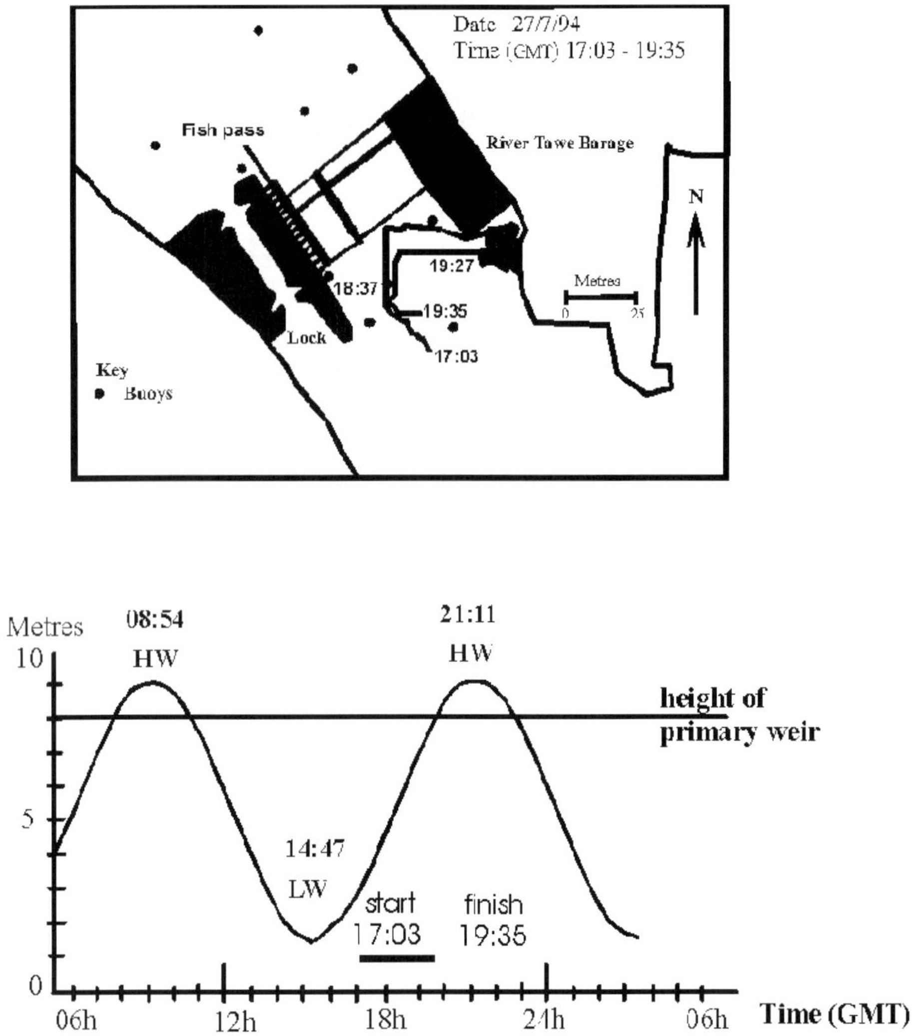


*Figure 3.* High resolution track of an adult salmon in the area immediately below the River Tawe barrage on the 27 July 1994. The fish approaches the barrage on a flood tide, before moving to the margins of the estuary where it holds station for 44 minutes before moving back downstream approaching high water. The fish does not subsequently cross the barrage. The dark spot on the track indicates a period of extended residence in a localised area. The horizontal line on the graph at the foot of the Figure indicates the duration of the track relative to the tidal cycle.

fish adopted a rotational pattern of movement or moved along the length of the barrage, occasionally holding station in shallower water at the margins of the estuary (e.g. Figure 3). Many of the fish recorded approaching the barrage (68%) were not subsequently recorded within the impoundment, immediately upstream of the barrage, within the period of acoustic life of the tag.

Passage over the barrage itself commonly occurred around high water on over-topping tides, with fish mostly migrating over the primary or secondary weirs (78% of those recorded passing the barrage). Fish appeared to either hold station below the weirs before moving upstream when conditions were suitable (Figures 4 & 5), or approached and crossed the barrage more rapidly on a flooding tide (Figure 6). However, two fish were also recorded moving through the fish pass when the barrage was not over-topped by the tide (Figure 7). Two of the fish which were recorded crossing the secondary weir briefly dropped below the barrage before immediately re-entering the impoundment, again over the secondary weir (Figure 6). The movement of fish over the barrage was significantly

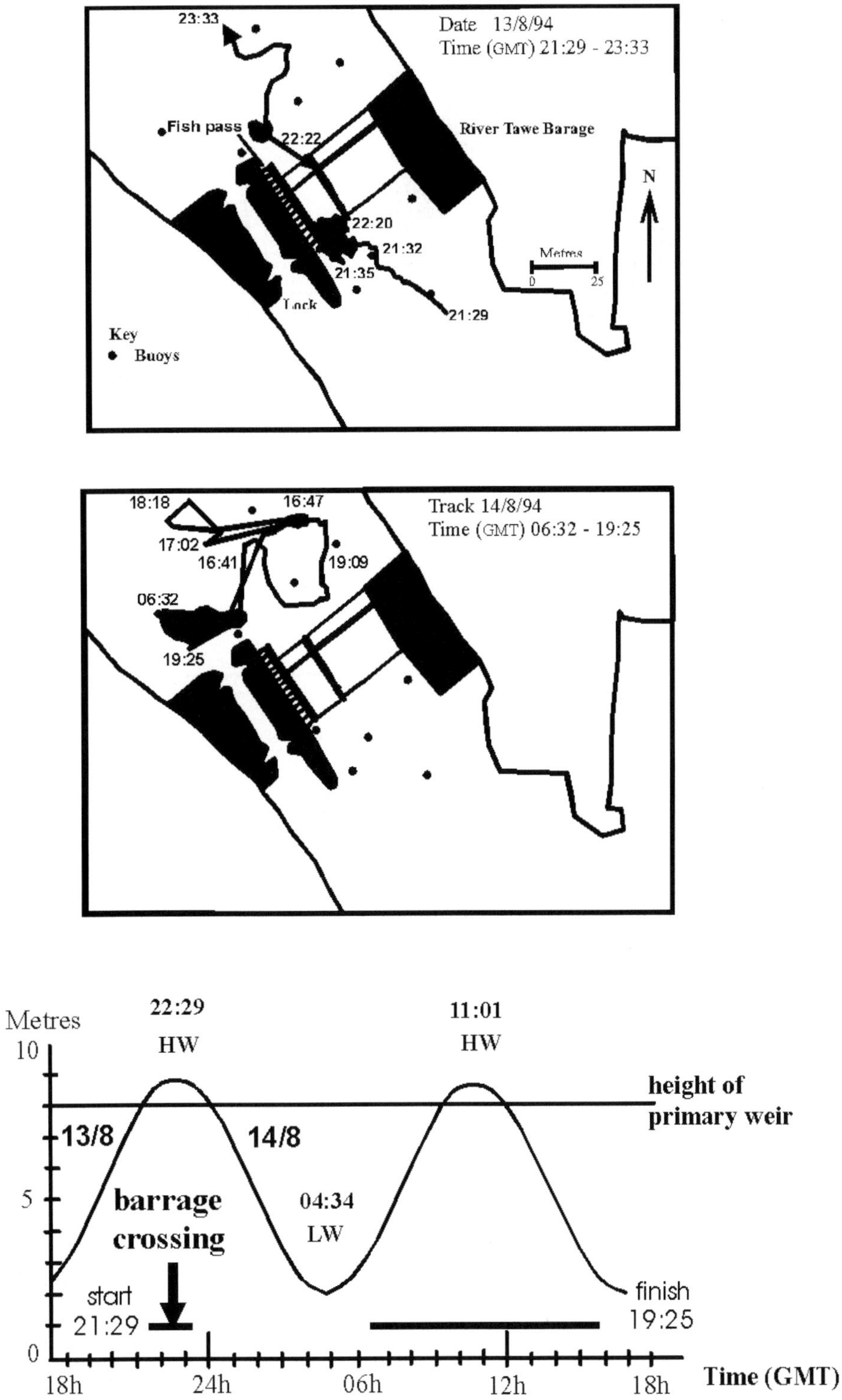


*Figure 4.* High resolution track of an adult salmon in the area immediately below and above the River Tawe barrage on the 13 to 14 August 1994. The fish approaches the barrage fairly rapidly on a flood tide, holds station adjacent to the plume from the fish pass for 45 minutes and then crosses the barrage over the primary weir at high water. It subsequently remains within $\sim$ 50 m of the barrage for 71 minutes before moving further upstream, but returns to a position close to the barrage ($<$ 50 m) for a further period of about 13 hours on the following day. The dark spots on the track indicate periods of extended residence in a localised area. The horizontal lines on the graph at the foot of the Figure indicate the duration of tracks relative to the tidal cycle.

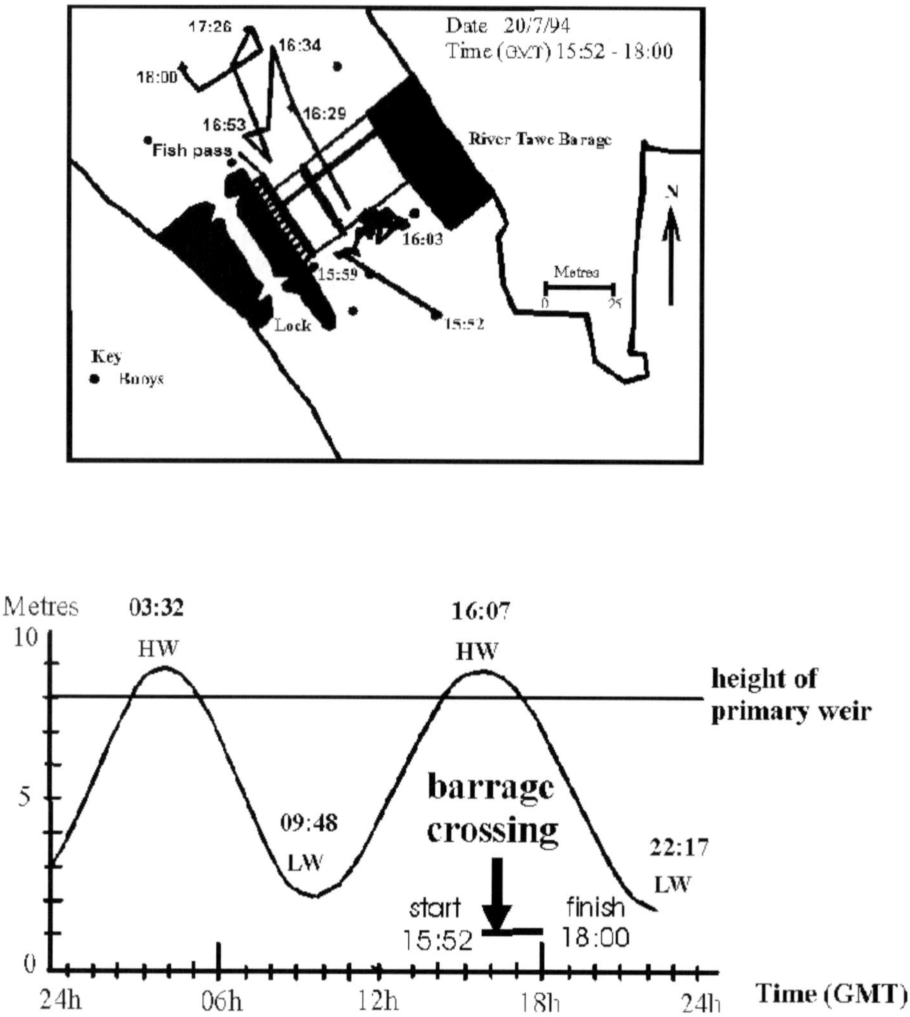


*Figure 5.* High resolution track of an adult sea trout in the area immediately below and above the River Tawe barrage on the 20 July 1994. The fish approaches the barrage fairly rapidly on a flood tide, holds station below for 29 minutes and then crosses the barrage over the secondary weir just after high water. It subsequently remains within $\sim$ 50 m of the barrage for a further 91 minutes before moving further upstream. The dark spots on the track indicate periods of extended residence in a localised area. The horizontal line on the graph at the foot of the Figure indicates the duration of the track relative to the tidal cycle.

unimodal with respect to the tidal cycle (Rayleigh Test $r=0.64$, $p < 0.05$, $n=9$; Batschelet, 1981), with a mean passage time of 4 minutes after high water.

On crossing the barrage into the impoundment, the movements of most fish appeared to become fairly random with quite long periods (an hour or two to several days) of residency relatively close to the barrage (< 100 m) (Figures 4 & 8). Three fish (43% of those tracked into the impoundment) subsequently dropped back over the barrage for varying periods, some momentarily, others for periods of a few hours to a few days, before either moving out to sea (1 fish) or re-entering the impoundment (2 fish). One fish traversed the barrage three times in this way before moving further up the impoundment.

*Smolt movements*

Full details of the study of smolt movements in the Tawe are reported by Moore et al. (1996). In brief, the movement of salmon and sea trout smolts through the upstream sections of the impoundment was found

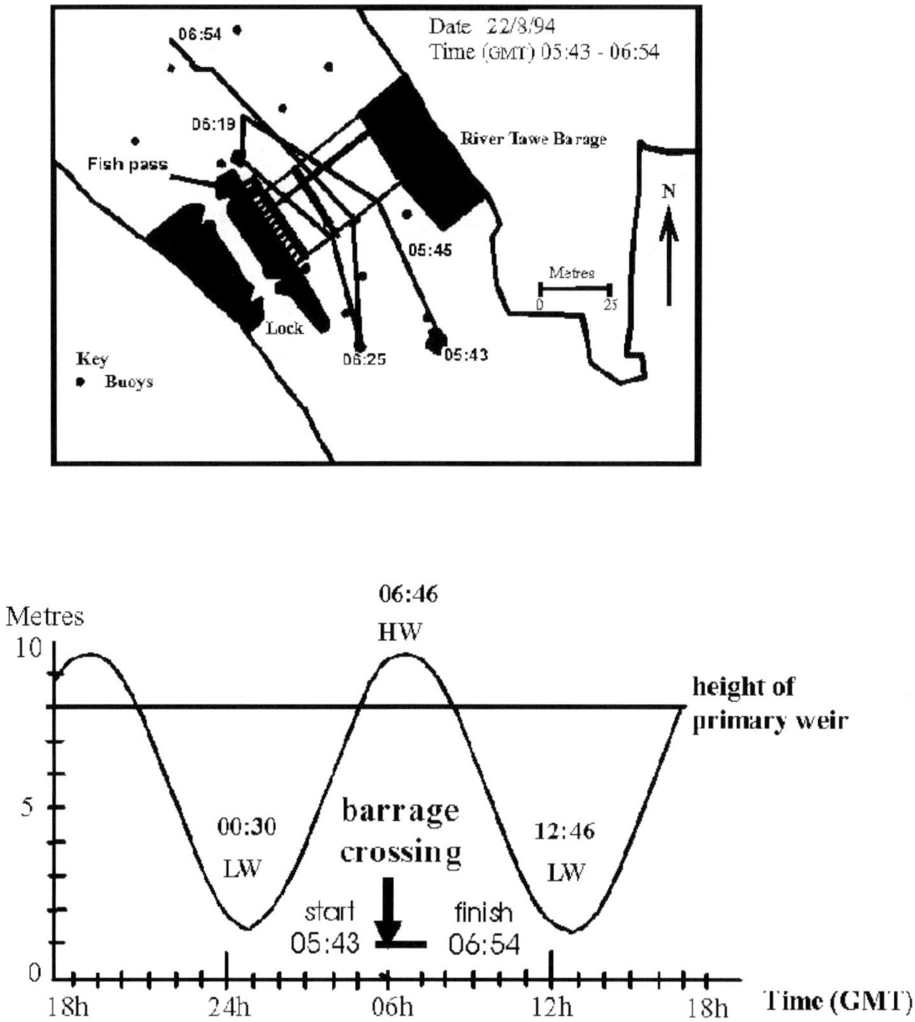


*Figure 6.* High resolution track of an adult salmon in the area immediately below and above the River Tawe barrage on the 22 August 1994. The fish approaches the barrage fairly rapidly on a flood tide and crosses the barrage over the secondary weir without significant delay about an hour before high water. It subsequently drops back over the barrage, re-entering the impoundment about 15 minutes later, again over the secondary weir, before moving further upstream relatively quickly. The horizontal line on the graph at the foot of the Figure indicates the duration of the track relative to the tidal cycle.

to be predominantly nocturnal. The mean time taken for smolts to move through the tidal and impounded sections of the River Tawe to the area immediately upstream of the barrage (5.8 ± 1.25 hours) was not significantly different to that of an inert object travelling during comparable flows within the main current (6.94 ± 4.33 hours). This suggests that the smolts may have emigrated passively with the current, but with some degree of orientation that maintained them in the main flow.

However, migration was discontinuous and, on reaching the barrage, further movements were delayed. The mean residence time (± S.E.M.) of smolts immediately above the barrage was 54.8±23.1 hours in 1994 and 42.9 ± 22.5 hours in 1995, with a range of 0.2 to 389.5 hours. No consistent pattern of smolt movement was observed during the period that they were in the vicinity of the Tawe Barrage. However, it was noted that some smolts moved in the direction of the lock during filling, suggesting that this localised increase in

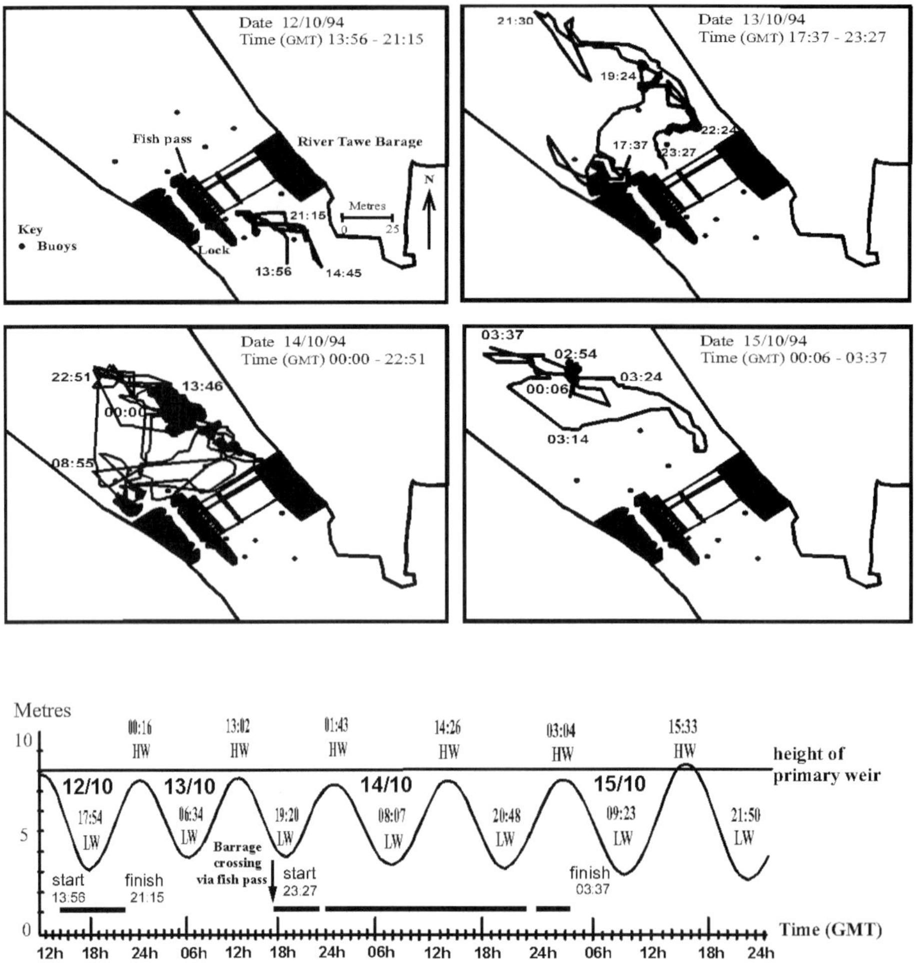


*Figure 7.* High resolution track of an adult salmon in the area immediately below and above the River Tawe barrage on the 12 to 15 October 1994 during a period of neap tides. The fish is recorded below the barrage for ~ 4 hours either side of low water on 12 October, but enters the impoundment, via the fish pass, just before low water about 24 hours later on 13 October (no downstream track was recorded at this time). The fish is recorded close to the barrage (< 100 m) for varying periods of time on each day from 13 to 18 October (see also Figure 8). The dark spots on the track indicate periods of extended residence in a localised area. The horizontal line on the graph at the foot of the Figure indicates the duration of the track relative to the tidal cycle.

flow acted as a stimulus for smolts to continue downstream emigration.

Smolts are believed to have migrated past the barrage via a number of possible routes: over the barrage weirs; via the fish pass; or through the lock. Two smolts (1 salmon and 1 sea trout) are known to have used the latter route, as this was verified by actively tracking the fish from a small vessel. In both years the movement of salmon smolts past the sonar buoys located immediately downstream of the barrage in the lower estuary was random with respect to the tidal cycle.

## Discussion

It is widely accepted that while the timing and duration of the passage of adult salmonids varies greatly with factors such as the dimensions, physical structure and hydrography of estuaries, movement in the main body of estuaries is predominantly tidally directed, with fish ascending on the flood tide and descending on the ebb (Milner, 1990). Such movements may last one tidal cycle or up to several weeks. However, fish may also hold up in deeper sheltered locations or drop out of an estuary altogether. It is also generally accepted that the entry of fish into the river is enhanced by increas-

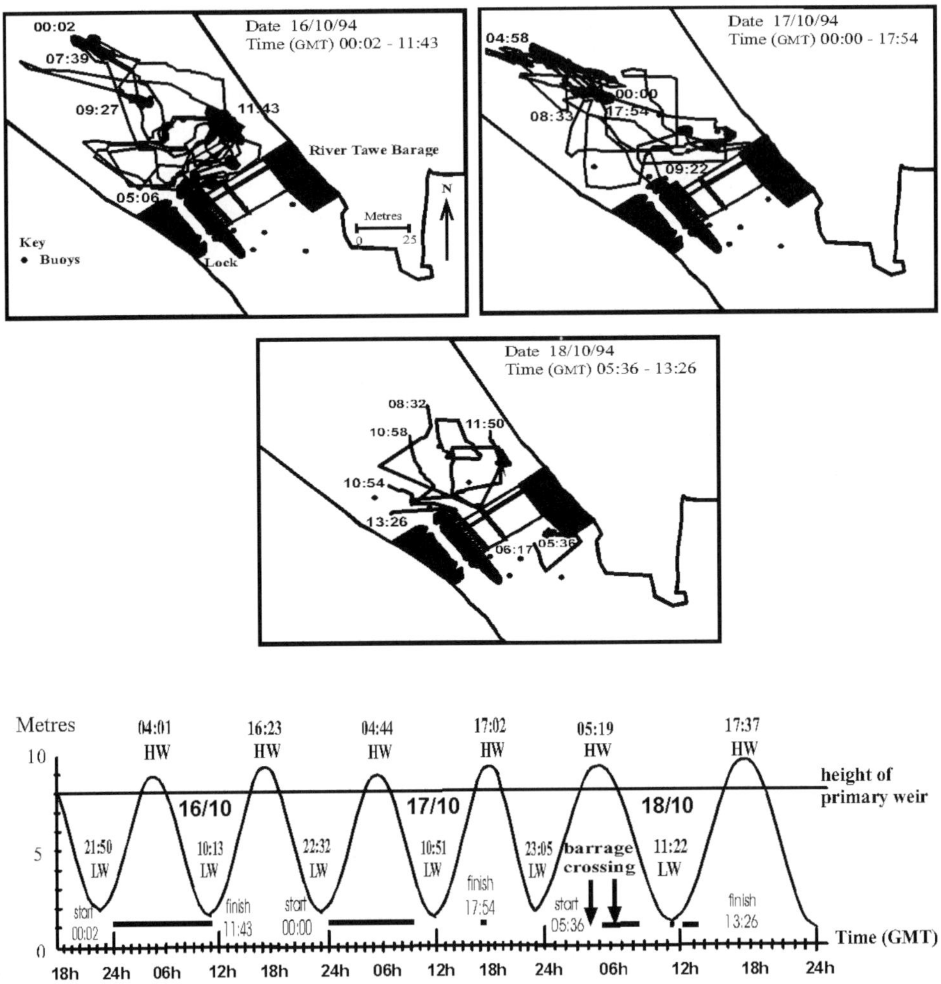


*Figure 8.* High resolution track of the adult salmon (as Figure 7) in the area immediately above the River Tawe barrage on the 16 to 18 October 1994. On the 18th October this fish briefly drops over the barrage just after high water, before re-entering the impoundment about 40 minutes later while the barrage is still over-topped. The dark spots on the track indicate periods of extended residence in a localised area. The horizontal line on the graph at the foot of the Figure indicates the duration of the track relative to the tidal cycle.

ing river flows (Potter, 1988; Milner, 1990; Potter et al., 1992; Clarke et al., 1994; Evans et al., 1994; Smith et al., 1994) and is largely independent of tidal phase (Potter, 1988; Potter et al., 1992; Priede et al., 1988; Smith & Smith, 1997), although the environmental stimuli that promote migration through estuaries and stimulate entry to freshwater are generally poorly understood (Gough, 1996).

In the absence of a baseline study, these observations of fish movements in unmodified river estuaries may allow some comparison with adult salmonid movements in the lower Tawe estuary. Mee et al. (1996) have observed that the time between tagging and river entry is longer for fish tagged on the Tawe than for other rivers in South Wales (Aprahamian et al., 1988; Clarke et al., 1994; Evans et al., 1994) and have suggested that river entry on the Tawe may therefore be delayed by the barrage. The topography of the Tawe estuary will have been modified by the construction of the barrage, and estuarine movements of salmon are thought to be strongly influenced by the topography and flow (Potter et al., 1992). However, Mee et al. (1996) have indicated that the probability of fish successfully migrating past the Tawe barrage does not appear to be correlated with river flow.

This study has enabled the movements of fish around the barrage to be described in some detail and enables tentative comparison with the expected movements in an unmodified estuary. However, in the absence of a baseline study this is inevitably somewhat speculative. It should also be noted that this study did not run for the entire duration of the tracking programme described by Stonehewer et al. (1995) and Mee et al. (1996) and that the tracks presented here represent a sub-set of the fish movements described by these authors.

The presence of the barrage greatly reduces the tidal velocity in the lower Tawe estuary and this, in turn, affects the ability of fish to migrate passively with the tide (Mee et al., 1996). From the HiRes tracks recorded below the barrage in both 1993 and 1994, it is evident that many fish are attracted and orientate with respect to the freshwater plume discharging from the fish pass, often maintaining position within such areas for lengthy periods. The apparent variation in fish orientation between 1993 and 1994 probably reflects the fact that the direction of the surface plume was changed due to the installation, in 1994, of a 'Chicken Pen' deflector below the lock and fish pass. However, despite the apparent attraction of the freshwater plume, relatively few fish ascended into the impoundment via the pass. This may be due to unfavourable conditions such as turbulent flow within the pass (Mee et al., 1996), inadequate fish pass design, or may indicate that fish are not yet adapted for the movement from salt to fresh water, or to switch from passive to active migration. The HiRes tracks also reveal that, where fish migrated over the barrage, this was commonly over the primary or secondary weir during an over-topping tide, suggesting a preference for tidally assisted movement and/or a salt water environment rather than directed swimming against a predominantly fresh water flow. The movement of adult fish past the barrage was thus generally limited to a relatively short period within the tidal cycle on larger spring tides.

Once over the barrage, adult salmonids move from saline water to predominantly fresh water, in which the directional tidal cues normally associated with this stage of migration are negligible. The tracks of fish movements immediately above the barrage appear to confirm this loss of orientation, with many fish moving around the impoundment in an apparently random and undirected manner. Mee et al. (1996) report a median residence time within the impoundment of 59.5 hours (range 3 hours to 68 days).

As for the adults, no baseline data are available for smolt movements on the Tawe prior to barrage construction. However, the patterns of movement of smolts in the Tawe estuary differed from the movements recorded in unmodified estuaries such as the Rivers Avon and Conwy (Moore et al., 1992; Moore & Potter, 1994; Moore, 1995; Moore et al., 1995). These previous studies demonstrated that salmon and sea trout smolts migrated seawards at night, making selective use of the ebb tide. In the lower portion of these estuaries, smolts migrated using active directed swimming and there was no apparent period of acclimation when smolts moved from fresh to salt water (Moore, 1995; Moore et al., 1995). It was concluded that the tidal currents within these estuaries were important cues for initiating smolt migration, and in enabling smolts to move rapidly out into coastal waters under the cover of darkness (Moore et al., 1992; Moore & Potter, 1994; Moore, 1995; Moore et al., 1995).

The modified tidal cycle in the River Tawe prevents smolts from using rapid ebb tide currents to assist movement through the estuary and into coastal waters. The barrage represents a partial barrier to smolt migration and delays their subsequent movement into coastal waters. Smolt migration through the estuary was not continuous and emigrating smolts remained immediately above the barrage for periods of up to 16 days. There was no consistent pattern of smolt movement within this period. Thus exposing fish to additional potential dangers arising from factors such as predation or water quality problems. Moore et al. (1996) speculated that the delays may result in smolts not moving out of the River Tawe at the optimal time. It is widely thought that Atlantic salmon smolts may have an optimum window of opportunity for migration into the marine environment (Harris, 1994) in much the same way as Pacific salmon (Bilton et al., 1982). In addition, smolts within the impoundment have few cues available to them on the tidal cycle below the barrage. Thus smolts moving past the barrage during a flooding tide within the lower estuary, are likely to be further delayed until the onset of the ebb, and further vulnerable to predation.

The results of this study indicate that the Tawe barrage does appear to physically constrain the migration of both salmon and sea trout smolts and returning adults and support the findings of Mee et al. (1996). This highlights the importance of safeguarding fish passage requirements and the need for constructing and operating barrages to minimise potential impacts on juvenile and adult salmonids, and for improved understanding of fish pass design in such structures.

## Acknowledgement

This study was undertaken collaboratively with the National Rivers Authority (now Environment Agency) Welsh Region. Thanks are due to D. M. Mee for his assistance.

## References

Aprahamian, M. W., C. D. Strange & C. Diamond, 1988. Migration of Atlantic salmon, *Salmo salar* L., into the River Usk, in relation to water quality. J. Fish Biol. 33 (A): 245–247.

Batschelet, E., 1981. Circular Statistics in Biology. London: Academic Press, 371 pp.

Bilton, H. T., D. F. Alerdice & J. T. Schnute, 1982. Influence of time and size at release of juvenile coho salmon (*Oncorhynchus kisutch*) on returns at maturity. Can. J. Fish. aquat. Sci. 39: 426–477.

Clarke, D. R. K., D. M. Evans, D. S. Ellery & W. K. Purvis, 1994. Migration of Atlantic salmon (*Salmo salar* L.) in the River Tywi estuary during 1988, 1989 and 1990. Cardiff: NRA (Welsh Region) Report RT/WQ/RCEU/94/7.

Evans, D. M., D. M. Mee, W. K. Purvis & D. R. K. Clarke, 1994. Migration of sea trout (*Salmo trutta* L.) in the River Tywi estuary during 1988, 1989 and 1990. Cardiff: NRA (Welsh Region) Report RT/WQ/RCEU/94/4.

Gough, P. J., 1996. Potential impact of estuarine barrages on migratory fish in England and Wales. In N. Burt & J. Watts (eds), Barrages – Engineering Design & Environmental Impacts, International Conference, 10–13 September 1996, Cardiff: 73–82.

Harris, G. S., 1994. The identification of cost-effective stocking strategies for migratory salmonids. National Rivers Authority, R&D Note 353, 150 pp.

Kell, L. T., I. C. Russell & M. J. Challiss, 1994. Development of a high resolution tracking system for monitoring the movement of migratory fish past obstructions. In I. J. Winfield (ed.), Proceedings of The Institute of Fisheries management 25th Annual Study Course, University of Lancaster, September 1994: 269–288.

Mee, D. M., A. J. Kirkpatrick & R. O. Stonehewer, 1996. Post impoundment fishery investigations on the Tawe Barrage, South Wales. In N. Burt & J. Watts (eds), Barrages – Engineering Design & Environmental Impacts, International Conference, 10–13 September 1996, Cardiff: 395–408.

Milner, N. J. (ed.), 1990. Fish movements in relation to freshwater flow and quality. Proceedings of the Atlantic Salmon Trust/Wessex Water Workshop, held at the University of Bristol, 4–6 April 1989, The Atlantic Salmon Trust, Pitlochry, 53 pp.

Mitson, R. B. & T. J. Storeton-West, 1971. A transponding acoustic fish tag. Radio-electron. Eng. 41: 483–489.

Moore, A., 1995. The movements of salmon (*Salmo salar* L.) and sea trout (*Salmo trutta* L.) smolts in the lower estuary of the River Conwy, North Wales. NRA R&D Note 349.

Moore, A., T. J. Storeton-West, I. C. Russell, E. C. E. Potter & M. J. Challiss, 1990a. A technique for tracking Atlantic salmon (*Salmo salar* L.) smolts through estuaries. ICES C.M. 1990/M:18, 1–5.

Moore, A., I. C. Russell & E. C. E. Potter, 1990b. The effects of intraperitoneally implanted dummy acoustic transmitters on the behaviour and physiology of juvenile Atlantic salmon, *Salmo salar* L. J. Fish Biol. 37: 713–721.

Moore, A., E. C. E. Potter & A. A. Buckley, 1992. Estuarine behaviour of migrating Atlantic salmon (*Salmo salar* L.) smolts. In I. G. Priede & S. M. Swift (eds), Wildlife Telemetry – Remote Monitoring and Tracking of Animals, Ellis Horwood Ltd., Chichester: 389–399.

Moore, A. & E. C. E. Potter, 1994. The movements of sea trout (*Salmo trutta* L.) smolts through the estuary of the River Avon, Southern England. Fish. Mgmt & Ecol. 1: 1–14.

Moore, A., E. C. E. Potter, N. J. Milner & S. Bamber, 1995. The migratory behaviour of wild Atlantic salmon (*Salmo salar* L.) smolts in the estuary of the River Conwy, North Wales. Can. J. Fish. aquat. Sci. 52: 1923–1935.

Moore, A., R. O. Stonehewer, M. J. Challiss, L. T. Kell, I. C. Russell, W. D. Riley & D. M. Mee, 1996. The movements of emigrating salmonid smolts in relation to the Tawe Barrage, Swansea. In N. Burt & J. Watts (eds), Barrages – Engineering Design & Environmental Impacts, International Conference, 10–13 September 1996, Cardiff, 409–417.

Potter, E. C. E., 1988 Movements of Atlantic salmon, *Salmo salar* L., in an estuary in south-west England. J. Fish Biol. 33 (A): 153–159.

Potter, E. C. E., D. J. Solomon & A. A. Buckley, 1992. Estuarine movements of adult Atlantic salmon (*Salmo salar* L.) in Christchurch Harbour, southern England. In I. G. Priede & S. M. Swift (eds), Wildlife Telemetry – Remote Monitoring and Tracking of Animals, Ellis Horwood Ltd., Chichester: 400–409.

Priede, I. G., J. F. de L. G. Solbé, J. E. Nott, K. T. O'Grady & D. Cragg-Hine, 1988. Behaviour of adult Atlantic salmon, *Salmo salar* L., in the estuary of the River Ribble in relation to variations in dissolved oxygen and tidal flow. J. Fish. Biol. 33 (A): 133–139.

Russell, I. C., L. T. Kell, E. C. E. Potter, T. Storeton-West & A. Moore, 1990. A telemetric tracking system for high resolution monitoring of fish movements. ICES C.M.1990/M:19.

Smith, G. W., I. P. Smith & S. M. Armstrong, 1994. The relationship between river flow and entry to the Aberdeenshire Dee by returning adult Alantic salmon. J. Fish. Biol. 45: 953–960.

Smith, I. P. & G. W. Smith, 1997. Tidal and diel timing of the entry by adult Atlantic salmon returning to the Aberdeenshire Dee, Scotland. J. Fish. Biol. 50: 463–474.

Solomon, D. J. & E. C. E. Potter, 1988. First results with a new estuarine fish tracking system. J. Fish. Biol. 33 (A): 127–132.

Solomon, D. J. & T. J. Storeton-West, 1983. Radio tracking of migratory salmonids in rivers: development of an effective system. Fish. Res. Tech. Rep., MAFF Direct. Fish. Res., Lowestoft No 75. 11 pp.

Stonehewer, R. O., A. J. Kirkpatrick & D. M. Mee, 1995. The results of the Tawe Barrage post impoundment fisheries impact studies, 1993–1994. National Rivers Authority, Welsh Region, Draft Report PL/EAW/95/4.

*Hydrobiologia* **371/372:** 335–337, 1998.
*J.-P. Lagardère, M.-L. Bégout Anras & G. Claireaux (eds), Advances in Invertebrates and Fish Telemetry.*



# Movements of adult sea trout *Salmo trutta* L. in the tailrace of a low-head dam at Włocławek hydroelectric station on the Vistula River, Poland

V. D. Linnik[1], L. K. Malinin[1], M. Wozniewski[2], R. Sych[2] & P. Dembowski[2]
[1] *Inst. of Biology of Inland Waters Russian Academy of Sciences, 152742 Borok, Nekouz, Yaroslavl, Russia*
[2] *Inland Fisheries Institute, Zabieniec 05-500 Piaseczno, Poland*

*Key words:* ultrasonic telemetry, sea trout, anadromous, migration, impoundment, fish ladder

## Abstract

Ultrasonic telemetry was used to study the movements and behaviour of 27 adult migratory sea trout, 60–90 cm in fork length. Fish were caught near the tailrace of a hydroelectric dam, tagged externally and released in October 1993, 1994 and 1996. Distribution of tagged sea trout depended on flow patterns in the tailrace. At high water discharge (maximum water current greater than 1 m $s^{-1}$) during the day, fish were observed about 250–500 m from turbines in the main stream or on the right of it. At lower water discharges, tagged fish were observed in the main stream both during the day and at night. Fish mainly moved upstream during the day and downstream at night. Fish were closer (about 50–250 m) to the turbines during the day than at night (about 300–500 m). They swam downstream to the right of main stream at night. The mean ground swimming speed of tagged sea trout was 1.98 cm $s^{-1}$ during the day and 8.33 cm $s^{-1}$ at night. Estimated swimming speeds correlated with time of day and fish length. Fish rarely entered the fishladder when water level near the entrance was low. Influence of water flow patterns on distribution of migratory sea trout is discussed. Fish transferred above the dam successfully migrated upstream through the reservoir under conditions of high water discharge.

## Introduction

Radio and acoustic telemetry techniques are currently used to study fish migrations and to survey fishway efficiency (McCleave et al., 1978; Travade et al., 1990; Ingendalh et al., 1996). Entry of migrating adult salmonids into fishladders is strongly influenced by the volume of discharge and flow patterns in the tailrace area (Leman & Paulik, 1966; Pavlov, 1979). The movements of adult sea trout during their spawning migration were investigated in the area close to Włocławek hydroelectric dam to identify the optimum conditions for entry to the fishladder.

## Materials and methods

Włocławek hydroelectric station is situated in the middle portion of the river Vistula. During the study, significant daily variations in the operation of the station resulted in maximum discharge rates during the day and at the beginning of night, while minimum rates occurred during the second half of night. The average daily discharge was $458 \pm 57$ $m^3$ $s^{-1}$ in 1993 and $511 \pm 51$ $m^3$ $s^{-1}$ in 1994 and 96. Water temperature was 7–10 °C. The chamber type fishladder is situated between the spillway and turbines with the entrance close to the turbines.

The station had 4 operating patterns. Patterns N1 and N2 were observed in 1993. Under Pattern N1, the last turbines on the right and left sides of the dam were operational. Water discharge was high, speed of the surface water was about 150 cm $s^{-1}$ and areas of back circulating currents were formed between the streams. Under Pattern N2 all the turbines were working but with low loading and the speed of the surface water in the area close to the turbines was 70–85 cm $s^{-1}$.

Patterns N3 and N4 were observed in 1994. Pattern N3 was used during the day and at twilight. The turbine adjacent to the fishladder entrance was shut, which resulted in low current speeds in the area close to the fishladder entrance, as observed by eye from a distance of 10 m. Pattern N4 was similar to N2, but at higher water discharge rates. There was no water discharge through the spillway during the period of investigations.

Sixteen male and eleven female sea trout with fork lengths between 60 and 90 cm and weights between 2.6 and 7.0 kg were tagged with ultrasonic transmitters placed on one side of the dorsal fin base (Gray & Haynes, 1979) without anaesthesia. The length of the transmitter is 35 mm, external diameter 17 mm and its operating frequency is 30–80 kHz (Saranchov, 1984). Tagged individuals were released in the tailrace area of the dam in 1993 and 94, and above the dam in 1996. The positions of tagged fish were located manually from a boat 2–3 times during the day and the first half of night. The tracking lasted for 1–10 days and was performed in October 1993, 1994 and 1996.

## Results

The operating patterns of the station had an effect on fish distribution. Under pattern N1 during the day, tagged fish were mainly 150–300 m downstream of the turbines or in the edge of the main stream where the current speed was about 70 cm $s^{-1}$. Fish constantly approached the operating turbines trying to go upstream. In the evening and the first half of night fish were concentrated in the area of circulating currents downstream the shut turbines. Sea trout in the tailrace area were located in the right-bank side of the main stream and periodically moved upstream to the dam. Fish remained in areas of high current speed for no longer than 4 hours before moving downstream to shallow water areas with low current speed. All tagged sea trout were never located together close to the dam.

Under pattern N2 tagged fish were usually located on the right-side of the main stream, 150–200 m from the dam, and were less often found close to the turbines than when pattern N1 was in operation. This may be explained by the lower water discharge which did not stimulate active searching behaviour of the migrants in the tailrace area.

Under Patterns N3 and N4 sea trout stayed to the right of the main stream in an area of low current speed. In the evening, when water discharge was less, they attempted to move upstream in the area of the lowest height of the fishladder. At night tagged fish more often moved downstream.

Spearman rank correlation coefficient showed no difference in fish movements between males and females. Speed of movement of tagged fish over the ground correlated positively with fish length and showed a significant association with time after release at the 10% level (Table 1). Speed of movement of tagged fish over the ground was significantly lower in the day-time ($2.0 \pm 0.8$ cm $s^{-1}$) than at the beginning of night ($8.3 \pm 3.0$ cm $s^{-1}$) (Student test 2.6, $P < 0.05$). Tagged fish moved mainly upstream in the daytime, and downstream at night.

At least 8 of the 12 tagged fish transferred above the dam swam upstream within the first four hours after release. Those fish showed rapid movement through the reservoir against the water flow 20–70 cm $s^{-1}$. Estimated ground speed of the upstream movement was about 20 km $day^{-1}$.

## Discussion

Tagged sea trout were never all found close to the dam at the same time, demonstrating the variability to be found between the behaviour of individuals. It should be noted that the signals of some transmitters disappeared after being located only a small number of times. Some of the tagged fish may have been caught by poachers. Even under strict control, the number of fish taken by poachers may equal those taken legally (Harris, 1980).

Both during the day and in the evening tagged fish moved upstream, while at night they moved downstream and swam to wide shallow waters with stony bottom and weak current. Fish with transmitters moved at slow speed in the edges of the stream caused by the working turbines, where current speeds were lower than in the central part of the main stream. Taking into account the counter current, the real swimming speed of fish was equal to 0.5–1.0 body length per second, i.e. it was close to optimum cruising swimming speed (Beamish, 1978, Figure 1). The correlation of the swimming speed with the length of tagged fish lends experimental support to the theoretical relationship between the swimming performance and the length of fish proposed by Weihs (1973). Reduction of fish swimming speed registered several days after tagging in 1993 may show the lessening of the shock caused by handling.

Quite often tagged fish moved into areas without current. Mainly, fish with transmitters were located in the area 150–500 m downstream from the turbines and

*Table 1.* Spearman rank correlation coefficient of fish swimming speed, their length and after the release time

| Year | Parameter | Spearman rank coefficient | The level of significance |
|---|---|---|---|
| 1993 | Fish length | 0.39 | < 0.05 |
| | Time after fish release | –0.35 | < 0.1 |
| 1994 | Fish length | 0.45 | < 0.05 |
| | Time after fish release | 0.27 | < 0.3 |

preferred the right edge zone of the main stream and the adjacent areas with low current speeds. The ability of sea trout to use local zones with low current speed for resting was discovered earlier (Linnik et al., 1993). Tagged fish did not enter the calm area situated down the spillway. The place where tagged sea trout were most frequently located depended considerably on the current speeds downstream of the turbines. Increases in the current speeds in the main stream from 70–80 cm $s^{-1}$ to 100 cm $s^{-1}$ and more caused a shift of the main area of fish presence downstream in which tagged fish were capable of maintaining themselves at sustainable swimming speeds.

The passage of migratory salmon through a fish-ladder greatly depends on hydraulic characteristics in the area of the fishladder entrance (Pavlov, 1979). During the observations of 1993 tagged fish were not located in the area of the fishladder entrance even under the most favourable conditions (the shut turbine adjacent to the entrance). This was probably because the small distance from which fish could react on the attractive stream of the fishladder. A single tagged sea trout was tracked moving directly to the entrance of the fishladder when the water discharge was high at the whole front of the turbine part of the dam. In the same area fish were observed leaping out of the water. Migrating fish were concentrated in the area of the circulating currents near the shut turbines.

## Acknowledgements

We are very thankful to Organizing Commitee of the Conference and Conference Secretariat, especially to Dr M. L. Bégout Anras, Dr G. Claireaux, Dr J. P. Lagardère for financial support of the visit of one of the authors to La Rochelle. Thanks are due to Dr E. I. Izvekov (IBIW, Russia) for editing the manuscript in English.

## References

Beamish, F. W. H., 1978. Swimming capacity. In W. S. Hoar & D. J. Randall (eds), Fish Physiology, Vol VII Locomotion: 101–187.

Gray, R. H. & J. M. Haynes, 1979. Spawning migration of adult chinook salmon (*Oncorhynchus tshawytcha*) carying external and internal radio transmitters. J. Fish. Res. Bd Can. 36: 1060–1064.

Harris, G. S., 1980. Ecological constrains on future salmon stocks in England and Wales. 'Proc. 2 Intern. Atlant. Salmon Symp., Edinburgh 1978', Farnham: 82–97.

Ingendahl, D., J. M. Bach, M. Larinier & F. Travade, 1996. The use of telemetry in studing downstream migration of Atlantic salmon smolts at a hydro-electric power plant in South-West France: preliminary results. In E. Baras & J. C. Philippart (eds), Underwater Biotelemetry: 121–128.

Leman, B. & G. J. Paulik, 1966. Spill pattern manipulation to guide migrant salmon upstream. Trans. am. Fish. Soc. 95: 397–407.

Linnik, V. D., A. Niemi, K. Poikola & T. Koskenala, 1993. Peremeschenija migrirrujuschih proizvoditelej kumzshy i siga v zaregulirovannoj sisteme r. Kymijoki (Finlandia). In D. S. Pavlov & A. G. Poddubny (eds), Biotelemetrija ryb. Metodicheskoe rukovodstvo. Hydroproekt, Moskva: 69–74.

Linnik V. D., A. Niemi, K. Poikola & T. Koskenala, 1993. Movements of migrating adults sea trout and whitefish in impoundment Kymijoki river system (Finland). In D. S. Pavlov & A. G. Poddubny (eds), Biotelemetry of Fish. Manual., Hydroproekt, Moscow: 69–74. (In Russian)

McCleave, J. D., J. H. Power & S. A. Rommel, Jr, 1978. Use of radio telemetry for studying upriver migration of adult Atlantic salmon (*Salmo salar*). J. Fish Biol. 12: 549–558.

Pavlov, D. S., 1979. Biologicheskie osnovy upravlenija povedeniem ryb v potoke vody. Nauka, Moskva: 319 pp.

Saranchov, S. I., 1984. Primenenije biotelemetricheskoj tehniki v rybohozjajstvennyh issledovanijah. Ministerstvo rybnogo hozjajstva SSSR. Tcentralny nauchno-tehnicheskij institut informatcii i tehniko-ekonomicheskih issledovanij rybnogo hozjajstva, Moskva, N 1: 76 pp.

Travade, F., P. Bomassi, J. M. Bach, C. Brugel, P. Steinbach, J. F. Luquet & G. Pustelnik, 1990. Use of radiotracking in France for recent studies concerning the E.D.F. fishways program. Hydroécol. Appl. 1/2: 33–51.

Weihs, D., 1973. Optimal fish cruising speed, Nature (London) 245: 48–50.

*Hydrobiologia* **371/372:** 339–346, 1998.
*J.-P. Lagardère, M.-L. Bégout Anras & G. Claireaux (eds), Advances in Invertebrates and Fish Telemetry.*



# Migration of adult Atlantic salmon (*Salmo salar*); the effects of artificial freshets

Eva B. Thorstad & Tor G. Heggberget
*Norwegian Institute for Nature Research, Tungasletta 2, N-7005 Trondheim, Norway*
*(e-mail: eva.thorstad@ninatrd.ninaniku.no)*

*Key words:* Atlantic salmon, *Salmo salar*, upstream migration, artificial freshets, radio telemetry

## Abstract

The upstream migration of Atlantic salmon tagged with radio transmitters was analysed on a 6 km long stretch of the River Mandalselva (southern Norway), where the residual flow was 3 $m^3$ $s^{-1}$. Over a period of 40 days six artificial freshets were released. Two of the freshets lasted continuously for 48 hours and four of the freshets consisted of successive pulses of increased water discharge. There were significant differences in mean migration distance per hour between the freshet period, the 11 hour post-freshet period and the residual flow period for three of the freshets. However, no clear relationship between increased water flow and stimulated salmon migration was found.

## Introduction

River flow is the factor most frequently reported to control the upstream migration of salmonids (Banks, 1969). Water flow appears to be an important factor stimulating adult Atlantic salmon (*Salmo salar*) to enter rivers from the sea (Huntsman, 1948; Hayes, 1953; Saunders, 1960; Brayshaw, 1967; Alabaster, 1970; Potter, 1988; Clarke et al., 1991; Smith et al., 1994). Upstream migration in rivers also appears to be stimulated by elevated flow (Huntsman, 1948; Hayes, 1953; Dunkley & Shearer, 1982; Laughton, 1989; Webb & Hawkins, 1989; Baglinière et al., 1990). However, the relationship between physical factors and migration is complex, and many other factors may also affect salmon migration (Banks, 1969; Jonsson, 1991).

In regulated rivers with possibilities of controlling the water discharge, artificial freshets may be used to stimulate the upstream migration of Atlantic salmon (Huntsman, 1948; Hayes, 1953; Banks, 1969). However, little is known about the effects of different water discharges and durations of freshets on migration. In rivers with hydroelectric operations, it is often the case that small amounts of water are available for residual flows and artificial freshets. The water, therefore, should be utilised as effectively as possible to maintain fish migrations. Knowledge of the response of fish to increased discharge and the effects of different types of freshets on fish migration will contribute to a more effective utilisation of the available water.

The aim of this study was to analyse the upstream migration of adult Atlantic salmon in the regulated River Mandalselva, Southern Norway, during different artificial freshets using radio telemetry. Previous studies on river migration and artificial freshets are based mainly on counting fences or catches in inland fisheries (e.g. Huntsman, 1948; Hayes, 1953; Swain & Champion, 1968; Dunkley & Shearer, 1982). Telemetry is a method that yields more detailed information on fish movements in response to changing water flow.

## Study area

The River Mandalselva is situated in Southern Norway (Figure 1). Mean annual flow at the outlet is 88 $m^3$ $s^{-1}$. The fish populations in the river system are affected by acidification due to acid precipitation. The native Atlantic salmon populations are extinct, while the sea trout populations are markedly depleted

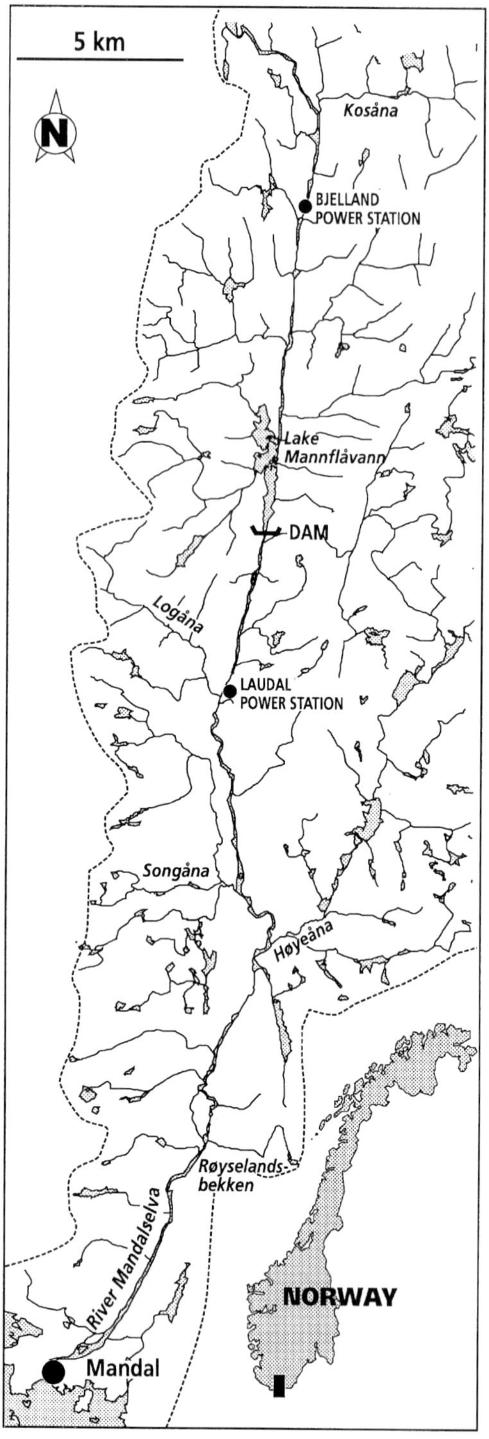


*Figure 1.* Location of the River Mandalselva, Southern Norway.

(Heggenes & Saltveit, 1992). Still, the annual catches of Atlantic salmon and sea trout were, on average, 778 kg during the period 1982–1993 (Flerbruksplan Mandalsvassdraget, 1994). The Atlantic salmon entering the River Mandalselva originate mainly from neighbouring rivers, salmon farm escapes and smolt releases in the estuary (Brabrand & Saltveit, 1995). The spawning period of these fish in the River Mandalselva is unknown, but for most Norwegian salmon populations peak spawning occurs after 20 October (Heggberget, 1988). The River Mandalselva is regulated. The study was located on a 6 km long stretch of river between the outlet of the Laudal power station and the dam at Lake Mannflåvann (Figure 1). Water is piped from Lake Mannflåvann to the Laudal power station, bypassing the study stretch. Residual flow on this stretch was 3 $m^3 s^{-1}$. The stretch mainly consists of artificial pools and weirs. Two weirs are constructed from concrete and ten weirs from wood and larger stones. Fish passes are constructed in the weirs to facilitate the passage of migratory fish. A fish trap is situated in the weir at Laudal. Before regulation, 80% of the spawning grounds were situated between Laudal and Kosåna (Figure 1) (Heggenes & Saltveit, 1992).

## Materials and methods

Upstream migrating Atlantic salmon were caught in the fish trap at Laudal, just after entering the river stretch with residual flow (Figure 1). Radio transmitters (Model 7PN, Advanced Telemetry Systems, ATS) were attached externally to the fish below the dorsal fin. The transmitters were flat with outline dimensions of 10 × 20 × 38 mm, weight in air of 11 g and weight in water of 6 g. Frequencies were restricted to the 142.010–142.500 MHz range. To recognise individual fish, each radio tag had a unique combination of frequency and pulse rate. The salmon were radio tagged and immediately released ($n = 9$), or kept in a net pen in the river up to eight days before tagging and release ($n = 26$).

During the period 9 August–8 September in 1996, 35 salmon (26 males and 9 females) were radio tagged. Mean total length of the fish was 63 cm (range = 54–96, SD = 8). The salmon were identified according to their external characteristics and from scale analysis (Lund et al., 1989). Of the 35 radio tagged salmon, 3% were identified as wild salmon, 31% as escaped farmed salmon, 11% as hatchery smolts, 14% as released as 0+/1+ and 40% as of unknown origin.

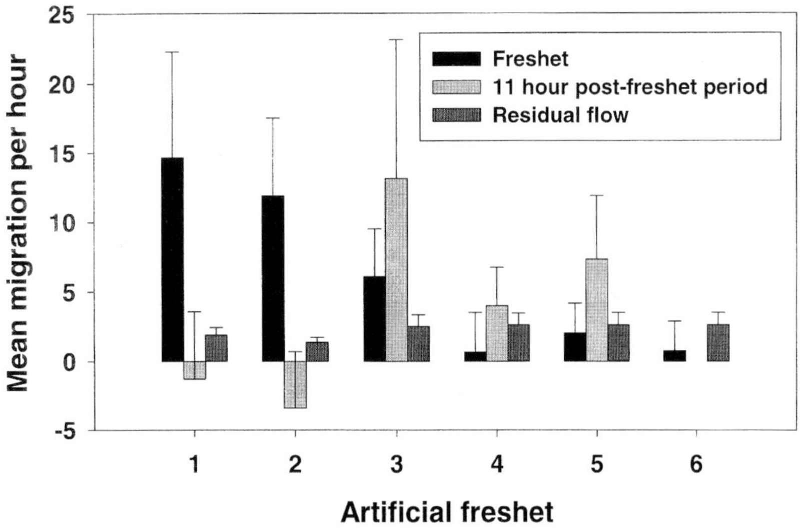


*Figure 2.* Mean migration distance per hour (+1 SE) of radio tagged Atlantic salmon during the artificial freshet, the 11 hour post-freshet period and the period of residual flow for the six artificial freshets on the river stretch between Laudal and Lake Mannflåvann in the River Mandalselva in 1996. The post-freshet period for freshet no. 6 is not included, because the residual flow was reduced from summer discharge to winter discharge immediately after the freshet finished, and migration during the post-freshet period was not recorded.

Six different freshets were released from the dam at Lake Mannflåvann during the period 26 August–4 October (Table 1). Two of the freshets lasted continuously for 48 hours and four of the freshets consisted of successive pulses of increased water discharge (Table 1). The design of the dam limited the maximum rate at which water could be released to 13 $m^3$ $s^{-1}$. Residual flow on the stretch was reduced from 3 to 1.5 $m^3$ $s^{-1}$ on 4 October.

The positions of radio tagged fish in the river between Laudal and Mannflåvann were determined to the nearest 300 m from a car every third day during the period 10 August–31 October, and every sixth day during the period 6 November–18 December (receiver Model R2100, ATS). In addition, the fish were located just prior to every freshet, twice a day during the freshets (morning and evening) and the morning after the freshets (except the morning after the last freshet). With this tracking scheme, the position of all fish was recorded, on average, 11 hours (range 9.5–12.5) after the end of the freshets, except in the case of the last freshet.

Of the radio tagged salmon, 7 individuals (4 males and 3 females, mean total length 63 cm, range = 55–69, SD = 5) migrated downstream from the area of catch and release, and thus, downstream from the stretch of residual flow. These fish were not included in the results. Before 4 October, 3 males passed the dam at Lake Mannflåvann and migrated further upstream. Numbers of radio tagged salmon present in the river between Laudal and Mannflåvann during the six artificial freshets were 11, 16, 27, 26, 25 and 25, respectively.

Non-parametric statistical tests were used because the data were not normally distributed and variances were not homogeneous (Kolmogorov-Smirnov One-Sample Test and Levene's test for homogeneity of variance).

## Results

The mean migration distance per hour differed among the 6 artificial freshets (Kruskal-Wallis test, df = 5, $x^2 = 17.5$, $P = 0.004$) (Figure 2). There were significant differences in mean migration distance per hour among the freshet period, the 11 hour post-freshet period and the residual flow period for freshets number 2, 3, and 4, but not for freshets number 1 and 5 (Friedman tests, test statistics given in Table 2) (Figure 2). There were no significant differences in mean migration distance per hour between the freshet period and residual flow period for any of the freshets (Wilcoxon signed ranks tests, test statistics given in Table 3) (Figure 2). There were significant differences in mean migration distance per hour between the 11 hour post-freshet period and the residual flow period for freshet number 3,

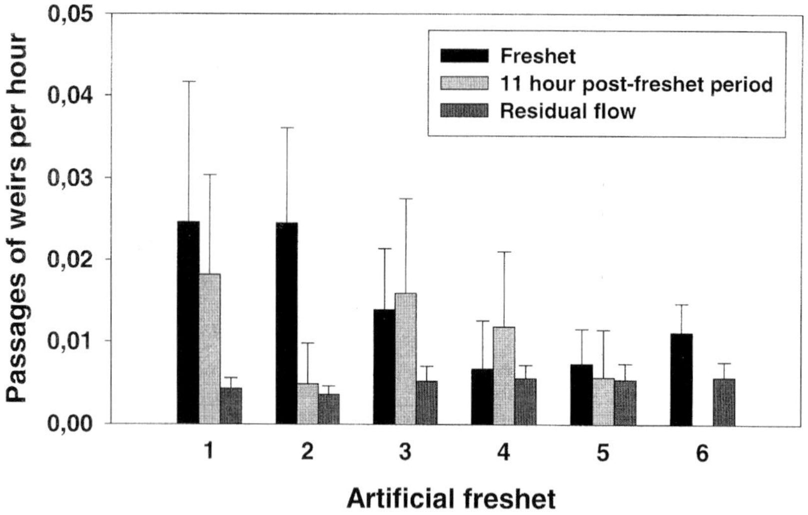


*Figure 3.* Number of passages of weirs per hour (+ 1 SE) of radio tagged Atlantic salmon during the artificial freshet, the 11 hour post-freshet period and the period of residual flow for the six artificial freshets on the river stretch between Laudal and Lake Mannflåvann in the River Mandalselva in 1996. The post-freshet period for freshet no. 6 is not included, because the residual flow was reduced from summer discharge to winter discharge immediately after the freshet finished, and migration during the post-freshet period was not recorded.

but not for the other freshets (Wilcoxon signed ranks tests, test statistics given in Table 4).

During the first three freshets, both newly tagged and released salmon (released 3–8 days before the freshets) and salmon that had stayed for some time in the river after tagging were tracked. Mean migration per hour of the newly released salmon during the freshets was not different from salmon that had stayed for some time in the part of the river with residual flow (Mann-Whitney *U*-tests, freshet no 1: $U = 12.0$, $P = 0.70$, freshet no 2: $U = 28.0$, $P = 0.83$, freshet no 3: $U = 56.0$, $P = 0.83$).

Number of passages of weirs per hour differed among the 6 artificial freshets (Kruskal-Wallis test, $df = 6$, $x^2 = 12.2$, $P = 0.032$) (Figure 3). There were significant differences in number of passages of weirs per hour among the freshet period, the 11 hour post-freshet period and residual flow period for freshets number 3, 4 and 5, but not for freshet number 1 and 2 (Friedman tests, test statistics given in Table 2) (Figure 3). There were no significant differences in number of passages of weirs per hour between the freshet period and residual flow period for any of the freshets (Wilcoxon signed ranks tests, test statistics given in Table 3) (Figure 3). There were significant differences in passages of weirs per hour between the 11 hour post-freshet period and residual flow period for freshet number 5, but not for the other freshets (Wilcoxon signed ranks tests, test statistics given in Table 4) (Fig-

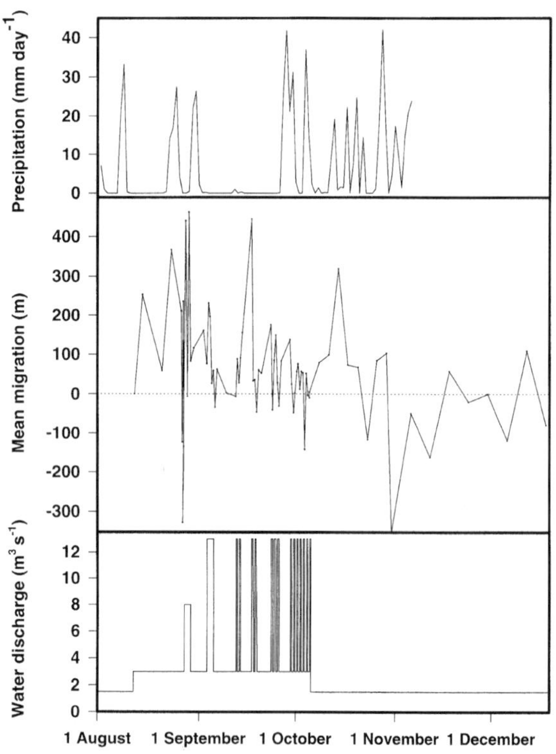


*Figure 4.* Upper: Daily precipitation recorded at Laudal Power Station daily during the period 1 August–6 November in 1996. Middle: Mean migration distance of radio tagged Atlantic salmon on the river stretch Laudal-Lake Mannflåvann in the River Mandalselva in Southern Norway during the period 10 August–18 December in 1996. Bottom: Water discharge from the dam at Lake Mannflåvann in the River Mandalselva in 1996.

*Table 1.* Artificial freshets released from the dam at Lake Manflåvann in the River Mandalselva in 1996. Residual flow on the river stretch was 3 $m^3 s^{-1}$ until 4 October and 1.5 $m^3 s^{-1}$ for the remainder of the year

| Start release of water | | Stop release of water | | Duration of release | Total water discharge during release |
|---|---|---|---|---|---|
| Date (dd.mm.) | Time | Date (dd.mm.) | Time | (hours) | ($m^3 s^{-1}$) |
| **Artificial freshet 1** | | | | | |
| 26.08. | 12.00 | 28.08. | 12.00 | 48 | 8 |
| **Artificial freshet 2** | | | | | |
| 02.09. | 12.00 | 04.09. | 12.00 | 48 | 13 |
| **Artificial freshet 3** | | | | | |
| 11.09. | 12.00 | 11.09. | 20.00 | 8 | 13 |
| 12.09. | 12.00 | 12.09. | 20.00 | 8 | 13 |
| **Artificial freshet 4** | | | | | |
| 16.09. | 08.00 | 16.09. | 16.00 | 8 | 13 |
| 16.09. | 16.00 | 16.09. | 18.00 | 2 | 11 |
| 16.09. | 18.00 | 16.09. | 20.00 | 2 | 9 |
| 16.09. | 20.00 | 16.09. | 22.00 | 2 | 7 |
| 16.09. | 22.00 | 16.09. | 24.00 | 2 | 5 |
| 17.09. | 08.00 | 17.09. | 16.00 | 8 | 13 |
| 17.09. | 16.00 | 17.09. | 18.00 | 2 | 11 |
| 17.09. | 18.00 | 17.09. | 20.00 | 2 | 9 |
| 17.09. | 20.00 | 17.09. | 22.00 | 2 | 7 |
| 17.09. | 22.00 | 17.09. | 24.00 | 2 | 5 |
| **Artificial freshet 5** | | | | | |
| 22.09. | 08.30 | 22.09. | 20.30 | 12 | 13 |
| 23.09. | 08.30 | 23.09. | 20.30 | 12 | 13 |
| 24.09. | 08.30 | 24.09. | 20.30 | 12 | 13 |
| **Artificial freshet 6** | | | | | |
| 28.09. | 10.00 | 28.09. | 18.00 | 8 | 13 |
| 29.09. | 10.00 | 29.09. | 18.00 | 8 | 13 |
| 30.09. | 10.00 | 30.09. | 18.00 | 8 | 13 |
| 01.10. | 10.00 | 01.10. | 18.00 | 8 | 13 |
| 02.10. | 10.00 | 02.10. | 18.00 | 8 | 13 |
| 03.10. | 10.00 | 03.10. | 18.00 | 8 | 13 |
| 04.10. | 10.00 | 04.10. | 18.00 | 8 | 13 |

*Table 2.* Test statistics for Friedman tests, testing the hypothesis that (1) mean migration distance per hour and (2) number of passages of weirs per hour was similar during the artificial freshets, during the 11 hour post-freshet periods and during periods with residual flow for radio tagged salmon in the River Mandalselva in 1996. Artificial freshet no. 6 is not included, because the residual flow was reduced from summer discharge to winter discharge immediately after the freshet finished, and migration during the post-freshet period was not recorded. Asterisks indicate significant *P*-values when the significance level is corrected according to sequentially rejective Bonferroni test procedures (Holm, 1979).

| Artific freshet no | Number of radio tagged salmon | Mean migration distance per hour | | Number of passages of weirs per hour | |
|---|---|---|---|---|---|
| | | $x^2$ | *P* | $x^2$ | *P* |
| 1 | 11 | 3.82 | 0.15 | 3.94 | 0.14 |
| 2 | 16 | 8.38 | 0.015* | 6.49 | 0.039 |
| 3 | 27 | 9.17 | 0.010* | 11.7 | 0.003* |
| 4 | 26 | 9.69 | 0.008* | 15.2 | 0.001* |
| 5 | 25 | 4.36 | 0.11 | 9.88 | 0.007* |

*Table 3.* Test statistics for Wilcoxon signed ranks tests, testing the hypothesis that (1) mean migration distance per hour and (2) number of passages of weirs per hour was similar during the artificial freshets and during periods with residual flow for radio tagged salmon in the River Mandalselva in 1996. None of the *P*-values are significant when correcting the significance level according to sequentially rejective Bonferroni test procedures (Holm, 1979).

| Artific freshet no | Number of radio tagged salmon | Mean migration distance per hour | | Number of passages of weirs per hour | |
|---|---|---|---|---|---|
| | | Z | P | Z | P |
| 1 | 11 | –1.87 | 0.062 | –0.18 | 0.86 |
| 2 | 16 | –1.81 | 0.070 | –1.41 | 0.16 |
| 3 | 27 | –0.27 | 0.80 | –0.62 | 0.54 |
| 4 | 26 | –2.20 | 0.028 | –1.93 | 0.053 |
| 5 | 25 | –1.52 | 0.13 | –0.36 | 0.72 |
| 6 | 25 | –1.22 | 0.22 | –0.89 | 0.37 |

ure 3). All of the weirs were passed by radio tagged salmon both during the freshets and during residual flow.

The position in the river on 28 October was not dependent on the size of the fish (linear regression analysis, $n=35$, $r^2=0.053$, $P = 0.18$). Mean mi-

*Table 4.* Test statistics for Wilcoxon signed ranks tests, testing the hypothesis that (1) mean migration distance per hour and (2) number of passages of weirs per hour was similar during the 11 hour post-freshet periods and during periods with residual flow for radio tagged salmon in the River Mandalselva in 1996. Artificial freshet no. 6 was not included, because the residual flow was reduced from summer discharge to winter discharge immediately after the freshet finished, and migration during the post-freshet period was not recorded. Asterisks indicate significant *P*-values when the significance level is corrected according to sequentially rejective Bonferroni test procedures (Holm, 1979).

| Artific freshet no | Number of radio tagged salmon | Mean migration distance per hour | | Number of passages of weirs per hour | |
|---|---|---|---|---|---|
| | | *Z* | *P* | *Z* | *P* |
| 1 | 11 | −1.07 | 0.29 | −0.87 | 0.39 |
| 2 | 16 | −1.91 | 0.056 | −1.96 | 0.050 |
| 3 | 27 | −2.73 | 0.006* | −1.91 | 0.056 |
| 4 | 26 | −1.51 | 0.13 | −1.76 | 0.078 |
| 5 | 25 | −0.73 | 0.47 | −2.27 | 0.023* |

gration distance of the fish, artificial freshets and precipitation at Laudal are shown in Figure 4.

## Discussion

In this study, the issue of how stored water can be used to facilitate salmonid migration was investigated. Detailed monitoring of fish migration in relation to changing water flow was possible by use of radio telemetry. Upstream migration of Atlantic salmon was recorded on a river stretch with a residual flow of 3 $m^3$ $s^{-1}$. During the study period, 6 different artificial freshets were released into the river stretch. There were significant differences in mean migration distance per hour between the freshet period, the 11 hour post-freshet period and the residual flow period for three of the freshets. However, no clear relationship between increased water flow and stimulated salmon migration was found. Similarly, Hawkins & Smith (1986) and McKinnell et al. (1994) did not find any simple relationship between salmon migration and changes in water flow. Hawkins & Smith (1986), however, reported a fairly high water discharge during their study period. The residual flow in the River Mandalselva was much lower, and increased upstream migration in relation to the artificial freshets was expected.

There may be several reasons that we did not record any clear effects of the artificial freshets on salmon migration. The design of the dam limited the maximum rate at which water could be released to 13 $m^3$ $s^{-1}$, and the results may have been influenced by the limitation on the size of the freshets. The artificial freshets in this study probably differed from natural spates in terms of discharge profile and water quality. It has been suggested that artificial freshets may be more effective in combination with rain (Baxter, 1961). The fact that most of the radio tagged salmon were of hatchery or farmed origin may also have influenced the results in the present study. In a radio telemetry study in the River Namsen in Norway, the number of movements of wild but not of farmed salmon increased when variation in water flow increased (Thorstad et al., 1998).

Another factor that should be considered is that the salmon in the present study may have found spawning areas in the study stretch, and therefore were not motivated to migrate further. However, the radio tagged salmon did not originate from the River Mandalselva, because the river has been too acidic for successful reproduction of Atlantic salmon (Heggenes & Saltveit, 1992). Salmon lacking river imprinting seem to distribute themselves higher up rivers than wild salmon (Heggberget et al., 1996; Økland et al., 1995; Thorstad et al., 1998). Trackings of the radio tagged salmon after 4 October also showed that seven of the salmon migrated above the study stretch (Thorstad & Heggberget, 1997). The dam at Lake Mannflåvann seemed to act as a migration barrier for the salmon, because 8 of the salmon reached the dam without passing it and instead migrated downstream again (Thorstad & Heggberget, 1997). Thus, most of the salmon in the present study were probably motivated to migrate upriver of the study stretch.

The results could be influenced by how long the fish had stayed in the part of the river with residual flow before release of the freshets. During the first three freshets some of the radio tagged salmon were newly tagged and released, whereas some had been in the river stretch with residual flow for some time. We found no differences in mean migration distance during the freshets between these two groups of fish.

Migration during the 11-hour post-freshet period was included as a variable in this study because several studies using counting fences or automatic counters have found that salmon ascent mainly occurs when a freshet is subsiding (Huntsman, 1948; Allan, 1966; Swain & Champion, 1968; Dunkley & Shearer, 1982;

Lawson et al., 1991). However, there has been some disagreement on this point. Jensen et al. (1986) found that most salmon ascended the River Vefsna before the maximum flow had been reached, regardless of the height of the flow. They suggested that the explanation for disagreement may be that increasing flows stimulate salmon to ascend, and that they continue to ascend over a period of time regardless of any change in river flow. When a spate only lasts for a few hours, some fish which are activated fail to reach the counting fence before the flood begins to subside. Allan (1966) suggested the same explanation. The present study gave no clear answer regarding this question, but has shown the advantage of using radio telemetry over counting fences or automatic counters when studying salmonid migration in relation to changes in water flow.

## Conclusion

Based on this study, no conclusive statements about how stored water may best be used to facilitate salmon migration can be made. The relationship between salmon migration and water discharge are probably not simple and straightforward, but may be influenced by several factors, as for example duration of the freshet, type of freshet, water discharge, water temperature, water quality, precipitation, time of the season, size of the fish, origin of the fish and motivation for migration. How stored water in a regulated river can be used to stimulate salmonid migration is an important issue, and further telemetry studies clarifying the effects of different artificial freshets under different conditions should be carried out.

## Acknowledgements

This study was financed by Office of County Governor in Vest-Agder county and Flerbruksplan Mandalsvassdraget. We wish to thank Jarl Fidje, Sigurd Haraldstad, Svein Haugland, Oddleiv Kleveland, Olav Kleveland and Dagfinn Laudal and their families who gave us great help with local organisation of the project and help with catch, tagging and tracking of the fish. We also wish to thank Lorraine Fleming for providing assistance with the English, Kari Sivertsen for graphic design of the map, Gunnel Østborg for the work with scale analyses, and two anonymous reviewers for helpful comments on an earlier draft of the manuscript.

## References


Alabaster, J. S., 1970. River flow and upstream movement and catch of migratory salmonids. J. Fish Biol. 2: 1–13.

Allan, I. R. H., 1966. Counting fences for salmon and sea trout, and what can be learned from them. Salm. Trout Mag. 176: 19–26.

Baglinière, J. L., G. Maisse & A. Nihouarn, 1990. Migratory and reproductive behaviour of female adult Atlantic salmon, *Salmo salar* L., in a spawning stream. J. Fish Biol. 36: 511–520.

Banks, J. W., 1969. A review of the literature on the upstream migration of adult salmonids. J. Fish Biol. 1: 85–136.

Baxter, G., 1961. River utilization and the preservation of migratory fish life. Proc. Instn. civ. Engrs. 18: 225–244.

Brabrand, Å. & S. J. Saltveit, 1995. Vandring av laks og sjøørret i Mandalselva mellom Mannflåvann og Laudal kraftverk. Laboratorium for ferskvannsøkologi og innlandsfiske (LFI), Zoological Museum, University in Oslo, Notat 3 – 1995. (In Norwegian).

Brayshaw, J. D., 1967. The effects of river discharge on inland fisheries. In P. G. Isaac (ed.), River Management. MacLaren, London: 102–118.

Clarke, D., W. K. Purvis & D. Mee, 1991. Use of telemetric tracking to examine environmental influence on catch effort indices. A case study of Atlantic salmon (*Salmo salar* L.) in the River Tywi, South Wales. In I. G. Cowx (ed.), Catch Effort Sampling Strategies: their Application in Freshwater Fisheries Management. Blackwell, London: 33–48.

Dunkley, D. A. & W. M. Shearer, 1982. An assessment of the performance of a resistivity fish counter. J. Fish Biol. 20: 717–737.

Flerbruksplan Mandalsvassdraget, 1994. Faggruppe for fisk og forurensing. Sluttrapport fra faggruppen, januar 1994: 1–28. (in Norwegian)

Hawkins, A. D. & G. W. Smith, 1986. Radio-tracking observations on Atlantic salmonascending the Aberdeenshire Dee. Scott. Fish. Res. Rep. 36: 1–24.

Hayes, F. R., 1953. Artificial freshets and other factors controlling the ascent and population of Atlantic salmon in the LaHave River, Nova Scotia. Bull. Biol. Bd Can. 99: 1–47.

Heggberget, T. G., 1988. Timing of spawning in Norwegian Atlantic salmon (*Salmo salar*). Can. J. Fish. aquat. Sci. 45: 845–849.

Heggberget, T. G., F. Økland & O. Ugedal, 1996. Prespawning migratory behaviour of wild and farmed Atlantic salmon (*Salmo salar*) in a North Norwegian river. Aquacult. Res. 27: 313–322.

Heggenes, J. & S. J. Saltveit, 1992. Reetablering av fiskebestanden i Mandalselva. Rapp. Lab. Ferskv. Økol. Innlandsfiske, Oslo 135: 1–77. (In Norwegian)

Holm, S., 1979. A simple sequentially rejective multiple test procedure. Scand. J. Statist. 6: 65–70.

Huntsman, A. G., 1948. Freshets and fish. Trans. am. Fish. Soc. 75: 257–266.

Jensen, A. J., T. G. Heggberget & B. O. Johnsen, 1986. Upstream migration of adult Atlantic salmon, *Salmo salar* L., in the River Vefsna, northern Norway. J. Fish Biol. 29: 459–465.

Jonsson, N., 1991. Influence of water flow, water temperature and light on fish migration in rivers. Nordic J. Freshwat. Res. 66: 20–35.

Laughton, R., 1989. The movements of adult salmon within the River Spey. Scott. Fish. Res. Rep. 41: 1–19.

Lawson, J. D., H. T. Sambrook, D. J. Solomon & G. Weilding, 1991. The Roadford scheme: minimizing environmental impact on affected catchments. Wat. Envir. Mgmt 5: 671–681.

Lund, R. A., L. P. Hansen & T. Järvi, 1989. Identifisering av oppdrettslaks og villaks ved ytre morfologi, finnestørrelse og skjelkarakterer. NINA Forskningsrapport 1: 1–54. (In Norwegian with English summary)

McKinnell, S., H. Lundqvist & H. Johansson. 1994. Biological characteristics of the upstream migration of naturally and hatchery reared Baltic salmon, *Salmo salar* L. Aquacult. Fish. Mgmt 25: 45–63.

Potter, E. C. E., 1988. Movements of Atlantic salmon, *Salmo salar* L., in an estuary in South-west England. J. Fish Biol. 33: 153–159.

Saunders, J. W., 1960. The effect of impoundment on the population and movement of Atlantic salmon in the Ellerslie Brook, Prince Edward Island. J. Fish. Res. Bd Can. 17: 453–473.

Smith, G. W., I. P. Smith & S. M. Armstrong, 1994. The relationship between river flow and entry to the Aberdeenshire Dee by returning adult Atlantic salmon. J. Fish Biol. 45: 953–960.

Swain, A. & A. S. Champion, 1968. Upstream movements of migratory salmonids in relation to river flows on the River Axe, Devon. I.C.E.S. Anadromous and Catadromous Fish Committee. CM 1868/M 9: 1–7.

Thorstad, E. B. & T. G. Heggberget, 1997. Oppvandring hos radiomerket laks og sjøørret i Mandalsvassdraget i forhold til minstevannføring, lokkeflommer, terskler og kalking. NINA Oppdragsmelding 470: 1–41. (In Norwegian with English summary)

Thorstad, E. B., T. G. Heggberget & F. Økland, 1998. Migratory behaviour of adult wild and escaped farmed Atlantic salmon, *Salmo salar* L., before, during and after spawning in a Norwegian river. Aquaculture Research 29: 101–110.

Webb, J. H. & A. D. Hawkins, 1989. The movements and spawning behaviour of adult salmon in the Girnock Burn, a tributary of the Aberdeenshire Dee, 1986. Scott. Fish. Res. Rep. 40: 1–42.

Økland, F., T. G. Heggberget & B. Jonsson, 1995. Migratory behaviour of wild and farmed Atlantic salmon (*Salmo salar*) during spawning. J. Fish Biol. 46: 1–7.

*Hydrobiologia* **371/372:** 347–353, 1998.
*J.-P. Lagardère, M.-L. Bégout Anras & G. Claireaux (eds), Advances in Invertebrates and Fish Telemetry.* 


# Survival of radio-tagged Atlantic salmon (*Salmo salar* L.) and trout (*Salmo trutta* L.) smolts passing a reservoir during seaward migration

Niels Jepsen, Kim Aarestrup, Finn Økland+ & Gorm Rasmussen
*Danish Institute for Fisheries Research Dept. of Inland Fisheries*
*+Norwegian Institute for Nature Research, Tungasletta 2, N-7005 Trondheim, Norway*
*Correspondence to: Niels Jepsen, Danish Institute for Fisheries Research. Dept. of Inland Fisheries, Vejlsøvej 39 DK-8600 Silkeborg Denmark. E-mail: nj@DFU.min.dk*

*Key words:* smolt, *Salmo salar, Salmo trutta*, radio-telemetry, predation

## Abstract

High mortality-rates of seaward migrating salmonid smolts when passing reservoirs and lakes have earlier been found in the Danish River Gudenå watershed. To reveal the causes of mortality of migrating smolts in Lake Tange, a 12 km long, shallow reservoir, 50 salmon smolts and 24 trout smolts were tagged with internal miniature radio-transmitters, and released in the river just upstream the reservoir on May 1, 1996. The salmon smolts were hatchery-reared, while the trout smolts were wild fish, caught in a smolt trap. The tagged smolts were tracked daily for 3 weeks, and when possible the cause of death was determined. During the 3-week period, 90% of the tagged smolts died. The main cause of death for both trout and salmon was predation from fish and birds. The most important predator was pike (*Esox lucius* L.), being responsible for 56% of the observed mortality. Avian predators were assumed to be responsible for 31% of the observed mortality. No trout smolts left the reservoir, but 5 salmon-smolts got out through the turbines. Others did traverse the reservoir, but were unable to enter the river downstream, and were later eaten. The present results suggest that mortalities for migrating smolts through Lake Tange are of such a magnitude, that stocking of juveniles in the river upstream is futile, and further, that the establishment of a natural population of salmon or sea-trout in river Gudenå, upstream Tange, is unrealistic under present conditions.

## Introduction

During seaward migration juvenile salmonids encounter habitats that are unfamiliar to them, while undergoing the physiological and behavioural changes associated with smoltification (Hoar, 1976). In this period smolts are often exposed to a high risk of predation (Larsen & Carl, 1994; Larsson, 1985; Rieman et al., 1991). Lakes and reservoirs have been shown to delay migrating smolts (Hansen et al., 1984). The delay may be dependent on the flow rates through the lake (Thorpe et al., 1981), which will influence patterns of smolt movement and the time spend traversing a lake or reservoir. A delay will further increase exposure to lake-predators and consequently decrease smolt survival rates.

Some studies have found high mortality-rates of smolts migrating through reservoirs and lakes on their way to the sea (Hansen et al., 1984; Rieman et al., 1991). However numerous natural salmon populations exists in river systems, where the juveniles after smoltification have to pass through lakes on their way to the sea, or where juveniles even use the lake as rearing habitat (Ryan, 1986). Obviously mortalities in these lakes must be moderate, and Bourgeois & O'Connell (1988) observed good survival of smolts traversing a large lake (> 18000 ha).

Mainly on basis of investigations of stomach contents, it seems widely accepted that predation from fish and birds are main causes of smolt loss during seaward migration (Larsson, 1985; Poe et al., 1991; Rieman et al., 1991; Ruggles, 1980), but attempts to quantify the actual causes of smolt-loss in lakes and reservoirs are scarce.

During the last decade substantial numbers of salmon smolts have been released in River Gudenå, as part of a program for reintroducing Atlantic salmon to the river. Trapping and tagging experiments revealed that only very few smolts succeed in traversing the lakes and reservoirs of the system. The largest reservoir, Tange, are very important because it is crucial for the establishment of a new self-reproducing salmon stock, and for the enhancement of migratory trout stocks, that smolts from upstream spawning and rearing sites can traverse the reservoir and reach the sea. The causes of the large (> 85%) smolt loss in the reservoir were unknown, but it was suggested that 1: The released smolts were unable to migrate through the reservoir, and subsequently became stationary and took residence in the river or in the reservoir, or 2: The released smolts did migrate but were eaten by predators. Radiotelemetry was used to test these hypotheses in the field, one study dealt with the causes of mortality and the other with the migratory behaviour.

## Materials and methods

### *Research area*

River Gudenå is from spring to outlet into a brackish fiord about 158 km. Mean annual discharge is 29 $m^3$ $sec^{-1}$. Elevation from mouth to spring is 69 m.

In 1920 a hydro-power station was build on the River Gudenå, 35 km upstream the tidal limit, creating a 537 ha, 12 km long, shallow (< 8 m.), eutrophic reservoir: Lake Tange. Due to high algae production, turbidity is rather high, with secchi-depths in May of 1–1.2 m. All spawning areas of the salmon, as well as important spawning and rearing areas for migratory trout, were situated above the power station. A few years later the salmon were extinct and the population of sea-trout reduced by at least 50%.

The variations in flow and temperature in the reservoir were small (9–13 °C and 11–14 $m^3$ $s^{-1}$) during the study period.

### *Tagging*

In May 96, 24 trout-smolts and 50 salmon-smolts, were implanted with miniature radio-transmitters. All 50 salmon-smolts were 1-year hatchery-reared, 16–18 cm in length. The trout smolts were caught in the river in a fyke-net type smolt-trap, and varied in length between 16 and 24 cm (total length), and were 1–3 years of age. All smolts selected for tagging were fully smoltified judged from morphological features (silvery appearance, enlarged eyes). Physiological tests ($Na^+$,$K^+$-ATPase activity and 24 h 35 ppt SW test) were carried out to check the smolting status of the salmon smolts. As the trout smolts were caught in the smolt-trap there are little doubt that they had started their seaward migration. They were probably all coming from upstream tributaries as the smolt production in the main river is very small.

The salmon-smolts were tagged at the hatchery. The smolts were anaesthetized in a 5 mg $l^{-1}$ solution of Methomidate (Marnil$^{TM}$, Wildlife Labs., Inc., Fort Collins, Colorado, USA) until operculum rate became slow and irregular (2–4 min). The fish were then placed in a V-shaped surgical table and the transmitter was inserted into the body cavity through a mid-ventral 8–10 mm incision, posterior to the pelvic girdle. The antenna was run through a hole from the body cavity, pierced with a blunt needle. The incision was closed with 2–3 separate sutures. The duration of the operation was between 1 and 2 min. Recovery time was 2–5 min. Within 24 hours after the implantation, the smolts were all transferred to River Gudenå, and released on a position just upstream Lake Tange.

The implantation of trout-smolts was done in a similar way at the riverbank by the smolt-trap. The trout-smolts were kept for 24–48 hours after implantation in a netpen in the river and released together with the salmon smolts.

The transmitters used were ATS internal smolt Model 377 and ATS int. smolt Model 384, weighing 1.4 and 1.7 gram. The transmitters had an expected operation time of 20 and 35 days. Each transmitter had a unique combination of frequency/pulse-rate, making each tagged fish individually recognizable. The reliability of these tags are very high, and according to the manufacturer failure rates are less than 1%. This is in accordance with our previous experience in the field and in the lab with this transmitter type.

All 50 salmon-smolts exhibited normal swimming- and feeding behaviour in the tank at the hatchery within few hours after the implantation. All tagged fish appeared to be in good condition when released.

### *Trapping experiments*

On two positions (2 and 6 km) upstream Lake Tange 8715 finn clipped salmon smolts were released on April 17. These were all 1 year smolts of the same

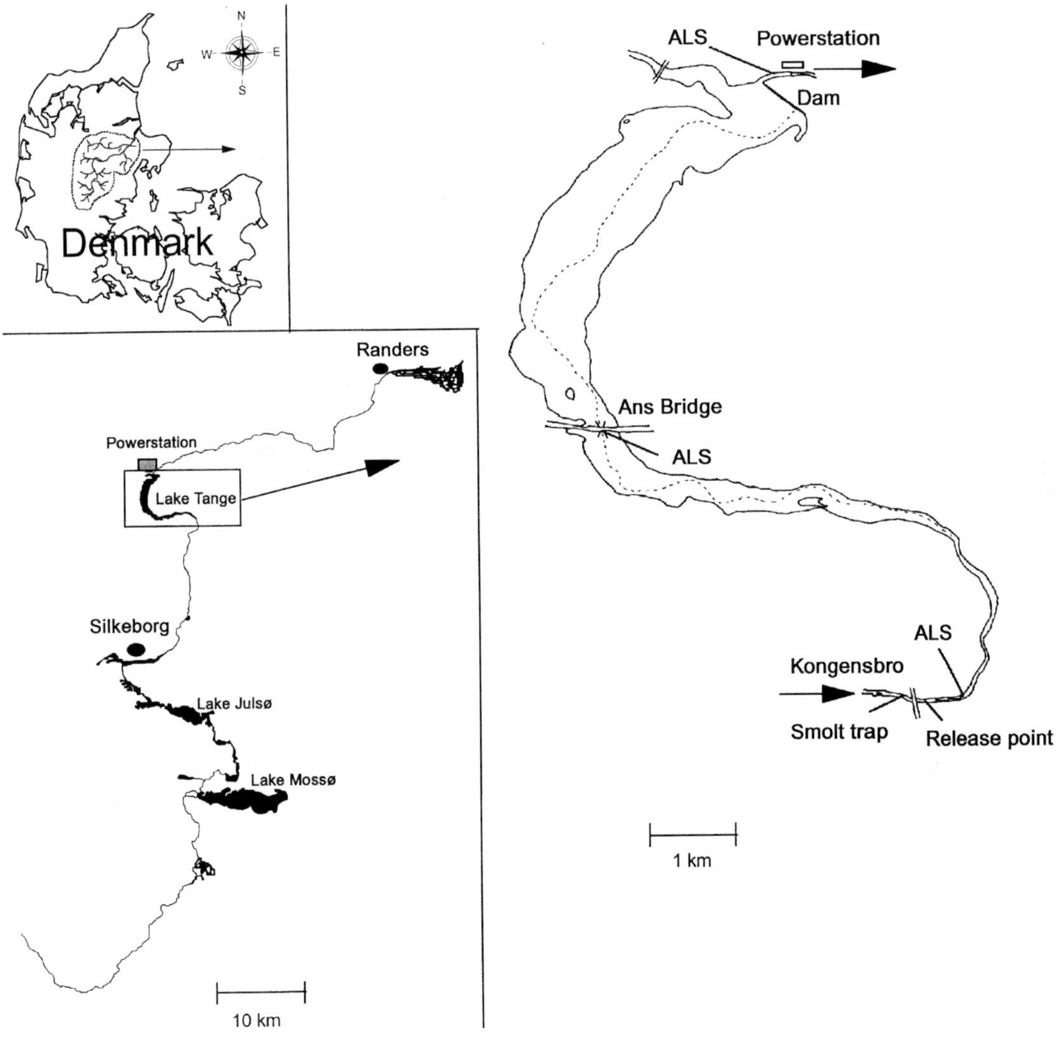


*Figure 1.* Map of study area, showing smolttrap, point of release and Automatic Listening Stations (ALS). Dotted line shows the old riverbed.

origin as the fish used for radio-tagging, and their success-rate (i.e. the fraction of smolts that succeed in traversing the reservoir and get out in the river downstream the dam) were estimated on basis of catches in a trap net situated just downstream the hydro power plant. The size of the wild trout smolt run was estimated from catches in the upstream trap, and their success-rate through the reservoir calculated on basis of the catches in the downstream trap. Finally 558 pan jet marked hatchery reared trout smolts were released upstream the reservoir to give estimates for their success-rate.

*Tracking and electric fishing*

The tagged smolt were tracked daily from boat with a mobile hand-operated receiver over a period of 3 weeks, and 3 automatic listening stations (ALS) were operating for five weeks, recording passage of tagged fish (see position on Figure 1).

When tagged fish did not migrate for long periods or were located in shallow water or weeded areas, the smolt was assumed to have been eaten, and, if possible, the predator was obtained by electro-fishing. Only on a single occasion was a tagged smolt caught. It was briefly examined and released.

*Table 1.* Different estimates of salmon- and trout smolt mortality in Tange reservoir 1996, based on 8715 released salmon smolts (from hatchery), 50 radiotagged salmon smolts, 4560 wild trout smolts, 24 radiotagged wild trout smolts and 558 released trout smolts (from hatchery)

| | Salmon (hatchery) | Trout (wild) | Trout (hatchery) |
|---|---|---|---|
| Mortality (calculated from trap catch) | 86.4% | 81.8% | 90.7% |
| Mortality (radiotagged smolts) | 87.5% | 87.5% | – |
| Mortality pr. km. (trap catch) | 15.2% | 13.2% | 18.0% |
| Mortality pr. km. (radiotagged) | 15.9% | 15.9% | – |

## Results

The physiological tests of the smolting status of the salmon, showed high $Na^{+}$,$K^{+}$-ATPase activity (10–12 $\mu$mol ADP mg protein$^{-1}$ s$^{-1}$), and saltwater tolerance (osmolality in freshwater; 310 mmol kg$^{-1}$ and after 24 h in SW; 348 mmol kg$^{-1}$). These values are normal for fully smoltified salmon.

Most smolts migrated downstream shortly after release, and after 24 hours only 18 tagged smolts (2 trout, 16 salmon) remained in the release-area. Some of these did stay in the area for up to 20 days before migrating. Mortality-rates on passing through the reservoir were high. No significant difference in mortality was seen between radio-tagged and untagged smolts (Table 1). Only 5 (7%) of the 74 radio-tagged smolts, successfully passed the reservoir during the study period. All 5 were salmon smolts. The first fish got out through the outlet 3 days after release. Time from release to death varied from 1–17 days (mean: 8.4).

As seen from Figure 2, no significant difference in predation pattern or survival was found between trout- and salmon smolts; no trout smolts got out during the study-period, but 3 were still alive in the reservoir, and may have got out later, or they could become resident in the reservoir as lake-trout. Because of this lack of difference, the trout- and salmon smolts are treated as one group in the further results.

Ten smolts were eaten or died in the vicinity of the release area (2 trout, 8 salmon), and never entered the reservoir. Electro-fishing was not carried out in this area before day 10, to avoid disturbance of live tagged smolts still present. When the area was electrofished, no fish were caught, and the transmitters remained on

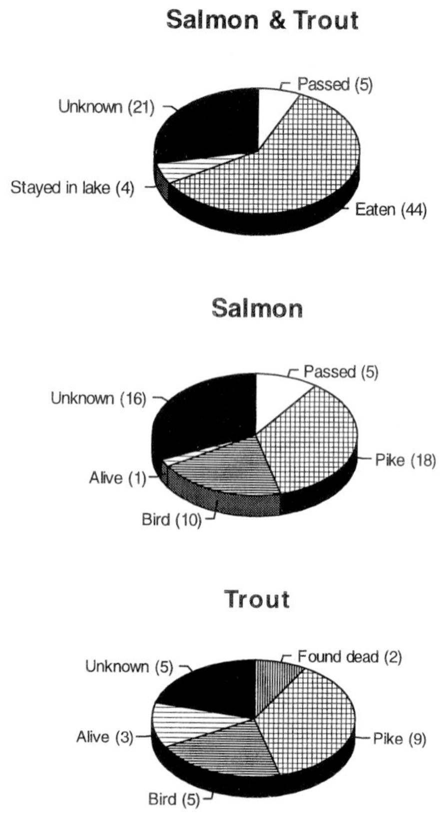


*Figure 2.* Fate of the 74 radiotagged smolts. Trout; $N = 24$, Salmon; $N = 50$. In brackets are numbers of smolts eaten by pike or bird or that was still alive in the reservoir after 3 weeks.

the river-bottom. Eleven smolts stopped moving in the reservoir and did not respond to electro-fishing.

Pike (*Esox lucius* L.) was the only predator caught containing transmitters. A total of 16 'transmitting' pike were obtained and opened to examine the stomachs for smolt-remnants and to remove the transmitters. In most cases the smolt was fully digested and only the transmitter was left, but also smolts in different stages of decay were found. One pike (62 cm.) contained 2 not fully digested smolts, 2 transmitters and 3 plastic-tags from smolts used for estimating trap efficiency. The pike, caught by electro-fishing, varied in size between 36 and 100 cm (mean: 60 cm). The smolts eaten were between 16 and 24 cm (mean: 17.7 cm). One pike containing a transmitter was caught by an angler and another was caught in a pound net by a commercial fisherman. Eight pike were tracked and observed, or seen when electro-fishing, but not obtained (in most cases because they moved

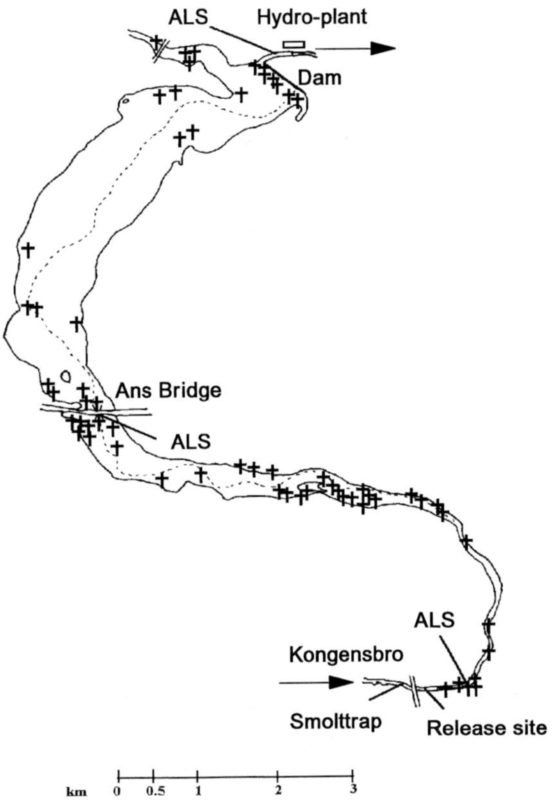


*Figure 3.* Geographical distribution of mortality. A cross denotes the last position of a smolt before it was classified as dead.

out into deeper water). So in total 27 tagged smolts were eaten by pike (9 trout- and 18 salmon smolts).

One trout smolt was obtained dying 7 days after release and another was found dead after 6 days. Presumably both had been attacked by predatory fish and had semicircular wounds on the anterior part of the body.

Fifteen tagged smolts were assumed to have been eaten by avian predators. Of these 5 were actually tracked while in the birds, 4 in grebe (*Podiceps griseigena* L.) and 1 in heron (*Ardea cinerea* L.). The remaining 10 suddenly disappeared from the area, and no signal was later received.

As seen from Figure 3 smolts were eaten all over the reservoir, but predation was highest on the following three locations:

- The very upper part from the point of release and 3 km downstream.
- The area around Ans Bridge, where the reservoir is only 25 m wide under the bridge.
- The lowest part of the reservoir, close to the outlet.

## Discussion

The implantation of radiotransmitters in salmonids is reported to have minor effects on mortality, growth, swimming performance and general behaviour (McCleave & Stred, 1975; Mellas & Haynes, 1985). In long term studies infections caused by the operation can be a problem (Lucas, 1989). However smolts are small and vulnerable creatures, and even though previous field and laboratory experiments with smolt implantation showed no increased mortality after tagging, tagged smolts may be expected to face an increased risk of predation. The behaviour of the tagged salmon smolts at the hatchery and the important fact that the mortality of the tagged smolts were not higher than that of the untagged smolts (Table 1), suggests that the results were not seriously affected by the adverse effect of handling and tagging the fish.

The fact that all trout and most salmon smolts started migrating shortly after release, confirms the conclusion (based on the results from the physiological tests and the morphological features), that the tagged fish were actually smolts and ready to migrate. However 3 trout- and one salmon smolt were still alive in the reservoir when the study was ended, and it is most likely that at least the trout smolts have lost their urge to migrate and become lake dwelling trout, if they survive long enough to reach a size that protects them against predation. Lake trout is frequently caught in the lakes and reservoirs of the Gudenå system.

The results does not indicate differences in ability of hatchery salmon smolts and wild trout smolts to avoid predation. Dellefors & Johnsson (1995) found that hatchery-reared trout was as capable as wild trout to avoid predation in a laboratory experiment, but the fact that no trout smolts got out during the study-period, is still surprising, and indicates that a wild nature in trout might not be advantageous when passing a reservoir

Pike were the most important smolt-predator in the reservoir, but the reservoir is also inhabited by other predators known to prey on migrating smolts; pike-perch (*Stizostedion lucioperca* L.), burbot (*Lota lota* L.) and eel (*Anguilla anguilla* L.). The predation by pike compared to other possible predators may, however be overestimated, because it is the most catchable predator, when using electricity, due to its preference for shallow vegetated areas (Diana et al. 1977; Chapman & Mackay, 1984), while other predators like pike-perch, burbot and eel, are known to take residence during daytime in the deeper parts of the lake (Ali et al., 1977; Nagiec, 1977; Carl, 1994). If a tagged smolt is eaten by a predator, situated on deeper water (> 2 m), the predator would be safe from electro-fishing.

When a radiotagged smolt was eaten by a pike, the transmitter would be withheld in the pike for 3–6 days, before being discharged, by regurgitation or through the rectum. The smolt itselves seemed to be digested rapidly (< 36 h), as found by Koed (1993). If the predator was not caught during these 3–6 days, the chance was passed, and the transmitter would be lying on the bottom, as was the case for the 21 tagged smolts classified as 'fate unknown'.

The results show that all pikes larger than 36 cm do eat smolt. The size of the smolts (salmon & trout) used in this study (16–24 cm) are slightly above the mean size of the natural trout smolt in the system. No direct correlation was found between predator and prey size. However, mean size of 575 wild trout smolts, caught in the smolt trap above the reservoir was 14.9 cm, while mean size of 260 wild trout smolts, caught in the trap downstream the power-plant was 18.4 cm., indicating size dependent mortality, as found by Bohlin et al. (1993), probably caused by size-selectivity of predators in the reservoir. An alternative explanation is that large fish can traverse the reservoir faster than small ones, and thereby decrease their time of exposure to predation.

Bird predation is usually hard to quantify, because of the high mobility of the predators. The assumption that sudden and permanent loss of a signal was due to avian predation, is based on the considerable range of the signals, even from transmitters at the deepest part of the reservoir, the intensive tracking and the reliability of the transmitters, making birdpredation the most plausible explanation. This was further validated by the fact that birds were directly tracked on 5 occasions. Furthermore one of the 10 disappeared transmitters were accidentally found in 1997 in a heron colony close to the reservoir.

The tracked birds were heron and grebe, but also cormorant (*Phalacrocorax carbo* L.), osprey (*Pandion haliaëtus*) and various gulls (*Larus* spp.) were present at the reservoir, and might have preyed on migrating smolt. The chance of actually tracking a signal to a certain bird are rather small, especially for species like cormorant and osprey, that are shy and very mobile. The grebe is a more stationary bird, that nests in the reservoir, and it does not flee, even when approached by boat. Therefore grebe has a high probability of getting 'caught in the act' in a study like this, and their importance as predators is easily overestimated. However at least 4 smolts were eaten by grebe, and that is 27% of the total number of smolts, classified as eaten by birds, and consequently this bird must be judged to have a not insignificant, though unexpected, impact on smolt survival in Lake Tange.

Predation pressure seemed to be highest at certain locations (Figure 3). There are no indications of a concentration of predators these places, but the smolts seemed to have problems in navigating these areas. The first high risk area is where the river becomes a lake, the current weakens and the smolts can not rely on passive displacement alone. Next problem is when having to pass under Ans Bridge. Most smolts got delayed here and were either unwilling to swim under the bridge or they had difficulties finding the hole. Finally the survivors got to the end of the reservoir, where they had similar problems getting into the power channel. Records showed that 13 smolts were registered by the ALS, covering the channel, but 8 turned around to reenter the reservoir. The overall picture seems to be that wherever a migrating smolt gets confused, it leaves the deeper middle part of the lake and comes close to the bank, where predation risk are very high. The smolts clearly preferred open waters far from the shores, and seemed to avoid shallow areas. Of more than 200 positionings in the broader part of the reservoir, only 3 were closer than 30 m. to the nearest shore. The few smolts that stayed alive in the reservoir for weeks, were never tracked close to the bank, but remained over deeper water close to the middle of the reservoir.

Considering that the mean time from release to death was more than 8 days, and that 4 smolts got out through the outlet only 3–5 days after release, the presence of predators might not be the major problem for the migrating smolt. If the smolts were not delayed and lost orientation, the mortality caused by predators would probably have been much lower. This could also explain why mortality rates of smolts in deep clear-watered lakes, like in Ireland, Scotland and Norway, are low (Bourgeois & O'Connell, 1988; Thorpe et al., 1981) compared to the ones found in shallow turbid reservoirs in Denmark. Many Irish lakes are known to hold large populations of pikes, but still have natural salmon populations, spawning in the lake-tributaries. Twomey (1965), however, reports of serious smolt-loss caused by pike predation in hydro-reservoirs on the Irish River Lee. This is in accordance with the findings of Buchanan et al. (1981), that Squawfish (*Ptychocheilus oregonensis*) predation rates were higher for smolts in watercourses where dams, water divisions or other obstructions disrupts water flow, than for smolts moving in streams with free-flowing water.

Schooling has been reported as a normal smolt response to threats like predation (Bakshtansky et al., 1981), but no schooling behaviour of the radiotagged smolts was observed in lake Tange, may be because of low visual contact between smolts due to high turbidity.

Two key factors to smolt survival may be turbidity of the water during smolt migration and the amount of 'obstacles', like Ans Bridge and the power channel, where smolts get confused and end up in areas with high predation risk.

Electro-fishing just downstream the power station, showed a very high density of large pike-perch, and their stomach-content indicated that smolts were the preferred prey, so even the few smolts that survived through both reservoir and turbines, were facing a high risk of being eaten by the pike-perches.

Obviously the efforts put into establishing a self-reproducing salmon population in Gudenå is wasted if more than 80% of the migrating smolts are lost in Lake Tange alone. This mortality, combined with river, estuary and marine mortalities, means that even intensive stockings of juveniles upstream Lake Tange, will fail to produce sufficient number of returning spawners.

## References


Ali, M. A., R. A. Ryder & M. Anctil, 1977. Photoreceptors and visual pigments as related to behavioural responses and preferred habitats of perches (Perca spp.) and pikeperches (*Stizostedion* spp.). J. Fish. Res. Bd Can. 34: 1475–1480.

Bakshtansky, E. L., V. D. Nesterov & M. N. Neklydov, 1981. The behaviour of Young Atlantic Salmon (*Salmo salar*), during Downstream Migration. J. Icthyol. 20: 93–100.

Bohlin, T., C. Dellefors & U. Faremo, 1993. Timing of Sea-run Brown Trout (*Salmo trutta*) Smolt Migration: Effects of Climatic Variation. Can. J. Fish. aquat. Sci. 50: 1132–1136.

Bourgeois, C. E. & M. F. O'Connell, 1988. Observations on the seaward migration of Atlantic salmon (*Salmo salar*) smolts through a large lake as determined by radiotelemetry and Carlin tagging studies. Can. J. Zool. 66: 685–691.

Buchanan, D. V., R. M. Hooton & J. R. Moring, 1981. Northern Squawfish (*Ptychocheilus oregonensis*) predation on juvenile salmonids in sections of the Willamette River Basin, Oregon. Can. J. Fish. aquat. Sci. 38: 360–364.

Carl, L. M., 1994. Sonic tracking of burbot in Lake Opeongo, Ontario. Trans. am. Fish. Soc. 124: 77–83.

Chapman, C. A. & W. C. Mackay, 1984. Direct observation of habitat utilization by northern pike. Copeia 1: 255–258.

Dellefors, C. & J. I. Johnsson, 1995. Foraging under risk of predation in wild and hatchery-reared juvenile sea-trout (*Salmo trutta* L.). Nordic J. Freshwat. Res. 70: 31–37.

Diana, J. S., W. C. Mackay & M. Ehrman, 1977. Movements and habitat preferences of northern pike in Lac St. Anne, Alberta. Trans. am. Fish. Soc. 106: 560–565.

Hansen, L. P., B. Jonsson & K. B. Døving, 1984. Migration of wild and hatchery reared smolts of Atlantic salmon, Salmo salar L., through lakes. J. Fish. Biol. 25, 617–623.

Hoar, W. S., 1976. Smolt Transformation: Evolution, Behaviour and Physiology. J. Fish. Res. Bd Can. 33: 1234–1252.

Koed, A., 1993. The feeding biology of pike (*Esox lucius* L.) and Zander (*Stizostedion lucioperca* (L.). M.Sc.-thesis, Dept. of Inland Fisheries and University of Aarhus.

Larsen, M. & J. D. Carl, 1994. Predation from pike (*Esox lucius* L.) and Zander (*Stizostedion lucioperca* L.) on down-migrating trout smolts in the Bygholm catchment area. M.Sc.-thesis, Dept. of Inland Fisheries and University of Aarhus.

Larsson, P. O., 1985. Predation on migrating smolt as a regulating factor in Baltic salmon, *Salmo salar*, populations. J. Fish Biol. 26: 391–397.

Lucas, M. C., 1989. Effects of implanted dummy transmitters on mortality, growth and tissue reaction in rainbow trout, (*Salmo gairdneri*). J. Fish Biol. 35: 577–587.

McCleave, J. D. & K. A. Stred, 1975. Effect of dummy telemetry transmitters on stamina of Atlantic salmon (*Salmo salar*) smolts. J. Fish. Res. Bd Can. 32: 559–563.

Moore, A., I. C. Russel & E. C. E. Potter, 1990. The effects of intraperitoneally implanted dummy acoustic transmitters on the behaviour and physiology of juvenile Atlantic salmon, *Salmo salar* L. J. Fish Biol. 37: 713–721.

Nagiec, M., 1977. Pike-perch in its natural habitats in Poland. J. Fish. Res. Bd Can. 34: 1581–1585.

Poe, T. P., C. H. Hal, S. Vigg, E. P. Douglas & L. A. Prendergast, 1991. Feeding of predaceous fishes on out-migrating juvenile salmonids in John Day reservoir, Columbia River. Trans. am. Fish. Soc. 120: 405–420.

Rieman, B. E., R. C. Beamsderfer, S. Vigg & T. P. Poe, 1991. Estimated loss of juvenile salmonids to predation by northern squawfish, walleyes, and smallmouth bass in John Day Reservoir, Columbia River. Trans. am. Fish. Soc. 120: 448–458.

Ruggles, C. P., 1980. A review of the downstream migration of Atlantic salmon. Can. Tech. Rep. Fish. aquat. Sci. 952: 39 pp.

Ryan, P. M., 1986. Lake use by wild anadromous Atlantic salmon, *Salmo salar*, as an index of subsequent adult abundance. Can. J. Fish. aquat. Sci. 43: 2–11.

Thorpe, J. E., L. G. Ross, G. Struthers & W. Watts, 1981. Tracking Atlantic salmon smolts, (*Salmo salar* L.) through Loch Voil, Scotland. J. Fish Biol. 19: 519–537.

Twomey, E., 1965. Predation by pike in the River Lee. ICES. Salmon and Trout Comm. C.M. 1965. No. 85. 3 pp.

*Hydrobiologia* **371/372:** 355–360, 1998.
*J. P. Lagardère, M.-L. Bégout Anras &G. Claireaux (eds), Advances in Invertebrates and Fish Telemetry.*




# The efficiency of a nature-like bypass channel for pike-perch (*Stizostedion lucioperca*) in the Marchfeldkanalsystem

S. Schmutz, C. Giefing & C. Wiesner
*Dept. of Hydrobiology, Fisheries and Aquaculture, Univ. of Agricultural Sciences Vienna, Max Emanuel-Strasse 17, A-1180 Vienna, Austria (e-mail: h524p2@edv1.boku.ac.at)*

*Key words:* nature-like bypass-channels, *Stizostedion lucioperca*, movement, colonisation, radio telemetry

## Abstract

This radio telemetry study is part of a large interdisciplinary research program on the colonisation and development of the Marchfeldkanal(MFK)-system, a man made channel. The immigration of fishes into the MFK is dependent on the effectiveness of fish bypass channels at several weirs. To investigate the efficiency of the lower most fishway we estimated the population densities along the MFK-system and below the weirs using electrofishing. In addition, the movements of 15 radio-tagged pike-perch at the fishway were observed. Although more than 57 000 fishes of 35 species passed the bypass channel, pike-perch (*Stizostedion lucioperca*) were under-represented in the fishway traps compared to their occurrence in the channel. The average daily movement of radio-tagged pike-perch was 108 m (range 6–333 m) and the maximum observed daily movement was 1050 m. The entrance to the bypass channel (280 m below the weir, and 100 m above the release site) was approached a number of times by 6 tagged fish, though none of them entered the bypass channel during the period of tracking. We conclude that although pike-perch migrate actively they do not utilise the bypass channel as much as most fish species of the MFK. Therefore the weir still represents a bottleneck for the immigration of pike-perch into the MFK.

## Introduction

Compared to the high number of fishways recently constructed throughout the world, very few have been evaluated (Francfort et al., 1994). Examples of the highly variable efficiency of upstream fish passage solutions are given in Schwalme et al. (1985), Monk et al. (1989), Baras et al. (1994), Jensen & Aass (1995), Schmutz et al. (1995), Jungwirth (1996) and Larinier et al. (1996). However, most of the documented evaluation studies are published only in reports or other grey literature, which are not easily accessible.

To assess the efficiency of fish passes, not only the number of fish passing the facilities but also a size estimation of the migratory population together with their migratory behaviour has to be analysed.

In Austria, nature-like fish bypass-channels have been developed during the last two decades and designed to provide fish passage for all species resident in the river (Jungwirth & Pelikan, 1989). These developments are useful in headwaters (Jungwirth, 1996) but their efficiency remains questionable in lowland rivers. Fish assemblages in lowland rivers are more diverse with respect to the migratory, swimming and leaping behaviours of individual species and this makes the design and construction of lowland river fishways more demanding.

The fish pass evaluation study shown in this paper is part of a large interdisciplinary research program on the colonisation and development of the Marchfeldkanal (MFK), a man made channel (Grubinger & Ernegger, 1994). The immigration of fish into the MFK is dependant on the effectiveness of nature-like fish bypass channels at several weirs. Within the first two years of the program up to 35 000 individuals of 34 species negotiated the fishways between April and July, demonstrating their overall high efficiency (Schmutz et al., 1995). However, when examining differences in species-specific passage into the channel, some species appear to be using the bypass less than expected. One of these is pike-perch, *Stizostedion lucioperca* (L.), which is expected to become the main predator in the MFK-system. Despite the fact that we detected pike-perch in the MFK, it is not certain if this migration from the Danube displays a spawning or a dispersive behaviour.

The main object of this study was to assess the efficiency of the lower most bypass channel at weir 4 (W4) of the MFK by observing the colonisation of pike-perch below and above this first obstacle, the movement behaviour below the weir, and the fishway passage.

## Study site

The MFK-system, a man-made irrigation channel, was filled in 1992 and corresponds to a branch or side channel of the Danube with strongly regulated flow conditions. The system is fed with water from the Danube in the vicinity of Vienna and is divided into two distinct sections. The upper section is the actual MFK and is about 19 km long with a newly created river bed. At Deutsch Wagram, the channel is joined by the Rußbach and then flows 37 km before re-entering the Danube (Figure 1).

The discharge of the MFK is dependent on the water level of the Danube and varies from 2 to 10 $m^3\ s^{-1}$. The channel is 10–20 m wide, with maximum water depths in a profile ranging from 0.7 m to 1.8 m, and a mean flow velocity of 0.3–0.9 $m\ s^{-1}$. Due to a low gradient (0.022%), the channel resembles a lowland river. The discharge and water surface elevation are regulated by three weirs, each has a fish bypass (F2, F3, F4). Below each weir is a pool which is about 2 m deep and 20 m long (W2, W3, W4) (Figure 1).

The fish bypass-channel at W4 consists of a sequence of 12 nature-like pools and is about 400 m long. The entrance of the pass is located 280 m below W4. The difference in water elevation between the pools ranges from 9 to 22 cm. The head at the weir is about 2 m and the discharge in the bypass is 0.25 $m^3\ s^{-1}$. More details about the physical characteristics of the bypass-channels in the MFK-system are given in Schmutz et al. (1995).

## Materials and methods

Between April and July we sampled 200 m-long reference sites (RS) in the MFK (RS1, RS3, RS4, RS5, RS6) and in the Rußbach (RS8, RS9) as well as the pools below the weirs (W2, W3, W4) three times a year (1993–95) with an equal effort using electrofishing gear as described by Schmutz et al. (1994). Fish passage at W4 was assessed through a daily monitored trap (T4) at the exit of the bypass from April to July as described by Schmutz et al. (1995).

For analysis, pike-perch were divided into two size classes, adults and juveniles. Pike-perch in the Danube are mature at a total length of about 370 mm (Zauner pers. com.). The ratio between the number of individuals found in the pool below weir 4 (W4) and those caught in the trap (T4) was calculated for all frequently caught taxa ($N > 40$). This ratio is a semi-quantitative indicator for a potential 'crowding effect' below the weir (see Table 2). In addition, fish caught in the pool, larger than 20 cm, were individually tagged with visible implant tags (VI-tags, Northwest Marine Technology Inc.).

For the tracking experiment we used LOTEK radio telemetry equipment (receiver SRX_400). Fifteen pike-perch (total length 345 mm to 475 mm) were caught in W4 by electrofishing, anaesthetised in a MS-222 solution (100 mg $l^{-1}$), measured to the nearest mm (total length) and weighed. LOTEK CFRT-3DM external transmitters with a theoretical lifespan of 65 days were attached through the dorsal musculature (Winter et al., 1978). Transmitters, with a weight in air of 3.3 g and a size of $11 \times 28$ mm were operating at a frequency of 142 MHz. The weight of the transmitter was 0.3–1.0% of the fish weight in air. Fish were returned to a holding cage in the MFK over night to allow recovery from the anaesthetic before being released on the next day 400 m below W4. Tracking was carried out with a hand-held receiver every morning from 18 May to 23 June. We started tracking at W4 and slowly drove with a car 2 km downstream along the channel.

## Results

The number of pike-perch caught by electrofishing in the MFK-system (RS1, RS3, RS4, RS5, RS6, RS8, RS9) as well as in the pools below the weirs (W2, W3, W4) increased from 8 individuals in 1993 to 29 in 1994, and 139 in 1995. While in 1993 and 1994, 75% and 45% of the total catch of pike-perch respectively were caught in W3, in 1995, 69% were found in W4. In 1995, the percentage of adult pike-perch above weir 4 was considerably lower (10%) than below (62%) (Table 1).

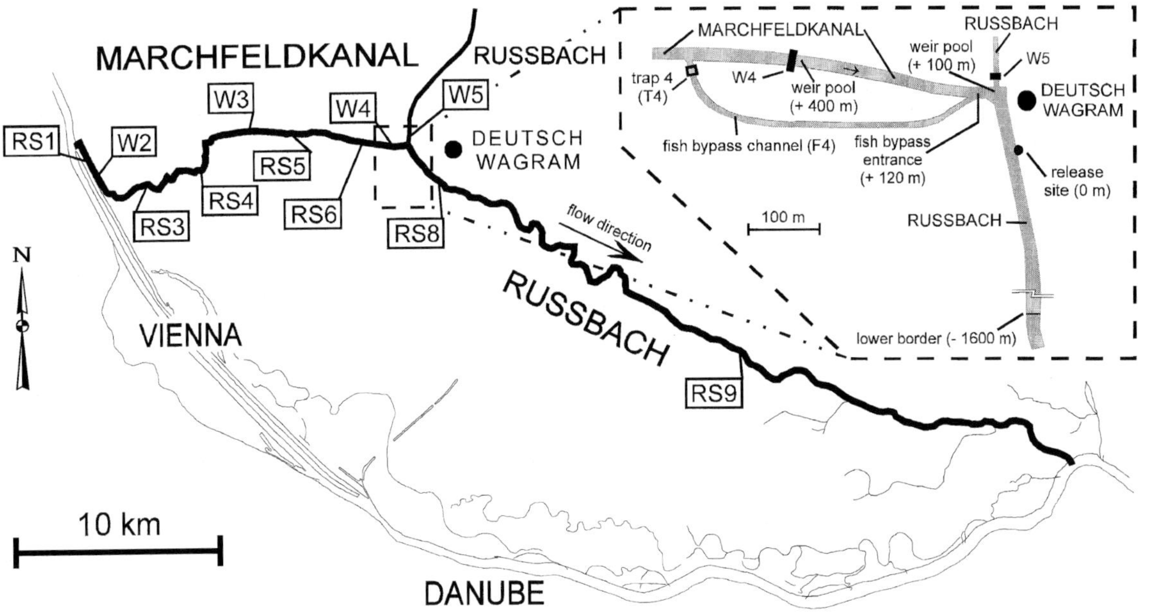


*Figure 1.* Map of the Marchfeldkanal-system with reference sites (RS1–RS9) and weirs (W2–W5) and description of the telemetry study site (distances from the release site in brackets).

*Table 1.* Number of pike-perch caught by electrofishing in the reference sites (RS1–RS9) and in the pools below the weirs (W2–W4) during the three sampling efforts in 1995

| Date | Life-stage | RS1 | W2 | RS3 | RS4 | W3 | RS5 | RS6 | W4 | RS8 | RS9 |
|---|---|---|---|---|---|---|---|---|---|---|---|
| 27–30 March | Juveniles | | | | 1 | 3 | 4 | | 18 | | |
| | Adults | | | | | | 1 | | 9 | 1 | |
| | Total | | | | 1 | 3 | 5 | | 27 | 1 | |
| 8–11 May | Juveniles | | | 4 | 4 | 10 | | | 18 | | |
| | Adults | | | | | | | | 26 | | 1 |
| | Total | | | 4 | 4 | 10 | | | 44 | | 1 |
| 26–29 June | Juveniles | | | 2 | 5 | 3 | 1 | | 3 | | |
| | Adults | | 3 | | | | | | 22 | | |
| | Total | | 3 | 2 | 5 | 3 | 1 | | 25 | | |

While in 1993 pike-perch did not pass through the bypass-channel, in 1994, four were recorded in the traps. In 1995, 35 pike-perch were continuously observed from 24 April to 16 July with a maximum of 3 individuals per day. This increase is due to progressive colonisation of the MFK-system.

In 1995, the ratio between the number of pike-perch caught in the trap and accumulating below weir 4 is 0.37. All other frequently caught species ($N > 40$), except *Barbus barbus*, (L.) and *Zingel zingel,* (L.), were caught in higher numbers in the trap than in the three fishing efforts in W4 (Table 2), demonstrating a significant difference to pike-perch (*G*-test, $G = 6436.4$, $p < 0.001$).

A significant difference between pike-perch and other species can also be found with tagged individuals from weir 4, recaptured in the pool and in the trap (*G*-test, $g = 13.0$, $p < 0.001$). Only 3 (4.5%) of the tagged pike-perch ($N = 67$) were recaptured in the trap and 10 (14.9%) in the pool itself, whereas individuals from other species ($N = 112$) were recaptured exclusively in the trap ($N = 10$, 8.9%).

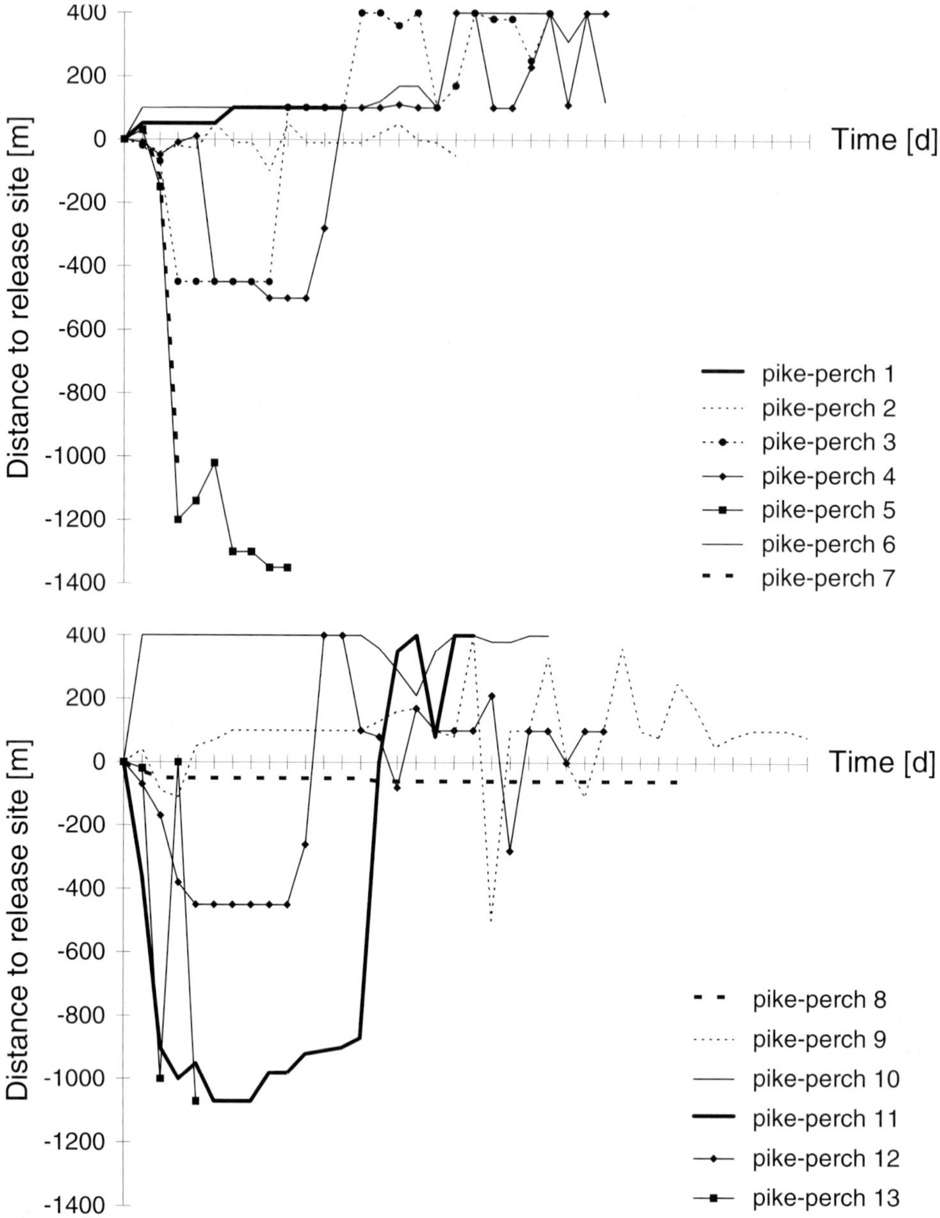


*Figure 2.* Position as the distance in meter from the release site of 13 radio-tagged pike-perch between 17 May and 23 June. Positive values indicate an upstream and negative a downstream position (pike-perch # 14 was only recorded one day, pike-perch # 15 was not recorded at all).

Fifteen radio-tagged pike-perch were released on 17 May 400 m below weir 4 (120 m below the entrance of the bypass-channel). One of the radio tagged pike-perch was not recorded after release. The other 14 pike-perch were tracked for 1 (pike-perch # 14) to 38 days (pike-perch # 9) with a mean observation duration of 18 days (Figure 2). Ten of the tagged individuals were observed more than 10 days.

Three pike-perch moved downstream more than 1000 m within the first three days and obviously left the study area in a downstream direction within the first nine days. The mean daily movement of radio tagged pike-perch calculated across the individual observation period varied between 6 and 333 m. The maximum observed daily movement was 1050 m (pike-perch # 5, 19 to 20 May). Six pike-perch approached the entrance

*Table 2.* Number of individuals for frequently caught species ($N > 40$) recorded in the trap (T4) and in the pool (three fishing efforts) below weir 4 (W4), and the T4:W4 ratio in 1995. Values in brackets indicate number of VI-tagged individuals, tagged in the pool (W4) and recaptured in the trap (T4)

| Taxa name | W4 | | T4 | | Ratio |
|---|---|---|---|---|---|
| Centrarchidae | | | | | |
| Lepomis gibbosus | 2 | | 40 | | 20.0 |
| Cyprinidae | | | | | |
| Abramis brama | 12 | (9) | 107 | (2) | 8.9 |
| Alburnus alburnus | 58 | | 35846 | | 618.0 |
| Aspius aspius | 1 | (1) | 756 | (0) | 756.0 |
| Barbus barbus | 57 | (9) | 24 | (0) | 0.4 |
| Blicca bjoerkna | 9 | (5) | 274 | (3) | 30.4 |
| Carassius auratus auratus | 38 | | 598 | | 15.7 |
| Carassius auratus gibelio | 23 | (2) | 529 | (0) | 23.0 |
| Chondrostoma nasus | 5 | (2) | 333 | (0) | 66.6 |
| Gobio albipinnatus | 2 | | 63 | | 31.5 |
| Gobio gobio | 101 | | 2068 | | 20.5 |
| Hybrids | 4 | | 172 | | 43.0 |
| Leuciscus cephalus | 146 | (61) | 462 | (2) | 3.2 |
| Leuciscus idus | 15 | (10) | 532 | (3) | 35.7 |
| Leuciscus leuciscus | 1 | | 392 | | 392.0 |
| Rutilus rutilius | 51 | (2) | 10764 | (0) | 211.1 |
| Scardinius erythrophthalmus | 6 | | 1806 | | 301.0 |
| Percidae | | | | | |
| Gymnocephalus cernua | 0 | | 85 | | – |
| Perca fluviatilis | 51 | | 2277 | | 44.6 |
| Stizostedion lucioperca | 96 | (67) | 35 | (3) | 0.4 |
| Zingel zingel | 39 | | 8 | | 0.2 |

of the fishway 28 times but none of them entered the pass. Tagged pike-perch preferred the pool below weir 4 (41 records) and the pool at the mouth to the MFK at weir 5 (W5) (59 records) but remained at one location on the average of only 2 days.

## Discussion

The MFK system was first filled in 1992 and the colonisation of pike-perch into the channel since then occurs via two routes. In 1993 and 1994, downstream drift of juveniles from the Danube was the most important colonisation mechanism (Schmutz et al., 1994; Schmutz et al., 1995). In 1995, upstream migration of adults from the Danube via the Rußbach increased which is shown in the increasing number of pike-perch in the pool of W4.

The migration of pike-perch from the Danube through the Rußbach up to W4 over a distance of 37 km indicates significant migratory behaviour. However, it is not clear if this behaviour represents a spawning migration or a dispersive movement. Pike-perch did not show a spawning migration peak through the bypass-channel at weir 4. Information on the migratory behaviour of pike-perch is very scare as few telemetry studies have been carried out (Baras, 1991). Therefore, it is not known if pike-perch show extensive spawning runs comparable to other potamodromous species. Travade et al. (in print) show that pike-perch migration in the Dordogne River peaks in September and October and only comparatively low migration activity is observed during the spawning season.

Although more than 57 000 fishes of 35 species passed the bypass-channel at weir 4 in 1995 only a small proportion of pike-perch, zingel and barbel found in the pool below the weir were recorded in the trap during spring and early summer. About 5% of the VI-tagged pike-perch in W4 were recaptured in the trap, compared with 9% mean recapture for other species.

Although 15% of tagged pike-perch were recaptured in the pool showing a higher 'crowding effect' compared to other species, most of them did not stay in W4.

In summary, the pike-perch was less likely to enter the bypass channel compared to other species as it is shown by the significant differences between weir pool and trap for tagged and non-tagged individuals. Typically cited causes of poor efficiency of fishways such as insufficient attraction flow and hydraulic limitations can be excluded as species with low swimming capabilities passed in high numbers. Although the entrance of the bypass channel is located 280 m below the weir most of the species had no obvious difficulty finding the entrance.

An explanation for the low efficiency of the bypass channel for pike-perch could be that larger fish avoid the comparatively small channel of the fishway. Bypass evaluation studies in Austrian headwaters showed a high efficiency for adult trouts *(Salmo trutta*, (L.) and *Oncorhynchus mykiss*, (WAL.)) and European grayling (*Thymallus thymallus,* (L.)) but the passage of large-sized Danube salmon (*Hucho hucho,* (L.)) is still questionable (Jungwirth, 1996).

We conclude that although pike-perch demonstrated considerable migratory activity the efficiency of the bypass-channel was low compared to other species. Despite the bypass channel, the weir still represents a bottleneck for the immigration of pike-perch into the MFK.

## Acknowledgements

Funds for this study were provided by the Austrian Science Foundation (project No. S 6406-B10). We also want to thank A. Lorber, E. Schager and G. Unfer for field assistance and S. Weiss for correcting the English.

## References


Baras, E., 1991. A bibliography on underwater telemetry, 1956–1990. Can. Tech. Rep. Fish. Aquat. Sci. 1819: 55 pp.

Baras, E., H. Lambert & J. Philippart, 1994. A comprehensive assessment of the failure of *Barbus barbus* spawning migrations through a fish pass in the canalized River Meuse (Belgium). Aquat. Living. Resour. 7: 181–189.

Francfort, J. E., G. F. Cada, D. D. Dauble, R. T. Hunt, D. W. Jones, B. N. Rinehart, G. L. Sommers & R. J. Costello, 1994. Environmental mitigation at hydroelectric projects. Volume II. Benefits and costs of fish passage and protection. Idaho Falls, ID, U.S. Department of Energy, Idaho Operations Office.

Grubinger, H. & T. Ernegger, 1994. Interdisziplinäre Forschung am Marchfeldkanal (Ein Schwerpunktprojekt des Fonds zur Förderung der wissenschaftlichen Forschung in Wien). Wetter und Leben, Z. angew. Meteorol. 46: 185–198.

Jensen, A. J. & P. Aass, 1995. Migration of a fast-growing population of brown trout (*Salmo trutta* L.) through a fish ladder in relation to water flow and water temperature. Reg. Riv. Res. & Mgmt 10: 217–228.

Jungwirth, M., 1996. Bypass channels at weirs as appropriate aids for fish migration in rhithral rivers. Reg. Riv. Res. & Mgmt 12: 483–492.

Jungwirth, M. & B. Pelikan, 1989. Zur Problematik von Fischaufstiegshilfen. Österr. Was. Abfallwirt. 41: 80–89.

Larinier, M., J. P. Porcher, F. Travade & C. Gosset, 1996. Passes à poissons. Expertise, conception des ouvrages de franchissement. Conseil supérieur de la pêche, Paris, 336 pp.

Monk, B., D. Weaver, C. Thompson & F. Ossiander, 1989. Effects of flow and weir design on the passage behaviour of American shad and salmonids in an experimental fish ladder. North am. J. Fish. Mgmt 9: 60–67.

Schmutz, S., H. Mader & G. Unfer, 1995. Funktionalität von Potamalfischaufstiegshilfen im Marchfeldkanalsystem. Österr. Was. Abfallwirt. 47: 43–58.

Schmutz, S., S. Matheisz, A. Pohn, J. Rathgeb & G. Unfer, 1994. Erstbesiedelung des Marchfeldkanals aus fischökologischer Sicht. Österreichs Fischerei 47: 158–179.

Schwalme, K., W. Mackay & D. Lindner, 1985. Suitability of vertical slot and denil fishways for passing north-temperate, nonsalmonid fish. Can. J. Fish. Aquat. Sci. 42: 1815–1822.

Travade, F., M. Larinier, S. Boyer-Bernard & J. Dartiguelongue, 1998. Experiences on four fishpass installations recently built on two rivers in southwest France. In M. Jungwirth, S. Schmutz & S. Weiss (eds), Fish Migration and Fish-Bypasses. Blackwell Scientific Publications: in press.

Winter, J., V. Kuechle, D. Siniff & J. Tester, 1978. Equipment and methods for radiotracking freshwater fish. Univ. of Minn. Agric. exp. Stn. Misc. Rep.